SECOND EDITION

THE INTEGRATED NERVOUS SYSTEM

A SYSTEMATIC DIAGNOSTIC CASE-BASED APPROACH

SECOND EDITION

THE INTEGRATED NERVOUS SYSTEM

A SYSTEMATIC DIAGNOSTIC CASE-BASED APPROACH

WALTER J. HENDELMAN, MD, CM

University of Ottawa
Ontario, Canada

PETER HUMPHREYS, MD, FRCP(C)

Children's Hospital of Eastern Ontario
Ottawa, Canada

CHRISTOPHER R. SKINNER, MD, FRCP(C)

University of Ottawa
Ontario, Canada

CRC Press

Taylor & Francis Group
Boca Raton London New York

CRC Press is an imprint of the
Taylor & Francis Group, an **informa** business

CRC Press
Taylor & Francis Group
6000 Broken Sound Parkway NW, Suite 300
Boca Raton, FL 33487-2742

© 2018 by Taylor & Francis Group, LLC
CRC Press is an imprint of Taylor & Francis Group, an Informa business

No claim to original U.S. Government works

Printed on acid-free paper

ISBN-13: 978-1-138-03742-7 (hbk)
ISBN-13: 978-1-466-59593-4 (pbk)
ISBN-13: 978-1-315-15676-7 (ebk)

DOI: 10.1201/9781315156767

Visit the Taylor & Francis Web site at
http://www.taylorandfrancis.com

and the CRC Press Web site at
http://www.crcpress.com

CONTENTS

PREFACE

You might be wondering why this textbook about neurology and neurologic problem-solving, using a case-based approach, is called the INTEGRATED Nervous System.

If you have ever attended an orchestral concert playing classical music, perhaps a symphony by Beethoven or Brahms or Stravinsky, you would have heard, and perhaps seen, the marvelous synchronization of the various musical instruments. It is incredible how the music is created as each member of the group starts or stops playing according to the musical score, resulting in the sound that we hear. Perhaps, you are fortunate enough to actually play in an orchestra or band and are part of that experience. Next, you might think about yourself, the listener. After all, your brain receives just sound waves via a complex series of connections. Just where in the brain is 'the music centre'? Finally, you might contemplate the genius of the composers whose brains produced this complex combination of sounds, the music we hear.

So it is with our brains. No part of the brain acts alone. All its component parts participate in a wondrous synchronization of thought/feeling/sensation/motor activity that characterizes our daily behaviour. You almost need to 'dissect' behaviour to recognize the contribution of the component parts. The mature nervous system integrates information from different sensory modalities, amalgamates experiences and information from the past and correlates it with the present context, resulting in responses (verbal and/or motor) that are (hopefully) appropriate for the situation.

The integration that we the authors are striving to achieve with this book is the approach that you as a non-neurologist will need in order to weave together your knowledge of neuroanatomy with the art and science of clinical neurology.

We, the coauthors of this book, include a neuroanatomist (WH), an adult neurologist (CS) and a pediatric neurologist (PH), all of whom have been involved in teaching a combined neuroscience and neurology course to second-year medical students. This is a five-week course, case-based, with lectures, organized on the principle of adult learning in small groups, with expert tutors (almost all practicing neurologists). Clinical disease entities (e.g. multiple sclerosis) are used as the 'problems' in the course.

One of the real challenges in a course that includes both neuroscience and neurology is the enormous scope of the subject matter; this leads to significant information overload when the process is compressed into a short period of block learning. A by-product of this compressed learning process has been the observation by both clinicians that most of their students – members of their tutor groups who 'knew' their material and passed the course, often with a good grade – when returning as clinical clerks or while taking a neurology elective, were unable to use the knowledge they once had mastered to solve clinical problems at the bedside.

This lack of integration of basic science and clinical information has been postulated as the basis of a syndrome called *neurophobia*, which apparently can affect one of every two medical students. This lack of ability to reason through clinical problems results in anxiety and dislike of the subject matter and, eventually, negative sentiments about and even fear of neurology (Jozefowicz, 1994; Fantaneanu et al., 2014). This book has been created to integrate these two worlds and to overcome this pedagogical deficit, using a problem-based approach with clinical disease entities. Our objective is to bridge the gap between the book and the bedside, in other words between the classroom and the clinic.

The second edition of this book has been revised and edited keeping you, the learner, in mind. We have rewritten and updated the text, as well as many of the tables. To help you understand the neurological disorders that you will encounter in this text and the challenges that patients and relatives face when such disorders occur, we have created a fictional 'person' for each of the clinical chapters as well as for the cases that are found on the accompanying Web site, called 'e-cases', with each of these chapters. The most significant addition is the abbreviated presentation of the history and neurologic examination of all the e-cases (now numbering over 50) within the text at the end of each of the clinical chapters; along with this, we have created 'maps' – a visual representation of the clinical motor, sensory and reflex findings. We hope that this inclusion will encourage the student to go to the Web site where each of the cases is presented in detail and where the 'expert' discusses the reasoning for the localization and etiological diagnosis.

We believe that this book and the associated Web site will be of practical value to all the professionals who deal with people who have neurological conditions, not only medical students and residents. This includes physiatrists (rehabilitation medicine specialists), physiotherapists, occupational therapists and speech therapists, and nurses who specialize in the care of neurological patients. We think that this text will also be of value for family physicians and specialists in internal medicine and pediatrics, all of whom must differentiate between organic pathology of the nervous system and other conditions.

Dr. Walter J. Hendelman
Dr. Peter Humphreys
Dr. Christopher Skinner
Faculty of Medicine
University of Ottawa
Ottawa, Canada

REFERENCES

Fantaneanu, T.A. et al. Neurophobia inception. *Can J Neurol Sci* 2014. 41: 421–429.
Jozefowicz, R.F. Neurophobia: The fear of neurology among medical students. *Arch Neurol* 1994. 51: 328–329.

ORGANIZATION OF THE BOOK
AND INTEGRATION WITH THE WEB SITE

The aim of this book is to enable you, the learner, to use your knowledge of the nervous system combined with a neurologically based, problem-solving clinical reasoning approach to neurology to help in the diagnosis and treatment, in the broadest sense, of those who suffer from a neurological disease or injury. We hope that this approach meets with success, insofar as it leads to an improvement in the diagnosis and care of persons afflicted by neurological problems.

To help you understand the neurological disorders that you will encounter in this text and the challenges that patients and relatives face when such disorders appear, we have created fictional persons with names. We meet the first of these people in the Introduction, which lays out the complexity of the nervous system and its capacity to multitask. The introduction exposes the reader to neuroanatomical pathways, which, by the end of the book, you should be quite familiar with in terms of both their function and their importance in neurology.

The first section is called 'The Basics of Neurological Problem Solving'. Chapter 1 is a review of clinically relevant neuroanatomy, enough to set the stage in terms of the basic knowledge of the nervous system needed for this book. The second chapter is devoted to the clinical neurological examination and the integration of the information garnered in terms of the normal functioning nervous system, for example, the assessment of reflex activity. Chapter 3 introduces the student to *neurological clinical reasoning* for the purpose of localizing the disease or lesion within the nervous system and determining the aetiology, the pathophysiological mechanism of disease. Patients with neurological problems present great challenges to their physicians (as well as to their relatives and caregivers) to accurately and completely gather the information required to make a working diagnosis and plan of investigation based on a single clinic visit. The approach used in this chapter and its accompanying worksheets is designed to provide students and non-neurologist clinicians with practical guidelines and tools with respect to diagnosis for the full range of neurological problems seen by a neurological generalist. This approach is applied throughout the book.

The next section is called 'Applying the Basics to Clinical Cases'. The succeeding 10 chapters deal with important specific clinical diseases or syndromes that have afflicted our fictional individuals, each with the focus on a different component of the nervous system, starting with the peripheral nervous system. The following chapters 'ascend' – spinal cord, brainstem, all the way to the cortex, with the last chapter in this section devoted to behavioural issues. For each of these cases, the history is presented, followed by the findings of the neurological examination. Additional neuroanatomical, neurophysiological and neurochemical information is added, where required, as it pertains to the clinical condition discussed in the chapter. Notwithstanding the information given, it is suggested that students review their knowledge of neuroanatomy, neurophysiology, neuropharmacology and neuropathology, using other resource books (given in the suggested readings and references sections at the end of each chapter and in the Annotated Bibliography at the end of the book). In each of these chapters, there is an application of the process of neurological reasoning to narrow the possibilities of *where* the lesion is located. This is followed by a systematic analysis in order to determine *what* disease (or diseases) should be considered. Relevant selected investigations are then presented and the results discussed. Finally, the diagnosis is made, with its prognosis, and an outline of the appropriate management is given, ending with the outcome of the case.

The text illustrations have been prepared with an emphasis on the functioning nervous system. In addition to neuroanatomical drawings related to the cases and tables with relevant clinical data, there are figures illustrating neurophysiological concepts, clinical findings (such as radiographic images and electroencephalograms) and microscopic neuropathological images. Again, the information is described in the context of the disease presented in that chapter. The glossary of terms also emphasizes clinical terminology.

The third section, named 'Supplementary Considerations', presents additional insights written by guest authors, Dr. Anna McCormick, a physiatrist (a specialist in physical medicine rehabilitation), and Dr. Robert Nelson, a senior neurologist with expertise in ethics. Both discuss other important dimensions of neurological problems: rehabilitation and ethics. Currently, rehabilitation has much to offer for those afflicted by disease or injury of the nervous system; there is now a certain air of hopefulness that there can be recovery of function following an insult to the nervous system, in adults as well as in children. The ethical principles and reasoning on the basis of which decisions (sometimes quite unique) are taken in neurological cases are presented in the context of an inherited disease of the nervous system.

There is a Web site associated with this textbook: http://www.integratednervoussystem.com.

This Web site contains the worksheets that have been developed to apply the clinical reasoning approach to neurological problem solving and the e-cases that are now part of each of the clinical chapters. The e-cases enlarge the scope of the book by adding additional commonly seen neurological diseases for each level of the nervous system. These are presented in a more sequential fashion, although once having learned the analytic approach, it is hoped (expected) that you, the learner, will work through each clinical case on your own, using the worksheets, before reading the case evolution, investigations and resolution. It is highly recommended that you, the non-neurologist student/learner, apply this approach when confronted with a neurological patient. The authors want you to learn how to think like a neurologist. In addition, the Web site has all the illustrations found in the book, with animation added to assist in the understanding of some reflex circuits and various pathways. It also includes the glossary. There is also a learning module to assist you, as a non-expert, in understanding neuroimaging, how the various modes of computed tomography and magnetic resonance imaging assist in localizing a lesion and defining the likely etiology. The Web site will also be utilized to provide updates on the cases presented as well as new cases, so it may be wise to check it periodically.

We sincerely hope that our system will work for you and wish you every success in the diagnosis and management of your neurological patients.

ACKNOWLEDGEMENTS

ILLUSTRATIONS

We wish to express our appreciation and respect for the illustrators who have helped shape this book: Perry Ng, the principal illustrator and Dr. Tim Willett, who is the principal illustrator for the *Atlas of Functional Neuroanatomy* (also published by CRC Press). Without their creative and conscientious efforts, we could not have achieved what is necessary to convey our message to you, the learner. Several illustrations from the Atlas have been included in this book, with the permission of the publisher.

CONTRIBUTORS

A special note of gratitude is extended to our chapter contributors, Dr. R. Nelson (Neurology, The Ottawa Hospital, now retired) and Dr. A. McCormick (Physical Medicine and Rehabilitation, The Children's Hospital of Eastern Ontario). Dr. Nelson is a neurologist's neurologist, highly regarded by his colleagues, with a special interest in ethical issues. Dr. McCormick, who carries with her an air of enthusiasm and hope, has successfully championed the cause of pediatric rehabilitation and is one of the few people in her field who actively treats both children and adults.

Many colleagues have contributed collegially and willingly to this book, with illustrations and clinical material. We are particularly grateful not only for their particular and unique contribution but also for the spirit in which it has been donated. In many cases, their staff helped with the preparation of this material and we thank them as well.

Dr. D. Grimes: Division of Neurology, The Ottawa Hospital
Dr. R. Grover: Neuroradiology, The Ottawa Hospital (resident-in-training)
Dr. M. Kingstone: Neuroradiology, The Ottawa Hospital
Dr. D. Lelli: Division of Neurology, The Ottawa Hospital
Dr. J. Marsan: Otolaryngology, The Ottawa Hospital
Dr. J. Michaud: Neuropathology, The Ottawa Hospital
Dr. M. O'Connor: Ophthalmology, The Children's Hospital of Eastern Ontario
Dr. C. Torres: Neuroradiology, The Ottawa Hospital
Dr. S. Whiting: Neurology, The Children's Hospital of Eastern Ontario, with S. Bulusu (chief technologist, Clinical Neurophysiology Laboratory, CHEO)
Dr. J. Woulfe: Neuropathology, The Ottawa Hospital

WEB SITE

We would also like to acknowledge the work of David Skinner, who diligently crafted the Web site to emulate the problem-solving methodology of the text.

SUPPORT

We also wish to thank the secretarial staff that we work with in our various offices, including the Department of Cellular and Molecular Medicine at the Faculty of Medicine, and particularly Orma Lester, who has assisted us in addition to her regular duties in a busy hospital office.

The Health Sciences library staff, particularly M. Boutet, is thanked for assistance in creating the Annotated Bibliography.

Computer support has consistently been available at the Faculty of Medicine at the University of Ottawa from Medtech (Information Management Services) and from staff at the Children's Hospital of Eastern Ontario (CHEO). Photographic services were also provided at CHEO.

Last, but not least, the authors gratefully acknowledge the assistance and cooperation of the production team of CRC Press, particularly our project editors.

AUTHORS

Dr. Walter J. Hendelman is a Canadian born and raised in Montreal. He did his undergraduate studies at McGill University in science with honours in psychology, where he studied under Dr. Donald Hebb, who is now recognized for his "cell assembly" theory, explaining how the brain manages information. He then proceeded to do his medical studies also at McGill. Following a year of internship and a subsequent year of pediatric medicine in Montreal, Dr. Hendelman chose the path of brain research and academia.

Postdoctoral studies followed in the emerging field of developmental neuroscience, using the "new" techniques of nerve tissue culture at the Pasadena Foundation for Medical Research. These studies continued, including electron microscopy, at Columbia University Medical Center in New York City; Dr. Richard Bunge was his research mentor and his neuroanatomy mentor was Dr. Malcolm Carpenter, author of the well-known textbook *Human Neuroanatomy*.

Dr. Hendelman then returned to Canada and has made Ottawa his home for his academic career at the Faculty of Medicine at the University of Ottawa. He began his teaching in gross anatomy and neuroanatomy and then concentrated on the latter, first assuming the responsibility of coordinator for the course and then becoming co-chair for the teaching unit on the nervous system in the new curriculum. His research focused on the examination of the development of the cerebellum and the cerebral cortex in organotypic tissue cultures.

Dr. Hendelman has dedicated his career as a teacher to assisting those who wish to learn functional neuroanatomy including medical students and trainees, as well as those in the allied health sciences. The first edition of his teaching ATLAS (1987, published by the University of Ottawa Press) was followed by other editions and subsequently was published by CRC Press (2000) with the title *Atlas of Functional Neuroanatomy*. The second edition of the Atlas (2006) is accompanied by a Web site (www.atlasbrain.com) with interactive features, including roll-over labelling and animations of pathways and connections. This edition has been translated into Italian (2009) and also into French (2013), with a Web site in French. The third edition has now been published (2016, CRC Press), and a Web site for this new edition will be available. An Italian translation of this edition has already been published and other translations are pending.

Additional learning resources developed by Dr. Hendelman include several narrated teaching videotapes on the skull and the brain using anatomical specimens; these are now available online on the Web site for the Atlas.

Dr. Hendelman is currently professor emeritus in the Faculty of Medicine, University of Ottawa.

Dr. Peter Humphreys, a graduate of the McGill University Faculty of Medicine in 1966, trained in pediatrics at Boston Children's Hospital and at St. Mary's Hospital in London, followed by training in neurology at the Montreal Neurological Institute. After a six-year stint on the neurology staff of the Montreal Children's Hospital, he became the founding head of the Neurology Division at the Children's Hospital of Eastern Ontario, Ottawa, a position he held for 23 years. Although now semi-retired, Dr. Humphreys continues an active outpatient clinical practice. His principal area of interest is in disorders of brain development. Ten years ago, he started the first Canadian hospital-based clinic devoted to the comprehensive care of girls and women with Rett syndrome. In this role, he is involved in clinical research related to movement disorders in Rett syndrome as well as collaborating with researchers in the University of Ottawa Faculty of Medicine in the investigation of the enteric nervous system of mouse models of Rett syndrome.

A full professor in the Department of Pediatrics at the University of Ottawa, Dr. Humphreys has been active at all levels of the medical curriculum. For many years, he was a tutor in small-group learning sessions for second-year medical students doing a problem-based introductory course on the nervous system. During the same course, he presently conducts a full-class lecture on brain developmental disorders as well as a live patient demonstration devoted to the pediatric neurological examination. He also does clinic instruction for residents in neurology and pediatrics. Finally, he participates in a teaching role in refresher courses for paediatric neurology residents at the University of Ottawa and for pediatricians in courses offered by the Canadian Paediatric Society.

Dr. Christopher R. Skinner earned his BEng (electrical) from the Royal Military College in 1970. He worked as a systems engineer with the Department of National Defence, implementing nationwide information systems until 1975. He earned his medical degree from Queen's University in 1979. He received his specialist certification in general internal medicine in 1986, in neurology in 1987, and qualified as a Diplomat of the American Board of Sleep Medicine in 2005. He was chief information officer at the Ottawa Hospital from 1996 to 1998.

He has been a clinical teacher and lecturer in the Faculty of Medicine at the University of Ottawa since 1993. He has taught clinical neurology, occupational neurology and sleep medicine to all levels of study, including medical students, residents, physician assistants and military flight surgeons. He was also involved in the design and implementation of the problem-based digital curriculum portal used for the teaching of medical students.

In 2011, Dr. Skinner did a three-month sabbatical with the Russian Space Agency in Moscow, studying the effects of long-duration space flight on the nervous system and its effect on sleep in space.

Dr. Skinner currently practices general neurology and sleep medicine at the Ottawa Hospital. He also has a practice in rural Quebec as well as patients in inner city shelters in Ottawa.

INTRODUCTION

Crash (as he is known to his colleagues from his call sign) had applied and been accepted into the international astronaut training program.

This was a gruelling four-year training program learning how to control both Russian and American spacecraft in order to go into space for near-earth and deep-space exploration.

He was selected to be the commander of a mission, which consisted of a three-person crew to rendezvous, land on and redirect a near-earth asteroid.

He had over 800 days of spaceflight under his belt. After his last mission, he had noticed a decrease in his vision and the flight surgeon had noted some papilledema. This could be a sign of visual impairment and increased intracranial pressure, a condition seen in astronauts with long duration in space.

His latest post-flight magnetic resonance imaging scan of the brain also had shown some new changes in the white matter of the brain. He wondered if this could be damage related to long exposure to cosmic radiation, which damages the white matter.

Learning the command and control systems of the Russian spacecraft was particularly difficult, not only because the instructions and instruments were all in Russian but also due to the fact that the controls required a much more direct physical touch than the fly-by-wire American systems. He welcomed having to learn a new language as he had heard that it prevents dementia.

Crash was due to go on a test flight of eight days in orbit to evaluate the systems required for the asteroid mission.

As he entered the elevator to the preparation room at the launch site in Kazakhstan prior to launch, his olfactory system sensed the odour of kerosene in the rocket fuel outside and transmitted these impulses to his mesial temporal cortex. He says to himself, 'I love the smell of rocket fuel in the morning'. It is amazing how smells can bring back memories, thinking back to his early flying career.

As he sat with his pressure suit on the pad in the Baikonur Cosmodrome, he could feel the weight of the hand control at 1G of gravity. The sensation of the position of his fingers travelled up through the posterior columns of the spinal cord, relaying once in the brain stem and again in the contralateral thalamus to terminate in the parietal cortex. This information would be relayed to areas of the prefrontal cortex to allow Crash's corticospinal tract to send messages to the spinal cord and then spinal motor units to make small adjustments in his hand muscles to control the spacecraft. The motor sequences that had been practiced thousands of times were stored in a network involving the prefrontal cortex and basal ganglia modulated by the cerebellum. His memories of the procedures and instructions had been encoded through the mammillothalamohippocampal system and then exported to populations of cortical neurons for long-term storage.

As he waited on the pad with his two crewmates, his amygdala kicked in, causing him to feel anxious; he could feel that there was spasm of his anal sphincter through the spinothalamic tract passing pain and temperature information to the thalamus and contralateral cortex. He says to himself, 'Come on – Let's light up the old vodka burner'.

Time for launch: 5, 4, 3, 2, 1! The five large rocket motors of the booster assisted by two solid rockets light up under him and the vehicle clears the tower heading down range at a velocity faster than the speed of sound. At 65 seconds into the flight, an alarm sounds, there is a problem with the left hand solid rocket booster!

Crash has to make a decision: abort or not to abort; he had exactly 3 seconds to make this decision. The executive function of his frontal cortex is in overdrive. He presses the abort button; the solid rockets and upper stages fall away. He mutters to himself, 'another bunch of guys at the rocket factory are going to go to jail for this'. The spacecraft is now 60 km in altitude going 10,000 km per hour in freefall. Although he had practiced for such a 'forward abort' situation, this requires Crash to make some hard decisions about where to land. He coolly turns to his crewmates and asks, 'Вода или Земля' ('Water or Land')? They both answer 'land' knowing that water meant the Arctic Ocean in January.

With his visual and auditory system fully engaged, Crash calmly takes hold of the controls, commands the spacecraft a series of S turns to bleed off the speed and heads for the frozen, snow-bound taiga. The parachute then deploys and the capsule lands with a thud in the Siberian taiga, 500 km from any habitation.

The crew members crawl out, pull out the survival kits and light a big fire with wood that they have gathered. Crash says casually to his crewmates: 'I guess we will have to cross-country ski out like they did in Voskhod 2'.

He calls Fifi on his satellite cell phone, 'Good news dear, we will not miss for our anniversary this year'. His Russian colleague asks: 'What are we going to call this mess?'

Crash answers: 'Сибирский Лебединое погружение' ('The Siberian Swan Dive').

THE BASICS OF NEUROLOGICAL PROBLEM SOLVING

Chapter 1

Synopsis of the Nervous System

Objectives

- To review the basic histological knowledge of the nervous system from a functional (neurological) perspective
- To organize the nerves, nuclei and tracts of the nervous system into functional systems

In the Introduction, Crash, pilot and astronaut, executes a number of intricate tasks in response to input from several sensory systems: tactile, muscle and joint, visual, vestibular and auditory. He reacts to all of these stimuli appropriately and performs highly accurate and skilled motor movements. Pathways were sketched for the sensory input and for executing the motor movements.

How does the brain process all this information? Which parts of the nervous system are involved in the exquisite motor control required to fly a jet aircraft or pilot a space craft? Where in the 'brain' are the integration and decision-making functions carried out?

A neurologist's view of the nervous system is one of functionality – are all the components operational in order to receive information, analyze and assess its significance and produce the appropriate action? If not, the task of the physician is to determine where the problem is occurring and what is its most likely cause – the localization of the lesion and its possible etiology.

In order to determine where, the *localization*, one needs to have knowledge of the anatomy and physiology of the nervous system. This chapter will provide that information from a functional perspective, but the student should expect to consult other resources – details of neuroanatomy, neurophysiology and neuropathology – to supplement this presentation (see Suggested Readings and the Annotated Bibliography). One needs this knowledge to understand the significance of the findings of the neurological examination, which is outlined in the next chapter.

Determining the likely cause, the *etiology*, requires knowledge of disease processes. This determination is based initially on the nature of the symptoms and the history of the illness – how long the problem has been occurring (acute, subacute, chronic) and how the symptoms have evolved over time. The task of the practitioner – physician, resident or student – will be to determine what disease process (e.g. infectious, vascular, neoplastic) is occurring and its pathophysiology, and to identify diseases that most likely account for the patient's signs and symptoms.

Laboratory investigations, including blood work, special tests (e.g. disease-specific antibody levels) and particularly neuroimaging, usually provide additional information to help pinpoint the localization of the disease and often limit the possible list of most likely diseases.

Lastly, the neurologist will synthesize the patient's history and the symptoms with the signs found on neurological examination as well as the additional information provided by the investigations to come up with the definitive diagnosis. This diagnosis allows for a therapeutic plan and some idea of the likely outcome: the prognosis. All of this must be communicated sensitively to the patient and family in a way that can be readily understood. (A sample case exemplifying this approach is presented in Chapter 3.)

1.1 NEUROBASICS

The nervous system is designed to receive information, analyse the significance of this input and respond appropriately (the output), usually by performing a movement or by communicating ideas through spoken language. In its simplest form, this process would require a minimum of three neurons, but as we come up through the animal kingdom, the complexity of analysis increases incredibly. This evolutionary development culminates in the human central nervous system (CNS), with all its multifaceted functions.

The nervous system consists of two divisions (Figure 1.1), a peripheral component, called the *peripheral nervous system*, the PNS, and a central set of structures, called the *central nervous system*, the CNS. The PNS consists of sensory neurons and their fibers, which convey messages that originate from the skin, muscles and joints and from special sensory organs such as the cochlea (hearing); it also carries the motor nerve fibers that activate the muscles. The *autonomic nervous system*, the ANS, is also considered part of the PNS; it is involved with the regulation of the cardiac pacemaker system and of smooth

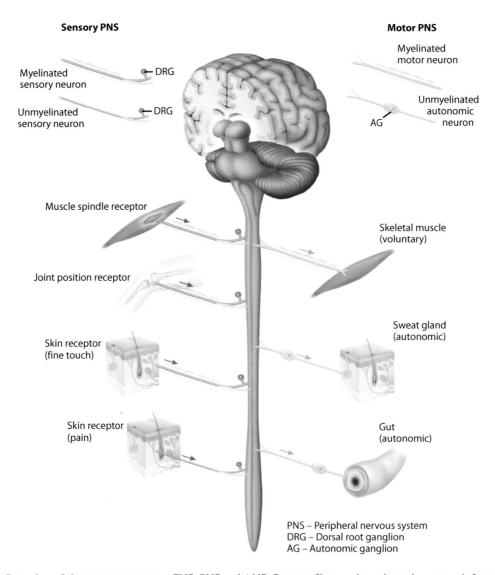

Sensory PNS

Myelinated
sensory neuron — DRG

Unmyelinated
sensory neuron — DRG

Muscle spindle receptor

Joint position receptor

Skin receptor
(fine touch)

Skin receptor
(pain)

Motor PNS

Myelinated
motor neuron

Unmyelinated
autonomic
neuron
AG

Skeletal muscle
(voluntary)

Sweat gland
(autonomic)

Gut
(autonomic)

PNS – Peripheral nervous system
DRG – Dorsal root ganglion
AG – Autonomic ganglion

FIGURE 1.1: Overview of the nervous system – CNS, PNS and ANS. Sensory fibers, coloured purple, convey information towards the CNS (afferent) from receptors in skin and muscle. Motor fibers, coloured green, carry instructions away from the CNS (efferent) to muscle and via autonomic ganglia to glands and viscera.

muscle and glands, including some control of bowel and bladder functions. The CNS consists of the spinal cord, the brainstem and the brain hemispheres. The CNS adds analytic functionality and varying levels of motor control, culminating in a remarkably intricate capacity for 'thinking' forward and backward in time, consciousness, language and executive functions, processes performed in different areas of the cerebral cortex.

1.1.1 THE NEURON

A *neuron* is the basic cellular element of the nervous system. In the most simplistic language of today's electronic world, each of the billions of neurons in the human CNS is equivalent to a unique microchip, possessing a specific information processing capacity. Like other

cells, the neuron has a cell body (the soma or perikaryon) with a nucleus and the cellular machinery to be its nutritive centre. Morphologically, it is the cellular processes – dendrites and axon – that distinguish a neuron from other cells. The electrochemical nature of its membrane, whereby the interior of the cell and its processes have a negative charge, is a characteristic feature of the neuron (see Figures 11.1, 11.2 and 11.3). The synapse, the electrochemical communication between neurons, is the other unique feature of nervous tissue (discussed in Section 1.1.4).

The typical neuron in the CNS (Figure 1.2a) has *dendrites* that extend from the cell body for several microns. Dendrites receive information from other neurons at specialized receptor areas, the synapses, some of which form small excrescences on the dendrites, called *synaptic*

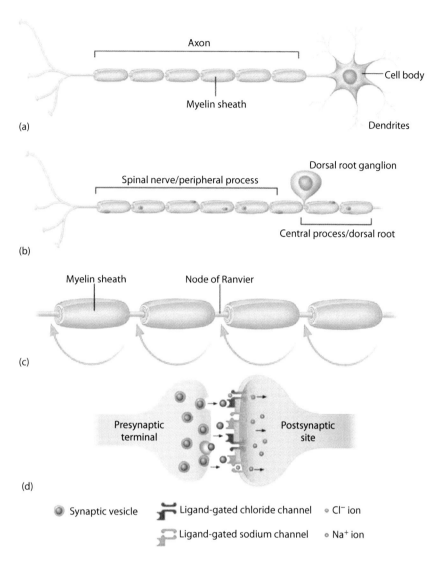

FIGURE 1.2: (a) A 'typical' CNS neuron has several dendrites and a single (myelinated) axon; its terminal branches communicate with other neurons. (b) A PNS neuron has both central and peripheral (myelinated) processes. The sensory endings of the peripheral process are located in the skin, joints and muscles. The cell body of the PNS neuron is located within the dorsal root ganglion (DRG). The central process enters the CNS via the dorsal root. (c) An axon of a nerve with its myelin sheath: each internode segment is the territory of a single Schwann cell in the PNS or an oligodendrocyte process in the CNS. A node of Ranvier separates each segment and the impulse 'jumps' from node to node, a process called saltatory conduction. (d) A 'generic' synapse is illustrated, including the presynaptic ending with the synaptic vesicles containing a neurotransmitter, the synaptic 'gap' and the postsynaptic site with its Na and Cl ion channels.

spines (discussed in Section 1.1.4). More complex neurons have an extensive arborization of dendrites and receive information from perhaps hundreds of other neurons. Neurons of a certain functional type tend to have a typical configuration of their dendrites and group together to form a *nucleus* (somewhat confusing terminology!) in the CNS, or a layer of *cortex* (e.g. the cerebral or cerebellar cortex). Brain tissue is traditionally fixed in formalin for the purpose of study, and neuronal areas (e.g. the cortex) become grayish in appearance when the brain is cut; hence, the term *gray matter* is used for areas of neurons, their processes and synapses.

The neurons in the PNS include both sensory neurons and those associated with the ANS. The cell body of a sensory neuron is displaced off to the side of its two processes: a peripheral process (e.g. to the skin) and a central process, which will form synapses within a nucleus in the spinal cord or brainstem (Figure 1.2b). The distal endings of sensory neurons (e.g. in the skin) are sensitized to receive information of a certain type and hence behave functionally as dendrites. Often, they are enveloped by specialized receptors (e.g. specialized touch and temperature receptors in the skin). Receptors for pain sensation are 'naked' nerve endings in the skin and are located within all tissues (except brain tissue of the CNS). For the special senses such as hearing, there are highly developed receptor cells (e.g. hair cells of the cochlea), which are in contact with the sensory neurons.

The cell bodies of sensory neurons congregate in a specific location, forming a peripheral *ganglion* (plural, ganglia), typically located along the dorsal root (the dorsal root ganglia; see Figure 1.1). Neurons of the ANS also form ganglia (see Figure 1.1), located alongside the vertebra (sympathetic) and within or closer to the end organ (parasympathetic; see Section 1.2.2.4).

1.1.2 AXONS (NERVE FIBERS)

Each CNS neuron has (with rare exception) a single *axon*, also called a nerve fiber, which is the efferent process of the neuron, acting like an electric wire to convey information from a neuron to other neurons, muscle or other tissue, with the possibility of many branches (collaterals) en route. Millions of axons course within the CNS providing extensive intercommunication within and between the CNS neuronal nuclei and cortical areas. The axons of functionally linked sensory and motor neurons usually bundle together and are called tracts or pathways. A neuron within the brainstem or spinal cord that sends its axon (via the PNS) to skeletal muscle fibers is called a motor neuron, also known clinically as the lower motor neuron (LMN).

A typical (observable, dissectible) nerve in the PNS usually has both motor and sensory nerve fiber bundles and, often, autonomic fibers. The postganglionic fibers of the ANS (sympathetic and parasympathetic) are distributed to smooth muscle and glands.

1.1.3 MYELIN

Myelin is the biological (lipid–protein) insulation surrounding axons; its function is to increase the speed of axonal conduction (Figure 1.2c). Since axonal conduction velocity increases in proportion to axonal diameter, an alternative to a myelin sheath would be a marked increase in axonal diameter; this, if present in most CNS axons, would cause the nervous system to be extremely bulky. Such an arrangement would require much larger axons that would be easily susceptible to transmission degradation, rendering the nervous system more inefficient. In the human brain, axons may travel for long distances, and the longer the distance, the more likely it is that the fibers are myelinated; faster transmission such as that required for certain sensory and motor functions requires axons with larger diameters, and these axons have thicker myelin sheaths.

Myelin is composed of segments, and between each segment is a very short 'naked' section of the axon, called the node of Ranvier; the segments are therefore called internodes. At each nodal section of an axon, there is a concentration of sodium ion channels within the axonal membrane. Impulse conduction along a myelinated axon 'jumps' from node to node, a process called saltatory conduction, which thus speeds the transmission of the impulse; the membrane potential is recharged at each node (discussed further in Chapter 4).

Myelin is formed and maintained by glia, supporting cells of the nervous system. In the PNS, a single cell, known as the Schwann cell, is responsible for each internode segment of myelin (see Figures 1.2c and 4.6). In the CNS, the equivalent glial cell is the oligodendrocyte, and each cell is responsible for several segments of myelin, on a number of axons. Areas of the CNS containing myelinated tracts have a whitish appearance with formalin fixation and are hence called the *white matter*.

1.1.4 SYNAPSE AND NEUROTRANSMISSION

In the CNS, the terminal end of each axon and each of its collaterals is a *synapse*, a specialized junction, the conduit by which one neuron communicates electrochemically with another. Synapses abut on the dendrites of other neurons, typically at synaptic spines; they are also located on the cell body of neurons, on the initial segment of the axon and sometimes on other synapses.

Synapses can be seen with light microscopic techniques but are best visualized with electron microscopy. A synapse (Figure 1.2d) consists of an enlargement of the terminal end of the presynaptic axon containing small (synaptic) vesicles, the presynaptic membrane and a postsynaptic receptor site (e.g. a dendritic spine), where the membrane is specialized for neurotransmission; in some cases, the postsynaptic membrane is thickened. In between is a space or cleft, the synaptic gap, which is sometimes widened compared to the usual space between adjacent cells in the CNS.

Biologic agents that can alter the membrane properties of neurons at the postsynaptic site are called *neurotransmitters*. These are synthesized in the cell body, transported down the axon and stored in the synaptic ending in packets, the synaptic vesicles. In some cases, the neurotransmitter may be synthesized within the nerve terminal. These endings are activated when the axonal electrical impulse, the action potential, invades the terminal, setting off a process whereby the transmitter is released into the synaptic cleft (see Figure 11.4).

Synaptic transmission is therefore both an electrical and a chemical event. In some instances, synapses have built-in mechanisms for recapturing the neurotransmitter (recycling); alternatively, enzymes in the synaptic cleft may destroy the active neurotransmitter. Glial cells (astrocytes) may be involved in this process, for example by removing the neurotransmitter from the synaptic site.

Receptors on the postsynaptic neuron are activated by the neurotransmitter, causing a shift of ions and a net change in the membrane potential of the postsynaptic neuron, leading to either depolarization or hyperpolarization of the membrane. This contributes to an increase in the likelihood of the neuron either to discharge more frequently (depolarization, excitatory) or to discharge less often (hyperpolarization, inhibitory).

1.1.5 NEUROTRANSMITTERS

A neuron may synthesize one or more neurotransmitters, and these are released at all of its synaptic endings. The action of any single neurotransmitter may differ at each site depending on the receptor type or subtype in the postsynaptic neuron.

Typical examples of neurotransmitter chemicals are simple amino acids, such as the inhibitory-acting gamma-amino butyric acid (GABA) or the excitatory-acting glutamate; these have an immediate but short-lasting (millisecond) effect on the postsynaptic membrane. More complex molecules may act long-term (seconds or minutes) to change the nature of the response (an effect called neuromodulation) or to alter the properties of the membrane of the postsynaptic neuron, thereby strengthening or diminishing the synaptic relationship, a process known as long-term potentiation (see Figure 13.7). Other neurotransmitters may cause the release of messengers, which enter the nucleus and bring about the activation or deactivation of genes in the nucleus, thereby producing a long-lasting effect on the cell and its synaptic relationships. (Neurotransmitters are discussed further in Chapter 11.)

1.1.6 MUSCLE

Although voluntary skeletal muscle is not part of the nervous system, the output of the nervous system most often includes some form of muscular activity. The neuromuscular junction between the motor neuron and its associated muscle fibers is an essential link in the chain, part of the PNS. Neurologists must therefore assess patients for muscle diseases (such as muscular dystrophy) and need to distinguish these entities from diseases that affect the synapse at the neuromuscular junction (e.g. myasthenia gravis), from diseases of the peripheral nerves, or from lesions of the spinal cord.

1.2 NERVOUS SYSTEM OVERVIEW

The perspective of this book is an understanding of the functioning nervous system as it goes about achieving its three essential tasks:

- To detect what is happening in the external (e.g. vision, hearing) or internal (e.g. within muscles and joints) environment

- To integrate this information with ongoing brain activity and, if possible, relate it to previous experience
- To act or react in an appropriate fashion, in order to accommodate to the new situation or perhaps to alter it

We detect changes in the external environment via the PNS, including particularly the special senses. The CNS is the integrative centre for analyzing the incoming information and organizing the output. The CNS consists of several distinct areas, each contributing a piece to this operation; all parts must function harmoniously in order to carry out complex tasks. Most of our responses include movements to adapt to these changes via the nerve fibers (of the PNS) that activate the skeletomuscular system.

One can discuss the nervous system as consisting of a set of *modules:*

- The periphery – sensory and motor nerves; neuromuscular junctions and muscles
- The spinal cord – the location of the LMNs and the site of sensorimotor reflex activity; pathways ascending and descending
- The brainstem – three divisions each with cranial nerve nuclei; reticular formation; pathways ascending and descending
- The diencephalon – hypothalamus for vegetative functions, and the thalamus for amalgamation with the cerebral cortex
- The cerebellum – a major modulator of the motor system in particular, but also systems involved in cognition
- The basal ganglia – several nuclei with both motor and nonmotor functions
- The cerebral cortex – integration, visuospatial orientation, language, memory, and executive function
- The limbic system – involved in the development and expression of emotional reactions

When disease or injury affects the nervous system, there is a disruption of function. It is the physician's (neurologist's) task to use his or her knowledge to diagnose where the nervous system is malfunctioning and the nature of the problem. It is the characteristic contribution of each part that permits the physician to determine the localization of any damage or lesion.

In order for the CNS to function collaboratively, pathways (tracts) are needed to carry information from the special senses, skin, muscle, joints, and viscera to higher 'centres' in the brain, including the cerebral cortex, as well as from these coordinating areas back down to the (lower) motor neurons producing actual movements. At the same time, there is a need for the various CNS modules to exchange information about the task that each is performing and what it is accomplishing. In fact, much of the substance of the hemispheres consists of nerve fibers interconnecting various parts of the brain and the two hemispheres with each other; these nerve fibers constitute the CNS white matter.

1.2.1 THE PERIPHERY/PNS

Strictly speaking, the PNS is the nervous system outside the brainstem and spinal cord. It includes the peripheral nerves, both sensory and motor, as well as the neuromuscular junctions. Muscle diseases are within the sphere of neurology; examples of muscle disease will be introduced in the online cases associated with Chapter 4, called *e-cases.*

Information from the skin and from receptors in muscles and joints is constantly needed for adaptation to a changing environment. Sensory information from the skin is detected by nonspecialized and specialized receptors for two main categories of sensation, called modalities:

1. Highly discriminative information such as fine touch and texture. Discriminative touch sensation is carried to the spinal cord by larger fibers with thicker myelin; therefore, the information is conveyed more rapidly (Figure 1.1).
2. Pain, temperature and non-discriminative (crude) touch. Fibers are mostly smaller and tend to be thinly myelinated or unmyelinated; impulse conduction along these fibers is therefore slower (Figure 1.1).

Motor nerve fibers, originating from motor neurons in the spinal cord, project to the muscles in order to initiate and control movements. Again, these nerves are well myelinated and carry information rapidly. The synapse on muscle cells is specialized as the neuromuscular junction, where the neurotransmitter acetylcholine is stored and released when the action potential invades the synapse.

Finally, intact muscle is required to produce the intended movements, either intentional (voluntary) or procedural, as well as for postural adjustments in response to changes in position or to the force of gravity (discussed further in the motor section in Chapter 2).

1.2.2 THE SPINAL CORD

The spinal cord is intimately connected with the PNS but is part of the CNS. Functionally, the spinal cord is responsible for receiving sensory (including muscular) input from the limbs and the body wall and for sending out the motor instructions to the muscles. It adds the functional capability for reflex activity, in response to information from both the muscles and the skin, and can organize some basic movements (e.g. walking). In addition, it carries ascending and descending axonal pathways (tracts).

The spinal cord is an elongated mass of nervous tissue with attached spinal roots that is located in the vertebral column; it ends normally at L2 (second lumbar vertebral level) in the adult (Figure 1.3a). Although the spinal cord is uninterrupted structurally (Figure 1.3b), it is organized segmentally, with each segment responsible for a portion of the body peripherally: a sensory area supplied by a segment is called a *dermatome*, and muscle supplied by a segment is called a *myotome*. The spinal cord segments are named according to the level at which their spinal nerves exit the vertebral column. There are 8 pairs of cervical spinal nerves (and spinal cord segments; C1–C8), 12 thoracic (T1–T12), 5 lumbar (L1–L5), 5 sacral (S1–S5) and 1 coccygeal.

Because the spinal cord is shorter than the vertebral column, a spinal cord segment responsible for a patch of skin and certain muscles in the limbs and the periphery does not correspond exactly with the vertebral level, with the exception of the upper cervical spinal cord. A lesion of the spinal cord is described as the level of the cord that has been damaged, not the vertebral level; therefore, knowing which part of the body is supplied by a spinal cord segment has clinical importance (see Table 2.1).

Each spinal cord segment has a collection of sensory and motor nerve rootlets that coalesce to form a single sensory (dorsal) and motor (ventral) nerve root on each side of the cord; the dorsal and sensory roots combine to form a spinal nerve (Figure 1.3c). Transverse sections (cross-sections) of the spinal cord reveal a core of gray matter (neurons and synapses) in a butterfly-like configuration, surrounded by white matter, consisting of tracts, also called pathways. The dorsal aspect of the spinal cord gray matter has sensory-associated functions; the sensory input is carried via the dorsal root. The ventral portion of the spinal cord gray matter has the motor neurons, known from a functional perspective as the LMNs (also called the alpha motor neurons). The axons of these motor cells leave the spinal cord (and change the nature of their myelin sheath as they do so) to be distributed via the ventral root to the muscles.

1.2.2.1 SENSORY ASPECTS

Peripheral nerves carry the sensory information from the two modalities to the spinal cord where the two systems follow quite different pathways (tracts). The modalities of discriminative (fine) touch along with proprioception and the 'sense' of vibration ascend and stay on the same side throughout the spinal cord but cross in the brainstem (Figure 1.4a), whereas crude touch, pain and temperature fibers synapse and cross to the other side in the spinal cord (Figure 1.4b) and then ascend. After the thalamic relay both reach the cortex, where further elaboration and identification of the sensory information occur, most of this being in the realm of consciousness (discussed in further detail in Chapter 2 and also in Chapter 5). Other information (known as proprioception), derived from special sensory

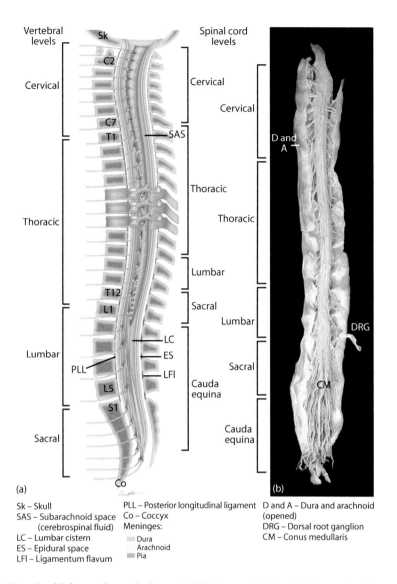

Sk – Skull
SAS – Subarachnoid space
          (cerebrospinal fluid)
LC – Lumbar cistern
ES – Epidural space
LFl – Ligamentum flavum

PLL – Posterior longitudinal ligament
Co – Coccyx
Meninges:
　　Dura
　　Arachnoid
　　Pia

D and A – Dura and arachnoid
(opened)
DRG – Dorsal root ganglion
CM – Conus medullaris

FIGURE 1.3: (a) The spinal cord, with its meninges, is situated within the vertebral canal. The vertebral levels are indicated on the left side of the illustration and the spinal cord levels on the right side. The cord terminates at the level of L2, the second lumbar vertebra (in the adult). The subarachnoid space (SAS) and lumbar cistern (LC) with CSF is shown (pale blue). Note the ligamentum flavum (LFl) and the epidural space (ES). (b) A photographic view of a (human) spinal cord with attached roots (dorsal and ventral) demonstrating its unsegmented appearance. The meninges (dura and arachnoid) have been opened and are displayed; note the dorsal root ganglion (DRG).
(Continued)

units responsive to stretch in the muscles (muscle spindles) and movement detectors in the joints, is also conveyed to the CNS via the discriminative touch pathway; much of this information does not reach the level of consciousness.

1.2.2.2 MOTOR ASPECTS

Neurons in the cerebral cortex (and brainstem) control the activity of the motor neurons in the spinal cord. One major pathway descends from the 'upper' levels of the nervous system (the cerebral cortex; see Section 1.2.7.2) to the spinal cord and is termed the *corticospinal tract* (Figure 1.5a; also discussed in Chapter 7); note that this is a crossed pathway (in the lower brainstem). The cortical neurons are known as the *upper motor neurons* (UMNs); the spinal cord neurons are called the *lower motor neurons*. The control of muscle activity is funnelled through the LMN, which is functionally the final common pathway for motor activity. The LMN, its motor axon and the muscle fibers it supplies are collectively known as the *motor unit* (see Figure 4.3).

1.2.2.3 REFLEXES

One of the most interesting aspects of muscle activity is a feedback mechanism whereby the muscle informs the nervous system about its degree of stretch. Receptors that gauge the degree of stretch are located among the regular muscle fibers; they are known appropriately as stretch

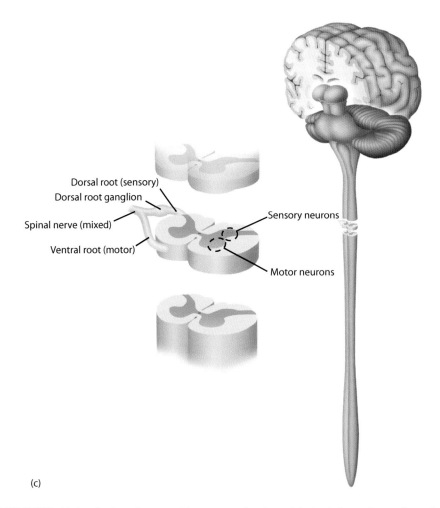

(c)

FIGURE 1.3 (CONTINUED): (c) A spinal cord segment in cross-section (an axial view) shows the configuration of the gray matter (neurons and synapses); sensory neurons are located in the dorsal horn and motor neurons in the ventral (anterior) horn. Ascending and descending tracts (pathways) are found in the surrounding white matter. The dorsal root with the dorsal root (sensory) ganglion and ventral (motor) root are also seen, forming the (mixed) spinal nerve.

receptors. These receptors are spindle shaped and are thus called *muscle spindles*. The afferent information from the muscle spindles is carried by a peripheral myelinated nerve fiber, which gives off a collateral branch in the spinal cord (see Figure 1.5b). This fiber synapses directly, by way of only a single synapse (i.e. monosynaptic), with an LMN supplying the muscle from which it originated. Activation of these receptors will normally lead to a reflex contraction of that very same muscle.

This reflex circuit, known as the (muscle) *stretch reflex*, the deep tendon reflex or myotatic reflex, is tested clinically by tapping on a tendon (e.g. the patellar tendon at the knee), which stretches the muscle and thereby activates the muscle spindles (Figure 1.5b) The muscle contracts (in this instance producing extension of the knee). This monosynaptic reflex requires the following elements:

- The muscle spindle with an intact functioning peripheral (myelinated) nerve carrying the afferent information

- The spinal cord, where afferent interfaces with efferent
- A motor neuron (a generic LMN) at the appropriate (lumbar) level of the spinal cord, with only a single synapse
- A (myelinated) motor axon travelling as a peripheral myelinated nerve, returning to the same muscle (with a neuromuscular synapse) to effect a reflex contraction

Reflex contraction of the muscle is graded in a standard way (see Table 2.3).

The sensitivity of this reflex circuit is influenced by neurons located in the brainstem reticular formation (see Section 1.2.4). Therefore, the assessment of reflex activity in the clinical setting not only is one of the most significant tests for motor functionality as well as spinal cord integrity but also is extremely important for overall assessment of the nervous system. Note that intact myelinated peripheral nerve fibers

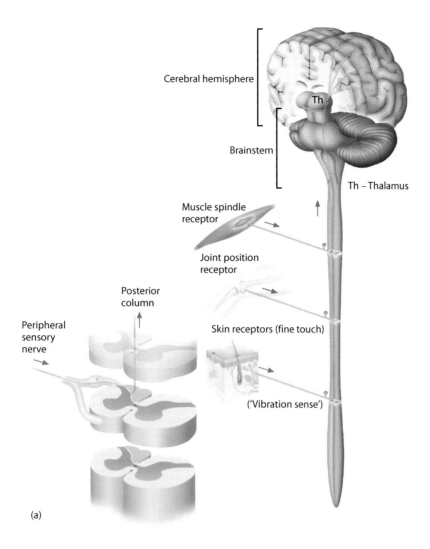

Cerebral hemisphere

Th

Brainstem

Th – Thalamus

Muscle spindle
receptor

Joint position
receptor

Posterior
column

Peripheral
sensory
nerve

Skin receptors (fine touch)

('Vibration sense')

(a)

FIGURE 1.4: (a) Sensory information from muscle spindles and joint receptors and for discriminative touch and vibration sensation enters the spinal cord and ascends on the same side, in the posterior (dorsal) column, crosses and is distributed via the thalamus to the postcentral gyrus. (*Continued*)

and functional neuromuscular junctions are also required for the reflex arc. Finally, healthy intact muscle, appropriate for the size and age of the individual, is needed for a response.

One of the remarkable features of the muscle spindles is their capability for resetting their sensitivity to the stretch stimulus; each spindle has within it a few muscle fibers that will reset the length of the spindle and thus alter its responsiveness. Specific neurons (called gamma motor neurons) located in the spinal cord supply the muscle fibers within each spindle.

Note that other reflexes that have a protective function, such as the response to touching a hot surface or stepping on a sharp object, involve more than one synapse.

To recapitulate, although there are many pathways in the spinal cord, both ascending (sensory) and descending (motor), from the clinical perspective, there are three that are highly relevant (note that each half of the cord has all three pathways):

- Two sensory pathways that ascend to higher levels – the posterior (dorsal) column (for discriminative touch, proprioception and vibration) and the spinothalamic pathway (for pain, temperature and crude [or non-discriminative] touch; Figure 1.4a and b)
- One major motor tract, the corticospinal (for voluntary motor actions), that descends from the cortex (Figure 1.5a)

1.2.2.4 The ANS

The ANS has two functional divisions, the sympathetic and parasympathetic. The sympathetic portion functions in circumstances of stress, for example, those requiring 'fight or flight' reactions (increased adrenaline, sweating, mobilization of glucose). The parasympathetic division is concerned with restoring energy and functions in quiet periods (such as after a big meal).

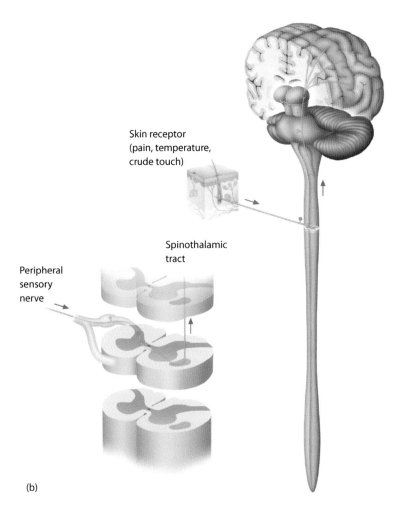

Skin receptor
(pain, temperature,
crude touch)

Spinothalamic
tract

Peripheral
sensory
nerve

(b)

FIGURE 1.4 (CONTINUED): (b) Sensory information for crude touch, pain and temperature enters the spinal cord, crosses and ascends, as the (lateral) spinothalamic tract and is distributed via the thalamus to the postcentral gyrus, as well as to other areas of the cortex.

Sympathetic outflow begins in the hypothalamus and descends through the brainstem (its fibers are located in the lateral aspect of the medulla). These fibers synapse with the preganglionic sympathetic neurons located in the 'lateral horn' of the spinal cord, a small gray matter excrescence, located from T1 to L2, between the dorsal (sensory) and ventral (motor) horns. Their axons exit with the ventral root and the fibers synapse next in the paraspinal sympathetic chain, located for the most part alongside the vertebral column. The sympathetic supply to the head region (particularly the pupil and the eyelid) is somewhat circuitous and will be discussed in Chapter 2 (with Figure 2.3).

Parasympathetic outflow starts in the lateral hypothalamus and descends to the brainstem for the control of heart rate, vasoconstriction, respiration, digestion and sweating. The parasympathetic outflow from the CNS is from the brainstem (see Section 1.2.3; with cranial nerves III, VII, IX and particularly with X, the vagus) and from the sacral division of the spinal cord (S2, 3, 4); the parasympathetic preganglionic neurons are found in the region of the conus medullaris, the lowest portion of the spinal cord (see Figures 1.3b and 5.3). Its ganglia are situated mostly nearer the target organ, including the gut, glands and the bladder wall (see Chapter 14).

1.2.3 THE BRAINSTEM

The brainstem, situated above the spinal cord and within the lower region of the skull, adds control mechanisms for basic movements, particularly in response to changes in position and to the effects of gravity. Most important, the nuclei controlling the vital functions of respiration and heart rate are found in the lower brainstem. In addition, there are groups of neurons located in the core of the brainstem, collectively known as the *reticular formation*; some of these neuronal groups have both a general effect on the level of activation of motor neurons of the spinal cord while others modulate the level of consciousness. The three major pathways continue through the brainstem – two

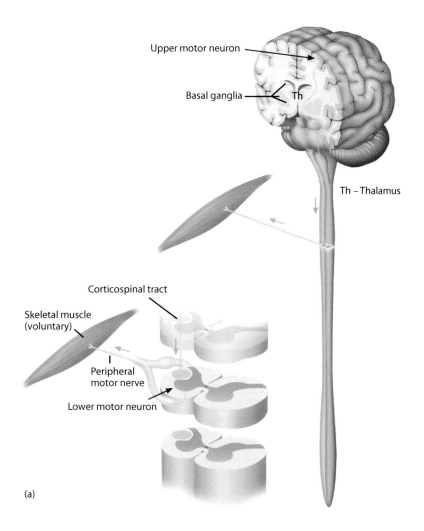

Upper motor neuron

Basal ganglia — Th

Th – Thalamus

Corticospinal tract

Skeletal muscle
(voluntary)

Peripheral
motor nerve

Lower motor neuron

(a)

FIGURE 1.5: (a) The motor system is organized at several levels, including the spinal cord, brainstem and cerebral cortex. The motor neurons of the cerebral cortex are the UMNs, while spinal cord neurons are the LMNs. The pathway descending from the cortical neurons to the spinal cord is the corticospinal tract. The axon of the LMN innervates skeletal (voluntary) muscle.

(Continued)

sensory (ascending) and one motor (descending). Finally, the brainstem is connected to the cerebellum.

The brainstem is divided into three parts, from above downward; each part is morphologically quite distinct (Figure 1.6a):

- The midbrain, demarcated by the cerebral peduncles, situated below the diencephalon
- The pons, distinguished by its prominent (anterior) bulge
- The medulla, with the pyramids on either side of the midline, which continues as the spinal cord

Attached to the brainstem are the sensory and motor nerves supplying the skin and muscles of the head and neck, known as the *cranial nerves* (usually abbreviated CN, Figure 1.6a). Within the brainstem are the nuclei,

sensory and motor, of these cranial nerves (Figure 1.6b and c):

- The nuclei of CN III (oculomotor) and IV (trochlear) are found in the midbrain; both of these nerves are involved with eye movements. In addition, the parasympathetic supply for constriction of the pupil runs with the CN III completing the connections for the pupillary light reflex (see Figure 2.2).
- The nuclei of CN V (trigeminal), VI (abducens), VII (facial) and VIII (vestibulo-cochlear) are all found within the pons. The trigeminal nerve supplies sensation to the skin of the face and is motor to the chewing muscles (mastication). The abducens nerve is responsible for lateral eye movement. The facial nerve supplies the

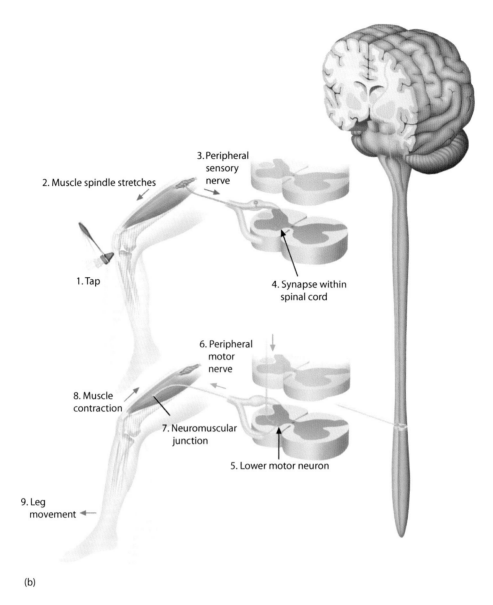

(b)

FIGURE 1.5 (CONTINUED): (b) The stretch reflex, a deep tendon monosynaptic reflex, is shown in this case as the patellar reflex. Upper figure: tapping on the tendon below the patella (1) causes a stretching of the muscle and firing of the muscle spindles (2); the afferents (3) enter via the dorsal root and synapse with motor neurons (4) in the spinal cord. This is a monosynaptic connection. Lower figure: the motor neurons (5) send axons via the ventral root (6) to the same muscle, and assuming a normal neuromuscular junction (7), the muscle contracts (8), causing the leg to extend at the knee (9). Note: This reflex circuit is animated on the text Web site.

muscles of facial expression (around the eyes and the lips); parasympathetic fibers supply some of the salivary glands and the lacrimal gland. The special senses of hearing and body motion are carried in the VIIIth nerve.

- The nuclei of CN IX (glossopharyngeal), X (vagus) and XII (hypoglossal) are all found in the medulla. The mucosa of the pharynx is supplied by the glossopharyngeal nerve, and the muscles of the pharynx and larynx, by the vagus. The vagus nerve provides the major parasympathetic innervation to the organs of

the chest and abdomen. The spinal accessory nerve, CN XI, is responsible for raising the shoulder and turning the head. (Its nucleus is actually in the spinal cord, but as it enters the skull, it has been included as one of the cranial nerves.) Movements of the tongue are controlled by the hypoglossal nerve.

Each cranial nerve emerges from the brainstem at approximately the level where its nucleus is located, except for CN V. Examination of these cranial nerves and the reflexes associated with each part of the brainstem is

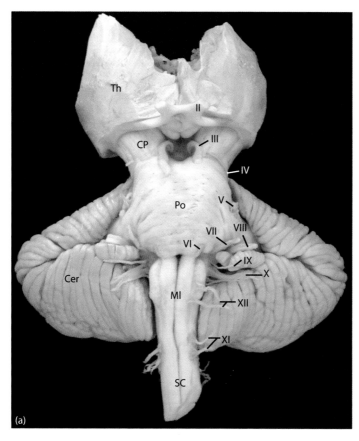

Th – Thalamus
CP – Cerebral peduncle (midbrain)
Po – Pons
Cer – Cerebellum
Ml – Medulla
SC – Spinal cord

Cranial nerves:
  II – Optic
  III – Oculomotor
  IV – Trochlear
  V – Trigeminal
  VI – Abducens
  VII – Facial
  VIII – Vestibulocochlear
  IX – Glossopharyngeal
  X – Vagus
  XI – Spinal accessory
  XII – Hypoglossal

FIGURE 1.6: (a) Photographic view of the (human) brainstem from the anterior (ventral) perspective, showing the medulla (Ml), the pons (Po) and the midbrain with its cerebral peduncles (CP). Cranial nerves III to XII (excluding CN XI) are attached to the brainstem. The optic nerves, CN II, are also present (along with the optic chiasm and tracts). The cerebellum is situated behind; the diencephalon (thalamus) is included, above. *(Continued)*

detailed in Chapter 2. Note that the neurons in the brainstem that innervate muscles of the head and neck are also considered LMNs.

The sympathetic supply to the head region follows a unique course (see Chapter 2 and Figure 2.3).

1.2.4 THE RETICULAR FORMATION

The reticular formation is found within the core area of the brainstem (Figure 1.6c). It exerts a diffuse effect on neuronal activity, both downward on the spinal cord and upward toward the diencephalon and cortex. It is best to think of this group of neurons as two systems: lower and upper.

Within the lower (medullary) component of the reticular formation are the previously mentioned neurons that control heart rate and respiration; lesions here may cause respiratory and cardiac arrest. Another role of the lower reticular system (via descending motor pathways) is to modify the excitability of the LMNs of the spinal cord; lesions or dysfunction of this neuronal pool will lead to changes in the responsiveness of skeletal muscles to passive stretch and altered stretch reflexes (further discussed in Chapters 2 and 7).

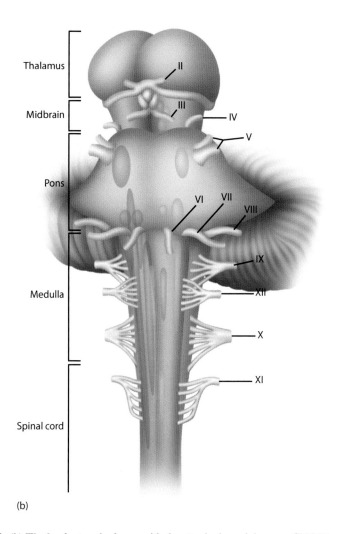

(b)

FIGURE 1.6 (CONTINUED): (b) The brainstem is shown with the attached cranial nerves CN III to CN XII (excluding CN XI), and their cranial nerve nuclei is shown from the same perspective. The nuclei of CN III and IV are located in the midbrain; CN V, VI, VII, and VIII in the pons; and CN IX, X, and XII in the medulla. The nucleus for CN XI is located in the upper spinal cord.   *(Continued)*

The projections of the neurons of the upper part of the reticular formation, the upper pontine and midbrain portions, are distributed widely to the next level, the thalamus, and beyond to the cortex. Lesions here will affect the level of arousal and consciousness (further discussed in Chapter 10).

1.2.5 THE CEREBELLUM

The cerebellum is situated at the back of the lower portion of the skull, in the posterior cranial fossa. It is positioned behind the brainstem (Figure 1.7a and b), with which it has many connections. The major contribution of the cerebellum is the facilitation of the smooth performance of motor activity and particularly voluntary motor movements (discussed further in Chapter 2).

The human cerebellum is fairly prominent, with easily observed narrow ridges of tissue, called *folia*.

Its cortex has a unique three-layered organization, with the prominent Purkinje neurons occupying the middle layer. A set of output nuclei is located deep within its structure (the deep cerebellar nuclei). The cerebellum is connected to the brainstem by three pairs of bundles of fibers called peduncles – the inferior cerebellar peduncles to the medulla (carrying mainly afferent information from the spinal cord and medulla), the middle ones to the pons (transmitting pontine input from the cerebral cortex) and the superior ones to the midbrain (carrying cerebellar efferents to the thalamus and from there to the cerebral cortex).

The midline (older, in evolutionary terms) portion of the cerebellum, the vermis, is connected to the vestibular system and is involved in the regulation of gait and balance. The lateral (newer) portions, the cerebellar hemispheres, are involved in coordination of motor activity of the limbs, receiving input from muscles and joints of the

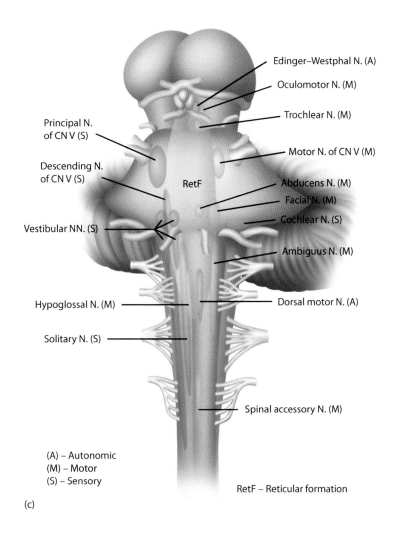

Edinger–Westphal N. (A)

Oculomotor N. (M)

Trochlear N. (M)

Principal N.
of CN V (S)

Motor N. of CN V (M)

Descending N.
of CN V (S)

RetF

Abducens N. (M)

Facial N. (M)

Cochlear N. (S)

Vestibular NN. (S)

Ambiguus N. (M)

Hypoglossal N. (M)

Dorsal motor N. (A)

Solitary N. (S)

Spinal accessory N. (M)

(A) – Autonomic
(M) – Motor
(S) – Sensory

RetF – Reticular formation

(c)

FIGURE 1.6 (CONTINUED): (c) The cranial nerve nuclei include sensory (S: CN V principal and descending, vestibular, cochlear, solitary), motor (M: oculomotor, trochlear, CN V, abducens, facial, ambiguus, spinal accessory, hypoglossal) and autonomic (A: Edinger–Westphal, dorsal motor). The reticular formation (RetF) is found within the core of the brainstem at all levels.

body and from motor regions of the cerebral cortex. (The connections of the cerebellum and its functional contribution to motor control are discussed in Chapter 7.) Recent evidence indicates that the cerebellum also contributes to a number of complex cerebral functions, such as learning, language, behaviour and mood stability.

1.2.6 THE DIENCEPHALON

The diencephalon is a small part of the brain located between the brainstem and the cerebral hemispheres (see Figure 1.1). The diencephalon is difficult to visualize as it is situated atop the brainstem (see Figure 1.6a and b; also see Figure 1.5a), deep within the hemispheres of the brain. One way of allocating its position within the brain is to locate it above the pituitary gland, which also happens to be the site of the optic chiasm.

The diencephalon has two major divisions, the hypothalamus and the thalamus, both consisting of a set of nuclei. The hypothalamus controls the activity of the pituitary gland; regulates autonomic functions, including temperature; and maintains water balance, as well as organizes certain 'basic' drives such as food intake; it is also involved in the regulation of sleep.

The thalamus consists of a number of relay and integrative nuclei with connections to and from (reciprocal) the cerebral cortex (Figure 1.8; see also Figure 13.3). With the exception of smell, all sensory pathways, including somatosensory, audition and vision, have their individual specific relay nucleus in the thalamus on the way to the cerebral cortex; this is the third-order neuron in the sensory chain. Other circuits of the thalamus involve the integration of motor regulatory data from the cerebellum and the basal ganglia (discussed next). Some nuclei of the

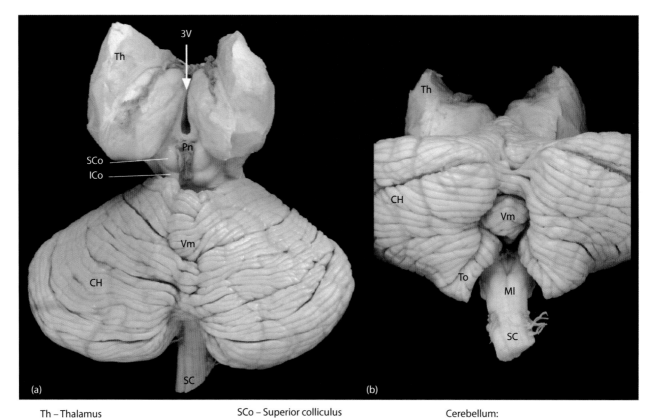

Th – Thalamus
3V – IIIrd ventricle
Pn – Pineal

SCo – Superior colliculus
ICo – Inferior colliculus
Ml – Medulla
SC – Spinal cord

Cerebellum:
   CH – Cerebellar hemisphere
   Vm – Vermis
   To – Tonsil

FIGURE 1.7: (a) Photographic view of the cerebellum and brainstem (and diencephalon), from the posterior (dorsal) perspective. The cerebellum includes the midline (vermis, Vm) and the lateral portions, the cerebellar hemispheres (CH). The posterior aspect of the midbrain includes the superior colliculus (SCo) and inferior colliculus (ICo); the IIIrd ventricle (3V) is also visible. (b) Photographic view of the cerebellum and brainstem, from the posterior and inferior perspective, showing the cerebellar tonsils (To) situated on either side of the lower medulla (Ml).

thalamus are reciprocally connected with other (association) areas of the cortex (see Figure 13.3), and still others have diffuse connections to wide regions of the cerebral cortex (see Figure 10.4). The latter projections are crucial for the maintenance of consciousness and the regulation of the different types of sleep. The details of the connections involving the thalamus will be explained with each of the systems in subsequent chapters.

1.2.7 THE CEREBRAL HEMISPHERES

The cerebral hemispheres, right and left, are the largest component of the brain; these are enormously developed in primates and more so in humans. This part of the CNS, often called simply the brain, fills most of the interior of the skull, including both the anterior and middle cranial fossae. Within the central portion of the hemispheres are large collections of neurons collectively called the basal ganglia. In addition, within the hemispheres are the cerebral ventricles, spaces with cerebrospinal fluid, the CSF.

1.2.7.1 THE BASAL GANGLIA

The basal ganglia are concerned with motor activity and other brain operations, functioning in collaboration with the cerebral cortex. The motor aspects involve particularly the organization of movement sequences, the control of agonist and antagonist muscle function and the storage of procedural memory (discussed further in Chapters 2 and 7). It is hard to describe what the basal ganglia actually do in motor control until their influence is altered by diseases such as Parkinson's disease and Huntington's chorea. Then one sees movements with too little or too much amplitude and abnormal movements such as tremor, chorea and athetosis. It is also clear that parts of the basal ganglia participate in circuits concerned with attention and executive function, as well as with emotion.

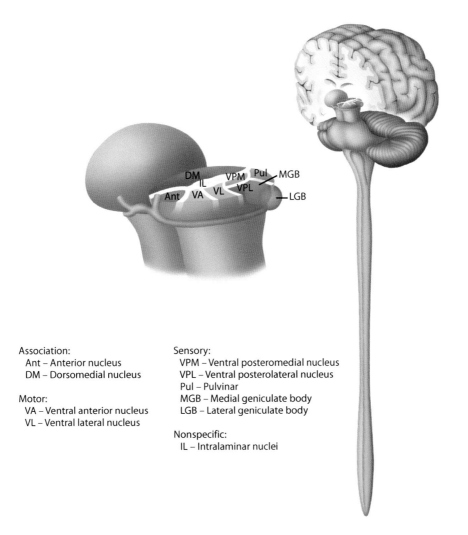

Association:
   Ant – Anterior nucleus
   DM – Dorsomedial nucleus

Motor:
   VA – Ventral anterior nucleus
   VL – Ventral lateral nucleus

Sensory:
   VPM – Ventral posteromedial nucleus
   VPL – Ventral posterolateral nucleus
   Pul – Pulvinar
   MGB – Medial geniculate body
   LGB – Lateral geniculate body

Nonspecific:
   IL – Intralaminar nuclei

FIGURE 1.8: The diencephalon, consisting of the thalamus and hypothalamus, is located between the brainstem and cerebrum (see Figure 1.4a). The thalamic nuclei are shown on one side: there are sensory relay nuclei (VPL, VPM, MGB, LGB, Pul), motor-related nuclei (VA, VL), reticular-related (nonspecific) nuclei (IL), as well as association nuclei (Ant, DM).

The basal ganglia include the following (Figure 1.9):

- The *caudate nucleus*, lying adjacent to the ventricle; it has the configuration of a large head and then a narrow body, with a trailing tail that follows the lateral ventricle into the temporal lobe of the brain (somewhat in the form of a comma; in the shape of an inverted letter C).
- The *putamen*, a globular nucleus located within the white matter of the hemispheres of the brain at the level of the lateral fissure (see Section 1.2.7.2).
- The *globus pallidus*, almost almond shaped, located medial to the putamen with two portions or divisions that are quite distinct and have different connections, external (lateral – usually referred to as the globus pallidus

externus) and internal (medial – the globus pallidus internus).

Terminology of the basal ganglia: The error in nomenclature referring to this CNS collection as ganglia should be noted; normally, *ganglia* refers to collections of neuronal cell bodies in the PNS. The alternative term for these nuclei is the *corpus striatum*, a term rarely used nowadays. The term *striatum* is used and refers to the caudate and putamen together because these are structurally linked; this portion of the basal ganglia is also called the neostriatum, based on its connections with the (neo)cortex. The putamen and globus pallidus together are called the lentiform (also called lenticular) nucleus.

Visualizing the configuration of the various parts of the basal ganglia is quite challenging, even after learning the names of the component parts. Different parts

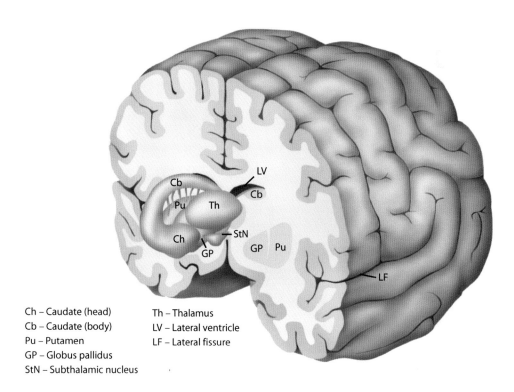

Ch – Caudate (head)
Cb – Caudate (body)
Pu – Putamen
GP – Globus pallidus
StN – Subthalamic nucleus

Th – Thalamus
LV – Lateral ventricle
LF – Lateral fissure

FIGURE 1.9: The basal ganglia are located within the white matter of the cerebral hemispheres. These include a number of nuclei: the caudate with its head (Ch) and body (Cb), the putamen (Pu) and the globus pallidus (GP). The subthalamic nucleus (StN), located below the thalamus, is functionally part of the basal ganglia. The cerebral lateral ventricle (LV) is also shown.

of the basal ganglia can be seen when the brain is visualized with computed tomography scans and with magnetic resonance imaging (see radiological examples on the Web site).

In order to complete the circuitry and connections of the basal ganglia, two nuclei are now added to the functional definition of the basal ganglia: the *substantia nigra*, a group of cells located in the midbrain, and the *subthalamic nucleus*, located below the thalamus. Input to the basal ganglia comes from all parts of the cerebral cortex; its output is funnelled through the thalamus and returns, in the case of motor activity, to the motor areas of the cortex. (Detailed connections of the basal ganglia will be elucidated in Chapter 7.)

1.2.7.2 THE CEREBRUM

The cerebrum, which refers to the cerebral hemispheres without the basal ganglia, is the part of the CNS that contains the billions of microprocessors and is what most people are thinking about when talking about 'the brain'; it occupies most of the interior of the skull. In fact, the two halves of the cerebrum in humans are by far the largest component of the CNS, dwarfing in size the brainstem, the cerebellum and the diencephalon. It is this part of the brain that is needed for 'thinking' (reasoning and planning), as well as language and emotion. In addition,

in higher mammals, activities of the cerebrum seem to dominate and control all aspects of CNS function.

Structurally, the cerebrum consists of an outer rim of billions of neurons and their dendrites and synapses, the *cerebral cortex* (gray matter), arranged as layers of different types of neurons (see Figure 10.5). It is the cerebral cortex with its intricate synaptic connections that is the biological substrate for the highest integration of nervous activity, what we call higher mental functions, including language.

The surface of the human cerebrum is characterized by sulci (deep indentations) and fissures (even deeper), with the cortex following the surface contour; this arrangement enormously increases the actual surface area of the cortical gray matter. Although there are variations in the configuration of sulci, there are certain major fissures that divide the hemispheres into cerebral lobes: the central fissure (of Rolando), the lateral fissure (of Sylvius) and the parieto-occipital fissure. Based on these fissures, there is a standard way of dividing the cortex into lobes (Figure 1.10a and b): frontal, parietal, occipital and temporal.

The *frontal lobe*, the area in front of the central fissure, is generally described as having chief executive functions, in other words the part of the brain involved with major planning and decision making. The *parietal lobe*, between the central fissure and the parieto-occipital fissure, functions mainly in the reception and integration

Lobes of the cerebral cortex

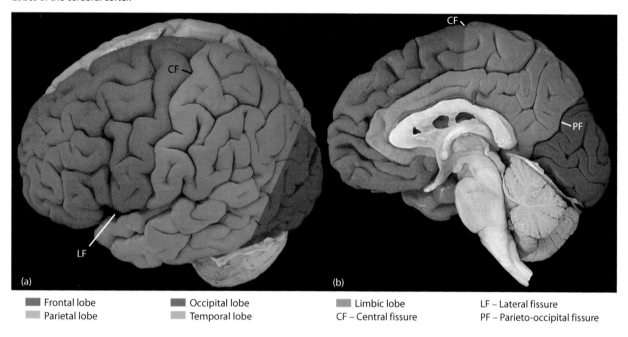

■ Frontal lobe	■ Occipital lobe	■ Limbic lobe	LF – Lateral fissure
■ Parietal lobe	■ Temporal lobe	CF – Central fissure	PF – Parieto-occipital fissure

Functional areas of the cerebral cortex

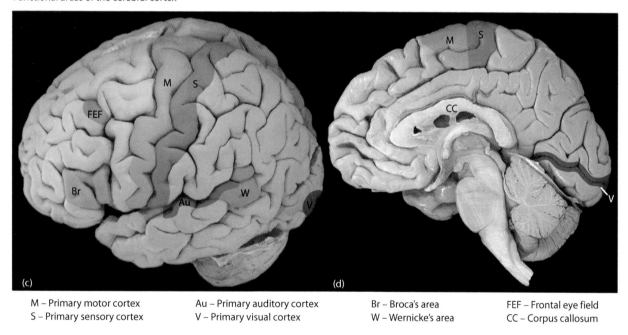

M – Primary motor cortex	Au – Primary auditory cortex	Br – Broca's area	FEF – Frontal eye field
S – Primary sensory cortex	V – Primary visual cortex	W – Wernicke's area	CC – Corpus callosum

FIGURE 1.10: Photographic views of the cerebral hemispheres are shown from the lateral (a and c) and the medial (mid-sagittal) (b and d) views. (a, b) The cerebral cortex of the cerebral hemispheres is divided into lobes: frontal, parietal, temporal and occipital. The major fissures are central (CF), lateral (LF) and parieto-occipital (PF). (c, d) Gyri within each lobe have specific functions: motor (M, precentral) and somatosensory (S, postcentral), auditory (Au), and visual (V). The two language areas in the 'dominant' (left) hemisphere are Broca's (Br) for expressive language and Wernicke's (W) for phonologic retrieval (see text). Voluntary eye movements are organized in the frontal eye field (FEF). Note the corpus callosum (CC).

of the various sensory inputs and contributes significantly to spatial orientation. The *occipital lobe*, behind the parieto-occipital fissure and mostly located on the medial and inferior aspects of the brain, has predominantly visual functions. The *temporal lobe*, located below the lateral fissure, has various sound-related association tasks on its lateral portion and structures important for memory formation and emotion in its medial portion (see Section 1.2.9).

The ridges of tissue between the sulci, called *gyri*, also vary somewhat but some of these have a reasonably constant location and are known to have an assigned function, motor or sensory. The most characteristic are as follows (Figure 1.10c and d):

- The *precentral gyrus*, in front of the central fissure, with motor functions related to the limbs and the face, the primary motor area
- The *postcentral gyrus*, located behind the central fissure, with sensory functions related to the limbs and the face, the primary somatosensory area
- The primary auditory area, in the upper temporal lobe (also known as Heschl's gyrus), located within the lateral fissure
- The primary visual area of the occipital lobe, located on the medial surface along both banks of the calcarine fissure (often called the calcarine cortex)

The portions of the cerebral cortex that are not related directly to sensory and motor functions are known as *association areas*. Areas adjacent to the primary sensory areas are involved in the elaboration and interpretation of the sensory input to that area.

Language: The left hemisphere in humans is most frequently the repository for language function and is called the dominant hemisphere. There are two distinct language centres, one for expressive functions, *Broca's area*, located in the inferior aspect of the frontal lobe (in the frontal opercular area), and the other known as *Wernicke's area*, located in the superior and posterior area of the temporal lobe and also within the lateral fissure. Wernicke's area has traditionally been assigned the function of language reception/comprehension. In a recent paper (by Binder, 2015), evidence is given that this area is involved in phoneme retrieval, used in all aspects of speech production; modern imaging and neuropsychological studies indicate that wide areas of the lateral temporal lobe and also the parietal lobe support speech comprehension, perhaps involving both hemispheres.

Memory: Memory for names, places and events is called *declarative memory*. One structure, the hippocampus, located in the medial portion of the temporal lobe, is necessary for the recording of memories of this type.

Working memory is a term used to describe memory 'in use', as a problem or task is being completed; portions of the frontal lobe are required for this type of memory function. Long-term memory is stored in widely distributed neuronal networks involving all parts of the brain (discussed further in Chapter 2).

There is a different type of memory system, called *procedural memory*, involved with performing motor acts, such as riding a bicycle, driving a car or various sporting activities. These motor sequences are usually learned after countless repetitions. Memories of this type are 'stored' in other areas of the brain such as the basal ganglia (also discussed in Chapter 2).

1.2.7.3 CEREBRAL WHITE MATTER

Beneath the cortex and within the cerebrum are large bundles of axons, mostly myelinated, interconnecting the various parts of the brain: the white matter (Figure 1.11). This elaborate interconnectivity underlines the fact that the brain acts as a totality, with all parts contributing to the whole even while some parts are more responsible for selective functions; this is its integrative role (hence the title). The fiber tracts (both sensory and motor) that project to and from the cerebral cortex funnel through a narrow channel between the basal ganglia and the thalamus, known as the *internal capsule* (see Figures 1.11 and 7.10b). This region is prone to infarction (discussed in Chapter 8). The large white matter bundle that interconnects the two hemispheres is the *corpus callosum* (see Figures 1.10 and 1.11).

1.2.8 THE CEREBRAL VENTRICLES

The CNS develops from the walls of the embryological neural tube. The original tube, filled with fluid, remains in each part of the nervous system. The largest spaces are found in the cerebral hemispheres; they are called the cerebral ventricles or lateral ventricles (also ventricles I and II). The fluid within the ventricular system is the CSF, which also envelops the exterior of the nervous system as part of the meningeal covering of the brain and spinal cord. (The ventricular system, the meninges and the CSF circulation are described in Chapter 9.)

1.2.9 THE LIMBIC SYSTEM

The areas of the brain involved in the reactions to emotionally laden stimuli and to events, situations and people are collectively called the limbic system. This system gives rise to our physiological responses (e.g. sweating, rapid pulse), our psychological reactions (e.g. fear, anger) and motor responses (e.g. fight, flight).

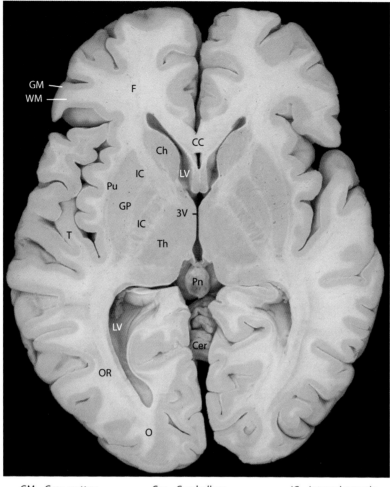

GM – Gray matter	Cer – Cerebellum	IC – Internal capsule
WM – White matter	CC – Corpus callosum	Th – Thalamus
F – Frontal lobe	Ch – Caudate (head)	LV – Lateral ventricle
T – Temporal lobe	Pu – Putamen	3V – IIIrd ventricle
O – Occipital lobe	GP – Globus pallidus	Pn – Pineal
		OR – Optic radiation

FIGURE 1.11: Horizontal section (photographic view) through the cerebral hemispheres shows areas of neurons (gray matter), including the cerebral cortex, the thalamus and the basal ganglia, and areas of axons (white matter), including the internal capsule (IC) and the optic radiation (OR). Ventricles are also seen (LV = lateral ventricle, 3V = IIIrd ventricle).

The limbic system includes cortical areas and nuclei in the anterior and medial aspects of the temporal lobe, including the hippocampus along with the amygdala, and subcortical nuclei within the diencephalon (both thalamus and hypothalamus) and the brainstem. (The limbic system is discussed in Chapter 13.)

1.2.10 THE VASCULAR SUPPLY

Brain tissue is dependent on a continuous supply of both oxygen and glucose for its function and viability. The blood supply of the nervous system will be discussed in Chapter 5 (spinal cord), Chapter 8 (cerebral) and Chapter 9 (the meninges).

REFERENCE

Binder, J.R. The Wernicke area. Modern evidence and a reinterpretation. *Neurology* 2015. 85: 2170–2175.

SUGGESTED READING

Haines, D.E., Ed. *Fundamental Neuroscience*, 4th ed. London: Churchill Livingstone, 2012.
Hendelman, W. *Atlas of Functional Neuroanatomy*, 3rd ed. Boca Raton, FL: CRC Press, Taylor & Francis, 2016.
Nolte, J. *The Human Brain*, 7th ed. St. Louis, MO: Mosby, 2015.
See also Annotated Bibliography.

Chapter 2

Neurological Examination

Objectives

- To integrate the anatomical information of the nervous system with the neurological examination
- To understand the significance of abnormal signs during the neurological examination for the localization of the lesion

2.1 INTRODUCTION

There is usually a routine sequence for carrying out a neurological examination in an adult. While taking the history, the neurologist is interacting with the patient and is already assessing language and cognition, as well as emotional facets. As the examination is proceeding, the neurologist attempts to localize the problem and is considering possible etiologies. Attention may be focused on a particular area of the nervous system, based on the history and as the neurologic signs are found during the assessment. Many physicians will choose to do the motor and sensory examination first, so as to get to know the person and allow him or her to feel more relaxed, considering the unfamiliar (and confined) space of most examining rooms and the not unexpected tension associated with a physical examination of any type.

The neurologic examination of young children requires a different and flexible approach, as discussed at the end of this chapter. Examination of the nervous system in newborns and young infants is beyond the scope of this book.

Formal testing begins with the cranial nerves, each of which is tested either individually or in groups. In the process, one is assessing the integrity of each portion of the brainstem. This is followed by an examination of the upper limbs, both motor and sensory aspects, including reflexes – comparing one side with the other; this is followed by examination of the lower extremities. Finally, coordination and gait are tested. Fundoscopic (ophthalmoscopic) examination can be postponed until the end when the room is darkened; it should be performed after testing of visual acuity and visual fields in order to avoid dazzling the patient with bright light. Formal testing of higher mental functions can also be carried out at the end.

Note: A neurological screening examination of a normal adult is available on the text Web site.

2.2 SYSTEMS

The nervous system is best examined from the perspective of functional systems according to the following template.

Cranial nerves:
- CN I – olfaction
- CN II – the visual system
  - Including fundoscopy
- CN III–XII
  - Including hearing (auditory) and balance (vestibular)

Motor systems:
- Voluntary (intentional, purposeful)
  - Precisional (skilled)
  - Programmed (procedural)
- Nonvoluntary (postural)

Sensory systems:
- Somatosensory – from the skin and muscles
  - Vibration, proprioception and discriminative touch
  - Pain and temperature

Mental status:
- Executive function
- Language
- Memory

Emotional status

Note: The integrity of the autonomic nervous system is typically screened by way of the assessment of bladder and bowel functions. (These are described in Chapter 14.)

2.3 THE CRANIAL NERVES

CN I and CN II: Although always included in the list of cranial nerves, both 'nerves' are in fact white matter extensions or tracts of the central nervous system (CNS) (with the myelin of CNS). Both carry a special sense.

2.3.1 CN I: OLFACTION (SMELL)

Olfaction is not highly developed in humans, yet smell is an important component of our lives and plays a significant part in the partaking of food and beverages. Any potent and distinctive aroma can be used (e.g. cinnamon or cloves). The patient is asked to 'sniff' the vial with one nostril closed and to identify the nature of the aroma; then the other nostril is tested.

Pathway: The sensory cells for smell are located in the roof of the nose, and active sniffing is needed to get the odour to activate them. After processing in the olfactory bulb, fibers travel along the base of the frontal lobe, as the olfactory tract, to end in the anterior temporal lobe and influence the limbic system (discussed in Chapter 13). This is the only sensory system whose pathway does not pass through the thalamus.

2.3.2 CN II: VISUAL SYSTEM

We humans are predominantly visual creatures and visual deficits are important to detect. There are many aspects to vision. Visual acuity (discrimination of shapes and borders) and colour vision are both found in the central foveal portion of the retina, where the cones are located. A Snellen chart is used for testing of visual acuity, with each eye examined separately (while the other eye is covered) and the best corrected visual acuity in each eye is recorded. Peripheral visual images, including movement as well as images seen under low levels of illumination, are detected by the rods, found throughout the non-foveal (peripheral) portions of the retina.

Recognition of movement is used at the bedside to detect a lesion in the visual system: retina, optic nerve, optic chiasm, optic tract, visual radiation and visual cortex. As the visual pathway spans the brain from the front to the back, its examination is most important in detecting any lesion within the cerebral hemispheres. This is done through examination of the visual fields.

Testing is done by the examiner facing the patient, with the patient covering one eye and both of them looking at each other 'straight in the uncovered eye'. The examiner, with arms extended, moves one or more digits in either or both hands and the patient is asked to report when he or she sees the fingers moving on one or the other side. (This requires a fair degree of cooperation and may have to be modified for the examination of young children or adults with dementia.)

Pathway: The visual fields for each eye can be charted using a perimetry apparatus (explained in Chapter 9, including the foveal region, the blind spot and the periphery; Figure 2.1). The visual field is then divided in half; deficits in vision are described for each eye in terms of the temporal or nasal half of the visual field: a temporal (lateral) or nasal (medial) hemianopia. If the visual loss involves the visual field on the same side in both eyes, the clinical term is *homonymous* hemianopia; if not, the term is *heteronymous* hemianopia. (The details of the visual pathway are explained in Chapter 9, which will help to clarify the usage of this terminology.) A defect in the central region of the visual field that does not fit the boundary of a quadrant is called a *scotoma*.

The visual cortex identifies objects on the basis of shapes and borders, movement and colour. Further elaboration of the visual information is performed in adjacent areas of the parieto-occipital lobe cortex. Higher mammals have a distinct region for facial recognition. Reading is accomplished by transferring the information to the

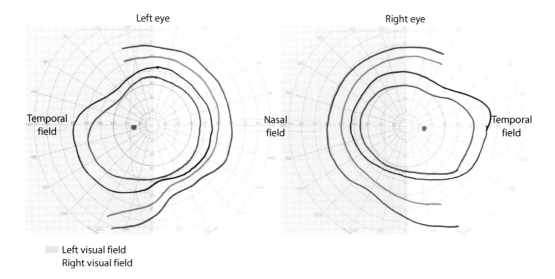

Left eye   Right eye

Temporal field   Nasal field   Temporal field

Left visual field
Right visual field

FIGURE 2.1: Perimetry chart. This depicts the (normal) visual fields of both eyes. The visual field of each eye is divided in half: lateral (temporal) and medial (nasal). The central area for vision is the fovea. The axons leave the eye at the optic disc to form the optic nerve; this is the site of the 'blind spot' (shown in red). (Courtesy of Dr. M. O'Connor.)

posterior temporal lobe and adjacent visual association cortex in the left (dominant) hemisphere (discussed in Chapter 1).

Fundoscopy: Examination of the retina and the optic disc with an ophthalmoscope usually requires the room to be darkened. The examiner employs the same eye as the one being examined (i.e. right eye for the patient's right eye) to visualize the light reflex of the patient through the ophthalmoscope, then moves closer to examine the optic disc, blood vessels and surrounding retina. A normal optic disc with the blood vessels in that area is shown in Figure 9.1.

2.3.2.1 PUPILLARY LIGHT REFLEX

The pupillary light reflex is an essential basic reflex. This reflex is tested by shining a pinpoint light on the eye and noting the reaction of the pupil in the eye stimulated as well as that of the other pupil – the consensual light reflex. Both eyes must be tested separately. The pupillary light reflex is used in anesthetized and comatose patients to test for the integrity of the midbrain portion of the brainstem.

Pathway (Figure 2.2): The visual image, light in this case, is carried from the retina into the optic nerve; after the partial crossing of fibers in the optic chiasm, the stimulus is carried in the optic tract. Some of these fibers leave the optic tract and go to the dorsal area of the midbrain, in the pretectal area, where they terminate in a group of cells called the *light reflex centre*; the information is relayed immediately to the same group of neurons on the other side.

Next, the neurons in the pretectal light reflex centre project to a group of cells known as the *Edinger–Westphal (E–W) nucleus*; these are parasympathetic neurons that form part of the nucleus of CN III. These cells send out parasympathetic fibers with CN III (the

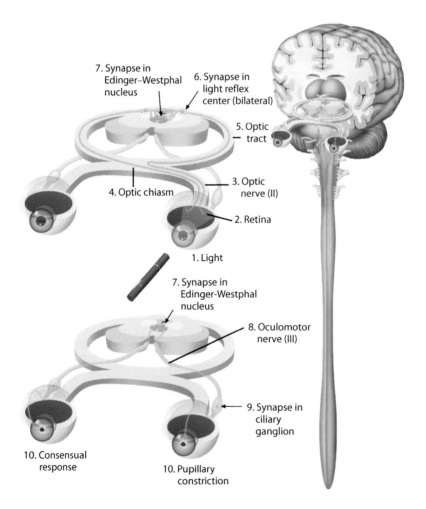

FIGURE 2.2: Pupillary light reflex. Upper illustration: Shining a light onto one eye (1) stimulates cones in the retina (2). Afferents from the retina travel in CN II, the optic nerve (3), through the optic chiasm (4) and via optic tract (5) to the light reflex center of the midbrain and synapse (6), with fibers going to the other side. Lower illustration: The reflex center relays next to the Edinger–Westphal (EW) nucleus (7), part of the IIIrd nerve nucleus in the upper midbrain. Parasympathetic fibers exit with CN III, the oculomotor nerve (8), synapse in the ciliary ganglion in the orbit (9), and supply the constrictor muscle of the pupil and causing constriction of the pupil (10). This is the direct pupillary reflex response; the response of the other eye is called the consensual light reflex (10). This reflex circuit is animated on the text Web site.

oculomotor nerve). After synapse in the ciliary (para-sympathetic) ganglion in the orbit, the postganglionic fibers cause release of the neurotransmitter Acetylcholine (Ach) to stimulate the muscle of iris to contract resulting in pupillary constriction.

2.3.3 SYMPATHETIC TO PUPIL AND EYELIDS

The sympathetic innervation to the eye results in pupillary dilatation and elevation of the eyelid via Mueller's muscle. These fibers follow a unique course (see Figure 2.3). The fibers descend from the hypothalamus through the brainstem where they are located in the lateral aspect of the medulla. Following a synapse in the upper spinal cord, the fibers exit and ascend in the extension of the sympathetic chain in the neck, synapsing in the superior cervical ganglion. The postganglionic fibers then travel with the carotid then the ophthalmic arteries to supply the dilator of the pupil and the smooth muscle that is needed to elevate the upper eyelid. A lesion anywhere along this pathway produces the Horner's syndrome (discussed with e-case 8e.1 [BJ]).

2.3.4 CN III, IV AND VI

These three nuclei, which innervate the muscles that move the eye, are collectively called the visuomotor nuclei. Testing is done by asking the patient to follow the tip of a pencil in eight specific positions: to both sides, up and down, inward down and up, and laterally down

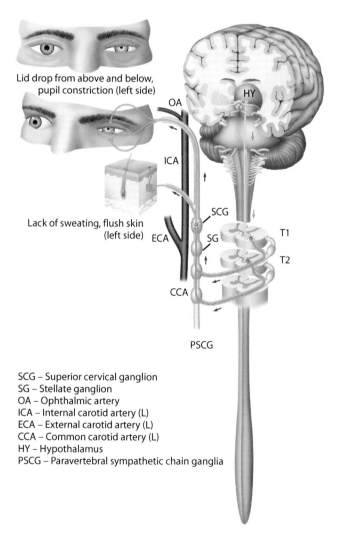

SCG – Superior cervical ganglion
SG – Stellate ganglion
OA – Ophthalmic artery
ICA – Internal carotid artery (L)
ECA – External carotid artery (L)
CCA – Common carotid artery (L)
HY – Hypothalamus
PSCG – Paravertebral sympathetic chain ganglia

FIGURE 2.3: Sympathetic supply to pupil and eyelid elevator. The sympathetic pathway to the head region begins in the hypothalamus and descends through the lateral aspect of the brainstem. The fibers synapse in the lateral horn and exit to the sympathetic chain (T1–T2) and then ascend to the superior cervical ganglion, where they again synapse. The postganglionic fibers travel with the arteries to the dilator muscle of the pupil and to the muscle that elevates the upper eyelid. The figure also shows the clinical findings in Horner syndrome.

and up, as well as in a rotatory manner. The examiner notes whether both eyes are moving symmetrically and whether there is any oscillation of the eyeball (called nystagmus) when the eye is held in position looking to either side. Movement of the eye inward (adduction) as well as upward is done exclusively by CN III; CN IV acts to move the eye down and inward; and CN VI is the nerve involved exclusively in moving the eye laterally (abduction).

Pathway: The oculomotor (III), trochlear (IV) and abducens (VI) nuclei all control eye movements. CN III and IV are located in the midbrain and CN VI in the lower pons. The eyes are yoked for carrying out movements in the horizontal plane, requiring the medial rectus muscle of the eye moving inward (adduction, CN III in the midbrain on one side) and the lateral rectus muscle of the other eye (abduction, CN VI in the pons on the other side). There is a distinct pathway that interconnects these nuclei, known as the medial longitudinal fasciculus, so that movements of the eyes are carried out in a coordinated manner.

2.3.5 CN V

Facial sensation (via the trigeminal nerve) is tested for the two principal types of sensation (discriminative touch; pain and temperature), as is done with other parts of the body. Three distinct areas are tested: the forehead region, the cheek and the jaw area, corresponding to the three divisions of the trigeminal nerve (ophthalmic, maxillary and mandibular). With eyes closed, the person is asked to compare the quality of the sensation on the two sides.

Asking the patient to open and close the mouth and to grind the teeth (from side to side) tests the motor portion of CN V (the muscles of mastication). The jaw reflex is tested (gently) by tapping on the point of the chin with the jaw relaxed, using a reflex hammer.

Pathways: CN V, the trigeminal nerve, carries the sensory fibers coming from the face and lips, as well as sensation (not taste) from the surface of the tongue. The discriminative touch pathway for facial sensation synapses in the pons, crosses the midline and ascends to the ventral posteromedial nucleus of the thalamus and then distributes to a separate face area of the postcentral gyrus (the sensory homunculus is discussed in Section 2.5.1; see also Figure 4.2).

Pain and temperature sensations form a distinct pathway starting in the pons, descending through the medulla and into the upper spinal cord; this is called the descending (spinal) trigeminal tract. After a relay in a nucleus on the same side, the pathway crosses and ascends, joining with the pain and temperature pathway from the body regions below the neck; the sensory

information reaches the cortex via the thalamus and is distributed to the facial portion of the postcentral gyrus as well as to other areas (like somatic pain; see Section 2.5.2).

2.3.6 CN VII

The facial nerve is tested by asking the person to look upward and wrinkle the forehead; one notes the presence or absence and symmetry of creases in the forehead. Then the patient is asked to close the eyes tightly, at first alone and then against resistance applied by the examiner. Other muscles of facial expression are tested by asking the patient to show his or her teeth and to smile, looking for any asymmetry, particularly at the corners of the mouth.

Pathway: The facial nucleus, which controls all the muscles of facial expression, is located in the lowermost pons. After an unusual loop within the brainstem, the fibers leave at the cerebello-pontine angle and exit the skull to be distributed to the upper and lower face. Part of this nucleus innervates the muscles of the forehead (tested by asking the person to look upward) and the eyelid (tested by the blink/corneal reflex and by asking the person to close the eyes against finger resistance); the other part of this nucleus innervates the muscles of the lower face (around the mouth, tested by asking the person to smile).

2.3.7 CORNEAL REFLEX

The corneal reflex is another basic reflex that is most often tested in unconscious patients to check for the integrity of the pons (Figure 2.4). The cornea is gently stimulated with a wisp of (clean) cotton or tissue; this is an irritant to the cornea and is picked up by nociceptive fibers of CN V (the ophthalmic division).

Pathway: The sensory information is carried by the pain and temperature pathway, in the descending tract and nucleus of V (in the medulla). The reflex circuit goes to CN VII (the nucleus of the facial nerve, in the pons) of both sides. The reflex consists of a closure of the eyelids, on both sides.

2.3.8 CN VIII

2.3.8.1 Auditory System

Hearing is one of the special senses and is one of the two afferents that comprise the VIIIth nerve, the vestibulocochlear. Hearing is tested by using a tuning fork (activated) or by whispering near one ear and asking whether the person can hear the sound(s), again comparing the two sides. A tuning fork is also used to test the auditory

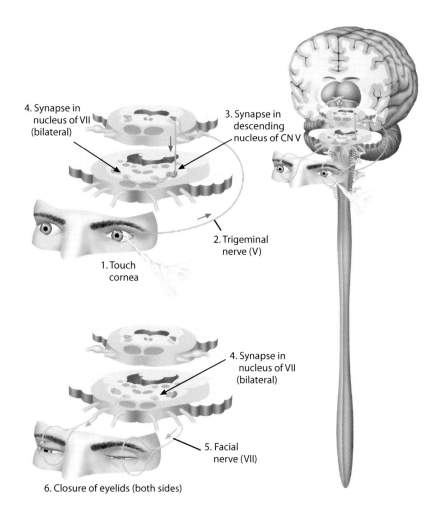

4. Synapse in nucleus of VII (bilateral)

3. Synapse in descending nucleus of CN V

2. Trigeminal nerve (V)

1. Touch cornea

4. Synapse in nucleus of VII (bilateral)

5. Facial nerve (VII)

6. Closure of eyelids (both sides)

FIGURE 2.4: Corneal reflex. Upper illustration: Any stimulus to the cornea (1) is carried via CN V, the trigeminal nerve (2). These enter the pons and descend, synapsing in the descending (trigeminal) nucleus of CN V (3); next there is a relay to the nucleus of the facial nerve, on both sides (4). Lower illustration: Fibers exit in CN VII (5) to supply the muscles of the eyelid, which results in eyelid blinking, i.e. closure of the eyelids (6), on both sides. This reflex circuit is animated on the text Web site.

input of the two ears, utilizing what are called the Weber and Rinne tests (fully described in Chapter 6). (The details of the auditory pathway are also described in Chapter 6.)

2.3.8.2 VESTIBULAR SYSTEM

The vestibular system is usually tested only if there is a history of vertigo (dizziness) or gait imbalance. The specific test is called the Romberg and its extension, tandem walking, as well as the ability to turn sharply and change direction (further discussed in Section 2.4.5.1 and in Chapter 6). (The details of the vestibular pathway, as well as dizziness, are also described in Chapter 6.) The examiner looks for nystagmus appearing with changes in head and body position and on testing of eye movements (examples of nystagmus are shown with e-case 6-1 [Erno] on the text Web site).

2.3.9 CN IX AND X

The throat and tonsillar areas are supplied by sensory fibers from CN IX; the vagus nerve (CN X) supplies the muscles of the soft palate and the constrictor muscles of the pharynx, as well as parasympathetic innervation to the upper gastrointestinal tract. The sensory supply to the laryngeal area and the innervation of the muscles of the larynx are both supplied by CN X.

2.3.9.1 GAG REFLEX

Stimulation of the area of the throat and tonsillar area with a tongue depressor leads to a gag reflex (Figure 2.5), which is a reflex to protect the upper airway. The response consists of several components, which include a 'lifting up' of the soft palate (which can be seen) and, in some individuals, a strong gag response (possibly including vomiting).

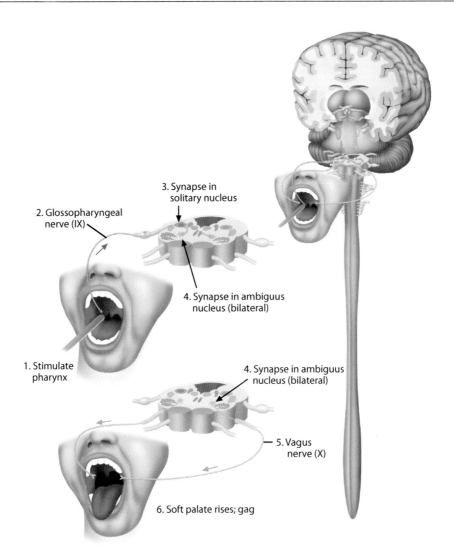

FIGURE 2.5: Gag reflex. Upper illustration: Stimulation of the pharynx with a tongue depressor (1) stimulates afferents, which are carried via CN IX, the glossopharyngeal nerve (2), to the solitary nucleus located in the medulla (3). The fibers relay to the nearby ambiguus nucleus of the vagus in the medulla, on both sides (4). Lower illustration: Fibers exit as part of CN X, the vagus nerve (5), to the muscles of the soft palate. The reflex action consists of a lifting of the palate (which is shown) and a constriction of the muscles of the pharynx, that is, the gag (6). Note: This reflex circuit is animated on the text Web site.

Pathway: The sensory fibers from CN IX terminate in the solitary nucleus in the medulla and connections are made with the motor nucleus of CN X, also located in the medulla, on both sides. The motor response is carried by CN X, the vagus nerve. Testing of the gag reflex helps to verify that the medullary level of the brainstem is functioning.

2.3.10 CN XI

The person is asked to raise the shoulder on each side and this can be tested against resistance. This manoeuvre tests the strength of the trapezius muscle on each side. The person is also asked to rotate the head to one side and then the other, against resistance. This manoeuvre tests the strength of the sternocleidomastoid muscle on each side.

Pathway: The spinal accessory nerve is in fact located in the upper part of the cervical spinal cord, but because of its circuitous route upward into and then downward out of the skull, it has traditionally been included as one of the cranial nerves. These fibers supply the upper portion of the trapezius muscle which elevates the shoulder, and the sternomastoid muscle which turns the head.

2.3.11 CN XII

The patient is asked to stick out his or her tongue and move it about. One looks for abnormal small worm-like movements (fasciculations). One also notes the surface of the tongue itself and whether there is any difference in appearance between the two sides, such as unilateral atrophy. Weakness of the tongue muscles can be tested by asking the patient to press the tongue inside the cheek with the examiner's finger on the outside resisting the movement. If one side of the tongue is completely paralyzed, the intact muscles on the opposite site will push (deviate) the tongue toward the weak side.

Pathway: The hypoglossal nerve supplies all the muscles of the tongue as a purely motor nerve; the innervation is to the muscles on the same side. Its nucleus is also located in the medulla.

2.4 MOTOR SYSTEMS

2.4.1 MOTOR REGULATION

Before elaborating on the examination of the motor system, it is necessary to outline the overall conception of motor regulation. Control of movements is complex and requires several systems, working in parallel and in harmony. We know that many components of the motor system are involved because we see a breakdown of smooth coordinated movements and/or abnormal movements as a consequence of disease and injury to different parts of the CNS. There may be a partial or one-sided weakness (paresis) or paralysis (assuming the absence of muscle disease); movements may be too slow, too rapid, imprecise, not done in the proper sequence (fragmented, uncoordinated), or there may be movements that are not controllable (involuntary) such as tremors, twitches or flinging movements (ballism).

Three sets of muscle groups need to be considered under the term *motor regulation:*

- Distal musculature (fingers and toes)
- Proximal joint musculature (elbow and shoulder of the upper limb and equivalent muscles of the lower limb)
- Axial musculature (the spine)

(The motor system is further discussed with the problems presented in Chapters 4 and 7.)

The control of the small muscles of the hand and fingers is the most recently acquired in evolution (in primates and humans). Intentional voluntary movements need precise, rapidly implemented control of specific muscle groups that necessitates a dedicated and speedy pathway from the motor cortex to the spinal cord, the *corticospinal tract*. The usual example given for skilled movements involves the distal musculature of the fingers and hand (for example, the work of a watchmaker, playing a musical instrument, writing a letter, performing robotic surgery).

In order to carry out these movements, associated adjustments of the larger joints by proximal muscle groups of the elbow and shoulder are necessary. Usually, postural adjustments (e.g. of the back while playing the piano) are also essential in order for the volitional movement to be carried out successfully.

Other types of skilled motor performances require the direct involvement of these proximal muscle groups (for example, using a pottery wheel, professional tennis playing, figure skating). It is important to note that feedback mechanisms (via the cerebellum; discussed in detail in Chapter 7) are required for adjustment and refinement so that the final movement is exactly as required.

Some purposeful voluntary action sequences have become routine, such as tying one's shoelaces, driving a car and riding a bicycle. These complex motor patterns are called into action and can be accomplished without any real conscious effort, seemingly done on 'automatic pilot', although learning them has usually required countless repetitions with progressive error correction. (Note that putting a necktie on someone else requires much more concentrated effort than putting it on oneself.) The basal ganglia, particularly the putamen and globus pallidus, are thought to be involved in this type of motor action. What to name these activities is problematic, and often the term used is *procedural movements*; we will also use the term *programmed movements*. These can involve fine hand movements as well as muscle activity of larger joints of the elbow and shoulder, in addition to postural adjustments.

The command instructions for the voluntary control of movements of all types are also sent to the cerebellum, which then feeds back via cerebello-thalamo-cortical connections for possible adjustment of the movement (further discussed in Chapter 7). In neurologically healthy persons, this is done without conscious involvement.

All these systems need to be connected (via tracts) in order for each contributor to inform the others of its participation in planning, organizing, coordinating and adjusting the movement(s). The eventual outcome is 'instructions' from the upper motor neuron, descending via various pathways – direct and indirect – to the spinal cord to connect with and control the activity of the lower motor neurons and their subsequent output to the muscles (see Figure 7.1; further discussed in Chapter 7). Any motor act requires functioning motor units; as was defined in Chapter 1, these consist of a lower motor neuron, its (myelinated) axon and all its collaterals, the intact neuromuscular junctions and normally acting muscle fibers.

The control of motor activity thus requires the involvement of some or all of the following areas of the CNS:

- The cerebral cortex, for planning the action
- The motor strip, to command the desired movement, site of the upper motor neuron

- The basal ganglia, to organize the muscle groups involved
- The cerebellum, to adjust the desired movements
- The brainstem, for associated movements and adjustments to gravity
- The spinal cord and its 'descending' pathways
- The lower motor neuron, as the final common pathway
- Intact peripheral nerves, neuromuscular junctions and muscle

2.4.2 MOTOR SYSTEM EXAMINATION

Assessing the motor system necessitates a variety of different tests; this approach reflects the complexity of the control of motor functions. These include the following:

- Muscle inspection and palpation for signs of atrophy, hypertrophy, tremor, involuntary twitching, fasciculations and abnormal consistency (e.g. induration)
- Passive movements of joints, both flexion and extension, performed both quickly and slowly by the examiner, for the assessment of muscle tone
- Voluntary rapid fine movements, carried out by the patient, of fingers and hand, as well as toes and foot
- Muscle strength, at various joints, usually tested against resistance
- Deep tendon reflexes

The examination of reflexes tests for the integrity of the peripheral nervous system (PNS) myelinated fibers (sensory and motor), the spinal cord with its upper motor neuron modulatory activity and the sensitivity of the motor neurons.

- Coordinated movements requiring the combined actions of various muscle groups (e.g. finger-nose test)
- Balance and gait (e.g. tandem walking)

The examination usually focuses on the activity of the limbs rather than of the trunk, unless the history indicates otherwise. Table 2.1 lists the levels of the spinal cord involved in the motor control of the upper and lower limbs, the trunk and the anal sphincter. The routine includes comparing one side with the other side for motor power and reflexes. All the while the examiner is on the lookout for abnormal and/or involuntary movements.

Starting with the upper limb, the muscles are palpated and the elbow joint is flexed and extended passively, and the examiner notes the tone of the muscles being stretched

TABLE 2.1: Motor and Sensory Levels

	Motor	Sensory
C2	Diaphragm, neck flexors, extensors	Occiput
C3	Diaphragm, neck flexors, extensors	Skin on neck
C4	Diaphragm, neck flexors, extensors	Upper shoulder
C5	Rhomboids, supraspinatus, infraspinatus, deltoid, biceps	Lateral shoulder
C6	Pectoralis major, supinator, pronator	Anterolateral forearm and thumb
C7	Triceps, latissimus dorsi, wrist, finger extensors, flexors	Middle finger
C8	Thenar, hypothenar, interossei	4th and 5th fingers and medial palm
T1	Thenar, hypothenar, interossei	Medial forearm
T4	Abdominal muscles	Nipple line
T10	Abdominal muscles	Umbilicus
L1	Abdominal muscles	Skin of lower abdomen
L2	Psoas	Skin of lower abdomen
L3	Iliopsoas, quadriceps, thigh adductors	Thigh
L4	Quads, tibialis anterior	Thigh
L5	Gluteus medius, hamstring internal, extensor hallucus longus, tibialis posterior	Lateral lower leg and foot
S1	Gastrocnemius, soleus, gluteus maximus, hamstring external, tibialis posterior	Posterior lower leg
S2	Hamstrings	Posterior upper and lower leg
S3	External anal sphincter	Buttock
S4	External anal sphincter	Perianal skin

Abbreviations: C = cervical; T = thoracic; L = lumbar; S = sacral.

and whether there is any resistance to movement. Then the patient is asked to perform a precise (voluntary) movement, such as apposing the finger and thumb rapidly.

Next, movements are done against resistance, testing shoulder abduction and adduction, elbow and wrist flexion and extension, all the time comparing the findings with what one might expect from the patient (depending on gender, body size and age). Finally, the hand muscles are tested by asking the patient to squeeze the examiner's hand and to spread the fingers against (the examiner's) resistance. Muscle strength is graded in a standardized way, as classified by the Medical Research Council (Table 2.2). A novel way of displaying and summarizing the findings on a muscle strength assessment has been

TABLE 2.2: Muscle Power Grading Scale

Grade 5	Normal strength in muscle being tested
Grade 4	Relative to normal, reduced muscle strength against resistance but no difficulty moving body part against gravity
Grade 3	Marked difficulty contracting muscle against resistance but just able to move body part against gravity
Grade 2	Visible movement of body part if gravity removed (e.g. movement in horizontal plane with limb supported by examiner)
Grade 1	Visible muscle contraction but no movement
Grade 0	No visible muscle contraction

Note: Classification according to the Medical Research Council (MRC).

devised, as shown in Figure 2.6; this muscle map will be used in the text for appropriate chapter cases and for those e-cases where there is a deficit of muscular strength. (An example of abnormal findings will be shown in the sample case in Chapter 3.)

The reflexes are tested again using side to side comparison: biceps to biceps, triceps to triceps and so on. A brisk tap is applied to the tendon attached to the muscle serving the reflex. Often, in the upper limb, the examiner places a thumb over the tendon being tested. The examiner not only looks for the reflex response but also feels for the output response. In the lower limbs, the examiner holds the lower leg or ankle when testing these reflexes not only to feel the output of the reflex but also to protect against being kicked in the event of extreme hyperreflexia. There is a standardized way of recording the reflex responses (Table 2.3). Reflex responsiveness is shown in patient charts using a 'stick figurine', and this will be used in the text to display the changes of reflex responsiveness for appropriate chapter and e-cases. (An example of abnormal findings will be shown in the sample case in Chapter 3.)

TABLE 2.3: Reflex Response Grading System

0+	No reflex
1+	Decreased reflex
2+	Normal reflex
3+	Increased reflex
4+	Abnormally brisk reflex with sustained clonus

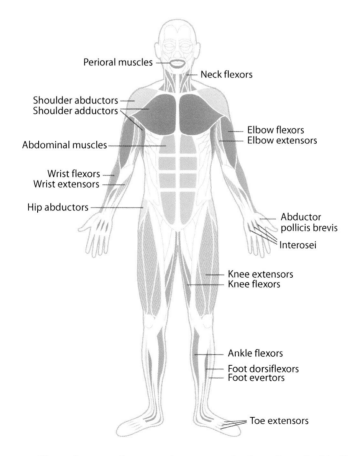

FIGURE 2.6: Muscle power map. The various muscle groups that are examined are shown in this display of musculature. The normal (following the MRC grading scale) would be shown as 5/5. (An example of an abnormal finding is shown with the sample case in Chapter 3.)

The examination follows the same pattern in the lower limb: palpating the muscles, flexing and extending the knee, testing for voluntary wiggling of the toes, strength of hip flexion (and extension), knee flexion and extension, and ankle dorsiflexion and plantar flexion. Reflex testing includes the patellar and the Achilles tendons at the knee and the ankle, respectively. Rapidly jerking the foot into dorsiflexion tests for a sustained rebounding rhythmic flexion–extension, referred to as clonus; this is an abnormal response usually accompanying increased tendon reflexes (hyperreflexia, see below).

Finally, the plantar reflex is tested by stroking the outside of the bottom of the foot firmly with a blunt object. The normal response in adults is a plantar flexion of the toes, often accompanied by the person 'withdrawing' the whole lower limb (so the examiner usually grasps the lower leg while doing this reflex examination). An abnormal plantar response consists of extension of the big toe and fanning of the other toes; this is called the extensor plantar response (formerly called the *Babinski sign* or *response*; oftentimes, clinicians will speak of a positive or negative Babinski). An abnormal plantar response invariably indicates damage to the corticospinal pathway (discussed in Section 2.4.3.1).

Spasticity is defined as a velocity-related increase in muscle tone. The combination of velocity-related increased resistance to passive stretch of the antigravity muscles (in particular the biceps brachii, hamstrings and plantar flexors) with increased reflex responsiveness (hyperreflexia) is a sign of upper motor neuron damage. This condition is usually accompanied by clonus and by an abnormal extensor plantar response. Increased constant resistance to passive movement of both flexors and extensors that is not velocity dependent is known as *rigidity*. Reflexes in this state are often normal and the plantar response is usually downgoing (normal). Rigidity usually occurs in diseases involving the basal ganglia.

Motor coordination is tested in the upper limb by asking the patient to perform the finger-nose test, in which the patient is asked to touch the tip of his or her own nose with the tip of his or her finger and then to touch the examiner's finger tip (which is moved each time). An additional test for coordination involves rapid alternating movements of the hand with palm up and palm down. The lower limb is tested by asking the patient to move the heel along the front of the shin bone (the tibia) down to the dorsum of the foot; an accompanying test is to have the patient tap the floor rapidly with the foot while leaving the heel on the ground. Deficits in coordination, with intact motor and sensory systems, are associated with disease processes involving the ipsilateral cerebellar hemisphere.

Balance and gait are tested by asking the person to walk along a straight line with heel touching the toe (tandem walking) and by asking him or her while walking normally to rapidly turn (change direction). In addition to assessing gait stability, one also notes whether there is a normal associated swinging of the arms. The testing should also include walking on the heels and the toes, and with the feet inverted and everted.

A loss of coordination of voluntary movements is known as *ataxia*. Ataxia can occur with upper and lower limb involvement as well as with walking.

2.4.3 PATHWAYS FOR VOLUNTARY (INTENTIONAL, PURPOSEFUL) MOVEMENTS

2.4.3.1 THE CORTICOSPINAL TRACT FOR COMPLEX, PRECISE MOVEMENTS

The cerebral cortex is where all planning of intended movement occurs, starting with the 'decision' to make a movement in the supplementary and premotor areas. The final command originates in the motor strip (the precentral gyrus) and usually also involves the premotor area. Control of the distal musculature is conveyed via a single uninterrupted pathway from the upper motor neuron in the motor cortex to the lower motor neuron in the spinal cord. This is the corticospinal tract, also referred to as the *direct* pathway (discussed further in Chapter 7).

Pathway (Figure 2.7): The corticospinal pathway descends through the white matter of the cerebral hemispheres. As the fibers funnel together between the thalamus and basal ganglia, the pathway becomes part of the internal capsule. It continues through the midbrain cerebral peduncle and the middle of the pontine bulge into the pyramid of the medulla; in the lower medulla, most of the fibers cross and continue down the spinal cord as the *lateral corticospinal tract* (see also Figure 1.5a). This tract is responsible for 'instructing' the lower motor neurons of the spinal cord regarding voluntary motor movements. For control of fine movements of the fingers, the pathway is thought to terminate directly onto the lower motor neurons. Such 'instructions' require both the activation of agonist muscles required for the movement and the inhibition of antagonist muscles.

Lesions of the cortex, white matter and brainstem will cause a deficit of voluntary movement on the opposite (contralateral) side; spinal cord lesions will affect movements on the same (ipsilateral) side. A lesion anywhere along this pathway will lead to hyperreflexia and an abnormal plantar response (discussed earlier), the extensor plantar reflex (previously called a positive Babinski sign).

The pathway for voluntary control of the cranial nerve motor nuclei is called the *corticobulbar pathway* (the older name for the brainstem is the bulb). This pathway is involved in the control of eye movements, chewing (mastication), facial expression (except for those accompanying emotions), swallowing, tongue movements, as well as shoulder elevation and neck rotation. A unilateral lesion of this pathway will lead to weakness of the lower facial muscles, on the opposite side (discussed further in Chapter 6).

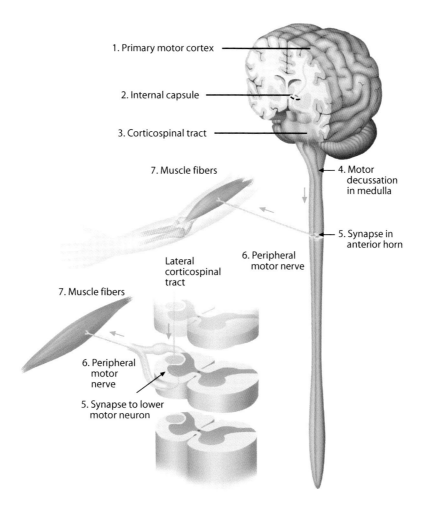

1. Primary motor cortex

2. Internal capsule

3. Corticospinal tract

7. Muscle fibers

4. Motor decussation in medulla

5. Synapse in anterior horn

6. Peripheral motor nerve

Lateral corticospinal tract

7. Muscle fibers

6. Peripheral motor nerve

5. Synapse to lower motor neuron

FIGURE 2.7: Voluntary direct motor pathway. The motor pathway begins in the primary motor cortex, the precentral gyrus (1), travels through the white matter of the hemispheres passing via the internal capsule (2), and becoming the corticospinal tract in the brainstem (3). Most of the fibers cross in the lower medulla (4) to form the lateral corticospinal tract in the spinal cord. Inset: The voluntary fibers in the lateral corticospinal tract synapse with a lower motor neuron in the anterior horn of the spinal cord (5); their axons exit in the peripheral motor (efferent) nerves (6) to activate the voluntary muscles (7). Note: This pathway is animated on the text Web site.

2.4.3.2 PATHWAYS FOR PROGRAMMED (PROCEDURAL) MOVEMENTS

The basal ganglia are definitely associated with learned motor patterns that can be called into action when needed, a phenomenon known as procedural memory. This system is likely involved in organizing the sequencing of muscle contractions and relaxations, agonist and antagonists, needed for a smooth action (starting and stopping muscle action), as well as the force necessary to perform the movement.

As was mentioned in Chapter 1, the substantia nigra (in the midbrain) and the subthalamic nucleus are, from the functional perspective, part of the basal ganglia and are also involved in this type of motor control; lesions involving these structures lead to problems with voluntary movements as well as the occurrence of involuntary movements (e.g. resting tremor, ballistic or flinging movements). The most common example of motor problems

associated with degeneration of the substantia nigra system is Parkinson's disease (discussed further in Chapter 7).

Pathway: The putamen is the part of the basal ganglia most involved with organizing the motor patterns. The signals are relayed via the globus pallidus and the thalamus to the motor areas of the cerebral cortex. The resulting actions are carried by the premotor area, and the fibers descend in the corticospinal tract; there are no direct connections from the basal ganglia to the spinal cord (discussed in Section 7.4.1).

2.4.4 PATHWAYS FOR NONVOLUNTARY (ASSOCIATED AND POSTURAL) MOVEMENTS

Most volitional movements are accompanied by complementary or associated movements, such as positioning of the elbow and shoulder (proximal) joints and adjustments in trunk posture.

2.4.4.1 CORTICO-BULBO-SPINAL PATHWAYS

Commands for associated movements are carried from the cortex to the spinal cord via motor nuclei of the brainstem. Because there is a second nucleus in between the cortex and the lower motor neuron, these tracts are called the *indirect pathways* (also discussed in Chapter 7), or the cortico-bulbo-spinal tracts.

Pathways: The nuclei of the brainstem that are involved in such actions include the red nucleus (in the upper midbrain) and the pontomedullary portion of the reticular formation. Tracts, named rubrospinal and reticulospinal, descend from their respective brainstem nuclei to the (lower) motor neurons of the spinal cord (see Figure 7.2). Thus, the motor cortex remains the origin for the proximal limb and trunk movement instructions, but through a two-step pathway via the brainstem.

The reticular formation is also responsible for setting the level of excitability of the motor neurons. For example, loss of this influence on the lower motor neurons because of a lesion of the motor areas of the cortex or of the descending pathways in the spinal cord (above the level of the lesion) causes increased excitability of the motor neurons below the level of the lesion, leading to muscles that have increased tone (spasticity) and hyperreflexia.

2.4.4.2 THE VESTIBULOSPINAL TRACT

The response to changes in position in relation to gravity or to acceleration and deceleration also requires postural adjustments; these occur without conscious involvement and are totally nonvoluntary in nature.

Pathway: The vestibular nuclei of the brainstem are responsible for these adjustments and give rise to pathways (lateral and medial vestibulospinal tracts) that descend to the spinal cord to control the axial musculature of the spinal column; this system is under the influence of the cerebellum. Other output from these nuclei controls the adjustments of the eye muscles, forming the basis for the vestibular-ocular reflex (discussed in Chapter 6, see Figure 6.3).

2.4.5 COORDINATION OF MOVEMENT

The command for a voluntary movement is also sent from the motor cortex to the cerebellum. As the movement is carried out, the cerebellum itself receives input from the periphery (the muscles and the joints involved) and detects errors in the performance of the movement that are fed back to the cerebral cortex for an appropriate 'online' adjustment.

Pathway: The lateral portions of the cerebellum, the cerebellar hemispheres (also called the neocerebellum), are involved in this aspect of motor control. (The detailed pathway for accomplishing this is explained in Chapter 7.)

2.4.5.1 GAIT AND BALANCE

Gait is a complex activity requiring the collaboration of the indirect motor pathways, the cerebellum, the visual system and the vestibular apparatus. We have already discussed the examination of gait and tandem walking earlier in this section. (It should be noted that children under the age of 5 or 6 are usually unable to tandem walk.)

Pathway: Medial (older) parts of the cerebellum are involved with balance and gait (i.e. the central portion or vermis). Lesions in these parts of the cerebellum or in the pathways from the cerebellum to the brainstem (medullary portion) may cause problems with balance and gait (ataxia).

The accepted test for balance is called the *Romberg Test*. The patient is asked to stand with feet together and told to look at a distant point. If able to accomplish this without beginning to lose his or her balance, the patient is then asked to close his or her eyes and balance is reassessed. The examiner should stand nearby and be prepared to assist should the person begin to fall.

Pathways: The two components of the Romberg test evaluate both the vestibulo-cerebellar apparatus and the posterior columns of the spinal cord. If the patient is unable to stand with feet together, eyes open, a vestibulo-cerebellar lesion is strongly suspected. On the other hand, if the patient has no trouble standing feet together with eyes open, then begins to fall with eyes shut, a posterior column lesion is likely. Should there be a lesion affecting the latter pathway, unconscious proprioception (for limb position) does not reach the cortex (discussed next), and therefore, visual guidance is needed for balance; when the eyes are closed this guidance is lost and the person 'loses his or her balance', that is, a positive Romberg.

2.5 SOMATOSENSORY SENSORY SYSTEMS

The sensory system is examined in a systematic fashion, including the two major sensory modalities – vibration, proprioception and discriminative touch on the one hand, pain and temperature on the other. Various levels of the spinal cord are examined (see Table 2.1).

2.5.1 POSTERIOR COLUMN PATHWAY: VIBRATION, PROPRIOCEPTION AND DISCRIMINATIVE TOUCH

The ability to sense vibration is tested by placing a vibrating tuning fork over bony prominences. Joint sensation can be evaluated by asking the patient to identify (with the eyes closed) which way the finger or toe joint is being moved (up or down). Discriminative touch is tested by asking the person whether he or she feels the examiner touch selected skin areas (usually with a wisp of cotton or tissue) and whether the sensation is the same on both sides.

Identifying objects placed in the hand with the eyes closed (stereognosis) and numbers stroked on the skin (graphesthesia) test whether the cortical areas involved with interpreting these sensations are intact.

Pathway (Figure 2.8a): These modalities of vibration, proprioception and discriminative touch are conveyed from their specialized receptors in the skin, joints and muscle and carried rapidly in thickly myelinated fibers to the spinal cord (see also Figure 1.4a). The fibers enter the cord and give off collateral branches for the deep tendon (myotatic) reflex (Figure 1.5b) and other reflexes, then turn upward, without synapsing, on the same (ipsilateral) side, forming the posterior (dorsal) columns in the spinal cord. A lesion of this pathway in the spinal cord interrupts the sensory information from the same side of the body.

The fibers ascend the spinal cord and synapse in the lowermost medulla in the posterior column nuclei. The fibers of the second-order neurons cross and form a new pathway, the *medial lemniscus*, which ascends through the brainstem, terminating in the *ventral posterolateral (VPL) nucleus* of the thalamus; from here, fibers are distributed to the postcentral gyrus, the primary somatosensory cortex. (The details of this pathway are also discussed in Chapter 5.) This gyrus of the brain has different portions dedicated to each part of the body, called the sensory homunculus (similar to the motor homunculus; see Figure 4.2). Cortical processing in this gyrus allows for the localization of the stimulus.

Identification of the nature of the stimulus or object is accomplished by further processing of the sensory

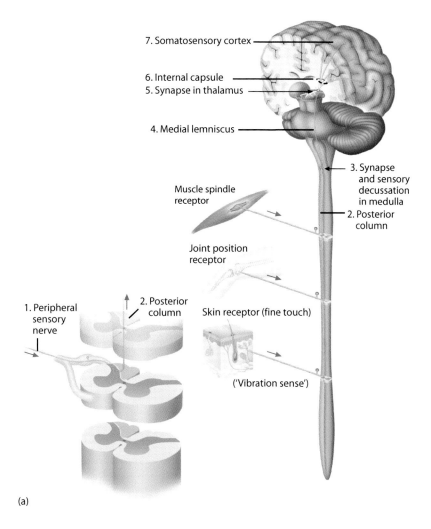

7. Somatosensory cortex

6. Internal capsule
5. Synapse in thalamus

4. Medial lemniscus

3. Synapse and sensory decussation in medulla

2. Posterior column

Muscle spindle receptor

Joint position receptor

1. Peripheral sensory nerve

2. Posterior column

Skin receptor (fine touch)

('Vibration sense')

(a)

FIGURE 2.8: (a) Pathway for discriminative touch, proprioception, and 'vibration'. Afferents from receptors for fine (discriminative) touch, muscle spindles and joint capsule receptors, and the 'sense of vibration', are carried from the periphery to the spinal cord. As shown in the inset, the fibers (1) enter the dorsal horn of the cord via the dorsal root and ascend, without synapsing, in the posterior columns, on the same side (2). The first synapse occurs in the lowermost medulla (3), then the fibers decussate (cross) and traverse the brainstem as the medial lemniscus (4). The second synapse occurs in the specific relay nucleus of the thalamus (5), following which the information is relayed via the internal capsule (6) to the somatosensory cortex (7), the postcentral gyrus of the parietal lobe. Note: This pathway is animated on the text Web site. *(Continued)*

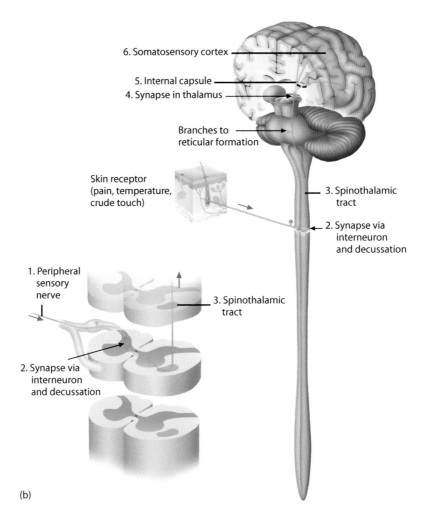

(b)

FIGURE 2.8 (CONTINUED): (b) Pathway for pain, temperature and crude touch. Afferents from pain, temperature and crude touch receptors are carried from the periphery to the spinal cord. As shown in the inset, the fibers (1) enter the dorsal horn of the cord via the dorsal root and synapse (2). The fibers then decussate (cross, 2) soon after entry, just anterior to the central canal in the ventral white commissure and ascend in the (lateral) spinothalamic tract on the opposite side (3). The tract traverses the brainstem, giving off collaterals to the reticular formation. The next synapse occurs in the specific relay nucleus of the thalamus (4), following which the information is relayed via the internal capsule (5) to the somatosensory cortex (6), the postcentral gyrus of the parietal lobe, and other areas. Note: This pathway is animated on the text Web site.

information in the adjacent areas of the superior parietal lobe, and from there, the patient, when asked, can name an object held and manipulated in the palm of the hand. This is called *stereognosis*. Connections with other association areas of the cortex may lead to memories about the nature of the object. Loss of ability to recognize the significance of sensory stimuli, even though the primary sensory system is intact, is called an *agnosia*.

2.5.2 SPINOTHALAMIC PATHWAY: PAIN AND TEMPERATURE

Pain is usually tested by using the tip of a new unused pin or safety pin. The examiner can press very lightly or more heavily on the pin so that the degree of sensation of pain can be graded. (One must never press so heavily as to draw

blood!) Again, the two sides are compared. Occasionally, the blunt side of the safety pin is used just to make sure that the person is responding correctly. The safety pin is to be used for that patient only and must be discarded in the designated needle box afterward.

It is often useful to calibrate the degree of sensation using a sensitive area on the face or nape of the neck and to ask the patient to close his or her eyes and image in the mind that the pinprick sensation is 100% pricky. Then as one tests areas of corresponding dermatomes in the upper and lower limbs from side to side, one asks the patient if the sensation is 100% pricky and, if not, what percentage of 'prickiness' is present. The examiner can then record the degree of sensation by location, which can then be used by future examiners to test for clinical changes.

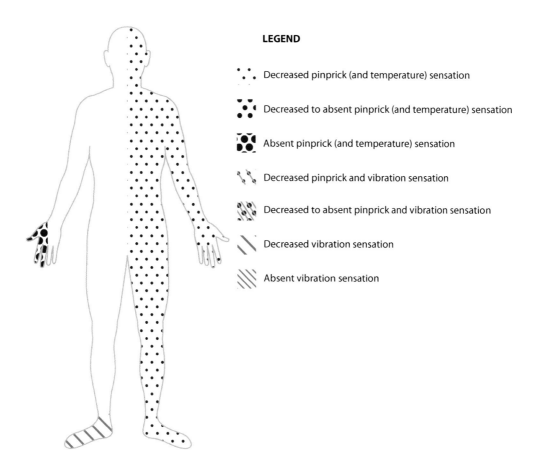

LEGEND

∴ Decreased pinprick (and temperature) sensation

Decreased to absent pinprick (and temperature) sensation

Absent pinprick (and temperature) sensation

Decreased pinprick and vibration sensation

Decreased to absent pinprick and vibration sensation

Decreased vibration sensation

Absent vibration sensation

FIGURE 2.9: Sensory examination map. Abnormal sensory finding will be shown using different patterns and densities for the various sensory findings; note particularly the figure legend for the various different sensory deficits.

Pathway (Figure 2.8b): These modalities are conveyed from their nonspecialized receptors, mostly free nerve endings in the skin and body wall, and are carried more slowly in thinly myelinated and non-myelinated fibers to the spinal cord (see also Figure 1.4b). At this level, reflex activity may occur, for example, withdrawal of the limb because of the pain of the pin. The fibers synapse soon after their entry in the gray matter of the dorsal horn of the cord and cross and ascend the spinal cord as the (lateral) spinothalamic tract (also called the anterolateral pathway). A lesion involving this pathway on one side of the spinal cord therefore interrupts these sensations from the opposite side of the body. The pathway continues through the brainstem, where it is joined by the pain pathway from the opposite side of the face. Collaterals from the pain and temperature pathway enter the reticular formation (see Figure 2.8b), where they may cause a general 'alerting' response.

The pain and temperature fibers again relay in the thalamus and are distributed not only to the postcentral gyrus but also to other areas of the cortex. It is at the cortical level where the source of the pain is localized and the nature of the stimulus is identified. Other connections, including the limbic structures, may lead to other associations and possibly to an emotional reaction.

In summary, the various sensory modalities from the skin and muscles follow two distinct routes in the spinal cord. Lesions of one-half of the cord will lead to a *dissociated sensory loss* involving loss of discriminative touch sensations on the same side and loss of pain and temperature on the opposite side. Knowing which areas of the body are supplied by what level of the spinal cord (Table 2.1) will help determine the level of the lesion (level of the cord, not the vertebral level). A lesion at the brainstem level may affect both pathways leading to a loss of all sensations on the opposite side, since both pathways have already crossed. (These lesions and others will be discussed further in Chapters 5 and 6.) A lesion may also involve the corticospinal tract, thereby causing a loss of voluntary movement.

A sensory map (Figure 2.9) will be used with appropriate chapters and e-cases (in the text) to display the changes in the sensory system (see legend with this figure).

2.6 HIGHER MENTAL FUNCTIONS

The assessment of higher mental functions includes the evaluation of attention, executive function, language, memory, visuospatial ability and personality (which embraces the unique emotional and reactive aspects of a person). It is often difficult to recognize and classify disorders associated with cortical dysfunction. Very often, the person affected also loses insight into this loss of functionality and it is usually the spouse, children, relatives and friends who become aware of these changes. The challenge has been to understand what changes in mental functionality, if any, should be associated with normal healthy aging, which indicate the recently described syndrome of mild cognitive impairment and which are indeed indicative of disease states, that is, a dementia. (Dementia is discussed with the case presented in Chapter 12.)

2.6.1 MENTAL STATUS

The mainstay for the screening of higher mental functions has been the Folstein Mini-Mental Status Examination with modifications for language, visuospatial and frontal lobe testing. This instrument, consisting of 10 questions with different point values, tests the basic elements of higher brain function: attention, orientation (time, place, person), short- and long-term memory, calculation (100 – 7 test), verbal reception and verbal output, reading and writing, sequential processing, a visuospatial task (drawing the face of a clock) and integration.

The score out of 30 gives an estimate of cortical function as well as subcortical connectivity. A score of lower than 27, provided that attention is intact (see Section 2.6.2.2), indicates a mild impairment; lower scores need more thorough testing depending on the examiner's assessment of the patient, the history and the concerns of the family.

The mental status examination has to be tailored to age, education and the developmental level of the patient. In the pediatric population, age-referenced milestones are available in chart form, such as the Denver Development Scale.

Recently, other tests of mental capacity have been devised as our population ages and with the emergence of newer concepts such as mild cognitive impairment or dysfunction. One that has been verified in the literature is the Montreal Cognitive Assessment (MOCA), designed as a rapid screening instrument for just such a condition. The various functions tested include visuospatial, executive, naming, memory, attention, language, abstraction, delayed recall and orientation (to time and place).

2.6.2 CORTICAL FUNCTIONS

The cerebral cortex, with its microprocessor array and exquisitely complex interconnections, is where our thinking occurs, as well as our language function; our memory for events and places; and our ability to plan, to react and to interact with others (see Figure 1.10). No part of the brain can function in isolation, without the participation of many other parts. Normal functional capacity therefore requires integration of all the contributing CNS modules (hence the title of this book). With the use of current technology, including functional magnetic resonance imaging (which shows changes in regional blood supply) and positron emission tomographic studies (using radioactive tracers), certain parts of the brain can be seen to 'light up' during task-related activation of specific areas. With all the 'background' brain activity removed, these images are now the conventional way of presenting information regarding regional brain functionality.

2.6.2.1 EXECUTIVE FUNCTIONS

In the normal course of our lives, we must decide on what has to be done and in what sequence. Decisions have to be made as to priorities, and judgements are made about the importance of the task and its ethical dimension or its social appropriateness. This ordering of our tasks is usually attributed to the function of the frontal lobes of the brain, the area in front of the motor regions, sometimes called the *prefrontal cortex*, particularly the dorsolateral aspects (see Figure 13.1). Often, this region of the brain is called its chief executive officer. It is this function of the brain that is impaired, progressively, in cortical dementias (e.g. frontotemporal dementia, see Chapter 12).

2.6.2.2 ATTENTION

A particularly important feature of the brain is the ability to focus on certain happenings, whether a conversation or a task. This ability to attend is very important while doing any task (e.g. reading, or performing a surgical procedure) or in any learning situation. Areas of the brain involved with attention include the frontal lobe as well as the anterior portions of the cingulate gyrus.

The assessment of attention has to be made first in the mental status exam as the degree of attentiveness will influence all of the subsequent testing. Often, an inattentive patient will score very poorly on a mental status exam. This result overestimates the patient's deficits in other areas.

In addition, deficits in vision, hearing, proprioception and motor function can also influence the score obtained on the mental status examination. It is hard to draw a clock face if you cannot see the result properly. It is therefore often useful to perform the other parts of the neurological examination first and then come back to the mental status testing.

2.6.2.3 LANGUAGE

Our ability to communicate using symbols is perhaps unique to humans. The left hemisphere is usually the hemisphere specialized for language – expression and comprehension, both written and spoken. As noted in Chapter 1, two areas have been identified (see Figure 1.10c):

- Broca's area in the lower frontal lobe for expressive language

- Wernicke's area in the posterior-superior temporal region (near the auditory area) (as discussed in Chapter 1)
- Association areas in the temporal and parietal cortex, mainly in the left (dominant for language) hemisphere

2.6.2.4 MEMORY

Memory for events, names and places is the memory that we use most frequently. Usually, attention to the event is important for recording a memory. The essential structure needed for memory formation is the hippocampus, a neuroanatomical structure located within the temporal lobe of the brain in its medial aspect (see Chapter 13). Once the memory is 'registered', if it is considered important, it is 'stored' in various areas of the cerebral cortex.

While one is engaged in a task or event (for example, going to fetch something, remembering a name), the information that is in use needs to be retained for short periods of time, a process called *working memory*. It is thought that this aspect of memory is carried out in the frontal lobes. Our long-term memory retains information, names, places and events; in order to use these data, the memory trace has to be retrieved in a timely fashion. Broad areas of the cerebral cortex are involved in the storage of this type of memory. Various memory disorders affect the registration, storage or retrieval of memory.

The second type of memory is called *procedural*, which involves wide areas of the brain, particularly the basal ganglia (also discussed in Section 2.4.3.2). We use this type of memory for all kinds of complex motor tasks, which, once learned, can be retained and used over a person's lifetime (e.g. riding a bicycle, performing a piece of music). This type of memory is not usually affected in the dementias, disorders that affect cognitive functions and memory for events, names and places (discussed in Section 2.6.2.6 and in Chapter 12).

2.6.2.5 VISUOSPATIAL

The association area of the lower parietal lobe in the nondominant hemisphere, situated between the somatosensory, visual and auditory areas, is thought to be the region of the brain most involved with visuospatial tasks. The task most often used for testing of this function is drawing a clock face within a prescribed period of time.

Lesions of this area may give rise to hemi-neglect (described with e-case 13e-4 [Eddie]). A visuospatial impairment is usually difficult to detect until formally tested in the manner described.

2.6.2.6 IMPAIRMENT OF FUNCTION

Impairment of function can occur in any of these domains, separately or in combination. One of the impairments most challenging to detect, particularly for a physician meeting

a patient for the first time, is executive dysfunction, problems with weighing information (the basis for decision making) or with attention. Changes in personality often accompany the deterioration of higher mental functions; sometimes the person is able to cover up deficits or lapses for a time by using his or her personal charm or social graces to slide through an awkward moment.

A language disorder is called an *aphasia*, defined as an acquired disruption, disorder or deficit in the expression or comprehension of spoken or written language. More specifically, there can be an expressive or receptive aphasia, or a global aphasia affecting all language functions. Difficulty with articulation or enunciation of speech is called a *dysarthria*. *Alexia* is the term used for word blindness; this is the acquired loss of the ability to grasp the meaning of the written word due to a central (not visual system) lesion. Inability to write owing to a lesion of higher centres, even though muscle strength and coordination are preserved, is called *agraphia*.

Difficulty with short-term memory is one of the most common concerns, particularly as people age (in fact, some older people call these lapses 'senior moments'). This concern is to be appreciated because short-term memory is one of the functions most affected in a common form of dementia, Alzheimer's disease. Dementia is defined as a progressive brain disorder that gradually destroys a person's memory, starting with short-term memory, and cognitive abilities (including judgement, discussed in Chapter 12).

A specific disorder of cortical function is the loss of ability to carry out a purposeful or skilled movement (e.g. buttoning a shirt), despite the preservation of power, sensation and coordination; this dysfunction is called an *apraxia*.

2.6.2.7 EMOTIONS

Although most of our cerebral functions are associated with analyzing our external environment, other (older) parts of the brain have been found to 'react' to these happenings. A reaction may include alterations in pulse, respiration and blood pressure and changes in other autonomic functions (e.g. gastrointestinal). Over the short and long term, there may be endocrine changes involving the hypothalamus.

The areas of the brain associated with these emotional reactions are located for the most part in the temporal lobe. The relevant structures include the amygdala as well as other nuclei, and the cortex of the insula. Parts of the frontal lobe are also involved in emotional reactions, mostly the ventromedial aspects (in particular, the cingulate gyrus) and the orbital portion (sitting above the orbits; see Chapter 13). These brain structures are collectively called the limbic system and are considered in more depth in Chapter 13 (see Figure 13.5). Some limbic system components such as the septal nuclei and the nucleus accumbens (beneath the anterior corpus callosum and caudate head, respectively – see Chapter 13) are also involved in behaviour associated with rewards and punishment.

2.7 HUMAN CONSCIOUSNESS

One particular aspect of higher mental function is consciousness, wherein the individual is in contact with those around him or her and reactive to outside events (beyond himself). A unique characteristic of humans, so it is thought, is self-awareness, considered by many as an aspect of consciousness. The exact structure or structures necessary for consciousness have not been clearly defined but involve the upper part of the reticular formation, the thalamus (the nonspecific nuclei) and the cerebral cortex. Perhaps, in fact, consciousness parallels total brain activity of a specific type, involving all these structures.

Consciousness can be impaired (e.g. a confusional state after a major seizure or with drug or alcohol toxicity) or lost after a major head injury. Prolonged loss of consciousness occurs after severe brain injury, most often resulting from a closed head injury, and after vascular ischemic lesions affecting the upper part of the reticular formation. Of course, general anaesthesia does away with consciousness temporarily. A prolonged state of unconsciousness is called a *coma*.

Consciousness is assessed by a widely accepted set of criteria, collectively called the Glasgow Coma Scale, involving ocular, verbal and motor responsiveness. Its scoring system (see Table 2.4) is important for instituting therapeutic measures and sometimes for prognosis. The verbal output responses are clearly limited if the patient is intubated. There is a more comprehensive discussion of coma scales and the clinical assessment of comatose patients in Chapter 10.

TABLE 2.4: Glasgow Coma Scale

Best ocular response	Opens eyes spontaneously	4
	Opens eyes to voice	3
	Opens eyes to pain	2
	No ocular response	1
Best verbal response	Oriented, conversing normally	5
	Confused, disoriented	4
	Says inappropriate words	3
	Incomprehensible sounds	2
	No sounds	1
Best motor response	Obeys commands	6
	Localizes painful stimuli	5
	Withdraws to pain	4
	Decorticate response to pain	3
	Decerebrate response to pain	2
	No response to pain	1
Maximum score		15
Minimum score		3

Other important items in the assessment of the comatose patient with a major cerebral lesion include the position of the limbs (see Section 10.7.4.4 and Figure 10.11). With so-called 'decorticate' posturing, the arms are in the flexed position and the legs extended; this is seen with bilateral cerebral pathology above the red nuclei (in the midbrain). With 'decerebrate' posturing, both arms and legs are in the extended position; this is seen with midbrain pathology at or below the level of the red nuclei. In addition, the reflex responses of the cornea and the pupil and the response to testing of the oculocephalic reflex for pontine integrity (the caloric test, described in Chapter 6) would be important aspects of the neurological examination of someone in coma.

Additional testing, including brain imaging and an electroencephalogram (EEG), may be needed to determine whether the comatose person is in epileptic status, whether there is brain death or a condition called persistent vegetative state.

2.8 SLEEP

Sleep, a normal brain function, is a reversible behavioural state of perceptual disengagement from and unresponsiveness to the environment. Most adults and older children enjoy an awake daytime and a night-time sleep, a diurnal rhythm. Neuronal networks located in the diencephalon (the hypothalamus) and the brainstem reticular formation control this diurnal rhythm.

The amount of time spent in sleep changes with age and differs among individuals, although there is a minimum amount of sleep time (6–7 hours) that almost all adults require. An infant spends most of its time in sleep; as she or he matures, sleep tends to occur during definite time periods, which eventually (by the age of about six months) become regularized such that one is mostly awake during the day and asleep during the night. There are different stages of sleep, and the proportion of these stages changes with age.

There are two principal states of sleep described in adults: *rapid eye movement (REM) sleep*, which has a waking form of EEG and occurs with dreaming, and *non-REM (NREM) sleep*, which has specific EEG patterns of sleep known as K-complexes, sleep spindles and slow waves. The restorative value of sleep depends on the ability of an individual to maintain four to five sequences per night, each lasting 90 to 120 minutes, of NREM followed by REM sleep. We usually remember our dreams only if our REM sleep phase is interrupted.

The biologic purpose for sleep has been hotly debated, although most agree that a good sleep is restorative and that some form of memory consolidation occurs during sleep. Restorative sleep is necessary for the normal function of an individual.

It is commonly agreed that up to 25% of the adult population suffers from some form of sleep disorder at some time or another. As people age, sleep generally becomes more disrupted and fractionated, and less restorative. This often spills over into daytime activities and can interfere with daytime function or cause irritability and social problems. Besides poor sleep hygiene, the most common sleep disorder is obstructive sleep apnea.

Loss of sleep, even for a few days (such as with airplane travel and jet lag), does affect a person's ability to perform normal daily activities and clearly affects the quality of higher mental functions, including attention and decision making. Shift work (e.g. in the case of nurses) can severely affect a person's life. Physicians in training are equally vulnerable as are physicians on call who need to respond to complex issues and questions when they are awakened in the middle of the night.

2.9 THE NEUROLOGICAL EXAMINATION OF YOUNG CHILDREN

Performing an adequate neurological examination on a young child is not an easy task. This is particularly the case for children aged 36 months and younger, as well as for older children (and adults) with significant cognitive impairment. At the developmental equivalent of 6–36 months in particular, children are wary of strangers, have their own agendas as to what they would like to do next and do not like having their heads touched. Typically, any attempt by the examiner to examine the fundi, ear canals or throat will be met with violent objection!

The following list of suggestions should help you perform a fairly complete neurological assessment most of the time:

1. *The neurological examination of a young child should commence, in part, the moment the child is brought into the examining room.* In other words, while taking the medical history from the parents, you need to keep an eye on what the child is doing. If an 11-month-old patient crawls over to a small table containing toys, pulls into a standing position and starts playing with blocks, you already know that gross and fine motor development is fairly normal and that the child has adequate vision in at least one eye.

2. *During the eventual 'formal' neurological examination, it is important to defer laying hands on the child until absolutely necessary.* A great deal of useful information can be obtained by playing games with the child using hand puppets, blocks and small pieces of food (such as breakfast cereal) – provided that the child's current diet includes such items. For example, grossly intact visual fields may be documented with hand puppets; visual acuity and fine motor function of both hands may be evaluated by offering the patient a piece of cereal held in the palm of your hand. In other words, always do what is least threatening first; i.e. save the ophthalmoscope and otoscope for last.

3. *Throughout the assessment, the adult neurological examination sequence that you have already learned must be kept in mind, with each item mentally filled in as circumstances permit.* Thus, for example, examination of the gag reflex would normally be performed during the cranial nerve assessment, early in the course of the neurological examination of an older child or adult. For a 12-month-old, this part of the examination should be done at the very end, when it matters less if the patient becomes upset. In other words, having internalized a logical examination sequence for adult patients, you have to be far more flexible in assessing a small child while still remembering to evaluate every aspect of the neurological examination sequence that is mandated by the parents' history.

4. The best examination table is the parent's lap, where the child feels most secure. The real examination table should only be used for aspects of the physical examination that have been saved for last and often require the patient's lying in a supine position with hands restrained, for example, in examining the throat or abdomen.

The neurological examination of newborn infants is even further removed from that of adults and is beyond the scope of this book.

SUGGESTED READING

Bickley, L.S., Ed. *Bate's Guide to Physical Examination and History Taking*, 11th ed. Philadelphia, PA: Lippincott Williams & Wilkins, 2012.

Blumenfeld, H. *Neuroanatomy through Clinical Cases*, 2nd ed. Sunderland, MA: Sinauer Associates, 2011.

Schapira, A.H.V., Ed. *Neurology and Clinical Neuroscience*. St. Louis, MO: Mosby, 2007.

Weiner, W.J., Goetz, C.G., Shin, R.K., and Lewis, S.L., Eds. *Neurology for the Non-Neurologist*, 6th ed. Philadelphia, PA: Lippincott Williams & Wilkins, 2010.

See also Annotated Bibliography.

Chapter 3

Clinical Problem Solving

Objectives

- To provide a framework for clinical reasoning when presented with a patient having neurological symptoms
- To provide a framework for localization of lesions within the nervous system
- To provide a framework to correctly identify etiological diagnoses
- To review how the data gathered by neurological history and physical examination are synthesized in a logical process to allow for localization and etiological diagnosis

3.1 INTRODUCTION

When confronted with a patient with neurological symptoms, the questions that a clinician must ask while conducting a neurological history and physical examination include the following:

1. Is the problem truly neurological? Or putting it another way, is the problem due to real organic pathology? Alternatively, could the patient's symptoms be due to a disease process located outside the nervous system that is indirectly affecting the nervous system?
2. Where is the lesion in the nervous system? (Where is the problem?)
3. What is the pathophysiological cause or etiology of the lesion? (What is the problem?)

The clinical encounter usually starts with history taking. The history or sequence of events is received from the patient directly or from other witnesses such as family, friends or emergency medical services. By means of a series of simple questions, the onset and evolution of each of the neurological symptoms can be traced and recorded.

One of the most important questions that needs to be asked is 'What exactly were you doing when the symptoms first appeared?' Another way of asking this question is 'What time was the patient last seen well?' This is especially important for stroke patients to determine the eligibility and type of acute thrombolytic treatment.

The history is probably most important in determining the underlying disease process by carefully retracing the steps through which the illness evolved. Did the symptoms come on suddenly or slowly? What was the patient's health like in the preceding days and weeks?

The various neurological symptoms can then be plotted in sequence on the History Worksheet and then correlations can be made between the various symptoms. Several examples illustrating various conditions are posted on the Web site under Chapter 3 under 'Printable Worksheets'.

After the history is taken, one then performs a neurological examination. Often in emergency situations the history and examination are performed concurrently.

The neurological examination is most important in determining the localization or location of the problem in the nervous system. Each element of the neurological exam tests the various systems involved in normal neurological function and therefore is used to detect dysfunction. Chapter 2 has provided the template for a basic screening neurological examination.

This chapter presents a system of sequential analysis of codified troubleshooting of the problems within the nervous system; it is focused on determining first the localization and then the etiological diagnosis.

Knowing where the problem is located often makes the determination of the etiological diagnosis easier. At first glance, it might seem that this approach is contrary to the classical clinical method of first eliciting a history for etiological diagnosis and then performing a physical examination for localization. Practically speaking, the order of operations is not very important as the history and physical are usually gathered in one session. As will be seen in this chapter, it is the use of the worksheet tools provided and the analytic sequence that are important and not necessarily the order in which the clinical data are gathered.

This process of analysis will now be illustrated using a typical sample case. The process includes gathering the history and performing a clinical examination, including the derivation of a minimum dataset for the neurological examination.

The data from the neurological examination will be codified in an Expanded Localization Matrix to assist in working out the location of the patient's lesion(s). Then, turning to the etiological question, the illness timeline is then plotted on a History Worksheet; this visualization of the history will then be applied to the completion of the Etiology Worksheet. In the process, a likely disease location and disease type will become apparent.

3.2 IS THE PROBLEM TRULY NEUROLOGICAL?

The usual approach of most clinicians is to assume that the symptoms have an organic basis until further information is gathered. Certain aspects will emerge during the course of the history and neurological exam, which will alert the physician to suspect that the symptoms may not have an organic basis and that other causes need to be explored.

The features that would make one suspect that the symptoms are not organically based would be neurological symptoms, which, as the history taking progresses, shift from side to side, and a lack of consistent hard neurological findings. One must approach all patients with neurological symptoms with an open mind. A known psychiatric patient can develop neurological symptoms and abnormal behaviour, which might turn out to be a brain tumor or herpes encephalitis and not merely another episode of psychiatric illness. There are some tricks during the examination that can be used to help determine whether there is real neurological disease. For instance, a patient presenting with 'leg paralysis' who, upon request, cannot lift his or her legs from the bed at the hip is then asked to sit up from the supine position with the hips stabilized by the examiner; in the latter test, the patient is able to sit up without difficulty. This result would indicate that the hip flexors are working properly and therefore the paresis is non-organic.

3.3 SAMPLE CASE

A 45-year-old man presented to the local emergency department having difficulty with speech and right-sided weakness.

He was overweight, with a body mass index of 35, was a smoker, and had previously elevated blood cholesterol levels. There was a strong family history of coronary disease and stroke.

Three weeks previously, he had experienced two attacks one week apart, during which he suddenly lost vision in his left eye. He described a pattern in which the upper part of the visual field in the left eye would become gray, then black (as in a curtain dropping down over his vision). This phenomenon lasted 10 minutes at its maximum, following which his vision would slowly return in reverse sequence over 5 to 20 minutes. He was sure that it was his left eye in that he remembered covering each eye separately and noticed that the disturbance was only in his left eye.

The present episode occurred after the evening meal, when he experienced a sudden-onset of difficulty speaking, which he described as knowing what he wanted to say but being unable to output more than single disjointed words in a telegraphic fashion.

At the same time, he noticed weakness of his right arm while his wife noticed that the lower part of his face was drooping on the right. By the time he reached the emergency department and was seen, over three hours had elapsed. He was now unable to output any speech and the weakness of his right arm and right side of the face was worse; there was also mild to moderate weakness of his right leg. However, he was able to understand questions and obey commands using gestures.

His neurological examination showed a blood pressure of 170/100 in both arms; he had normal heart sounds and a left carotid bruit. His heart rate was 84 and regular.

Mental status examination was limited due to the speech difficulty, but he was alert and able to answer questions (primarily using gestures and head nodding) and to obey commands. He could not name any of the pictures on the National Institutes of Health scoring card but would nod if the names were suggested. He could not repeat simple phrases.

Cranial nerve examination showed he had a gaze preference to the left side, which could be overcome by moving his head briskly from side to side. Otherwise, the extra-ocular movements and visual fields were normal. There was weakness of the lower part of the face on the right. Motor exam showed a severe 2/5 weakness of the right upper limb and less weakness of the right lower limb 3/5 with hyperreflexia of the right upper and lower limbs with an extensor plantar response on the right ('positive' Babinski sign). Sensory examination was normal on both sides; coordination testing was limited on the right due to weakness and normal on the left. Gait testing was not performed because of weakness.

With the exception of the expressive language disturbance and eye movements, the findings on neurological examination are summarized in Figure 3.1.

Where is the lesion in the nervous system?

This is the most important question in that the location of the lesion in the nervous system will often give context to the underlying neuropathological process.

One approach to answering this question is to use an established troubleshooting technique used in electrical engineering. This process relies on the principles of

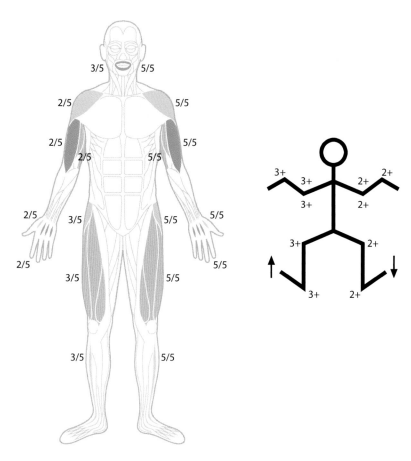

FIGURE 3.1: Neurological examination findings motor and reflex map in the sample case.

testing the sequential function of each element and its hardware connection. The system is simple: one starts where the system exerts its final output and works backwards towards the source of the input to determine whether each element is functioning or not. If there is a disconnection of communication between one element and another, then one is able to localize the problem. One can think of the nervous system as a series of nodes and links. If one utilizes a systematic program of analyzing each node and link, one can determine where the fault is in the nervous system.

For the remainder of this chapter, as we proceed with a series of discrete steps in the analysis of the sample case and develop a rational explanation for the location and type of disease seen in the patient, we will be utilizing a sequence of text sections that will be seen (with some necessary variations) in all 10 of the clinical case chapters that are to follow.

3.4 CLINICAL DATA EXTRACTION

As we have already mentioned in the Introduction to this chapter, the data gathered in the neurological examination

provides the most important information for determining the location of the problem in the nervous system. Each element of the neurological exam tests the various systems involved in normal neurological function and therefore is used to detect dysfunction. From Chapter 2, you will recall that the neurological examination includes the following components:

Mental status
Cranial nerves
Motor
Reflexes
Sensory systems
Coordination
Gait

3.5 MAIN CLINICAL POINTS

- High-risk factors for vascular disease
- Transient left eye disturbance two weeks prior
- Sudden-onset right paresis face, arm greater than leg
- Expressive aphasia
- Gaze preference to the left side
- Right-sided hyperreflexia

3.6 RELEVANT NEUROANATOMY

Each chapter will provide a synopsis of the relevant neuro-anatomy that will assist in the localization process.

3.7 LOCALIZATION PROCESS

The Basic Localization Matrix consists of a knowledge cube with the neurological examination on one axis and the major functional elements of the nervous system on the other axis, as shown in Table 3.1.

The process of determining the location of the disease process causing the nervous system to fail requires certain basic knowledge of the principal functional parts of the nervous system. This knowledge includes the various tissue types present in each neuroanatomical region, their function and their connections. These functional elements are listed across the top of the matrix and include the following:

Muscle
Neuromuscular junction (NMJ)
Peripheral nerve
Spinal cord
Brainstem
Deep white matter and thalamus
Cortex
Basal ganglia
Cerebellum

This matrix can be expanded to provide further detail of both the neurological examination and the functional elements. A more detailed model of the matrix, incorporating the data from the sample case, the Expanded Localization Matrix, is shown in Table 3.2.

For instance, the mental status examination can be expanded to include language, memory and visuospatial testing, each corresponding to different areas of the cerebral cortex. The structures relevant to the mental status examination include the frontal, temporal, parietal and occipital cortices, as well as subcortical structures (basal

ganglia, thalamus) with which the cortical areas are connected. The basal ganglia and cerebellum are modulating systems that depend on the primary systems and can only be isolated with respect to localization by exclusion.

Localization of damage to cerebral white matter exclusively requires elimination of cortical, ganglionic and cerebellar localizations with features of weakness, sensory loss and hyperreflexia.

Many disease processes involve the destruction of a specific component of the nervous system or its connections to or from other parts of the nervous system. For many nervous system functions, there are several parallel pathways that can be sequentially tested; this is what we do when we examine the human nervous system.

On completion of the neurological examination as described in Chapter 2, the pertinent findings can be inserted into the appropriate rows and columns of the matrix using other correlative factors to help determine the most likely localization.

For the sample case, the clinical findings included an episode of transient visual disturbance in the left eye, followed several weeks later by an episode of expressive aphasia (difficulty of expression of language) and right-sided weakness and hyperreflexia including a positive Babinski sign.

The spaces in the localization matrix that correspond to positive findings are denoted by a circle ○. When, as all the localization data are considered together, a potential localization becomes impossible or unlikely, the circle is then crossed out ⊘.

The full matrix is completed first; then, starting from muscle and moving up to NMJ, peripheral nerve, spinal cord, brainstem, deep white matter, cortex, basal ganglia and cerebellum, each potential location is either confirmed or eliminated.

This general approach is especially important when there may be more than one localizations: a typical example is multiple sclerosis (MS), where one might have a lesion in the spinal cord, causing damage to a posterior column (and sensory loss on one side) with another lesion in the deep cerebral white matter, causing sensory loss

TABLE 3.1: Basic Localization Matrix

	Muscle	Neuromuscular Junction	Peripheral Nerve	Spinal Cord	Brainstem	White Matter Thalamus	Cortex	Basal Ganglia	Cerebellum
Mental status									
Cranial nerves									
Motor									
Reflexes									
Sensory									
Coordination									
Gait									

TABLE 3.2: Expanded Localization Matrix

			MUS	NMJ	PN	SPC	BRST	WM/TH	OCC	PAR	TEMP	FR	BG	CBL
Attention	N													
Memory	U													
Executive	U													
Language	Exp													
Visuospatial	N													
	Right	Left												
CRN I	N	N												
CRN II	N	N												
CRN III, IV, VI	Gaze to left Able to overcome with head movement													
CN V	N	N												
CRN VII	Lower face	N												
CRN IX, X	N	N												
CRN XI	↓	N												
CRN XII	Dev R													
Tone	↑	N												
Power	↓	N												
Reflex	↑	N												
Involuntary Mvt														
PP	N	N												
Vib/prop	N	N												
Cortical sensation	N	N												
Coord UL	Abs	N												
Coord LL	Abs	N												
Walk	NA	NA												
Toe/heel/invert/evert	NA	NA												
Tandem	NA	NA												
Romberg	NA	NA												

Abbreviations: Abs, absent; Dev R, deviated to the right; Exp, expressive aphasia; N, normal; NA, not attempted; U, untestable.

on the opposite side. Some diseases such as vitamin B12 deficiency cause concurrent localizations with a combined myelopathy and neuropathy. In these cases, it is important to keep an open mind and leave the various localization options open; these then can be confirmed or refuted by imaging, neurophysiology or laboratory testing.

The final localization hypothesis – or top two localizations – is made based on the remaining anatomic structures most likely to explain the patient's signs and symptoms.

Making a localization based on the clinical findings and on this system of clinical reasoning, then having it confirmed by the neuroimaging or neurophysiology test results, is extremely satisfying for students, teachers and, of course, the patient. When assessing new patients during morning rounds, it is important not to cheat by looking at the imaging first before going through the process of localization using this system.

We will now review the localization implications for each anatomical level in the matrix, considering the findings of the neurological examination and utilizing the information in the sample case.

3.7.A MUSCLE DISEASE

As the function of muscle is specifically motor output, the correlative factors for muscle disease relate to motor function or dysfunction. Motor dysfunction can originate either in a lower motor unit (which includes the alpha motor neuron, its axon, NMJ and muscle fibers) or in an upper motor neuron unit (which includes the pyramidal

TABLE 3.3: Muscle Localization: Clinical Correlations

History/Symptoms	Observations	Investigations
Medication toxic exposure	Statins, colchicine	Elevated CK Myopathic EMG Genetic markers Muscle biopsy
Weakness	Decreased motor power atrophy/ hypertrophy Decreased deep tendon reflexes	
Sensory changes	Absent	
Muscle pain	May be present	
Medical comorbidities	Malignancy	
Family history	May be present	

cell in the cortex, its axon and the latter's variably complex connection to the lower motor neuron in the spinal cord).

The important clinical distinction is that lower motor neuron unit disease causes weakness, muscle wasting and decreased amplitude of reflexes (hyporeflexia), whereas upper motor unit disease causes weakness, spasticity and increased amplitude, velocity and spread of reflexes (hyperreflexia). The upper motor neuron tracts, the corticospinal tracts, are somatotopically organized such that the pattern of muscle weakness can give information pertaining to the precise localization of the lesion within the corticospinal tract.

The term *somatotopic* means that, for example, upper cervical cord area corticospinal tract axons destined for the lumbosacral cord region travel together in the outer portion of the tract while axons destined for the cervical cord anterior horn cells travel in the inner portion, closer to the gray matter column.

Careful observation will determine if there is wasting or hypertrophy of muscle. The pattern of wasting is important in terms of etiology. Is it located in the face or limb girdles; is it symmetric? The combination of wasting, weakness, hyporeflexia and lack of sensory findings represents the most important clinical features of muscle disease. The presence of tenderness on direct palpation of muscle is useful in inflammatory disorders of muscle. Many primary muscle diseases have a genetic basis and therefore careful family history can be very useful (Table 3.3).

3.7.A.1 INVESTIGATIONS

Investigations for muscle disease include serum assay for creatine kinase (CK) and aldolase. These are enzymes released from muscle when it has been damaged. A high serum level of either of these enzymes thus implies muscle tissue damage. Different isoforms of CK are contained in skeletal and cardiac muscle, thus making it possible to differentiate skeletal muscle damage (elevation of the

MM isoform of CK) from heart muscle damage (raised MB isoform levels or Troponin I).

The electromyogram (EMG) is a technique for investigating peripheral neuromuscular disease by recording electrical activity directly from muscle, using either needle or surface recording electrodes. Using this technique, it is often possible to distinguish between myopathic and neuropathic causes of muscle weakness.

In the past, the gold standard test for muscle disease was to perform a biopsy, which then showed the nature of the muscle pathology. More recently, DNA probes have been developed for many of the most common genetically inherited muscle diseases.

In the sample case outlined in Table 3.2, muscle disease was considered due to weakness. It would be annotated in the Expanded Localization Matrix with a ◯ but ruled out and changed to ⦸ because the weakness was unilateral and associated with hyperreflexia. An extensor plantar response, a classic sign of upper motor neuron dysfunction, also makes the localization of pure muscle disease unlikely.

3.7.B NMJ

The NMJ is the second last step in the process of motor output between nerve and muscle. It requires intact presynaptic, synaptic and postsynaptic function. The presence of a decrease in muscle strength with sustained effort (fatigue) in the absence of reflex changes or sensory signs is highly suggestive of a postsynaptic NMJ disease such as myasthenia gravis (Table 3.4). In contrast, facilitation of muscle strength with repeated activity suggests a presynaptic disorder such as 'Lambert–Eaton Myasthenic Syndrome'.

TABLE 3.4: NMJ Localization: Clinical Correlations

History/Symptoms	Observations	Investigations
Medication toxic exposure	Aminoglycosides, hydroyxchloroquine, recent anaesthesia, nerve gas	Normal CK Positive Tensilon test Abnormal single-fiber EMG Positive anti-Ach receptor or MuSK antibodies Thymoma on CT chest Genetic markers Muscle/nerve biopsy
Double vision	Opthalmoplegia not conforming to CRN III, IV, VI	
Dysphagia	Abnormal CRN IX, X, XII	
Weakness	Fatigability with continued effort (myasthenia)	
Atrophy/ hypertrophy	Absent	
Reflexes	Normal	
Sensory changes	Absent	

3.7.B.1 INVESTIGATIONS

The investigations for abnormalities of the NMJ include the Tensilon test, repetitive nerve stimulation, single-fiber EMG and assays for serum autoantibodies, antibodies against acetylcholine receptors or antibodies to muscle specific kinase, on the muscle side of the NMJ. In specialized clinics, biopsies of the NMJs can be performed and analyzed.

The Tensilon (edrophonium) test is a technique that temporarily increases the amount of acetylcholine present in the synaptic cleft to overcome the block at the NMJ. This test uses a medication called Tensilon or edrophonium bromide, a mild, reversible acetylcholinesterase inhibitor. The test is positive if the Tensilon reverses the neurological deficit for 5 to 10 minutes.

Acetylcholinesterase is an enzyme produced in the postsynaptic folds of the NMJ; its function is to break down acetylcholine molecules released from the presynaptic site. Were acetylcholine permitted to remain in the synapse after its initial action on the postsynaptic receptor, the result would be an effective blockade of the junction.

Repetitive nerve stimulation is a technique by which a repetitive electrical stimulus is applied to a peripheral nerve and the output is measured over the muscle it innervates (a compound muscle action potential [CMAP]), noting whether there are changes in CMAP amplitude as the stimulation continues. Normally, there are virtually no changes in CMAP amplitude after 10 to 15 seconds of rapid (2 to 3 Hz) stimulation. A decremental response (steadily decreasing CMAP amplitude with repetitive stimulation) suggests a fatigable NMJ and, therefore, a postsynaptic problem. An incremental response (increasing CMAP amplitude with repetitive stimulation) suggests a presynaptic problem.

The single-fiber EMG is a test of the integrity of the NMJ. The ability of an NMJ to faithfully transmit impulses to all the constituent muscle fibers within a given motor unit is tested and expressed as a parameter called 'jitter'. The test is based on measuring the variance of jitter between two NMJs within the same motor unit.

In this case, disease of the NMJ was considered due to weakness but was ruled out for the same reasons as muscle disease.

3.7.C PERIPHERAL NERVE

Peripheral nerve includes motor and sensory neuronal cell bodies, their axons including myelin, their terminal boutons and the metabolic machinery that supports them.

As the peripheral nerve serves sensory, motor and autonomic functions, destruction of the cell bodies, axons or myelin of this tissue leads to loss of function in any or all of these modalities. Sensory findings are often the presenting symptom, with motor and autonomic symptoms and signs coming later. Therefore, a combination of

TABLE 3.5: Peripheral Nerve Localization: Clinical Correlations

History/Symptoms	Observations	Investigations
Medication toxic exposure	Check history vincristine, cisplatin	Glucose, B12, folate, thyroid, protein electrophoresis
Weakness	Usually late, except for acute conditions such as AIDP	Nerve conduction study/EMG
Sensory changes	Stocking and glove, radicular	Tilt Table Test Genetic markers – CMT
Reflexes	Decreased	
Autonomic	Decreased sweating, postural hypotension	
Metabolic disorders	Diabetes, thyroid, vitamin B12 deficiency	
Family history	May be present	

sensory, motor and autonomic dysfunction with normal or decreased reflexes suggests peripheral nerve localization (Table 3.5).

It is important to recognize that, with diffuse peripheral nerve disease, motor and sensory symptoms and signs typically appear earlier in, are more severe in and are often confined to the distal extremities. The reason for this phenomenon is that the longest nerve fibers, typically from the lower spinal cord to the feet, are the most metabolically at risk and therefore more sensitive to any form of damage. Therefore, it is common that symptoms of peripheral nerve dysfunction occur in the feet and hands first. This is the 'stocking and glove' pattern of peripheral neuropathy.

One must beware of combined lesions such as cord and peripheral nerve; these can give a combination of lower and upper motor neuron findings. Typical examples include amyotrophic lateral sclerosis (ALS) and a syndrome caused by vitamin B12 deficiency (subacute combined degeneration).

3.7.C.1 INVESTIGATIONS

The investigations for peripheral nerve lesions include nerve conduction studies (NCSs), EMG, metabolic studies for the detection of various specific nerve disorders and in some cases biopsies of peripheral nerves.

NCSs are a test of the electrical response of peripheral nerves to faithfully transmit electrical stimuli. These tests help to determine the speed of conduction and the amplitude of the electrical response from peripheral nerves. This information aids in localization of dysfunction to specific peripheral nerves, plexi or roots. Slowing of conduction suggests disruption of myelin (a demyelination disorder), whereas low amplitude suggests axonal dysfunction or destruction.

EMG has been described in the section on muscle disease.

Metabolic studies useful in determining the etiology of some peripheral nerve problems include serum assays for glucose (in diabetes), vitamin B12, folic acid, thyroid hormone, protein electrophoresis and specialized DNA assays for genetic forms of peripheral nerve disorders.

There are over 20 different hereditary neuropathies that can be expressed with different levels of penetrance. Family history is important if positive (see Table 3.14).

In the sample case, disease of the peripheral nerves was considered due to weakness but ruled out due to the presence of unilateral findings, expressive aphasia and hyperreflexia. The Expanded Localization Matrix would show the peripheral nerves as being considered but ruled out as less likely, that is, ○ overwritten with ⊘s.

3.7.D SPINAL CORD

Motor disturbances due to lesions in the spinal cord depend on which tracts are involved and at what spinal cord level. Patterns of spinal cord injury relate to the cross-sectional anatomy of the cord. From the viewpoint of clinical localization, the three most important tracts in the spinal cord are the spinothalamic tract, the posterior columns and the corticospinal tract. These have been reviewed in Chapter 1 (Figures 1.4 and 1.5).

The clinical picture in spinal cord injury depends on which tracts are damaged. Table 3.6 details the differences in the clinical findings for each of the principal spinal cord syndromes (these will be discussed in detail in Chapter 5).

For instance, damage to the anterior two-thirds of the spinal cord will lead to loss of function of the gray matter containing the lower motor neurons, the spinothalamic tracts and the corticospinal tracts on both sides but spare the posterior columns (as will be discussed in Chapter 5). The clinical picture would consist of flaccid paraplegia with loss of sensation to pin prick and temperature at the spinal cord level and below but preserved sensation to vibration and proprioception as the posterior columns are still intact.

3.7.D.1 INVESTIGATIONS

The single most important investigation for spinal cord localization is neuroradiological imaging. If available, MRI of the spine at the appropriate level is optimal; however, CT scan, CT myelogram or plain X-ray films are useful whenever MRI is unavailable or (for medical reasons such as the presence of a pacemaker or other ferrous object embedded in the body) cannot be performed.

The clinical level of the spinal cord lesion needs to be transmitted to the neuroradiologist so that the appropriate level can be scanned with several segments above and below the clinical level. Other neuroradiological investigations include plain or CT myelography, in which, after a lumbar puncture has been performed, an iodinated dye is injected into the subarachnoid space. The spread of the dye in the subarachnoid space is followed up to the neck as the patient is tilted head down to make the dye flow with gravity.

Somatosensory evoked potential (SSEP) is a test that measures the response to an electrical stimulus applied to either upper or lower limb nerves in order to determine the speed of conduction from peripheral nerves through the spinal cord through the posterior columns to the contralateral somatosensory cortex. An electrical stimulus is usually applied to the median nerves at the wrist or to the posterior tibial nerves at the ankles. Depending upon the site of stimulation (leg versus arm), the passage of the stimulus-induced volley through the central nervous system (CNS) is detected with recording electrodes placed over the lumbar spine, in the midline neck and in the appropriate region of the contralateral scalp. Each side and limb are tested separately. This test assists in localization by assessing at what level (and structure) there is slowing or cessation of transmission of the evoked potential.

In this case, spinal cord was considered due to limb weakness and hyperreflexia, but isolated damage to the spinal cord would not explain the right facial weakness, expressive aphasia and other cortical deficits. The Expanded Localization Matrix would show the spinal cord as being considered but ruled out as less likely, that is, ○s overwritten with ⊘s.

TABLE 3.6: Spinal Cord Localization: Spinal Cord Syndromes

	Weakness	Spasticity	Pain/ Temp Loss	Vibration/ Proprioception Loss	Investigation
Transection	Below	Below	Below	Below	MRI, CT myelogram/ SSEP
Brown–Sequard[a]	Ipsilateral below	Ipsilateral below	Contralateral below	Ipsilateral below	Search for source of compression
Anterior spinal artery	Below	Below	Below	Preserved	
Syrinx	None unless large	None unless large	Depends on level	Preserved	

[a] Common clinical spinal cord syndrome named after the nineteenth century French-American physician who first described it.

3.7.E BRAINSTEM

The brainstem includes the midbrain, pons, medulla, cranial nerve nuclei and the ascending and descending axonal tracts, which pass through the brainstem. Brainstem lesions produce motor and sensory disturbances in the face and body, along with localized cranial nerve findings.

The location of the cranial nerves originating in the midbrain, pons and medulla was reviewed in Chapter 2. The combination of isolated cranial nerve findings with long tract sensory and motor findings with or without ataxia suggests a brainstem localization.

Due to the compact nature of the brainstem nuclei and the adjacent tracts running through the brainstem, it is important when there is dysfunction in one cranial nerve or tract to look at the neighbours at the same level. For instance, a patient with unilateral facial weakness or Bell's palsy may have less obvious findings in the other cranial nerves at the same level in the pons: CN V, VI or VIII. Involvement of these other cranial nerves would suggest a central, more widespread lesion such as a pontine tumor, stroke or demyelination. The signs and symptoms are summarized in Table 3.7.

It is also important to look for dissociated sensory loss (i.e. isolated discriminative touch or pain and temperature)

TABLE 3.7: Brainstem Localization: Clinical Correlations

Level	Symptoms	Signs	Investigations
	Midbrain		
CN III	Horizontal or vertical diplopia	Loss pupillary reflex Loss of elevation, depression and adduction	MRI, CT
CN IV	Oblique diplopia	Head tilt, decreased movement of eye down and inward	
Corticospinal tract	Weakness	Hyperreflexia to opposite side	
Spinothalamic	Numbness, paresthesias	Loss of pin prick and temperature sensation to opposite side	
Medial lemniscus	Numbness, paresthesias	Loss of vibration and proprioception to opposite side	
	Pons		
CN V	Loss of facial sensation Difficulty with mastication	Decreased pin prick CN V – I, II, III Weakness of masseter, temporalis, pterygoid muscles	MRI, CT Cold calorics
CN VI	Horizontal diplopia	Decreased abduction	Blink reflex BSEP
CN VII	Facial weakness	Decreased facial movement to upper, mid and lower face	
CN VIII	Decreased hearing, vertigo, loss of balance	Loss of hearing, nystagmus, loss of balance	
Corticospinal tract	Weakness	Hyperreflexia	
Spinothalamic	Numbness, paresthesias	Loss of pin prick and temperature sensation to opposite side	
Medial lemniscus	Numbness, paresthesias	Loss of vibration and proprioception to opposite side	
	Medulla		
CN IX CN X	Difficulty swallowing or controlling secretions	Loss of gag reflex	MRI, CT
CN XI	Inability to turn head with sternocleidomastoid muscles or lift shoulders using trapezius muscles	Weakness of sternocleidomastoid and trapezius muscles	
CN XII	Inability to move tongue to one side	Deviation to side of weakness	
Corticospinal tract	Weakness	Hyperreflexia	
Spinothalamic	Numbness, paresthesias	Loss of pin prick and temperature sensation	
Medial lemniscus	Numbness, paresthesias	Loss of vibration and proprioception to opposite side	
Sympathetic tract	Drooping of eye, small pupil, lack of facial sweating	Ipsilateral, miosis, ptosis, loss of sweating	
Respiratory centre	Apnea	Loss of CO_2 drive	Apnea test

and motor signs in the face and body. Similar to the spinal cord, damage to one side of the brainstem can give facial sensory loss on the same side with loss of contralateral sensation to the body due to damage of tracts that have already crossed and are damaged in the brainstem.

The presence of coma or a progression of accumulating deficits from midbrain to pons to medulla – the rostrocaudal sequence of deterioration – suggests a lesion compressing the brainstem from above.

3.7.E.1 INVESTIGATIONS

The best investigation to confirm brainstem localization is MRI of the brain. Imaging the brainstem using CT scanning suffers from artefact caused by the petrous ridges of the temporal bones, and therefore, MRI is preferable.

Brainstem evoked potentials (BSEPs) involve testing the auditory portion of CN VIII using an auditory clicking sound and measuring the responses from CN VIII and the various brainstem structures that relay auditory information through the brainstem and eventually to the cortex. This test might show a delay or slowing of the signal evoked by the sound on one or both sides, allowing for precise localization of the lesion.

The blink reflex is an electrical equivalent of the corneal reflex: it measures the latency of the input of an electrical stimulus to CN V and the resulting blink output of CN VII on both sides. If the reflex response is slow or absent, this can help localize the lesion to CN V, the pons or CN VII on either side.

In the case under discussion, brainstem damage was considered due to unilateral right-sided weakness, speech difficulties and hyperreflexia; however, isolated damage to the left side of the pons would also produce left-sided weakness of the upper, mid and lower face along with right-sided hemiplegia. This patient had a conjugate gaze preference and not an isolated abnormality of extra-ocular movements. This would suggest a lesion at a higher level.

Finally, brainstem damage would also not explain the presence of aphasia. Brainstem lesions can produce dysarthria (a disturbance of articulation of speech) by involvement of CN V, VII, X and XII or their upper motor neuron connections. Therefore, localization of the disease to the brainstem is less likely. The Expanded Localization Matrix would show the brainstem as being considered but ruled out as less likely, that is, Os overwritten with Øs.

3.7.F DEEP WHITE MATTER AND LATERAL THALAMUS

Ipsilateral lesions of the deep white matter and lateral thalamus tend to give hemi-motor or hemi-sensory symptoms and signs on the opposite side of the body, including the face, arm trunk and leg. For larger lesions, both motor and sensory findings can occur. The trunk does not tend to be spared for sensory loss as occurs with cortical lesions (Table 3.8).

3.7.F.1 INVESTIGATIONS

CT scan with CT angiography and MRI scanning with angiography of the brain are the preferable modes of investigation to localize the lesions in this area. Clinically it is impossible to differentiate a small hemorrhage from a small infarction. Treatment decisions with respect to thrombolysis need to ensure that there is no evidence of hemorrhage prior to the initiation of thrombolytic therapy (for further explanation of this intervention, see Chapter 8).

In this case, white matter damage in the left hemisphere was considered due to unilateral right-sided weakness and hyperreflexia. Isolated white matter damage usually does not cause Broca's aphasia. White matter injury usually causes dysarthria (difficulty in articulation) but not deficits in the formulation of language.

Large areas of damage can obviously affect both cortex and white matter in this area, at which point therefore it is sometimes difficult to distinguish between the two pathology locations. The Expanded Localization Matrix would show the deep white and thalamus as being considered as a possibility but as less likely; that is, Os are left without change.

3.7.G CORTEX

Cortical lesions cause dysfunction of complex processing of information, the type of deficit depending on the cortical region involved. Lesions of the speech centres can impair input (Wernicke's area) with involvement of parietal cortex or output (Broca's area) or conduction of speech (arcuate fasciculus) or all of these. Lesions of a specific area such as the angular gyrus can affect calculation and left/right discrimination (Gerstmann syndrome).

Often, there is underlying white matter dysfunction causing long tract sensory or motor dysfunction from regions adjacent to the cortical lesion.

TABLE 3.8: Deep White Matter and Lateral Thalamus: Clinical Correlations

History/Symptoms/Signs	Observations	Localization	Investigations
Loss of motor function on contralateral side	Hemiplegia face/arm/leg	Internal capsule	MRI, MRA CT/CTA
Loss of sensory function on contralateral side	Decreased pinprick, vibration without sparing the trunk	Lateral thalamus	

TABLE 3.9: Cortex: Clinical Correlations

Location	Function	Effects of Damage	Dominance (Right Handed)
Frontal Cortex			
Frontal poles	Executive function, attention	Loss of judgement, planning, emotional, disinhibition, inattention	Both
Frontal eye fields	Conjugate eye movements	Loss of conjugate movement to contralateral side	Both
Frontal operculum Broca's area	Expression of language	Expressive aphasia[a]	Left hemisphere
Precentral gyrus	Motor output	Loss of motor activity on contralateral side	Both
Cingulate gyrus	Motor planning, attention and execution	Apraxia, attention deficits	Both
Temporal Cortex			
Superior temporal gyrus	Reception of speech	Impairment of comprehension of speech[a]	Left hemisphere
Hippocampus	Short-term memory	Loss of short-term memory	Slight left hemisphere
Amygdala	Generation of fear and sexual responses	Inappropriate sexual behaviour	Both
Insular cortex	Reception of sound and vestibular input	Loss of auditory attention and discrimination	Both
Parietal Cortex			
Parietal lobe	Integration of and attention to sensory information, speech recognition	Loss of ability to recognize visuospatial relationships, sensory inattention, speech recognition[a]	Right hemisphere for sensory inattention
Postcentral gyrus	Reception of somatosensory information relayed from thalamus	Loss of contralateral sensation	Both
Angular gyrus	Calculation, left–right discrimination, concept of finger position	Inability to calculate, recognize left/right, finger agnosia	Both
Occipital Cortex			
Occipital poles	Reception of visual information from ipsilateral lateral geniculate body (LGB)	Hemianopia, cortical blindness	Both
Parieto-occipital cortex	Recognition of visual objects such as faces	Hemi-attention, hemianopia, inability to recognize faces, cortical colour blindness	Both

[a] Understanding and production of the meaning and non-semantic aspects of language that conveys meaning and emotion are thought to be localized to homolgous areas in the non-dominant hemisphere and are called prosody.

For instance, the processing of visual information for the purpose of generating visual imaging (such as visual acuity and visual fields) localizes above the tentorium involving both cortex and white matter tracts transmitting visual information from the eyes to the occipital cortex.

Table 3.9 lists the various cortical functions by location and the effects of damage to these areas.

3.7.G.1 Investigations

Some clinical tools used for testing cortical function include the Mini Mental Status Exam and the Montreal Cognitive Assessment, both of which are discussed in Chapter 2. These bedside tools help with localization in areas of attention, memory, executive function, calculation, praxis and language but are not as useful in cases of mild to moderate cognitive decline.

Detailed neuropsychological testing can be performed by a licensed neuropsychologist, who can administer a battery of cognitive tests to help in localization.

High-resolution CT scan or MRI scanning and angiography of the brain are the preferable modes of investigation to localize lesions in this area of the nervous system.

The electroencephalogram (EEG) is a useful tool to assess the background rhythms generated by the interplay of the cortex and thalamus. Localized slowing on the EEG can reflect focal cortical damage, whereas diffuse slowing can represent widespread injury such as in dementia or metabolic/toxic injury, which affects the whole brain. More specific EEG patterns can be found such as the periodic 1 per second slow wave discharges seen in Creutzfeldt–Jakob disease (Figure 3.2). For more information concerning the neurophysiological basis of normal and abnormal EEG patterns, see Chapters 10 and 11.

In the case under discussion, cortical damage in the left hemisphere was considered due to unilateral right-sided weakness, and hyperreflexia. Left frontal lobe damage would explain the expressive aphasia, lower right facial weakness, right upper limb hemiparesis and hyperreflexia with an extensor plantar response (Babinski sign).

The expressive aphasia is due to destruction of Broca's area in the operculum of the left frontal lobe in a right-handed person.

The right hemiplegia is due to damage to the left precentral gyrus (and to adjacent cortical areas); this region is the source of the upper motor neuron input to the facial nerve nucleus on the right side of the brainstem and right half of the spinal cord that controls movement on the right side of the body. Damage to these upper motor neurons leads to weakness and hyperreflexia. Left temporal and parietal lobe damage would be suspected in the case of a receptive aphasia. The functions of speech and word recognition are widely distributed in area including the dominant left temporal and parietal lobes. Recent literature has shown that there are several functions required in speech and word recognition that include phoneme and phonological perception. The term *Wernicke's aphasia* localized to the posterior third of the superior gyrus of the left temporal lobe should now be modified to reflect the new information based on functional MRI data.

Sensation on the right side is intact, which implies that the left parietal cortex and underlying white matter connections have been spared.

The Expanded Localization Matrix would show the frontal cortex as being considered as most likely, and the localizations populated with Os.

3.7.H CENTRES OF INTEGRATION

3.7.H.1 CEREBELLUM

The cerebellum has widespread connections to and from the brainstem and cerebral cortex, which results in clinical deficits not only in balance and coordination but also in the cognitive domains of executive function, reasoning, calculation, affect and visual spatial function.

Lesions in the cerebellum cause either truncal unsteadiness for midline lesions involving the vermis or ipsilateral limb incoordination for hemispheric lesions. The output of the cerebellum produces damping of motor movements; therefore, loss of damping will cause movements to overshoot. This is the explanation for the clinical signs consisting of ataxia, past-pointing and hyporeflexia. The diagnosis of a cerebellar lesion has to be made only by exclusion. An ipsilateral hemispheric cerebellar localization can be made only when the lower motor neuronal system, upper motor system and the sensory systems are intact.

The cerebellar cognitive affective syndrome may not be initially clinically obvious and be overshadowed by the difficulties with articulation, incoordination and balance.

A common mistake made by students seeing a patient with a lesion of the right internal capsule is to localize it as a left cerebellar lesion due to the presence of apparent ataxia in the left upper limb. In this case, the left arm ataxia results from weakness caused by dysfunction of the upper motor neuron unit damaged by the right hemisphere lesion.

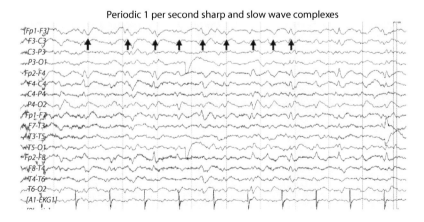

FIGURE 3.2: EEG in Creutzfeldt–Jakob disease. Arrows indicate periodic slow wave discharges.

TABLE 3.10: Cerebellum: Clinical Correlations

History/Symptoms	Observations	Localization	Investigations
Difficulty with balance, falls	Wide-based gait, nystagmus, dysarthria	Vermis, midline lesion	MRI, CT
Vertigo, difficulty with articulation, loss of coordination	Dysarthria, nystagmus, past-pointing, pendular reflexes, decreased muscle tone	Ispilateral hemispheric lesion if motor and sensory exams are normal	

The clinical picture clarifies the situation in that the right hemisphere lesion will often be associated with an upper motor neuron pattern of weakness and hyperreflexia of the left upper limb.

Table 3.10 details the common signs and symptoms associated with midline and hemispheric cerebellar lesions.

3.7.H.2 BASAL GANGLIA

Localization to the basal ganglia is suggested by hemi body signs of either increased (Huntington's chorea) or decreased (Parkinson's disease) activity of agonist and antagonist muscle groups involved in postural and voluntary motor control. If both basal ganglia are involved, then the symptoms and signs are bilateral. Dyskinesia, dystonia, tremor, rigidity, bradykinesia and difficulties maintaining a standing posture all suggest basal ganglia

dysfunction. Table 3.11 details the signs and symptoms of disorders of modulating structures at several levels.

In the present case, damage to the modulating structures (in particular the basal ganglia) is difficult to determine as there is clinical evidence of dysfunction of the corticospinal tracts originating from the left hemisphere.

The general rule is that localization of dysfunction to the modulating structures is made by exclusion in that all the primary motor and sensory modalities must be found to be normal before one can localize primarily to a modulating structure.

The Expanded Localization Matrix shows the cerebellum and basal ganglia as being considered but ruled out as less likely, that is, O overwritten with Ⓞ.

The completed Expanded Localization Matrix is shown in Table 3.12. Although pathology of the white matter/thalamus, parietal lobe and temporal lobe has

TABLE 3.11: Basal Ganglia: Clinical Correlations

History/Symptoms	Observations	Investigations
	Basal Ganglia	
Tremor, slowness, rigidity, falls	3–4 Hz tremor Cogwheel rigidity Akinesia Micrographia Postural instability Festinating gait Flexor posturing	MRI – tumor, infarct Cerulosplasmin – Wilson's disease Heavy metals – manganese Genetic – LKKK2, PARK2, PARK7, PINK, SNCA
Involuntary movements, family history	Dystonia, hemiballismus	Medication Inflammatory markers Genetic markers
	Caudate	
Involuntary movements, psychiatric disturbance, cognitive decline, family history	Choreothetosis Dancing gait Dementia Pathologoical damage to posterior putamen and lateral thalamus	MRI – caudate atrophy Genetic – chromosome 4p16 excessive CAG repeats encoding glutamine for the gene encoding for huntingtin
	Midbrain	
Falls	Loss upper motor neuron control of eye movements Extensor posture	MRI – 'mouse ear' midbrain Heavy metals Genetic – DYT5, EOTD, DYT11, TOR1A

Note: The basal ganglia consist of the putamen, globus pallidus, subthalamic nucleus and substantia nigra. These nuclei interact with other subcortical structures such as the thalamus and upper midbrain to modulate the initiation of thought and movement.

TABLE 3.12: Completed Expanded Localization Matrix (Sample Case)

		MUS	NMJ	PN	SPC	BRST	WM/TH	OCC	PAR	TEMP	FR	BG	CBL
Attention	N												
Memory	U												
Executive	U												
Language	Exp	⊘	⊘	⊘			⊘			⊘	○		
Visuospatial	N												
	Right	Left											
CRN I	N	N											
CRN II	N	N											
CRN III, IV, VI	Gaze to left Able to overcome with head movement						⊘	⊘				○		
CN V	N	N	⊘	⊘	⊘		⊘	⊘						
CRN VII	Lower face	N	⊘	⊘	⊘		⊘	⊘				○		
CRN IX, X	N	N	⊘	⊘	⊘		⊘	⊘						
CRN XI	↓	N	⊘	⊘	⊘		⊘	⊘				○		
CRN XII	Dev R		⊘	⊘	⊘		⊘	⊘		⊘		○		
Tone	↑	N	⊘	⊘	⊘		⊘	⊘		⊘		○		
Power	↓	N	⊘	⊘	⊘		⊘	⊘		⊘		○		
Reflex	↑	N	⊘	⊘	⊘		⊘	⊘		⊘		○		
Involuntary Mvt	N	N												
PP	N	N	⊘	⊘	⊘		⊘	⊘		⊘		○		
Vib/prop	N	N	⊘	⊘	⊘		⊘	⊘		⊘		○		
Cortical sensation	N	N	⊘	⊘	⊘		⊘	⊘		⊘		○		
Coord UL	Abs	N	⊘	⊘	⊘		⊘	⊘		⊘		○		
Coord LL	Abs	N	⊘	⊘	⊘		⊘	⊘		⊘		○		
Walk	NA	NA												
Toe/heel/invert/evert	NA	NA												
Tandem	NA	NA												
Romberg	NA	NA												

Abbreviations: Abs, absent; BG, basal ganglia; BRST, brainstem; CBL, cerebellum; CRN, cranial nerve; Dev R, deviated to the right; Exp, expressive aphasia; FR, frontal lobe; MUS, muscle; N, normal; NA, not attempted; NMJ, neuromuscular junction; OCC, occipital lobe; PAR, parietal lobe; PN, peripheral nerve; PP, pin prick; SPC, spinal cord; TEMP, temporal lobe; U, untestable; WM/TH, white matter/thalamus.

remained as possibly explaining the motor findings, only the option of frontal lobe pathology also explains the patient's right hemiparesis, expressive aphasia and eye deviation. Thus, among the four remaining possible locations for the patient's pathological process, only the left frontal lobe location explains all of the clinical findings.

3.7.1 LOCALIZATION SHORTCUTS

Patients with isolated signs and symptoms can sometimes be quickly localized using the following shortcuts. These rules are generalizations that have to be put into the context of a given patient.

Findings of sensory loss rule out muscle disease and disorders of the NMJ as the primary localization.

Disturbances of visual acuity and colour vision due to neurological disease can be localized to those structures above the brainstem that transmit information from the retina to the occipital cortex areas dedicated to the production of visual imagery.

Seizures nearly always arise from the cerebral cortex or thalamus; this means that seizures without other deficits can be localized to the cortex or to the related subcortical modulating structures in the thalamus.

Deficits of cognition without disturbances of consciousness or attention localize above the brainstem.

3.8 ETIOLOGY

What is the pathophysiological cause or etiology of the lesion?

Now that we have worked out the probable location of our patient's pathology, we must go back to the medical history to begin the process of determining the most likely cause of the problem.

The first step is to extract the key elements of the patient's history and enter them in the History Worksheet (Figure 3.3).

The relevant historical information has been added to the sheet based on time zero being the onset of the major presenting complaint of right-sided weakness and speech disturbance. It is then important to try and correlate the sequence and evolution of the symptoms with the etiological possibilities. To help identify the appropriate category of etiology, it is important to ask the status of the patient's health over the last weeks to months, for example, if there have been any toxic exposures, previous medical illnesses or foreign travel.

It is often very useful to ask patients exactly what they were doing at the time of onset of defining symptoms, what

time of day it was, their sleep and nutritional status as well as the names and dosage schedule of current medications and when these medications were last taken. Any information related to recent addition of medications and changes in dose or timing can often explain changes in neurological status.

For instance, a 40-year-old man who presents with new-onset weakness, muscle pain, and elevated CK two weeks after starting a statin medication can be easily diagnosed as a statin-induced myopathy and the treatment is simple – stop the statin.

Determination of a range of etiologic possibilities can be accomplished by referring to the grid shown in Figure 3.4 that correlates the duration of a given patient's set of symptoms with the type of disease process that is most likely to present with such a time course.

Before we proceed to use the grid to work out the etiology of the sample case, the information in the grid requires further explanation. Similar to the approach taken for the localization matrix, a checklist approach to rule in or out each of the possible causes or etiologies provides the student with a consistent process of thought, which avoids the need to memorize long lists of diseases and their causes.

Symptom	Time zero	
Loss of consciousness		
Cognitive impairment		
Pain		
Weakness		Right hemiplegia – Face, arm > leg
Eye movement		Conjugate gaze to left
Loss of vision	Left eye 3 weeks	
Loss of sensation		
Loss of hearing		
Loss of balance		
Loss of coordination		Due to weakness of right side
Memory loss		
Speech disturbance		Expressive aphasia
Incontinence		
Sleep disturbance		
Loss of milestones		

FIGURE 3.3: History worksheet.

Disease type	Secs	Mins	Hrs	Days	Weeks	Months	Years
Paroxysmal (seizures, faints, migraine)							
Traumatic							
Vascular (ischaemic stroke, bleed)							
Toxic							
Infectious							
Metabolic							
Inflammatory/autoimmune							
Neoplastic							
Degenerative							
Genetic							

← Acute →
← Subacute →
← Chronic →

FIGURE 3.4: Etiology matrix.

The etiology matrix uses the time sequence of the onset and duration of the principal symptoms to generate a focused list of the most probable generic causes of the patient's symptoms. This list, combined with the most probable localization, narrows down the diagnostic possibilities significantly. The same system of symbols using a ○ for possibilities that have not been ruled out and ⊘ for possibilities that have been ruled out is used.

When listening to a resident presenting a case, it is much preferable to listen to a precise presentation of the possible localization and etiology than to a long list of differential diagnoses memorized from a textbook with no context. For clinical teachers, it also provides a method to explore the resident's process of clinical reasoning rather than the ability to memorize lists.

With this introduction, the time course of the symptoms of a given case can be plotted on the etiology matrix sheet and then the most common etiologies determined. There will always be exceptions to these rules especially if two disease processes are in play: for example, a patient with an acute stroke and long standing diabetic neuropathy – both causing sensory loss. In this case, two different sheets would be necessary to plot the different time courses.

As a rule, disorders with extremely rapid onset in seconds to minutes with loss of neurological function are paroxysmal, vascular or traumatic in nature, but can also be due to the acute effect of medications or nonmedical drugs.

Disease categories evolving over a few weeks to months include toxic, infectious, metabolic, inflammatory and neoplastic disorders.

Chronic disease processes evolving over months to years tend to be metabolic, neoplastic, degenerative or genetic in nature.

Let us consider each disease category in more detail with reference to the sample case.

3.8.A PAROXYSMAL*

This disease category includes disorders that transiently disrupt metabolic and electrical function of the brain, spinal cord or peripheral nerves.

3.8.A.1 ELECTRICAL

Focal electrical discharges in a specific part of the brain can cause focal or partial epileptic seizures. These focal discharges can then spread to involve the entire brain, giving rise to generalized seizures. Primary generalized seizures can occur without focal onset. Epileptic seizures will be discussed in depth in Chapter 11.

Seizures generally are due either to an underlying genetic or metabolic/toxic abnormality, with the actual events triggered by environmental factors, or to focal brain

injury. The brain injury can be from a variety of causes, such as a developmental abnormality, stroke, trauma or tumor.

All seizures have an acute timeline, recurrent seizures occurring on many occasions, often over many years.

Global and focal deficits can occur after any type of seizure. Seizures cause hypermetabolism, which can then exhaust cellular substrates. This metabolic exhaustion can manifest clinically as brain dysfunction either globally with generalized seizures or focally with partial seizures. In the latter instance, there may be unilateral weakness, a phenomenon known as Todd's paralysis. This metabolic exhaustion and neurological dysfunction can last up to 24 hours and can mimic stroke or transient ischemic attack (TIA).

3.8.A.2 SYNCOPE

Syncope is a frequent cause of transient neurological but reversible deficits. Syncope is defined as loss of consciousness associated with absence of postural tone, followed by complete recovery. This accounts for approximately 5% of all emergency room visits and hospital admissions. There is often a prodrome of presyncopal symptoms such as dizziness, nausea or palpitations. Main cardiac etiologies include cardiac arrhythmia (bradycardia or tachycardia), structural heart disease and outflow tract obstruction (aortic stenosis, pulmonary stenosis and hypertrophic obstructive cardiomyopathy). Vasovagal syncope accounts for 20% of all syncope cases in both genders and is the most commonly identified etiology of syncope.

3.8.A.3 MIGRAINE

Migraine is a form of paroxysmal transient disturbance of control of regional cerebral metabolic activity and the region's associated cerebral vasculature due, at least in part, to inflammation and dysregulation of cranial blood vessels supplied by sensory branches of the trigeminal nerve. On first presentation, migraine patients having focal deficits followed by headache cannot be distinguished from those with stroke or TIA. The availability of vascular neuroimaging is extremely helpful to rule out underlying vascular abnormalities in these patients.

In the case under discussion, the focal right-sided deficits could fit with a Todd's paralysis or migraine, but the transient loss of vision in the left eye four weeks earlier would not be consistent with a cortically mediated seizure-related phenomenon or migraine. Also, there is no history of any rhythmic twitching on the right side of the body preceding the development of right-sided weakness. Abnormal electrical discharges in the spinal cord can cause localized myoclonus, usually abnormal involuntary shock-like movements in a single limb or group of muscles. Generalized discharges of the spinal cord involving the lower motor neurons can occur, resulting in a clinical syndrome called Stiff Person Syndrome. Electrophysiologically, this is known as neuromyotonia; it

---

* A lettering system for Etiological Diagnosis is used for Chapters 4–13 to provide a classification framework to the reader.

is believed to be an autoimmune disorder and may occasionally be triggered by occult neoplasia.

3.8.B TRAUMA

This disease category includes all forms of damage to the brain and spinal cord as a result of excessive forces applied directly or indirectly to the nervous system.

Trauma can cause damage (usually in the form of hemorrhage) at every level of the nervous system. This is best remembered by tracing the structures and layers that a knife or projectile must pass through in order to enter the brain.

The outer layer is the skin and muscle of the scalp, followed by the outer skull periosteum, the skull and then the inner skull periosteum. Traumatic hemorrhage to these structures is usually externalized and causes local damage related to nearby structures such as the eyes, ears and external branches of cranial nerves.

Trauma under the inner periosteum and above the dura may cause laceration of the middle meningeal artery under the squamous portion of the temporal bone, resulting in epidural hemorrhage (discussed further in Chapter 9, case 9e-2). This type of intracerebral hemorrhage is caused by arterial bleeding, which is under much higher pressure than venous bleeding and therefore can cause rapid expansion of the hematoma with compression of adjacent structures including brain herniation.

The next space into which blood can accumulate is the space between the arachnoid and the dura, the subdural space. Injury is usually to the veins bridging the space between the arachnoid and the dura causing venous bleeding. Since venous pressure is relatively low, blood in the subdural space may accumulate over days to weeks, causing subacute deficits due to pressure on the underlying brain.

Trauma to the major arteries that course in the subarachnoid space can cause subarachnoid hemorrhage. This can be due to either direct or indirect trauma causing shearing forces on the blood vessels and their attachments. Traumatic subarachnoid hemorrhage is clinically similar to that due to rupture of aneurysms: there is severe headache – 'thunderclap headache' – and meningsmus (irritation of the meninges). If a hematoma forms that extends into the brain, then focal deficits may ensue. The mass effect of a large hematoma may result in decreased level of consciousness and brain herniation phenomena (see relevant e-cases in Chapters 9 and 10).

The most serious situation is that of open head injury, in which there is a skull fracture with direct damage to brain and blood vessels causing crush injury and hemorrhage to cortex and white matter. If the injury has been associated with hypoxia and hypotension as with a gunshot wound or motor vehicle accident, then there is potential for the metabolic injury due to hypoxia and hypotension to make the direct brain injury worse.

Excessive forces caused by either acceleration or, more commonly, deceleration can cause damage at various locations within the nervous system without actually breaching any of the protective structures. The damage caused in closed head injury includes intraparenchymal hemorrhage and shearing injury to axonal tracts: diffuse axonal injury.

The term *concussion* is used for a closed head injury with momentary loss of consciousness with no focal signs. CT and MRI scans of these individuals often show small hemorrhages in the cortex and subcortical white matter, which resolve spontaneously over several weeks. Neuropathological studies have identified a disorder characterized by cumulative brain injury related to sports such as ultimate fighting, boxing, hockey, football and other contact sports. This has led to the use of the diagnostic term *chronic traumatic encephalopathy* (CTE) to describe this condition.

Immediate neuroimaging is the principal form of investigation for traumatic injuries. In the context of trauma, the appropriate area of the neuraxis needs to be stabilized prior to being assessed by neuroimaging to rule out any fractures or instability.

In our sample case, the history and findings do not fit with a traumatic injury.

3.8.C VASCULAR

This category includes disorders involving disease of large and small blood vessels supplying the brain and spinal cord.

Blood vessels usually malfunction in one of two different ways: they can burst or block. Both of these events are usually acute. Often, there is a small warning blockage or leak before the complete blockage or rupture occurs. It is therefore very important to intervene at an early stage to prevent a major blockage or rupture of a blood vessel supplying the brain or spinal cord.

In general, acute vascular occlusions or ruptures occur in blood vessels on one or other side of the brain, resulting in focal symptoms on the opposite side of the body. The only exception to this rule is the cerebellum, in which an ipsilateral cerebellar hemorrhage or infarction will cause clinical symptoms of ataxia and decreased coordination on the same side of the body (see Chapter 7 for explanation).

The time course of blockage or rupture of blood vessels is always acute, sometimes in a recurrent fashion, with recovery between events, at least initially. Because of the potentially devastating consequences of a vascular event, an unexplained acute focal neurological deficit must be taken seriously and the precise cause identified as quickly as possible in order to prevent eventual serious permanent damage.

The term *TIA* or *transient ischemic attack* refers to the situation of a temporary interruption of blood supply to an area of the brain, eye or spinal cord, causing a neurological deficit lasting less than 24 hours but usually less than 30 minutes. In neuroanatomic terms, this usually represents the effect of a small embolus that has become lodged in cerebral vessels, causing ischemic changes and neurological deficits related to the area of brain supplied by that vessel; subsequently the embolus breaks into smaller fragments, resulting in resumption of blood flow and reversal of the deficits. In the sample case, the two episodes of transient loss of vision in the left eye occurring two to three weeks before the main clinical event could certainly be explained by embolic occlusions of the left ophthalmic artery, a branch of the left internal carotid artery.

The term *ischemic stroke* refers to the situation as described earlier except that the embolus does not move or fragment, causing irreversible ischemic damage to the brain supplied by the vessel.

With the availability of CT scanning, CT angiography or MRI angiography in most major centres, immediate neuroimaging of the cerebral vascular anatomy of patients with cerebrovascular disease has become the standard of care. With these modalities of imaging, it is possible to visualize the cerebral vessel presumed to be occluded on clinical grounds. Depending on the location of the clot in the cerebral vasculature, the attending neurologists and neuroradiologists can decide whether systemic thrombolysis with tissue thromboplastin activator (tPA) versus localized thrombolysis or clot removal by interventional therapy is the best treatment for a given patient.

Recent studies have shown that branch occlusions of the first branch of the middle cerebral artery or M1 branch are best treated by removal of the clot by a clot retrieval device. Depending on the clinical situation, initiation of systemic therapy by 4½ hours or 6 hours for interventional therapy is essential to maximize the chances of neurological recovery.

Other acute causes of ischemic injury include sustained system hypotension. Sustained systemic hypotension that is eventually reversed may cause differential ischemic damage to areas of the brain and spinal cord located between the territories of major supplying vessels (the watershed or arterial border-zone areas).

A complete discussion of cerebrovascular anatomy as well as ischemic and hemorrhagic stroke is detailed in Chapter 8.

The timeline for the case under discussion (with symptoms related to the left eye followed three weeks later by symptoms related to the left cerebral hemisphere) fits with a vascular etiology most likely due to a clot originating from an ulcerated plaque in the left internal carotid artery. The emboli were originally small (capable of occluding only the smaller diameter left ophthalmic artery) and then grew larger as the clot on the left internal carotid artery grew in size.

Venous occlusive disease such as sagittal or transverse sinus thrombosis should also be considered. The time course of these disorders is often over days, in contrast to arterial occlusions, and is often associated with non-focal headache. Often, venous occlusive disease presents with small to moderate-size hemorrhages adjacent to areas of venous occlusion.

3.8.D DRUGS AND TOXINS

In general, toxic and drug-induced neurological problems cause non-focal deficits, meaning that there is no difference in the degree of functional disturbance from one side of the body to the other. For instance, the pupillary reaction to light may be abnormal, but in a symmetrical fashion.

Another general rule specifically concerning 'recreational' drug abuse is that drugs that stimulate the nervous system (amphetamines, cocaine) tend to produce large but equal pupils during acute exposure, whereas drugs that depress the nervous systems (narcotics) tend to produce small but equal pupils.

A common cause of acute neurological symptoms is a change in medication or dosage. It is extremely important to review every medicine that the patient is taking by inspecting the actual bottles of each medication and reviewing doses, times and recent changes. In considering the possibility of drug exposure, it is important to have relatives of patients to check the medicine cabinet and the trash can, as well as to ask specific questions about hobbies, for example, gardening (cholinesterase inhibitors from insecticides) or soldering (lead solder).

There is a wide list of possibilities (see Table 3.13); thus, the medication and recreational drug histories are essential in honing the list down to a few agents that are most likely. For nonmedical or recreational drug exposure, it is often necessary to obtain collateral history, with the patient's consent if possible, from other family members or friends.

Chronic exposures to substances such as heavy metals can lead to slow deterioration of neurological function of both brain and spinal cord. When these are suspected, blood and urine assays are available to confirm the presence of heavy metal poisoning such as lead or mercury.

Screening tests of urine and blood for medications and recreational drugs involved in overdose situations are available in most emergency departments.

In terms of the matrix, drugs and medications can cause acute, subacute and chronic effects. In the sample case, the patient was not taking any medications and recreational drugs are an unlikely cause of the patient's focal deficits.

Delirium must be considered as an acute event, but often, it is a combination of an acute drug intoxication, metabolic insult or encephalopathy related to infection with a background of a neurodegenerative disorder such as Alzheimer's disease (AD). It is therefore important to identify and reverse the acute component but then follow-up to evaluate a possible more chronic neurodegenerative disorder.

TABLE 3.13: Common Toxic Exposures

	Acute Symptoms	Acute Signs	Withdrawal Features	Withdrawal Signs	Investigations	Treatment	Long Term
Ethanol	Loss of coordination and balance, slurred speech, cognitive impairment	Ataxia, cognitive impairment	Irritability, tremors, seizures	Seizures 18–24 hours, delirium tremens	Etoh level	Thiamine, folate, benzodiazepine, abstinence	Wernick–Korsakoff, cerebellar degeneration, dementia
Marijuana	Loss of coordination and balance, slurred speech, cognitive impairment	Ataxia, cognitive impairment	CNS irritability	CNS irritability	Blood THC levels	Thiamine, folate, benzodiazepine, abstinence	Cognitive impairment, potential for schizophrenia
Ecstasy (methamphetamine)	Hallucinations, euphoria	Combative behaviour, seizures	CNS depression	Seizures, CNS depression	Toxin screen for methamphetamine	Thiamine, folate, benzodiazepine, abstinence	Trauma, medical co-morbidities
Angle dust (phencyclidine)	Hallucinations, euphoria	Combative behaviour, seizures	CNS depression	CNS depression	Toxin screen for PCP	Thiamine, folate, benzodiazepine, abstinence	Trauma, medical co-morbidities
Heroin	Euphoria, loss of balance and coordination	Miosis, decreased level of consciousness	CNS irritability	CNS irritability, seizures	Toxin screen for opioids	Naloxone for coma, long-term rehab, methadone	Trauma, medical co-morbidities
Cocaine	Euphoria	Mydriasis, hypertension, stroke	CNS depression	CNS depression	Toxin screen for cocaine	Thiamine, folate, benzodiazepine, abstinence	Stroke, trauma, medical co-morbidities
Methanol	Loss of coordination and balance, visual disturbance, slurred speech, cognitive impairment	Ataxia, cognitive impairment	CNS irritability	CNS irritability	Acidosis, increased osmolar gap	Ethanol, dialysis	Depends on exposure, blindness, pancreatitis
Ethylene glycol	Loss of coordination and balance, visual disturbance, slurred speech, cognitive impairment	Ataxia, cognitive impairment	CNS irritability	CNS irritability	Acidosis, increased osmolar gap	Ethanol, dialysis	Depends on exposure, blindness, renal failure
Inhaled solvents	Euphoria, loss of balance and coordination	Smell of solvent on breath	CNS depression	CNS depression	None readily available	Time	Trauma, medical co-morbidities
Heavy metal (Hg, arsenic, lead)	Loss of sensation, cognitive decline	Neuropathy, decreased mental status	None	None	Blood, urine for toxic level, hair sample	Chelation	Neuropathy, dementia, death

Abbreviations: PCP, phencyclidine; THC, tetrahydrocannabinol.

3.8.E INFECTIONS

The most common bacterial infections of the nervous system include meningitis and cerebral abscess; the latter is typically a metastatic infection from the heart, from the paranasal sinuses or from penetrating injuries. Encephalitis, which is usually viral in origin, refers to infection of the brain parenchyma itself. Meningitis is infection of the arachnoid layer of the meninges. Infected material can also collect in the epidural and subdural spaces; these infections are referred to as empyemas. The history for most of these disorders usually shows an onset with systemic features of fever, lethargy and signs of meningeal irritation over several days.

Meningeal irritation refers to inflammatory changes to the meninges due to a noxious agent such as blood, infection or sometimes a chemical such as contrast material that has been introduced into the subarachnoid space. A clinical sign of meningsmus is extreme neck stiffness (due to irritation of the meninges) when flexing either the head or the legs. Brudzinski's sign occurs with meningeal irritation, causing reflex flexion of the legs when the head is flexed on the neck by the examiner. Similarly, Kernig's sign is an inability for the examiner to fully extend the flexed knee because of reflex resistance from the hamstring muscles. Neither of these signs is specific but if present should guide the clinician to look for causes of meningeal irritation by performing a lumbar puncture if there are no contraindications.

If the brain parenchyma is involved, then focal signs and symptoms evolve. The commonest bacterial organisms are staphylococcus, streptococcus, H influenza and Neisseria meningitis. Listeria and tuberculosis can affect the brainstem. Immunocompromised individuals can harbour unusual bacterial organisms.

Syphilis and Lyme disease are bacterial infections of the nervous system caused by spirochetes that usually have three phases. There usually is a rash or lesion at the point of entry; days to weeks later, there is involvement of joints or skin, followed by the phase of involvement of the nervous system. Not all patients follow this sequence.

Brain abscesses usually evolve over a period of several weeks, typically with fever, mental status change, unilateral signs and epileptic seizures.

Acute viral infections of the nervous system include Japanese Encephalitis, West Nile, Ebola, Zika, coxsackie, echovirus and herpes viruses. The most important for rapid diagnosis is Herpes Simplex type I, which can cause a devastating hemorrhagic necrosis of the temporal lobes and has specific effective treatment with anti-viral agents. Herpes zoster usually manifests as painful eruptions of vesicles, similar to chickenpox but within the confines of one or two dermatomes; this represents a reactivation of latent virus in sensory ganglia from a primary chickenpox infection earlier in life. Herpes zoster can also cause diffuse encephalitis and myelitis, usually in immunocompromised individuals.

Encephalitis caused by mosquito-born viruses such as Japanese West Nile, Ebola, Chikungunya, Zika and Dengue viruses are seasonal and depend on the current epidemiology in the affected regions.

Some encephalitogenic viruses such as Zika have been associated with the development of severe microencephaly in the offspring of infected women. The mechanism by which Zika virus causes microencephaly *in utero* is not known. The virus is known to infect neural progenitor cells and produces cell death and abnormal growth and therefore is probably a form of pre-natal encephalitis occurring during critical phases of embryogenesis affecting the forebrain. Pathological studies have revealed significant histopathologic changes that were limited to the brain and included parenchymal calcification, microglial nodules, gliosis and cell degeneration and necrosis.

Protozoan infections of the nervous system such as malaria or trypanosomiasis have a time course of onset of days to weeks. There is often a history of overseas travel or previous exposure in immigrants from countries where these diseases are endemic. Naegleria, a waterborne protozoan organism often found in stagnant swimming holes, can cause fulminant encephalitis and death.

Prions, a form of chronic infective agent, can cause destruction of brain by the accumulation of insoluble prion protein in the extracellular spaces. Prionic infections include Creutzfeldt–Jakob disease (wild type or variant), kuru, fatal familial insomnia and Gerstmann–Straussler–Scheinker disease; all of these disorders have timelines of years, but the phase of decompensation may be in the order of days to weeks. These disorders are uniformly fatal but have significantly different time courses.

Neuroimaging of patients with suspected infections of the nervous system should be performed before lumbar puncture if there are any focal signs or decreased level of consciousness, or if the infection has a duration of 48 hours or more. This is to prevent the possibility of a mass effect caused by an infection such as a brain abscess to result in downward herniation of the brain onto the brainstem.

Investigations for these disorders include appropriate cultures of any purulent discharge from ear, nose or throat, blood cultures, as well as gram stain and cultures of cerebrospinal fluid (CSF). Multiple blood cultures can be very useful if CSF is not available for culture. Specialized rapid assays for bacterial infection using immunofixation and enzyme-linked immuno assay tests are available to guide clinicians with respect to the offending organism.

The timelines for infections of the nervous system do not apply in the sample case under discussion.

3.8.F METABOLIC

This category of disorders includes disturbances of all of the basic metabolic substrates, electrolytes, hormones and vitamins necessary for life.

Acute systemic hypotension and hypoxia cause cellular metabolic failure and the start of irreversible neuronal cell damage within four minutes. These disorders cause acute neurological global dysfunction with loss of consciousness. Acute hypercarbia causes narcosis and coma and is often associated with hypoxia.

Acute hypoglycemia also causes CNS dysfunction in a global fashion, but sometimes can present with focal neurological features that are indistinguishable from those of ischemia (e.g. unilateral focal weakness).

The timeline for these events is a matter of minutes; they are often reversible in the same time frame.

Other metabolic disturbances such as electrolyte abnormalities typically evolve over several days; hyponatremia is a classic example and can lead to seizures and coma. Hypokalemia usually causes muscle weakness and is relatively slow in onset unless caused by a potassium wasting medication. Hyperkalemia affects the heart long before it affects the nervous system. Hypocalcemia produces muscle irritability and tetany, whereas hypercalcemia causes an encephalopathy; both usually have time courses of days to weeks but can present acutely.

Hormonal disturbances affecting the nervous system include hyper- and hypo-thyroidism, Addison's disease and Cushing's syndrome (typically from a functioning pituitary adenoma); all of these take weeks to months to produce dysfunction of the nervous system and are usually diagnosed by primary care physicians, internists or endocrinologists. Vitamin deficiencies may also cause neurological dysfunction, often with widely varying time courses. Vitamin B12 deficiency can cause peripheral nerve and spinal cord dysfunction. Vitamin B12 stores in the liver usually require three years to deplete to the point that the individual becomes symptomatic, after which the onset of symptoms may be very subtle and gradual. Strict vegans with marginal stores or intake of vitamin B12 are at particular risk. Patients taking medications to suppress stomach acid can become vitamin B1 or B12 deficient. Folate stores are much more volatile, becoming depleted in weeks to months if dietary deficiency or malabsorption develops.

Thiamine deficiency has deleterious effects on the mamillothalamic tract and several brainstem structures as well as on peripheral nerves. Thiamine deficiency in alcoholics (and in others who become malnourished for other reasons) is still often overlooked in acute settings. Previous gastric surgery can interfere with thiamine absorption and lead to chronic CNS and peripheral nervous system dysfunction.

Investigations of these disorders include blood tests for glucose, electrolytes, calcium, phosphate, vitamin B12, folate; iron studies; liver, thyroid and kidney function; and blood gases if clinically indicated.

Metabolic encephalopathies can also occur from organ failure elsewhere in the body, such as kidney, liver and lung.

With kidney failure, when creatinine levels get above 500 to 700 mmole/L, decreased levels of consciousness are seen, with tremor. The serum levels of sodium, calcium or phosphate will often determine the degree of impairment.

In liver failure, depending on the cause and rapidity, encephalopathy can manifest as progressively altered mental status, tremor, asterixis and hyperventilation, followed by, if untreated, coma and death.

Respiratory failure leads to altered mental status due to hypercarbia and hypoxia. Anxiety and agitation due to hypoxia are often misinterpreted as a psychiatric disturbance and treated with medications such as benzodiazepines, which then make the hypoxia worse to the point of respiratory arrest.

Treatment of these entities obviously targets the underlying cause of the primary organ failure.

In summary, the time course of metabolic disorders is typically days to years and, with the possible exception of hypoglycemia, is too long for consideration in the sample case being discussed.

3.8.G INFLAMMATORY/AUTOIMMUNE

The non-infectious inflammatory disorders of the nervous system include such entities as Guillain–Barré syndrome, MS, systemic lupus erythematosus, sarcoidosis, Sjögren's syndrome, rheumatoid arthritis, post infectious vasculitis and isolated CNS vasculitis. These disorders generally have timelines of weeks to months and often have concomitant systemic symptoms of vasculitis. Acute events such as seizures, stroke and encephalopathy can occur as complications to these disorders. The timeline in the history would then show a subacute-on-chronic picture. In the case of relapsing remitting MS, there will be attacks over many years.

There has been recent identification of several types of non-infectious autoimmune encephalitis such as anti-N-methyl-D-aspartate receptor encephalitis (the most common) and voltage-gated potassium channel-complex antibody encephalitis, the second most common. These disorders can present with psychiatric presentations such as psychosis or depression and, in adults but not in children, are often paraneoplastic in origin.

These disorders do not fit with the timeline of the case under discussion.

Most of these disorders are investigated with appropriate serological tests for autoimmune diseases. MRI scanning is also important for MS and gadolinium enhanced magnetic resonance arteriogram (MRA) for the various forms of vasculitis.

3.8.H NEOPLASTIC

Neoplasms cause neurological dysfunction by destruction of neural tissue secondary to direct expansion of the tumor mass, sometimes with concomitant necrosis and hemorrhage. Mass effect refers to the displacement of normal brain tissue by a space-occupying lesion such as a tumor.

Mass effect can also be caused by edema surrounding the tumor. The compression of normal structures can lead to altered function and ultimate destruction of these structures. The location of the tumor and its mass effect will determine the nature of the neurological deficits. Many tumors are initially slow growing and are not discovered until there is a defining event such as a seizure or hemorrhage into the tumor.

Primary brain tumors originate from basic neural tissue elements: astrocytomas and gliomas from astroglia, oligodendrogliomas from oligodendrocytes, meningiomas from meninges and ependymomas from the ependyma. Medulloblastomas of the cerebellum are common tumors in children. These tumors arise from cell lines, which occur in embryonic development; they often originate in the midline cerebellum and have the propensity to metastasize within the craniospinal space, often to the lumbar sac resulting in some cases in a cauda equina-like syndrome.

Metastatic tumors usually arise from hematogenous spread of primary tumors elsewhere in the body. The common malignancies that metastasize to the brain originate in the breast, lung and bowel. Prostate cancer tends to metastasize to the spinal vertebral bodies, occasionally causing spinal cord or compression of the cauda equina.

In terms of time frames, there may be a subacute to chronic onset of a focal neurological deficit reflecting the slow growing mass effect of the tumor, with an acute event such as a seizure or stroke-like event; the latter represents a sudden expansion of the tumor mass due to hemorrhage or necrosis. While hemorrhage into a previously silent tumor mass might explain the sudden right hemiparesis in the chapter case, this mechanism would not account for the previous, short-duration episodes of loss of vision in the left eye.

3.8.H.1 PARANEOPLASTIC

Paraneoplastic disorders of the nervous system are defined as pathological processes that affect the nervous system due to the remote effects of a neoplasm elsewhere in the body.

A common scenario is that of the formation and growth of a solid tumor (such as small cell lung carcinoma), which stimulates an immune antibody response that cross-reacts to tissue elements in the nervous system, thereby causing dysfunction and destruction in those elements. This is the case in Lambert–Eaton syndrome, in which the small cell lung tumor stimulates antibodies that are directed against presynaptic calcium channels in terminal boutons of peripheral nerves, resulting in proximal muscle weakness and autonomic dysfunction. This is an example of a presynaptic disorder of the NMJ.

Paraneoplastic autoimmune syndromes may also produce a polyneuropathy, other neuromuscular transmission disorders, cerebellar degeneration and occasionally cognitive decline.

There are now panels of antibody testing that are available to assist the clinician in finding the primary tumors responsible for the neurological symptoms.

Other paraneoplastic syndromes affect the nervous system by producing hormone-like compounds. A typical example is a type of lung cancer that produces an anti-diuretic hormone-like hormone that induces hyponatremia with secondary effects on CNS function (see the earlier section on metabolic disorders).

Like brain tumors, paraneoplastic disorders evolve over weeks to months. In our sample case, therefore, the history and findings do not fit with a paraneoplastic disorder.

3.8.I DEGENERATIVE DISEASES

Neurodegenerative diseases take many forms and are often associated with other comorbid conditions that amplify their effect. Some tissues within the nervous system are relatively immune to neurodegenerative processes, for example, muscle, the NMJ and peripheral nerves. Most neurodegenerative processes affect neurons in the cerebral cortex, basal ganglia, cerebellum and spinal cord.

Many of the degenerative disorders are of genetic origin and get worse over time. In other instances, the primary cause remains unknown, but the course of the disease may be modified by genetic factors.

Most of the cortical neurodegenerative diseases represent a failure of CNS neurons or glia to recycle breakdown protein products into soluble forms that can be either reutilized in synthetic processes or removed from the cell. The presence of excessive insoluble proteins leads to the accumulation of those proteins and eventual dysfunction and death of the cell.

In the case of AD, there is a failure of metabolism of amyloid and tau protein; in frontotemporal dementia (FTD), a failure of metabolism of tau protein; in Parkinson's disease, synuclein protein. The mechanisms may involve disorders of metabolism within the neurons at several levels related to the control and metabolism of RNA products involved in protein synthesis.

Research has revealed that many of what were thought to be sporadic degenerative disorders of the nervous system have contributory genetic cofactors.

For instance, ALS was thought to be a purely sporadic degenerative disorder until enzyme studies uncovered kindreds with superoxide dismutase deficiency (SOD1). A further subgroup of ALS is SOD negative but is related to recycling disorders of cellular proteins such as ubiquitin.

Examples of cerebrocortical degenerative disorders include AD, FTD and dementia with Lewy bodies. Degenerative diseases of white matter include microvascular leukoencephalopathy; those of the dopaminergic systems in the basal ganglia include Parkinson's disease and corticobasal degeneration. Multi-system atrophy is a neurodegenerative disorder associated with the accumulation of synuclein protein, which can affect cortex, white matter, basal ganglia, cerebellum, brainstem and peripheral nerves in various degrees and at various rates of progression.

Investigations for these disorders include neuroimaging, mental status and neuropsychological testing. In some cases, such as patients with suspected prion disorders, brain biopsy can be considered in order to establish a definitive diagnosis.

In terms of timeline, all of these disorders are chronic, each with a characteristic collection of symptoms and signs that suggest the ultimate diagnosis. These are not applicable in the case under discussion.

3.8.J GENETIC DISEASES

Genetic diseases affecting the nervous system often overlap with most other disease types in terms of timeline. With the increasing knowledge derived from genetic studies, the distinctions between the categories of degenerative and genetic disorders are becoming increasingly blurred. One can consider many degenerative disorders as late-onset genetic disorders with possible environmental triggers. Alternatively, genetic disorders could be considered as the cause of most degenerative disorders but with varying clinical expressions and longer time courses. For instance, Tay–Sachs disease has always been considered a genetic

disorder of sphingolipid metabolism, but it could be considered as a genetic neurogenerative disorder of early onset.

The genetic diseases may express themselves in terms of timeline from as early as *in utero* (diagnosed on ultrasound) to late adult life. They may present in an abrupt fashion (rapid onset coma in neonates due to urea cycle defects), over weeks to months (6 to 12 months for infants with Tay–Sachs disease), or over many years (Huntington's disease).

The defining event for the severe forms of genetic disorders is often birth and the inability of the neonate to support itself metabolically in the absence of the placental circulation and the mother's normal metabolic support. Other genetic degenerative disorders declare themselves in infancy or early childhood with parental concerns about slowness of development such as a missed milestone which then prompts an investigation for the cause.

Using the localization system to classify the genetic disorders is helpful. Table 3.14 gives some examples of how some of the common genetic disorders show different timelines throughout life.

In terms of timing of onset, all of these disorders have their own characteristics. However, the basic principle is that the more severe and life-threatening disorders will present earlier in life.

Blood tests exist for the various genetic markers for many of these disorders, which allows families to make

TABLE 3.14: Common Genetic Disorders by Localization and Time Course

Localization	Neonatal	Infancy	Childhood	Adolescent	Early Adult	Mid Adult	Late Adult
Muscle	CMTD	Metabolic myopathies	DMD, MELAS	BMD	MERRF, MELAS		
Neuromuscular junction	Congenital myasthenia					Myasthenia gravis	
Peripheral nerve			CMT	CMT	HNPP		
Spinal cord		SMA	SMA	SCA			ALS
Brainstem		LD					
White matter	Neonatal leukodystrophies, CHS	Leukodystr	Leukodystr	Leukodystr	LD Leukodystr	MVL	
Cortex	Lissencephaly and other cortical dysplasias	Tay–Sachs, Gaucher's CLF	CLF	CLF	HC, CLF	FTD, MSA	AD, DLB
Basal ganglia		LD	Generalized dystonias	Generalized dystonias	HC	PD	
Cerebellum	NMD	NMD	Dandy Walker		Chiari Malformation		

Abbreviations: AD, Alzheimer's disease; ALS, amyotrophic lateral sclerosis; BMD, Becker's muscular dystrophy; CHS, congenital hypomyelination syndromes; CLF, ceroid lipofuscinosis; CMT, Charcot–Marie–Tooth disease; CMTD, congenital myotonic dystrophy; DLB, dementia with Lewy bodies; DMD, Duchenne's muscular dystrophy; FTD, frontotemporal dementia; HC, Huntington's chorea; HNPP, hereditary neuropathy with liability for pressure palsies; LD, Leigh's disease (this is mostly a gray matter disease); MELAS, mitochondrial myopathy with lactic acid and stroke-like symptoms; MERRF, mitochondrial myopathy with ragged red fibers; MSA, multisystem atrophy; MVL, microvascular leukoencephalopathy; NMD, neuronal migration disorders; PD, Parkinson's disease; SCA, spinocerebellar atrophy; SMA, spinomuscular atrophy.

informed reproductive decisions, especially for those disorders with long latencies such as Huntington's disease, which does not manifest until the usual reproductive period of a couple's life has passed.

3.9 CASE SUMMARY

To summarize the process and conclusions for the case under discussion, let us refer back to the Expanded Localization Matrix to produce a concise summary of the case under discussion.

The completed Expanded Localization Matrix (Table 3.12) shows that the elements of the neurological exam (including conjugate gaze to left, right-sided weakness of the face and upper limb, hyperreflexia on the right side and difficulty with expression of speech) all point to localization in the left frontal area. Combining these findings with the sequence of events in the history and the correlative factors in the etiology checklist points to an acute vascular event involving the left internal carotid artery as the common element needed to explain all of the findings.

The history indicates recurrent episodes of visual disturbance in the left eye followed by a major disturbance to the area of the nervous system that controls speech as well as strength and coordination of the face and right upper limb. The History Worksheet (Figure 3.3) and the completed Etiology Matrix (Figure 3.5) suggest an acute etiology, such as a paroxysmal, traumatic or vascular disease. The unilateral nature of the symptoms and lack of history of seizures or trauma favour a vascular etiology.

The most likely explanation for the patient's story is repeated vascular ischemic events caused by emboli emanating from the left carotid artery. These initially caused disturbances in vision in the left eye through embolization of the ophthalmic artery, a branch of the left internal carotid. At the time of the major symptomatic event, the emboli caused disruption of flow to the left frontal cortex and white matter, resulting in expressive aphasia and right-sided weakness. The conjugate gaze to the left is explained by the unopposed conjugate movement of the eyes to the left by the unaffected right frontal eye field due to damage to the left frontal eye field.

The neuroradiological investigations can then be targeted to these structures. This would include an urgent CT scan with CT angiogram of the carotid arteries and intracranial vessels. An MRI and MRA of the brain would provide similar information, but this modality is usually not accessible on a 24-hour, 7-days-per-week basis.

Figures 3.6 and 3.7 show the CT scan and CT angiogram images that show the findings for this case. They confirm the clinical analysis of a left frontal infarction, high-grade stenosis of the left internal carotid artery and occlusion of an M2 branch of the left middle cerebral artery.

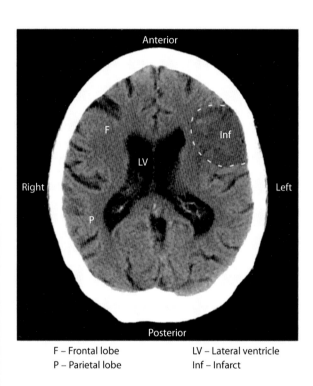

F – Frontal lobe   LV – Lateral ventricle
P – Parietal lobe   Inf – Infarct

FIGURE 3.6: CT scan brain for the sample case.

Disease type	Secs	Mins	Hrs	Days	Weeks	Months	Years
Paroxysmal (seizures, faints, migraine)	Ø	Ø					
Traumatic	Ø	Ø					
Vascular (ischaemic stroke, bleed)	O	O					
Toxic	Ø	Ø					
Infectious	Ø	Ø					
Metabolic	Ø	Ø					
Inflammatory/autoimmune	Ø	Ø					
Neoplastic	Ø	Ø					
Degenerative	Ø	Ø					
Genetic	Ø	Ø					

← Acute →
←Subacute→
←Chronic→

FIGURE 3.5: Completed etiology matrix (sample case).

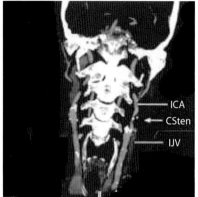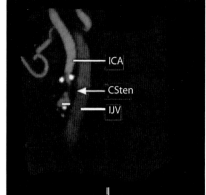

CSten – Stenosis
ICA – Internal carotid artery
IJV – Internal jugular vein

FIGURE 3.7: CT angiogram for the sample case.

3.9.1 CASE RESOLUTION

The patient arrived in the emergency room three hours after the onset of the new deficits and therefore was considered a candidate for intra-arterial interventional therapy. He was started on tPA and then taken to the angiography suite and the left M1 clot was removed without complication. The eye deviation and speech difficulty immediately improved.

The patient was admitted to the hospital and treated with intravenous heparin for five days; he then underwent a left carotid endarterectomy without complications seven days after the acute event. At six weeks after the surgery and neuro-interventional procedure, he showed only minor word-finding difficulties and slowness of rapid alternating movements on his right side.

3.10 FAIL SAFE: LOCALIZATION/ ETIOLOGY CHECKLIST

It is important to review the Fail Safe Localization/ Etiology Checklist. At each level of localization, there are several etiological diagnoses, which represent serious and often life-threatening neurological disorders that must always be considered so as not to be missed. Table 3.15 lists these disorders according to their level of localization.

3.11 IMPLEMENTATION

Now that a completely worked example of how the system of localization and etiological diagnosis works has been completed, readers, especially medical students and

TABLE 3.15: Fail-Safe Localization/Etiology Checklist

Localization	Diagnosis
Muscle	Congenital muscular dystrophy Polymyositis Polymyalgia rheumatica
Neuromuscular junction	Medications Myasthenia gravis Lambert Eaton myasthenic syndrome
Peripheral nerve	Guillain–Barré syndrome Diabetic neuropathy Entrapment/traumatic
Spinal cord	Acute spinal cord compression Transverse myelitis Infarction
Brainstem	Infarction/hemorrhage Vertebral dissection Meningitis (tuberculosis, listeria) Tumor Fisher variant Guillain–Barré
Deep white matter	Infarction/hemorrhage Multiple sclerosis Progressive multifocal leukoencephalopathy
Cortex	Infarction/hemorrhage Herpes simplex encephalitis Subarachnoid hemorrhage Carotid dissection
Cerebellum	Expanding hematoma Alcoholic- or drug-induced degeneration
Basal ganglia	Drug-induced movement disorder Parkinson's disease Huntington's disease Wilson's disease

residents, are encouraged to use the system by practicing it for each of the cases in the rest of the text and for the e-cases, on the text Web site.

It is recommended that prior to reading each chapter and e-case, the reader print out a blank copy of the Expanded Localization Matrix, Etiology Matrix and History Worksheet, which are available on the Web site. As the reader works through the case in each of the case-based chapters and online cases, he or she fills in the blanks for the localization and etiology using the method and criteria detailed in this chapter and also fills in the timelines of the symptoms on the History Worksheet. The conclusions reached in each case can then be compared against the discussion in the text and the completed sheets that appear on the Web site. The Web site has been constructed with workflow logic that does not allow the user to open the completed sheets and case discussion before the history, physical exam and investigations have been reviewed.

This will assist the reader in developing the skills to use the system on patients on the ward and in the clinic.

SUGGESTED READING

Paraneoplastic syndromes
> http://www.ncbi.nlm.nih.gov/pmc/articles/PMC 3299076/

Chronic traumatic encephalopathy (CTE)
> McKee, A.C., Cantu, R.C., Nowinski, C.J., Hedley-Whyte, E.T., Gavett, B.E., Budson, A.E., Santini, V.E., Lee, H.S., Kubilus, C.A., Stern, R.A. 2009. Chronic traumatic encephalopathy in athletes: Progressive tauopathy after repetitive head injury. *J Neuropathol Exp Neurol* 68(7): 709–735. doi:10.1097/NEN.0b013 e3181a9d503. PMC 2945234. PMID 19535999.

Huntington's disease
> http://www.omim.org/entry/143100?search=hunting ton&highlight=huntington

Section 2

APPLYING THE BASICS TO CLINICAL CASES

Chapter 4

Fifi

Objectives

- Learn the main anatomical components of the peripheral nervous system and their roles in the production and control of movement
- Review the microscopic anatomy and physiology of peripheral nerves
- Understand the symptoms and signs resulting from a disease process affecting the peripheral nervous system
- Understand the localization characteristics, as revealed by history and physical examination, of lesions of the peripheral nervous system affecting movement control, as distinct from lesions of the central nervous system (CNS)

4.1 FIFI

Currently 30 years old, Françoise (or Fifi to her friends and relations) works as a registered nurse in the emergency department at the University Hospital.

As often happens to medical personnel working on the 'front lines', notwithstanding constant hand-washing technique between patients, Fifi fell briefly ill with cough, sore throat, muscle aches and low-grade fever about two weeks ago. After three days at home, she recovered and went back to her regular shift in the emergency room (ER). Several other staff members had recently fallen ill with similar symptoms.

Yesterday morning, Fifi awoke with peculiar sensations in her legs: her feet felt as if they had 'gone to sleep' and her calves as if there were bugs crawling under the skin. Although she felt distinctly strange, Fifi (possessing the typical type 'A' personality so characteristic of health professionals) got ready to go to work. On the way downstairs to the kitchen, however, she noted that her legs seemed unusually heavy, and she tripped twice on the hall carpet. While not exactly sure what was going on, Fifi thought it might be better if she called in sick.

As the day progressed, things went from bad to worse. Fifi began to have trouble walking upstairs, her legs feeling like lead, and she had to proceed one step at a time, using one hand on the railing to help propel herself upward. She thought about going to a local walk-in clinic – of course it was Saturday and her family practitioner's office was closed for the weekend – but decided to stick it out for the day in the hope that tomorrow would be better. Nevertheless, Fifi had the foresight to ask her children's babysitter to stay overnight.

Today brought a marked deterioration in Fifi's condition. Her legs had become so weak that she could scarcely roll out of bed. She found that her fingers had begun to tingle and that she had difficulty lifting her arms to put them in the sleeves of her dressing gown. When she tried to head for the bathroom for a glass of water to take her birth control pill, she discovered that her legs would not support her properly; she had to shuffle slowly, holding on to the furniture. The physical effort involved made her extremely short of breath. Worse, when she tried to swallow water, she choked badly, some water exiting through her nose.

Now thoroughly alarmed, Fifi had the babysitter call an ambulance and was brought to the very emergency department where she normally works.

Concerned with Fifi's appearance, the triage nurse immediately transported her to the resuscitation area and called for one of the ER physicians – this means *you*.

On your examination, you find Fifi to be anxious and short of breath. Her blood pressure is 130/80 and her heart rate is 110/minute. She appears mentally intact and answers all questions appropriately, but her voice sounds nasal, almost a whisper. Her visual acuity is grossly normal; her fundi are also normal. Pupillary size and reaction to light are equal and symmetric. Extra-ocular movements are full, with no nystagmus; she has no ptosis (droopy eyelids). She is unable to keep her eyes shut against resistance; her smile is but a weak grimace. She can protrude her tongue on request but has a poor gag reflex.

The motor examination reveals diffuse extremity weakness, more severe proximally than distally, and more severe in the legs than in the arms (see Figure 4.1a) She has no difficulty initiating or stopping movement on request but cannot sustain a strong contraction of the involved muscles. Even though very weak, Fifi is still able to move her fingers and toes quickly and to perform rapid alternating hand and foot movements. Passive stretch of her limbs reveals a reduction in muscle tone. Her tendon reflexes are

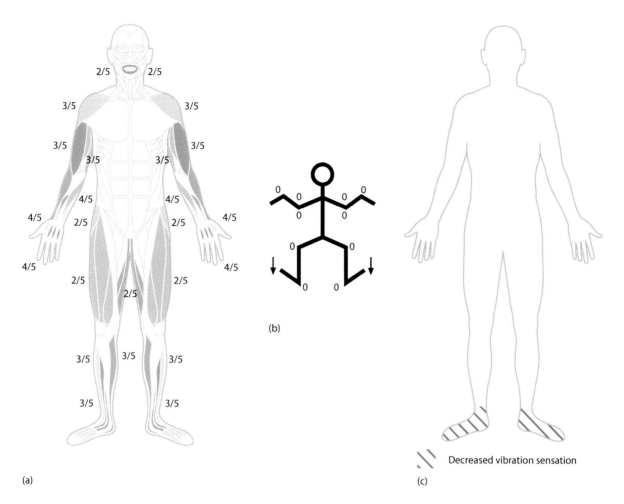

(b)

(a)

(c)

NNN Decreased vibration sensation

FIGURE 4.1: Motor examination (a), reflex (b) and sensory examination maps (c) for Fifi, the chapter case (see Figure 2.6 for explanation).

completely absent in all four limbs (Figure 4.1b). Brisk stroking of the soles of her feet with the handle of your reflex hammer produces a normal plantar flexion response.

Despite the history of sensory disturbances in hands and feet, the only abnormal sensory finding is a decrease in vibration sense at the ankles and toes (Figure 4.1c). Other sensory modalities (touch, position, pain, temperature) are normal.

Finally, Fifi's abdominal examination shows no bladder enlargement; she is able to provide a urine specimen on request without difficulty.

4.2 CLINICAL DATA EXTRACTION

Before proceeding with this chapter, you should first carefully review Fifi's story in order to extract key information concerning the history of the illness and the physical examination. On the Web site that accompanies this text, you will find a sequence of tools to assist you in this process.

Note: it is essential that you complete the Web site worksheets at this stage in the chapter and that you NOT skip through to the next section. The purpose of this textbook is to help you learn to navigate your way around the nervous system in clinical encounters. The only way to learn this properly is by trying to solve clinical problems and, in the process, by making mistakes. It is obviously preferable to make *virtual* mistakes now rather than *real* mistakes later on.

4.3 MAIN CLINICAL POINTS

Fifi's chief complaints are that she *cannot walk, is unable to swallow liquids and is having trouble breathing.*

The essential elements of Fifi's medical history and neurological examination are as follows:

- A respiratory illness two weeks prior to the onset of symptoms
- The development of severe muscle weakness and mild sensory symptoms over a period of two days

- On examination, the presence of diffuse, symmetrical muscle weakness involving the face, swallowing musculature and all four limbs (proximally and distally), the legs more than the arms
- Difficulty breathing and a faint voice
- Complete absence of muscle tendon reflexes
- Relatively minor abnormalities on sensory examination despite the history of sensory disturbances.

Fifi's key examination findings are illustrated in Figure 4.1a, b, and c.

With respect to localizing the disease process, it is also important to note several findings that Fifi does *not* have:

- Despite her respiratory distress, she is alert, gives a coherent history and shows no sign of confusion or memory loss.
- She has no visual difficulties or gaze problems.
- She does not have difficulty in initiating movements, but rather difficulty in carrying out the movements with sufficient force.
- Her limbs are hypotonic, not stiff or rigid.
- Despite the presence of severe weakness, she has no difficulty in controlling her bladder and has no signs of urinary retention and inability to void.
- Her plantar responses are flexor, not extensor, hence normal.

In brief, therefore, Fifi has a neurological disorder primarily affecting the maintenance of muscle strength, developing over a period of just two days. In consequence, we must redirect our attention to the parts of the nervous system that are involved in carrying out voluntary movements.

4.4 RELEVANT NEUROANATOMY

Before proceeding to attempt to localize Fifi's disease process, we will briefly review and, where necessary, expand upon the anatomy of the motor system outlined in Chapter 2.

Based on the results of functional magnetic resonance imaging studies, it appears that the process of planning a movement prior to its performance takes place in the supplementary motor area of the cerebral cortex, anterior to the cortical areas that actually send the signals downwards to eventually result in the desired movement. The actual decision to move appears to reside within the complex circuitry linking the motor cortex, the basal ganglia and the thalamus (see Chapter 7). Following this preprocessing phase, the message to commence a specific movement is transmitted to the motor cortices, located bilaterally in the precentral gyrus, anterior to the central fissure.

The cortical area directly initiating movement of the right arm and leg is located in the left cerebral hemisphere, and vice versa for the left side of the body. The area controlling movement of the opposite leg is located near the midline, at the vertex of the brain, while arm movement originates more laterally (over the convexity of the hemisphere), and face/tongue movements originate where the inferior motor cortex dives into the lateral fissure (see Figure 4.2). This anatomical distribution of movement control constitutes the so-called motor homunculus originally derived by Penfield and colleagues in the context of direct motor cortex stimulation during cortical excision for the treatment of medically intractable epilepsy.

Once initiated, a specific movement, in order to be carried out in a precisely controlled fashion, requires moment-by-moment sensory feedback from movement detectors in the muscles and joints of the part of the body being moved. This information, as we saw in the introductory exposition of Crash McCool's rocket-ship flying activities, is transmitted unconsciously to the cerebellum, where it is constantly compared and integrated with parallel information originating in the motor cortex. The ongoing integrated information is then fed back to the motor cortex to permit performance of the next epoque of the limb movement required.

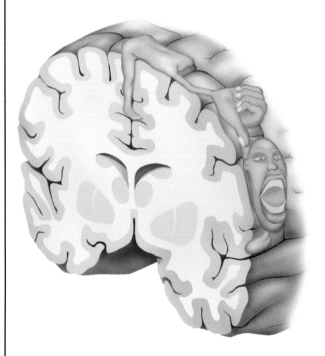

FIGURE 4.2: Three-quarter view of the cerebral hemispheres sectioned coronally just anterior to the precentral gyrus and illustrating the relative areal distribution of motor cortex devoted to movement of different parts of the body (the motor 'homunculus'). Note the relatively large areas devoted to movement of the hand, fingers, face and tongue.

As we saw in Chapter 2, the actual message, elaborated with the assistance of the basal ganglia and cerebellar motor circuitry, is transmitted to the spinal cord, thence to the muscles, by two parallel descending routes. One route is direct to the cord by single neurons without intermediate relays, the other an indirect route involving relay centres in the brainstem. These pathways, along with the contributions of the basal ganglia and cerebellum, will be considered in more detail in Chapter 7. Cortical motor neurons in both pathways are commonly referred to as upper motor neurons.

The final step in the descending pathways involved in carrying out a movement command is the recruitment of motor neurons in the area of the spinal cord responsible for a specific limb – the anterior horn cells (AHCs). These cells (or lower motor neurons), once excited by descending motor pathways, direct action potentials down their axons, which leave the spinal cord via the anterior (ventral) roots (Figure 1.5a) and travel down peripheral nerves to specific muscles participating in the desired movement. A given motor neuron, its axonal branches and terminal synaptic boutons and its associated cluster of muscle fibers constitute a motor unit (Figure 4.3). Any movement, whether of a finger, arm, tongue or eye, requires the graded and coordinated activity of a number of motor units.

Finally, no voluntary movement can be carried out in an organized fashion without constant feedback concerning muscle length, muscle tension and joint position via sensory neurons originating respectively in muscle spindles, tendon organs and joint capsules. In addition to sending sensory information to higher motor centres in the cerebellum (and elsewhere), as previously mentioned, the same sensory fibers send branches which synapse directly or indirectly with AHCs busy generating the movement (Figure 1.5b). These feedback loops form monosynaptic and disynaptic reflex arcs that are crucial in the peripheral circuitry involved with movement; they are also the anatomical basis of the reflexes tested in the basic neurological examination outlined in Chapter 2.

4.5 LOCALIZATION PROCESS

Having briefly surveyed the system presumably involved in the pathogenesis of Fifi's muscle weakness, we must now return to the clinical information provided and attempt to localize the lesion. Is the cause of her weakness located in the cerebral hemispheres, brainstem, spinal cord, peripheral nerves, neuromuscular junctions or muscles?

If we begin at 'the top', there are a number of reasons why the problem is unlikely to be located in the cerebral hemispheres or involve the command centre, as it were, of the motor system. First of all, even when profoundly weak and in respiratory distress, Fifi is alert and cooperative, with no evidence of a disturbance of consciousness or attention. Second, even if the disorder were confined to the cerebral motor system components, Fifi is not having any difficulty initiating or coordinating a movement: when she attempts to flex her elbow, for example, she does so quickly and in a smooth fashion. What she cannot do, however, is generate sufficient force to flex the elbow vigorously while you, as part of your examination, attempt to prevent her from so doing. She simply cannot generate sufficient force in the biceps brachii.

Since Fifi has weakness of facial movement and swallowing, it is conceivable that her disease could be located in the brainstem, the origin of nerves destined for facial and pharyngeal muscles. On the other hand, this possibility is made less likely by the fact that Fifi has no sign of a dysfunction clearly implicating the brainstem, e.g. nystagmus or a disturbance in consciousness. (The brainstem's contributions to consciousness will be reviewed in Chapter 10.)

In addition, as was mentioned in Chapter 3, clinical experience has taught us that lesions involving the descending motor control pathways above the spinal cord characteristically produce an *increase in muscle tone* below the level of the lesion. This tone increase is secondary to a disinhibition of segmental monosynaptic reflex arcs from muscle spindle to AHC back to muscle (Figure 1.5b). Without suppression from higher centres,

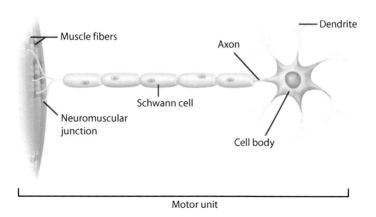

FIGURE 4.3: Drawing of the component parts of the motor unit.

these reflex pathways are normally set to produce considerable reflex contraction of a muscle when it is passively stretched: the phenomenon of *hypertonia*. The presence of hypertonia in a weak muscle is a cardinal sign, therefore, of a lesion in the descending motor pathways above the level of the AHC: the so-called upper motor neuron lesion.

A second characteristic of a motor system lesion at this level is the extensor plantar response (or Babinski sign) mentioned in Chapter 2. This response is a complex reflex involving plantar cutaneous sensory receptors and spinal cord circuitry that is normally present, at times, in newborn infants and is modified by the developing corticospinal tract into a plantar flexion response. A lesion anywhere in the descending motor pathways will result in the reappearance of the primitive toe extensor response (the Babinski sign).

Thus, in the absence of hypertonia in weakened muscles and of extensor plantar responses, it is very unlikely that Fifi has a lesion in the motor system above the level of the AHCs.

This conclusion applies just as well to the spinal cord as it does to the brainstem. There are a number of other reasons, however, why a high spinal cord lesion is unlikely. Most obviously, a spinal cord lesion could not explain the weakness of the muscles of the face and pharynx (the latter producing her swallowing problems). In addition, spinal cord lesions typically interrupt the descending control pathways for voluntary control of urination. If this pathway is suddenly disrupted for any reason, the bladder becomes incapable of releasing urine and becomes extremely large, a finding Fifi does not have at the time of your assessment.

Having travelled this far down the motor control pathways without encountering a mixture of clinical findings that accounts for Fifi's symptoms and signs, we are left with a problem involving the motor units: AHCs, their axons, neuromuscular junctions and the muscle fibers they control.

In general, a dysfunction of a proportion of motor units in a given muscle will produce difficulty in generating normal amounts of tension in the muscle when an attempt is made to contract it: exactly the problem Fifi has developed. Nevertheless, it is unlikely that Fifi has a specific problem of either AHC or muscle fiber function for the simple reason that she also has sensory symptoms. While it is possible that her disease process might be simultaneously affecting AHCs, muscles and sensory nerve fibers, as a general rule, such a scenario is unlikely. In large part, the most economical explanation for the location of a disease process is also the most likely.[a] This being the case, the most probable explanation for the muscle weakness is a process affecting motor axons, with the sensory axons also being involved to a lesser extent.

There is a key clinical observation that also supports the hypothesis of motor axon pathology: the complete *absence*

of muscle tendon reflexes. The anatomical basis of tendon reflexes was reviewed in Chapter 1. Any disease process that interrupts the function of a large percentage of large (muscle spindle afferent) sensory axons or motor axons participating in a given tendon reflex will abolish the reflex. If a disease process were to stop the function of the AHCs themselves, as distinct from their axons, the muscle tendon reflexes would also be abolished. Again, however, this explanation would not account for the sensory symptoms.

Now that we have examined the reasons why Fifi's muscle tendon reflexes are absent, we can go back and explore what would have happened to the reflexes had the disease process been located at the level of the upper motor neuron (cerebral hemispheres, brainstem or spinal cord). Dysfunction in the upper motor neurons results in a relative lack of input from the highest levels of the motor system to the AHCs. This lack of input, largely inhibitory in nature, results in the muscle tendon reflexes being exaggerated or abnormally brisk (hyperreflexia).

At this point, because of its crucial importance in neurological localization, we repeat the mantra already chanted in Chapter 3: *in a patient with muscle weakness, the three cardinal signs of an upper motor neuron lesion are hypertonia, hyper-reflexia and an extensor plantar response (Babinski sign). In contrast, lower motor neuron lesions are characterized by hypotonia, hypo- or areflexia and a flexor plantar response.* Clearly, Fifi's clinical picture corresponds to the latter description.

We have not devoted much attention to the possibility that Fifi might have an acute disorder of neuromuscular junction function, or of muscle itself, primarily because of her sensory symptoms. As we noted in Chapter 3, neuromuscular junction dysfunction results in progressively increased muscle weakness with sustained or repeated contractions over a short period of time, the phenomenon of muscle fatigability. Although Fifi is very weak, repeated contraction of given muscles do not make them any weaker than they were at the first contraction. In addition, unless they are very severe and protracted, neuromuscular junction problems do *not* typically decrease muscle tendon reflexes. Finally, neuromuscular junction disorders tend to specifically target the face and extra-ocular muscles, producing facial weakness (which Fifi has) and ptosis plus eye movement paresis (which she does not).

Finally, we again note that muscle diseases produce pure muscle weakness, often associated with a modest amount of muscle wasting if the disease process has been present long enough (months). Muscle diseases (myopathies) may decrease muscle tendon reflexes but rarely abolish them.

In summary, our descending voyage through the motor control system has indicated that the disease process acutely affecting Fifi probably resides in large fiber cranial and spinal motor nerve fibers and, to a lesser extent, sensory nerve fibers. Table 4.1 shows the completed Localization Matrix for Fifi's disease, indicating all

---

[a]　This line of reasoning dates back to a dictum of the medieval scholastic William of Ockham; it is known as 'Ockham's razor' and, in most circumstances, applies very well in clinical medicine.

TABLE 4.1: Completed Localization Matrix for Fifi

	Right	Left	MUS	NMJ	PN	SPC	BRST	WM/TH	OCC	PAR	TEMP	FR	BG	CBL
Attention	N													
Memory	N													
Executive	N													
Language	N													
Visuospatial	N													
CRN I														
CRN II	N	N												
CRN III, IV, VI	N													
CN V	↓	↓	⊘	⊘	O		⊘	⊘				⊘		
CRN VII	↓	↓	⊘	⊘	O		⊘	⊘				⊘		
CRN IX, X	↓	↓	⊘	⊘	O		⊘	⊘				⊘		
CRN XI														
CRN XII														
Tone	↓	↓	⊘		O	⊘	⊘	⊘				⊘		
Power	↓	↓	⊘	⊘	O	⊘	⊘	⊘				⊘		
Reflex	Absent	Absent	⊘		O	⊘	⊘	⊘				⊘		
Involuntary Mvt														
PP	N	N			O									
Vib/prop	↓F	↓F			O	⊘	⊘	⊘		⊘				
Cortical sensation														
Coord UL														
Coord LL														
Walk	U	U	⊘	⊘	O	⊘	⊘	⊘				⊘		
Toe/heel/invert/evert														
Tandem														
Romberg														

Abbreviations: BG, basal ganglia; BRST, brainstem; CBL, cerebellum; F, foot; FR, frontal lobe; MUS, muscle; N, normal; NMJ, neuromuscular junction; OCC, occipital lobe; PAR, parietal lobe; PN, peripheral nerves; SPC, spinal cord; TEMP, temporal lobe; U, unable; WM/TH, white matter/thalamus.

of the possibilities we have just considered along with our final decision as to the most likely location.

4.5.1 PERIPHERAL NERVES

Before we address our second basic question (i.e. *what* is Fifi's disease process?), it is important to consider in more detail the structure and function of the part of the nervous system involved: peripheral nerves.

In essence, peripheral nerves are cable systems consisting of motor and sensory axons, as well as various supporting structures. The latter include *myelin sheaths*, highly organized protein-lipid layers surrounding the thicker axons, *Schwann cells* (which produce the myelin), collagen fibers and nutrient blood vessels (see Figure 1.2). Most peripheral nerves are mixed, containing a combination of motor, sensory and autonomic axons; a few are purely motor or sensory.

Motor axons carry electrical impulses in an outward direction from cell bodies located in the brainstem, spinal cord and autonomic ganglia. The axons terminate, usually with numerous branches, in skeletal muscle, cardiac muscle, smooth muscle of intestinal and bronchial walls, selected glands or, in the case of the sympathetic and parasympathetic systems, peripherally located autonomic neurons.

Sensory axons carry electrical impulses in an inward direction from simple and complex receptors located in skin, joint capsules, muscle spindles, tendons, oral cavity, intestines, heart, lung, meninges or the walls of large arteries (see Figure 1.4a,b). For the most part, their cell bodies are located in ganglia present in spinal dorsal roots, cranial nerves and viscera. There are central extensions from these sensory ganglia, either from the cell body or from the sensory axon itself, entering the spinal cord and brainstem to make synaptic contacts with second-order sensory neurons that convey information to higher regions of the CNS. The organization of special senses (such as vision, hearing and olfaction) is highly specialized and will not be considered here (see Chapters 6 and 9).

A magnified microscopic view of a peripheral nerve in cross-section is shown in Figure 4.4. Individual axons are of varying diameters, with a range of about 0.5–20 μm. All of the larger axons, whether motor or sensory, have sheaths of myelin, the sheaths being thicker for larger axons (8–20 μm) and thinner for smaller axons (1–7 μm). Many of the smaller axons (0.5–1 μm) have no myelin sheaths whatever. As will be seen, there are significant functional differences between the large myelinated axons and the small unmyelinated fibers.

The development of a myelinated axon is illustrated in Figure 4.5. The myelin sheath itself is generated by adjacent Schwann cells and consists of tightly apposed lamellae of Schwann cell plasma membrane. In essence, the plasma membrane has been wrapped repeatedly around the axon like a jelly-roll. A given Schwann cell is responsible for the creation and maintenance of a segment of myelin for a single axon (unlike oligodendroglial cells in the brain and spinal cord which provide myelin segments for a number of axons). A section of myelin sheath provided by a Schwann cell is called an internode and is separated from the internode provided by an adjacent Schwann cell by a short section of unmyelinated axon, the node of Ranvier.

The node of Ranvier is a specialized structure whose function is to facilitate conduction of action potentials along the axon. Unmyelinated axons have sodium channels distributed evenly along their axonal membranes and conduct action potentials in a contiguous fashion by spread to adjacent membrane. Myelinated axons, on the other hand, cluster their sodium channels primarily at the nodes of Ranvier. As a result, action potentials are propagated

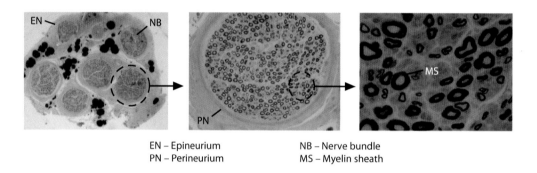

EN – Epineurium   NB – Nerve bundle
PN – Perineurium   MS – Myelin sheath

FIGURE 4.4: Photomicrographs of a cross-section of a typical mixed nerve, at different magnifications, showing myelinated axons of varying sizes, as well as unmyelinated fibers. (Courtesy of Dr. J. Michaud.)

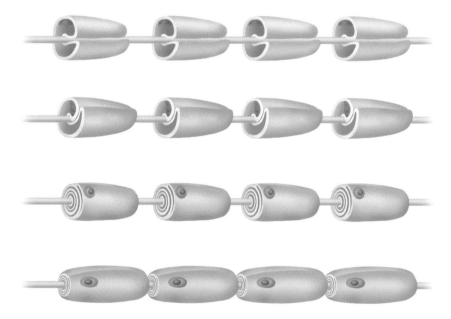

FIGURE 4.5: Cartoon illustrating the formation of myelin segments around a peripheral nerve axon. Each Schwann cell has a leading edge that gradually wraps around the axon to form multiple membranous layers, or lamellae. The mature myelinated axon is at the bottom.

rapidly along the axon by leaping from node to node, the process of *saltatory conduction* (see Figure 1.2c).

Saltatory conduction is obviously far more rapid than contiguous conduction: a large myelinated axon conducts action potentials at rates of 45–60 m/sec along its main trunk, while an unmyelinated axon can manage only a rate of 1 m/sec or less.

Large, rapidly conducting myelinated axons include muscle spindle afferents, sensory fibers from tendons and joint capsules, fibers carrying vibration data and motor axons destined for skeletal muscle fibers (α *motor neurons*). Intermediate-size (8–12 μm) myelinated axons include many sensory fibers for touch and motor axons destined for muscle spindles (γ *motor neurons*). Small myelinated axons (1–2 μm) include sensory fibers for fast ('sharp') pain and first-order efferent autonomic neurons (synapsing in sympathetic and parasympathetic ganglia). Finally, unmyelinated axons include those for slow ('burning') pain and second-order (post-ganglionic) autonomic fibers.

Thus, if we return to our patient, Fifi's peripheral nerve pathology appears to primarily involve motor axons originating in the brainstem and spinal cord and destined for muscles of the face, palate, pharynx, diaphragm, intercostal muscles and limbs. To a lesser extent, based on Fifi's symptoms, the disease process also must be affecting some of the large myelinated sensory axons (e.g. those for touch and vibration sensation); clearly, this involvement is not sufficiently severe to allow for detection of a sensory deficit on clinical examination.

When considering disorders of movement in this text, we will occasionally, in an attempt to relate a complex organization to a structure encountered in daily life, use the analogy of the automobile. For example, we can liken the motor planning centre in the frontal cortex to the driver of the automobile. With respect to the 'mechanical' side of the driver/automobile collaboration, we can likewise consider the skeletal muscles as the motor, and muscle energy substrates such as glycogen as one of the fossil fuels used by the engine. In Fifi's illness, we can consider the problem as residing in the cable linkage between the accelerator pedal and the fuel injection system, as well as the injection system itself.

4.6 ETIOLOGY – FIFI'S DISEASE PROCESS

If we examine Fifi's history, we find an evolution of symptoms from first inkling to severe disability over just two days. Referring back to Figure 3.4, we can see that she is unlikely to have a degenerative or neoplastic disease because the timeline is far too short. As well, in the case of a tumor, the process seems too dispersed in space (head, all four limbs to an equal extent) to allow for such a diagnosis. On the other hand, a vascular etiology is equally unlikely as the onset is too slow.

Between these extremes, we find several categories of disease that could, in principle, create a symptomatic evolution over two days: (accidental) ingestion of a neuro-active drug or environmental neurotoxin; an infectious disease with an affinity for the nervous system; an inflammatory, autoimmune disorder directed at the peripheral nerves.

For the sake of completeness, we will touch on all 10 of the disease categories listed in Figure 3.4, directing most of our attention to the four most likely categories.

4.6.A PAROXYSMAL

Paroxysmal symptoms in peripheral nerve do occur but, with one exception, are relatively rare.

The common exception is *myokymia*, a typically transient phenomenon in which there are irregular, focal muscle twitches that appear to derive from irritability in the motor axons that control the twitching region. A typical example is eyelid myokymia, the intermittent twitching of one eyelid that may develop when one is overtired or has consumed too many caffeinated beverages.

An inflammatory or degenerative process of a proximal nerve may produce brief, spontaneous discharges of motor units, manifested as irregular focal muscle twitches – usually referred to as *fasciculations*. A multiple sclerosis plaque in the brainstem or spinal cord at the exit point of the motor nerve roots may manifest itself in this way, without measurable muscle weakness.

A more generalized or multifocal form of uncontrollable muscle contraction that derives from nerve fiber irritability is *neuromyotonia*. This rare disorder manifests primarily as muscle stiffness, but muscle quivering or twitching may be seen. The disorder may be of hereditary or autoimmune (including paraneoplastic) origin.

4.6.B TRAUMA

Focal acute traumatic lesions of one or several peripheral nerves, including the brachial or lumbosacral plexus, are unfortunately all too common. Such lesions usually only affect one limb and are seen in the context of motorcycle accidents, sports injuries and war-related wounds. A generalized traumatic injury to peripheral nerves is not a phenomenon that is encountered in clinical practice.

4.6.C VASCULAR

An abrupt arterial occlusion in an arm or leg may produce an acute-onset regional ischemic multifocal neuropathy with predominantly distal weakness and numbness. An embolic clot from a fibrillating left atrium may lodge in a brachial artery or at the bifurcation of the abdominal aorta into the femoral arteries (a so-called saddle embolus). Patients with *diabetes mellitus* often develop small blood vessel obstructive disease in peripheral nerves; in consequence, they may develop a fairly abrupt focal motor neuropathy manifested as pain, weakness and muscle atrophy, typically in the quadriceps (diabetic amyotrophy). Generalized peripheral nerve ischemia is a theoretical possibility in severe systemic hypotension, but the patient would likely also lose consciousness.

4.6.D TOXIC

There are a large number of environmental toxins capable of selectively damaging peripheral nerves. Examples include inadvertent inhalation of volatile organic solvents such as toluene and ingestion of heavy metals such as lead (contained in lead-based paints, pottery glazes and automobile batteries). In addition, particularly for a patient employed in a hospital, one must consider accidental exposure to neurotoxic drugs such as the anticancer drugs vincristine and cisplatin.

All of these agents tend to damage peripheral nerves by disrupting the transportation of nutrients from the cell body to the peripheral axonal branches by axoplasmic flow. Nutritional deprivation of the distal portions of axons results in degeneration (sometimes in just a few days) of the terminal portion of the axon, a process known as *dying-back*. The dying-back phenomenon affects longer axons earlier than shorter ones and thus will impair movement and sensation in the hands and feet, the latter more than the former. Loss of sensation in hands and feet is commonly referred to as a 'glove and stocking' distribution and is a typical finding in peripheral neuropathies of the dying-back type.

What this means is that a toxin or drug-induced dying-back neuropathy will selectively produce weakness of distal musculature, leading to grip weakness, wrist drop and drop-foot, a phenomenon characterized by a tendency to trip constantly because of an inability to dorsiflex the feet while walking. Thus, quite apart from the fact that most toxic neuropathies require many days, if not weeks, to develop, Fifi is unlikely to have such a disease process for the simple reason that she also has severe proximal muscle weakness.

Another important cause of acute-onset severe, diffuse muscle weakness is *botulism*, the result of the consumption of food contaminated with *Clostridium botulinum* and its extremely dangerous toxin – for example, improperly home-canned vegetables. Patients with this disorder have peripheral neuromuscular blockade, hence no sensory symptoms; they typically also have dilated, poorly reactive pupils due to the fact that the toxin also blockades muscarinic – as well as nicotinic – cholinergic synapses.

4.6.E INFECTION

A number of infectious agents are capable of producing fairly acute peripheral nerve symptoms. Of these, the most notorious is probably poliovirus, the causative agent of acute poliomyelitis. This neurotropic virus selectively affects motor neurons in brainstem and spinal cord, leading to rapidly progressive weakness of both proximal and distal muscles, along with complete loss of muscle tendon reflexes. Respiratory failure, such as we have encountered in Fifi, is not uncommon.

There are several reasons, however, why Fifi is unlikely to have poliomyelitis. First, the disease is rare in North America because of childhood immunization programs. Second, the virus exclusively affects lower motor neurons, so sensory symptoms would be unlikely beyond pain secondary to parallel inflammation of the meninges. Third, poliovirus (as well as other more ubiquitous viruses that occasionally mimic it – Coxsackie, West Nile) tends to produce markedly asymmetric disease. One leg may be profoundly weak while the other is virtually intact. As we have seen, Fifi's weakness is quite symmetric.

Other micro-organisms capable of involving peripheral nerves include the spirochetal agents responsible for syphilis and Lyme disease (respectively, *Treponema pallidum and Borrelia burgdorferi*) and *Mycobacterium leprae*, the causative agent of leprosy. All three organisms tend to involve sensory axons more than motor. Furthermore, the spirochetal agents both have a constellation of non-neurological symptoms such as skin lesions and arthropathy, neither of which Fifi experienced. Leprosy follows a chronic course and has been virtually eliminated from North America.

4.6.F METABOLIC

Subacute-onset generalized weakness of skeletal muscle is not unusual in a number of metabolic disorders, in particular hypokalemia, hypermagnesemia, hypothyroidism and Cushing's disease (hypercortisolemia). A generalized neuropathy may occur in the context of chronic renal failure but typically evolves over a period of weeks to months rather than 48 hours. Similarly, diabetes mellitus may be accompanied by an insidiously progressive distal polyneuropathy of a predominantly sensory type.

4.6.G INFLAMMATORY/AUTOIMMUNE

We now come to an acute-onset, fairly common disorder that, as will be seen, best explains Fifi's time course, symptoms and signs: acute post-infectious polyradiculoneuropathy, better known as *Guillain–Barré syndrome* (polyradiculoneuropathy means a generalized disease process affecting both spinal nerve roots and peripheral nerves). This disorder is believed to be generated by the body's humoral and cellular immune systems in response to infection with a variety of micro-organisms. The most common trigger agents are Epstein–Barr virus (human herpesvirus 5), cytomegalovirus (human herpesvirus 4), influenza A virus, hepatitis E virus, *Mycoplasma pneumoniae*, *Hemophilus influenza* and *Campylobacter jejuni*;

these cause, variously, infectious mononucleosis (glandular fever), flu-like illnesses, hepatitis, pneumonia and acute diarrhoea. You will recall that Fifi developed flu-like symptoms two weeks before her neurological illness.

The mechanism of disease in Guillain–Barré syndrome is believed to be molecular mimicry. One of the surface components of the triggering organism shares a common molecular configuration with a component of the peripheral nervous system contained either in myelin or in the axonal membrane at the node of Ranvier (or both). An increasing number of ganglioside sub-types, all important components of nerve cell membranes, are strong contenders as the presumed molecules that are being mimicked. In attempting to combat the infectious agent, the immune system begins to attack peripheral nerves by at least two mechanisms, one or the other of which may predominate in individual patients (see Figure 4.6).

The first mechanism of attack is an antibody-mediated obstruction of sodium channels at the nodes of Ranvier, leading to a block in saltatory conduction. The second is a T-cell-mediated attack on myelin itself, with myelin being damaged by cytokines emanating from the T-cells, then stripped off the axon and subsequently digested by macrophages. In the latter case, the 'nude' axonal membrane, lacking appropriate numbers of sodium channels, is largely incapable of transmitting action potentials in the contiguous fashion of a normal unmyelinated axon.

The net effect is an inability to transmit signals to skeletal muscle, thus weakness of acute onset.

This autoimmune attack on peripheral nerves has several peculiar features which help explain the specific constellation of symptoms developed by Fifi. Presumably because the distribution of the potentially antigenic compounds varies from nerve to nerve, Guillain–Barré syndrome tends to involve motor axons more than sensory axons yet spares motor axons in select nerves altogether. The best examples are the three cranial nerves that contribute to eye movement (III, IV, VI) and the nerves for voluntary control of bladder and bowel sphincters. Thus, even though profoundly weak, with rare exceptions, patients with Guillain–Barré syndrome are typically able to look in all directions and have no loss of bladder or bowel control.

The second important peculiarity of the peripheral nerve pathology in Guillain–Barré syndrome is the fact that motor axons – let us say for a proximal anti-gravity leg muscle – are usually not involved homogeneously throughout their courses. The antibody- and cell-mediated disease processes tend to affect those portions of the axon where the so-called blood–nerve barrier is relatively deficient or leaky.

The blood–nerve barrier consists of tight junctions between endothelial cells that prevent the passage of large molecules from the nerve's blood vessel lumina to the extracellular space surrounding the axons. This barrier is relatively leaky at the level of the spinal nerves, the dorsal

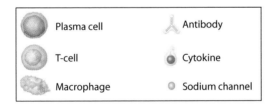

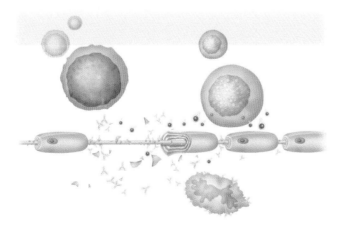

FIGURE 4.6: Illustration of the pathology in acute inflammatory demyelinating polyneuropathy. Some antibody molecules are attaching themselves to sodium channels at the nodes of Ranvier, leading to channel failure and a block in axon potential conduction. T-cells are actively damaging and stripping away myelin lamellae in one of the internodes; macrophages are present to ingest the myelin fragments. The myelin from one internode has completely disappeared.

and ventral roots, small distal nerve branches and in the region of potential compression points such as the carpal tunnel for the median nerve.

In the case of disease of the ventral roots, the 'radiculopathy' component of the acute post-infectious poly-*radiculo*neuropathy, we find the explanation for the severe *proximal*, as well as distal weakness in Guillain–Barré syndrome. Since motor axons for both proximal and distal muscles must all pass through the ventral roots, the subsequent length of the axon is relatively immaterial if the axon has already been blockaded at the root level. Clearly, this is a very different disease process from the one we encountered in dying-back neuropathy and accounts for the significantly different clinical picture.

Disease involving dorsal spinal roots is likely an important contributing factor in the production of the severe back pain that often accompanies the motor symptoms in Guillain–Barré syndrome.

There are other autoimmune disorders that may present with acute or subacute onset diffuse weakness. *Transverse myelitis*, an autoimmune disease affecting the spinal cord, may present with a 48-hour history of rapid-onset weakness; depending on the location of the myelitis, the pattern of weakness may involve all four limbs, legs more than arms or just the legs. Typically, there is loss of bladder control as well. Physical examination tends to reveal bilateral extensor plantar responses and sensory level in the mid-to-upper trunk. *Myasthenia gravis*, an autoimmune disorder of neuromuscular transmission, may present with diffuse weakness but usually over a period of weeks to months; there are no sensory symptoms. *Dermatomyositis* and *polymyositis*, both autoimmune disorders involving skeletal muscle, present with trunk and limb weakness (but no sensory symptoms other than muscle pain) also evolving over weeks to months.

4.6.H NEOPLASTIC

Primary neoplastic lesions affecting peripheral nerves usually involve a single nerve, although multiple peripheral nerve neurofibromas are seen in type 1 neurofibromatosis, multiple schwannomas (Schwann cell tumors) in type 2 disease. The latter disorders are multifocal but not generalized. Metastatic spread of non-neurological malignancies is unfortunately a common phenomenon but, again, only affects nerves or plexi in a regional fashion.

Some malignant neoplasia (for example, lung cancers) may be accompanied by the development of a multifocal or generalized polyneuropathy. In such situations, however, the neuropathies are of autoimmune origin, the antigenic trigger being chemical components in the neoplastic cells (so-called *paraneoplastic syndromes*); the neoplasia do not involve the peripheral nerves in these syndromes by direct invasion. In essence, the pathologic process in paraneoplastic sensorimotor polyneuropathy resembles the mechanism responsible for Guillain–Barré syndrome, the trigger for the autoimmune attack being a type of malignant tumor cell rather than an infectious agent.

4.6.I/J DEGENERATIVE/GENETIC

Generalized degenerative polyneuropathies are nearly always genetic in origin, hence our fusion of the last two etiologic categories. In general, these neuropathies are of the dying-back type, thus tending to produce progressive distal muscle weakness and sensory loss, more marked in the legs than in the arms. These diseases evolve over many years and, especially if the large sensory fibers are affected, may result in sensory ataxia and eventual loss of ambulation. Inheritance patterns may be dominant, recessive or X-linked; some forms are predominantly motor in nature, others purely sensory or even autonomic. A common and very imprecise eponym employed for hereditary motor and sensory neuropathies is *Charcot–Marie-Tooth disease*.

Our etiological diagnostic discussions in this section are summarized in the finalized Etiology Matrix in Figure 4.7. In completing this table, we have confined our decisions to the columns with disease durations that fit

Disease type	Secs	Mins	Hrs	Days	Weeks	Months	Years
Paroxysmal (seizures, faints, migraine)			Ø				
Traumatic			Ø				
Vascular (ischemic stroke, bleed)			Ø				
Toxic			Ø	Ø	Ø		
Infectious				Ø	Ø		
Metabolic			Ø	Ø	Ø		
Inflammatory/autoimmune				O	O		
Neoplastic					Ø		
Degenerative							
Genetic			Ø	Ø	Ø		
			← Acute →		← Subacute →		← Chronic →

FIGURE 4.7: Completed etiology matrix for Fifi.

Disease type	Secs	Mins	Hrs	Days	Weeks	Months	Years
Paroxysmal (seizures, faints, migraine)	+	+	+				
Traumatic	+	+	+				
Vascular (ischemic stroke, bleed)		+	+				
Toxic		+	+	+	+	+	
Infectious				+	+	+	+
Metabolic				+	+	+	+
Inflammatory/autoimmune				+	+	+	+
Neoplastic					+	+	+
Degenerative						+	+
Genetic	+	+	+	+	+	+	+
	← Acute →				← Subacute →		
						← Chronic →	

FIGURE 4.8: Etiology matrix showing the most commonly encountered symptom durations for each disease category.

best with the clinical history, i.e. a disease duration of days. In order to allow for unusual exceptions, we have included consideration of disease categories that typically evolve over hours, on the one hand, and weeks on the other.

In many clinical situations, based on symptom duration alone, it is possible to take a relative short cut in working out the disease categories that are most likely, without having to consider every disease category as we have done in this chapter. This short cut will be used in some of the later clinical chapters as well as in the e-cases. Figure 4.8 shows the most probable symptom durations that one will encounter with the various disease categories. As will occasionally be seen, there are exceptions to these generalizations. In most clinical encounters, however, Figure 4.8 will be found to correctly identify the disease categories that should be considered first.

4.7 SUMMARY AND SUPPLEMENTARY INFORMATION

Fifi has a relatively rapid-onset disease evolving over 48 hours; it is characterized by severe limb, facial and pharyngeal weakness accompanied by distal extremity numbness and paresthesiae. The presence of proximal and distal limb weakness, complete areflexia and flexor plantar responses is consistent with a generalized polyradiculoneuropathy. Given the recent history of a respiratory illness, the most likely diagnosis is an acute inflammatory demyelinating polyneuropathy, otherwise known as Guillain–Barré syndrome.

4.7.1 GUILLAIN–BARRÉ SYNDROME

Fifi's illness was first described in detail, in the French literature, by Guillain, Barré and Strohl during the First World War. In most, but not all, cases, there is an apparent inciting event, typically an infectious process as we have already seen, but occasionally a surgical procedure or an injury.

The onset of symptoms is usually over a period of several days, often beginning with spontaneous sensory symptoms (tingling, burning, crawling sensations) in the extremities, the legs more than the arms, as well as – in some cases – by the development of low back pain. There follows a rapidly progressive weakness, proximally and distally, at first in the legs, then the arms, sometimes the face and pharyngeal muscles. If nerves to intercostal and diaphragmatic muscles are also significantly involved, respiratory failure may ensue. The progression from first symptoms to greatest degree of weakness takes place over a maximum of six weeks, with the degree of disability extending from mild extremity weakness through inability to walk to complete paralysis. Sometimes the autonomic nervous system is also involved, the presenting features being one or more of arterial hypertension, postural hypotension, tachycardia and bowel or bladder dysfunction. Recovery occurs in most patients, usually over a period of weeks to an entire year, with functionally complete recovery in about 80% of patients.

In recent years, Guillain–Barré syndrome has evolved from a unitary diagnosis to a group of related, autoimmune disorders. In North America and Europe, the most common pattern is acute inflammatory demyelinating neuropathy (AIDP); in contrast, in Asia and Central/South America, the most common type is acute motor axonal neuropathy (AMAN). Clinically, these two disorders are quite different, although there is some overlap in presentation. With AIDP, there are mixed motor and sensory deficits; cranial nerve involvement is common as is autonomic dysfunction. In AMAN, on the other hand, motor weakness is largely confined to the extremities, sparing the cranial nerves; sensory symptoms are unusual. AIDP is predominantly a demyelinating process and is triggered by a variety of infections, in particular Epstein–Barr, cytomegatic inclusion and influenza viruses. AMAN, as the name suggests, involves antibody-mediated conduction block at nodes of Ranvier in motor nerves; the most common trigger by far is a *C. jejuni* infection. There are at least three other, much less common, clinical patterns of Guillain–Barré syndrome whose description is beyond the scope of this text. Interested readers should consult the recent review article on the disorder cited at the end of the chapter. Of the two most common clinical patterns, it is immediately clear that Fifi's story fits much better with AIDP.

The diagnosis of Guillain–Barré syndrome (Strohl's contribution has, unfortunately, been largely forgotten) is typically based on the characteristic presentation of ascending weakness, absence of muscle tendon reflexes and preservation of flexor plantar responses. There are two useful diagnostic tests that help confirm the diagnosis. An examination of cerebrospinal fluid (CSF) characteristically shows an elevated protein level but a normal or near-normal cell count, a combination referred to as *albuminocytological dissociation*. The protein elevation results from the disease process involving ventral spinal roots that are in direct contact with CSF in the spinal subarachnoid space. A high CSF protein level in a patient with Guillain–Barré syndrome may take several days to develop. Thus, if a CSF study done early in the disease course shows a normal protein level, yet the clinical picture is typical, it is worth repeating the lumbar puncture after a few days.

The second diagnostic test consists of nerve conduction studies. In a patient with AMAN, these usually reveal a marked reduction in the amplitude of compound motor action potentials, a result of extensive conduction block at nodes of Ranvier; motor nerve conduction velocities may be normal, as are sensory nerve conduction velocities. In AIDP, there is typically a striking decrease in motor conduction velocities and a marked increase in distal motor latencies, all due to patchy demyelination of motor axons; sensory nerve conduction is also slowed.

In those patients with almost exclusive involvement of spinal nerve roots, the peripheral nerve conduction studies may be initially normal. In such situations, the presence of a demyelinating radiculopathy can be detected by means of F-wave conduction studies, a technique that utilizes the fact that an electrical stimulation of a peripheral nerve will cause the nerve fibers to conduct the combined action potential back up the nerve (antidromically in the case of the motor fibers) as well as orthodromically down to the distal muscle recording site. A small percentage of the motor neurons respond to the antidromic stimulus by sending a second, smaller action potential (the F-wave) down to the recording site. In the presence of a demyelinating radiculopathy, the latency of the second wave will be abnormally prolonged.

4.7.2 INVESTIGATIONS

Pulmonary function studies carried out in the emergency department revealed a 75% reduction in vital capacity and forced expiratory volume.

A lumbar puncture done shortly after admission showed clear CSF with 1 lymphocyte/μL, 0 red blood cells, glucose 2.6 mmol/L and protein 1.26 g/L (normal <0.45 g/L).

Motor and sensory nerve conduction studies were carried out a few days later. Conduction velocities in main nerve trunks were significantly slowed; compound motor action potentials tended to be small in amplitude and dispersed (rather than smooth) in shape; sensory nerve action potentials were also smaller than normal. Distal motor latencies were 2–3 times normal.

4.7.3 TREATMENT/OUTCOME

Fifi was admitted to the intensive care unit (ICU) for close monitoring. Increasing respiratory distress and declining pulmonary function necessitated airway intubation and mechanical ventilation within 12 hours of admission.

Within 48 hours of admission, Fifi demonstrated grade 1/5 muscle strength proximally and distally in all four limbs. She was given two infusions of intravenous immunoglobulin 1 g/kg on succeeding days beginning the day after admission. The role of the immunoglobulin infusion was to obstruct access of the circulating pathogenic antibodies to their peripheral nerve targets. Although the presence of an autoimmune disorder might suggest to you that she should have been started on methylprednisolone or prednisone as well, this was not done. Large controlled studies have repeatedly demonstrated that steroids have no useful role to play in the management of Guillain–Barré syndrome, regardless of clinical sub-type. Beginning about one week after admission, Fifi began to improve steadily and was extubated on the 15th day.

Throughout the ICU admission, Fifi had an aggressive physiotherapy program, at first with passive range of movement and the use of ankle and wrist splints to prevent contraction deformities. At the same time, she was frequently repositioned to prevent the development of bed sores. Notwithstanding her medical background, she was understandably terrified by the severe nature of her disease, her resulting sense of helplessness and her complete reliance on the continuous efforts of the ICU staff. She required constant reassurance and psychological support from her colleagues in the ICU and rehabilitation teams.

Once extubated and stabilized, Fifi's program of physiotherapy was intensified to work on re-establishing independent sitting, then ambulation. She continued to improve and was walking without help six weeks after admission. By the time of discharge at eight weeks, she was functioning independently, her only deficit being a mild foot drop gait.

4.8 E-CASES

The following clinical cases are intended to broaden the scope of the material concerning the peripheral nervous system that we have introduced in this chapter. Only abbreviated case histories and physical examination findings are given here; more complete versions are found on the Web site. To work through these cases, please follow the same routine as you have just done for Fifi's story. The required materials, the discussions of localization and etiology for each case, relevant illustrations and summative comments are all available on the Web site.

4.8.1 CASE 4E-1: OSKAR, AGE 52

Background: Your patient lives with his wife, a fashion designer. They have three children, two away at university and one still at high school. He and his wife are moderately heavy cigarette smokers. He has smoked one pack of cigarettes a day for 30 years.

Chief complaint: Three-month history of increasing fatigue and weakness

History:

- Increasing difficulty walking upstairs for the past three months; increasing problems in brushing his hair; aching pain in his thighs and upper arms
- Two-week history of difficulty standing up from a sitting or lying position
- Constant non-productive cough for the past six months; unexplained 5 kg weight loss

Examination:

- Blood pressure (BP) 145/90
- Decreased air entry over right lower lobe of the lungs
- Pinkish-violet rash on the eyelids, cheeks, elbows, knuckles, knees and in the nail-beds (see Figure 4.9)
- Grade 3–4/5 weakness of the deltoids, latissimus dorsi, psoas, quadriceps femoris, hamstrings, gluteus medius and gluteus maximus; grade 2–3/5 weakness of neck flexors and rectus abdominis muscles (Figure 4.10)
- Tendon reflexes 1+; plantar responses flexor
- Normal sensory examination

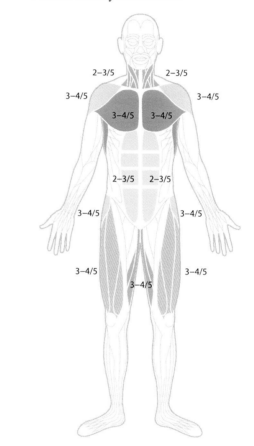

FIGURE 4.10: Motor examination map.

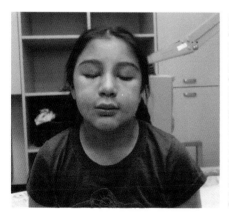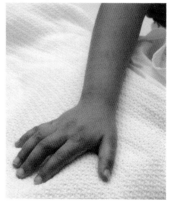

FIGURE 4.9: Typical rash seen in a patient with Oskar's disorder. (Patient's release and with permission; courtesy of Dr. H. McMillan.)

4.8.2 CASE 4E-2: KIYOMI, AGE 29

Background: Your patient has been too busy working on plans for a new civic concert hall and has delayed making an appointment for her increasing medical problems. She is married and has no children. Both her mother and her mother's sister suffer from rheumatoid arthritis; her mother is also hypothyroid. A 6-year-old niece has been diagnosed with type 1 diabetes mellitus.

Chief complaint: One-year history of double vision

History:

- Double vision developing later in the day, at first transient, then persistent; becomes worse as the day progresses; has not seen her own physician about this ('too busy!') and has been wearing an eye-patch
- Two-month history of difficulty chewing food and reduced voice volume, again worse later in the day; still too busy!
- Two-week history of difficulty swallowing liquids with occasional choking episodes – panic button finally pushed!

Examination:

- Normal vital signs; mild thyroid gland enlargement

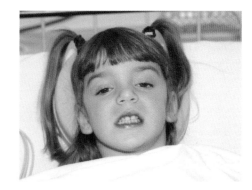

FIGURE 4.11: Typical facial appearance for a patient with her disorder. (Patient's release and with permission; courtesy of Dr. P. Humphreys.)

- *Pronounced drooping of eyelids (ptosis)*, progressively worse with sustained up-gaze
- Right eye deviated inwards (esotropia); *impaired abduction of both eyes*, inability to sustain upward eye movement
- *Marked facial weakness* with inability to smile (see example in Figure 4.11)
- *Nasal voice*; inability to keep mouth open against resistance
- Mild deltoid weakness; otherwise normal muscle strength in limbs
- 2+ tendon reflexes; flexor plantar responses

4.8.3 CASE 4E-3: FELIX, AGE 5

Background: Your patient is the oldest of three children. His 3½-year-old sister walked at 10 months and is now already reading simple words; the 2-year-old brother is more like our patient having learned to walk at 19 months and hardly talks at all. The mother is adopted and has no information about her biologic family.

Chief complaint: Two-year history of increasing difficulty walking

History:

- Increasing toe-walking over a two-year period, leading to tripping and falling; once fallen, increasing difficulty getting himself back into a standing position
- At present, no longer able to run; waddles when walking; trouble walking upstairs
- Delayed language development and toilet-training in comparison with three-year-old sister; younger brother, aged 2, is similar to Felix at that age

Examination:

- Easily distracted for someone his age
- *Abnormally large tongue and calves* (see Figure 4.12)

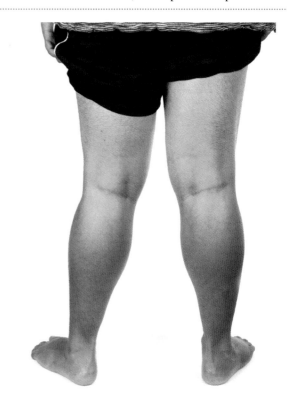

FIGURE 4.12: Appearance of calves in a patient with his disorder.

- Stands with exaggerated lumbar lordosis; tight, contracted heel cords
- Normal cranial nerve examination
- *Difficulty raising his arms above his head*
- In order to stand up from lying supine, Felix *rolls over into prone, gets on his hands and*

knees, plants feet on the floor and climbs up himself using his arms, in order to completely straighten his back (see video-clip on the text Web site)

- 1+ tendon reflexes; flexor plantar responses

4.8.4 CASE 4E-4: NASREEN, AGE 7

Background: This has been an extremely difficult year for the family. The child, our patient, has been treated for acute lymphoblastic anemia and is now in remission. The father is extremely distraught and in addition has recently been suspended from his job.

Chief complaint: Two-week history of progressive gait deterioration

History:

- *Acute lymphoblastic leukemia* diagnosed four months ago, treated aggressively with prednisone, vincristine and cytarabine; disease now in remission
- Two-week history of increasing tendency to trip over her feet when walking and going upstairs

- Unable to elevate toes during gait cycle – lifts whole leg and slaps foot forward; has worn out the toes of brand new shoes in two weeks due to friction with the floor

Examination:

- Sparse hair (due to chemotherapy); bright and chatty
- No lymphadenopathy; liver and spleen of normal size
- Weak hand grip; *grade 3/5 weakness of inter-ossei and wrist extensors*; proximal arm strength normal (Figure 4.13a)
- *Grade 2/5 weakness of foot dorsiflexion; grade 3–4/5 power in toe extensors/flexors, ankle evertors, plantar flexors*; proximal leg strength normal (Figure 4.13a)

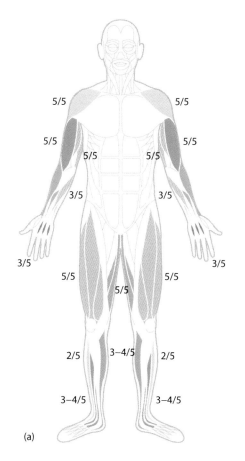

(a)

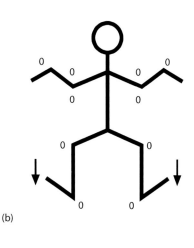

(b)

FIGURE 4.13: Motor examination (a) and reflex (b) maps.

- Foot-drop or 'steppage' gait, meaning that, in order not to trip while walking, she has to lift each foot well off the floor before putting it down (see video-clip on Web site)

4.8.5 CASE 4E-5: ANTONIA, AGE 62

Background: Your patient, a grandmother, is spending the summer with her daughter at the family cottage for quality time with her grandchildren (the son-in-law is chained to his work) at the family cottage and quality time with her grandchildren.

Chief complaints: Three to four days of headache and fever; two days of rapidly progressive right arm weakness

History:

- Four-day history of increasingly severe headache, with aching pain in the neck and back
- Three days of fever and chills followed by vomiting
- Two days of increasingly severe pain and weakness in the right arm
- On the day of assessment, Antonia is now confused and disoriented

Examination:

- Pulse 106/min; BP 150/95; respirations 25/min; temperature 39°C
- *Disoriented in time and place* but not person
- Marked *neck stiffness* on passive flexion
- Cranial nerve examination normal except weakness of the right sternocleidomastoid muscle; normal fundi

4.8.6 CASE 4E-6: CAROLINE, AGED 51

Background: Caroline is almost always busy with her hands knitting or crocheting. She has insulin-dependent diabetes mellitus. There is a family history of coronary artery disease, and both her mother and her sister have pernicious anemia. Your patient admits to smoking a half-pack of cigarettes per day.

Chief complaint: Pain and numbness in the hands for the past three months

History:

- Intermittent pain and numbness in the thumb, index and middle finger of both hands, right more than left
- Often wakes during the night or in the morning with these symptoms; they improve with rubbing or shaking the hands
- Caroline also develops the same symptoms in the right hand when doing repetitive activities like knitting or crocheting

- *Tendon reflexes*: see Figure 4.13b
- Sensory examination findings unreliable due to patient's age and developmental status

- *No movement (grade 0/5) of all muscles of the right arm above the wrist*; grade 3/5 weakness of right hand grip
- Normal power in the left arm except grade 3/5 weakness of the deltoid muscle
- *Profound hypotonia of the right arm*; normal tone in the left arm and in both legs
- Mild reduction of withdrawal of right leg to foot tickle compared with left leg
- *Tendon reflexes*: see Figure 4.14

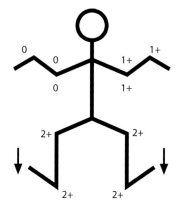

FIGURE 4.14: Reflex map.

- Has also had difficulties with vision in the left eye requiring laser treatments; her family physician has noted some changes in kidney function
- Current medications are insulin, hydrochlorothiazide, Altace, Norvasc and Lipitor

Examination:

- Wasting of the right thenar eminence; weakness of the right abductor pollicis brevis muscle
- Decreased touch and pinprick sensation over the first, second and third digits of both hands (Figure 4.15a)
- Tapping the palmar aspect of the wrists and forced wrist flexion both elicit paresthesias in the palms of the hands (positive *Tinel* and *Phalen* signs, respectively)
- Tendon reflexes: see Figure 4.15b
- Decreased sensation to vibration at 256 Hz over both medial malleoli (Figure 4.15a)

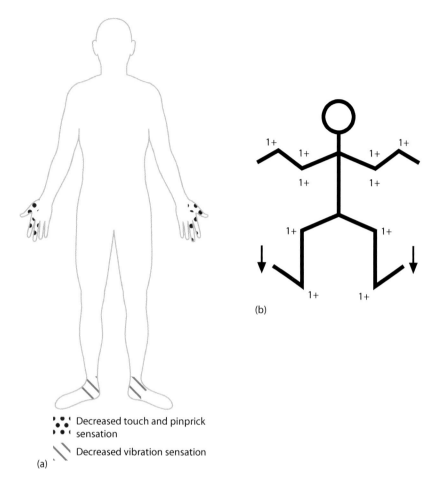

Decreased touch and pinprick sensation

Decreased vibration sensation

(a)

(b)

FIGURE 4.15: Sensory examination (a) and reflex (b) maps.

4.9 SUMMARY OF KEY NEUROANATOMICAL AND NEUROPHYSIOLOGICAL INFORMATION

- Motor neurons in the *precentral gyrus* (primary motor cortex, area 4) are known as the upper motor neurons. These are geographically distributed according to the region of the body they control. Proceeding from the most medial area of the motor cortex, there are areas that control, in sequence, the *feet* and *legs*; and then laterally along the precentral gyrus, beginning rostrally, the *trunk, arms, hands, face* and *tongue*, with more cortex devoted to areas of fine motor control. This more-or-less contiguous anatomical pattern for motor control is known as the *motor homunculus* (Figure 4.2).

- The next part of motor control is the pathway to the lower motor neurons – the cranial nerve motor nuclei (cortico-bulbar) and the anterior horns of the spinal cord (cortico-spinal). (The myelin sheath of these CNS axons is formed by oligodendrocytes.)

- The lower motor neuron and its axon (which travels in peripheral nerves) and the terminal synaptic boutons, plus the variable number of muscle fibers these innervate, collectively comprise the *motor unit* (Figure 4.3).

- *Peripheral nerves* usually consist of a mixture of motor and sensory axons, as well as axons of presynaptic and postsynaptic autonomic neurons (Figure 4.4). Sensory axons carry information from *muscle spindles, tendons, joint capsules* and *skin receptors*; this information is crucial to movement control in that it provides moment-by-moment *feedback* to the higher levels of the motor system.

- Motor and sensory axons vary in diameter from 0.5 μm to 20 μm. Large-diameter axons are encased in *myelin sheaths* and conduct action potentials at *rapid rates* (45–60 m/sec); the smallest axons are unmyelinated and conduct at much *slower rates* (<1 m/sec).

- Myelin sheaths of PNS axons consist of rolled-up, stacked cell membranes derived from *Schwann cells*. A single Schwann cell provides a segment of myelin sheath for a single axon, that segment being referred to as an *internode*. The narrow, regularly spaced bare axon segments between internodes contain concentrated collections of sodium channels and are known as *nodes of Ranvier* (Figure 4.5).
- In a myelinated motor or sensory axon, the action potential jumps extremely rapidly from node to node, bypassing the myelinated segments, in a process known as *saltatory conduction*.

4.10 CHAPTER-RELATED QUESTIONS

1. If you had direct access to a patient's cerebral cortex in the operating room and wished to stimulate the brain in order to produce a twitch of the left leg and foot, what region of the cortex would you stimulate?

2. List the main abnormalities on *motor system* examination that you would expect to find in a patient with a long-standing lesion in the thoracic spinal cord at T8.

3. In contrast, what would you expect to see on *motor and sensory* examination in a patient with long-standing damage to both lumbosacral plexi?

4. Imagine that you are an axonal growth cone sprouting form the cell body of an immature motor neuron in the left precentral gyrus of a second-trimester foetus. As you progressively increase in length, describe the route that you would have to take through the brain and spinal cord in order to eventually make synaptic contact with an AHC in the lower cervical spinal cord (e.g. to move the right index finger). Mention some of the adjacent anatomical structures that you would encounter (and ignore!) on your way down.

5. You are asked to see a 20-year-old man with a history of progressive difficulty walking over the last five years. He tends to trip over his feet, particularly when walking upstairs. In addition, on flat, uneven ground, he tends to lose his balance and fall down.

   Physical examination shows symmetric wasting of the small muscles of the hands and feet. Muscle strength assessments are as follows: deltoids 5/5; biceps brachii 5/5; wrist extensors 4/5; first dorsal interossei 3/5; hip flexors 5/5; knee extensors 5/5; foot dorsiflexors 3/5; toe extensors and flexors 2/5. Tendon reflexes are completely absent in all four limbs. Position sense is impaired in the toes and in the ankle joints. Touch, pinprick and vibration sensations are decreased in both legs below the knees and in both arms below the wrists. The plantar responses are absent.

   a. What is the most likely location of the disease process? Give the reasons for your choice
   b. What is the most likely category of disease process in this man?

6. Your next patient is a six-month-old girl. Her parents are concerned because their daughter has not yet developed control of her head; she is also unable to roll over by herself and seems much 'floppier' than was their older son at the same age. Apparently, she was a normal, vigorous infant until the age of two months, after which she seemed to be going 'backwards'. On examination, she is alert, smiles frequently and makes good eye contact. Her tongue is atrophic and seems to be fasciculating. She also has marked wasting of the extremity muscles and severe generalized hypotonia and is unable to lift her arms or legs off the bed against gravity. She is completely areflexic and has weakly flexor plantar responses.

   a. What is the most likely category of disease process in this girl?
   b. What are the two most likely locations of her disease process? Give the reasons for your choice.
   c. What simple neurophysiological investigation could you perform that would help you decide which of the two disease locations is the correct one in this child.

7. A two-year-old boy was found this morning with his right leg jammed between the bars of his crib. The parents suspect he may have tried to climb out of the crib during the night. When they got his diaper changed and put him on the floor, they noticed that he was tripping over his right foot while attempting to walk. On examination, he has no active dorsiflexion of either his right foot at the ankle or any of his toes (so-called drop foot); he is also unable to evert (move laterally) his foot at the ankle. He has no difficulty going up on the toes of both feet and can invert both ankles normally. The knee and ankle tendon reflexes are normal and the plantar responses are flexor. Finally, he seems to have little or no pinprick sensation on the

dorsum of his right foot while the sole of the foot reacts normally.

   a. How would you explain the clinical findings in this child? What process caused them to appear?

   b. How could you prove your hypothesis?

   c. What prognosis would you give the parents?

REFERENCE

Penfield, W., Roberts, L. 1959. *Speech and brain mechanisms*. Princeton, NJ: Princeton University Press.

SUGGESTED READING

Hadden, R.D.M., Karch, H., Hartung, H.-P., Zielasek, J., Weissbrich, B., Schubert, J., Weishaupt, A., Cornblath, D.R., Swan, A.V., Hughes, R.A.C., Toyka, K.V., the Plasma Exchange/Sandoglobulin Guillain-Barré Syndrome Trial Group. Preceding infections, immune factors, and outcome in Guillain-Barré syndrome. *Neurology* 2001. 56: 758–765.

Hughes, R.A.C., Wijdicks, E.F.M., Barohn, R., Benson, E., Cornblath, D.R., Hahn, A.F., Meythaler, J.M., Miller, R.G., Sladky, J.T., Stevens, J.C. Practice parameter: Immunotherapy for Guillain-Barré syndrome. *Neurology* 2003. 61: 736–740.

Hughes, R.A., Swan, A.V., Raphaël, J.C., Annane, D., van Koningsveld, R., van Doorn, P.A. Immunotherapy for Guillain-Barré syndrome: A systematic review. *Brain* 2007. 130: 2245–2257.

Oomes, P.G., Jacobs, B.C., Hazenberg, M.P.H., Bänffer, L.R.J., van der Meché, F.G. Anti-GM1 IgG antibodies and campylobacteria in Guillain-Barré syndrome: Evidence of molecular mimicry. *Ann Neurol* 1995. 38: 170–175.

Prineas, J.W. Pathology of the Guillain-Barré syndrome. *Ann Neurol* 1981. 9(suppl): 6–19.

Van den Berg, B., Walgaard, C., Drenthen, J., Fokke, C., Jacobs, B.C., van Doorn, P.A. et al. Guillain-Barré syndrome: Pathogenesis, diagnosis, treatment and prognosis. *Nat Rev Neurol* 2014. 10: 469–482.

Yuki, N., Takahashi, M., Tagawa, Y., Kashiwase, K., Tadokoro, K., Saito, K. Association of campylobacter jejuni serotype with antiganglioside antibody in Guillain-Barré syndrome and Fisher's syndrome. *Ann Neurol* 1997. 42: 28–33.

WEB SITES

Anatomy of the movement control system: http://www.neuro.wustl.edu/neuromuscular/index.html (University of Washington)

Guillain-Barré syndrome: http://www.jsmarcussen.com/gbs/uk/profs.htm

Chapter 5

Cletus

Objectives

- Learn the principal anatomic components of the spinal cord
- Learn the anatomy of the vascular supply of the spinal cord
- Learn the effects of damage to these spinal cord components and their clinical manifestations
- Learn the major spinal cord syndromes and their clinical manifestations
- Become sensitized to the medical, physical and psychosocial implications of spinal cord injury

5.1 CLETUS

Cletus, a 23-year-old man who makes his living as a hunting and fishing guide, lives in the bush country of North Fountainbleau County.

One Saturday night, Cletus and his buddies were peeling up and down County Road 1, the only one in the County, in their 1969 Duster. Ajutor, the driver of the old rattletrap, was travelling too fast and missed the corner at the end of the road, hitting a large boulder of four-billion-year-old pre-Cambrian rock, thus causing the car to come to a sudden stop. Cletus was in the back seat with no seatbelt and never knew what hit him.

At the scene of the accident, the paramedics found Cletus to be unresponsive to commands or questions but moving all limbs to pain; his blood pressure was very low. He was transported to the nearest university hospital.

The next thing that Cletus remembered was that he was in a hospital room with lots of beeping sounds, a tube down his throat and a big pain on the left side of his chest.

'Cletus, squeeze my fingers', asked the nurse: he squeezed her fingers – 'Cletus move your legs and feet' … What legs and feet? … He could not feel them, let alone move them. He suddenly became terrified and started to grab at his tubes: then the lights went out again.

When he woke up, there was this geeky looking old guy with blond hair at his bedside, making him do a lot of weird stuff – 'Cletus, follow the pen with your eyes; wrinkle your forehead; smile; stick out your tongue' – he could do all these things ok.

He was able to squeeze the doctor's fingers, bend and straighten his arms at the elbows and shoulders. When asked to move his legs and feet again, he was unable to feel or move his lower limbs. The examiner took out a safety pin and tested him on the face, arms, trunk and legs to the belly button. He could feel it on both sides down to the belly button but not below, on either side.

Then a strange thing happened – the guy took out this fork thing that vibrated and tested him over his body. He could feel the vibration not only over his face and arms but also over his toes, ankles and knees!

Then he could hear them talking in the background in muffled tones. 'Time is going to tell…' He heard the surgeon talking to this mother. 'The operation in his chest went very well, we had to put a patch on his aorta … the plastic tube in the side of his chest should come out in a couple of days … you will have to ask the neurologist whether he will walk again'. Cletus could hear his mother sobbing at the end of the intensive care unit bed.

5.2 CLINICAL DATA EXTRACTION

Before proceeding with this chapter, you should first carefully review Cletus' story in order to extract key information concerning the history of the illness and the physical examination. On the Web site accompanying this text, you will find a sequence of tools to assist you in this process.

The clinical findings as described earlier allow us to localize the lesion within the nervous system. The other piece of important information, of which Cletus was not aware, is the presence and characteristics of the deep tendon and superficial reflexes. For the sake of discussion, let us assume that the reflexes were normal (2+) in the upper limbs and absent in the lower limbs. The plantar responses were essentially absent: neither flexor nor extensor. Cletus was also unaware that his bladder was being drained by a urinary catheter as he had no feeling of bladder fullness or control. With this information, the neurological examination section of the Extended Localization Worksheet can be filled out.

The History Worksheet (Figure 5.1) is fairly straightforward: a sudden onset of lower limb weakness in the context of an automobile accident suggests that the etiology is probably trauma with complete transection of the

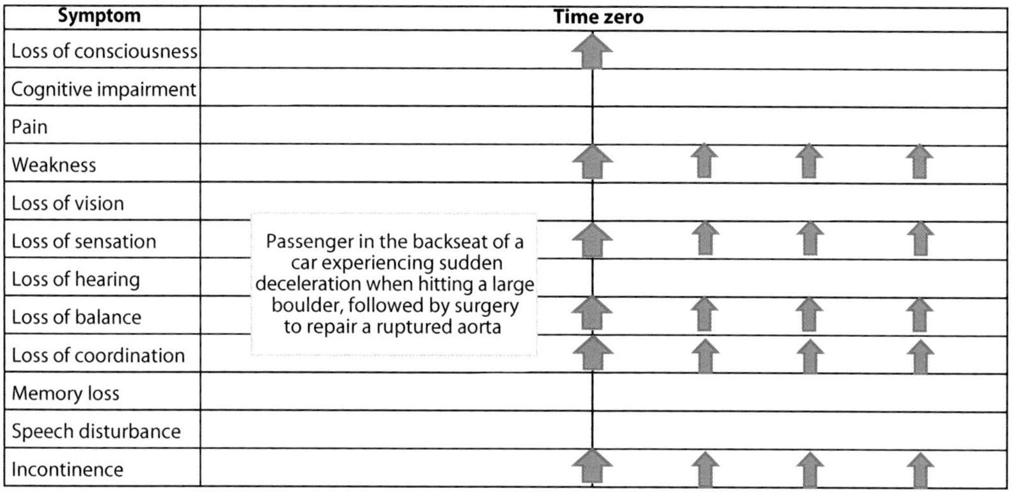

Symptom	Time zero			
Loss of consciousness	⬆			
Cognitive impairment				
Pain				
Weakness	⬆	⬆	⬆	⬆
Loss of vision				
Loss of sensation	⬆	⬆	⬆	⬆
Loss of hearing				
Loss of balance	⬆	⬆	⬆	⬆
Loss of coordination	⬆	⬆	⬆	⬆
Memory loss				
Speech disturbance				
Incontinence	⬆	⬆	⬆	⬆

Passenger in the backseat of a car experiencing sudden deceleration when hitting a large boulder, followed by surgery to repair a ruptured aorta

FIGURE 5.1: History worksheet.

spinal cord. Other things to consider are collateral etiologies that occur as a result of trauma, such as systemic hypotension due to blood loss or hypoxia due to hemo/pneumothorax. However, follow the localization and etiology protocols carefully; you might be in for a surprise.

5.3 MAIN CLINICAL POINTS

- Cletus was involved in a severe traumatic deceleration injury.
- Cletus was hypotensive but noted to be moving all four limbs at the scene of the accident.
- Cletus had a surgical procedure performed to his chest to repair a ruptured thoracic aorta.
- Following the procedure, Cletus was found to have lower limb paralysis and sensory loss to pinprick but not to vibration.
- Cletus also had no sensation of or voluntary control of bowel or bladder function.
- His mental status, cranial nerves and upper limb power, sensation and reflexes were all normal.

In summary, Cletus has suffered a neurological injury related to power and sensation in the lower limbs, including loss of bowel and bladder control; these deficits developed after the surgical procedure to correct a ruptured thoracic aorta. His neurological examination findings are summarized in Figure 5.2.

5.4 RELEVANT NEUROANATOMY

The spinal cord consists of a central gray matter H-shaped column surrounded by white matter tracts, which transmit information from either the peripheral nerves to the brain or information with respect to motor and autonomic function from the brain to the spinal cord. Figure 5.3 shows the anatomy of the spinal cord, its blood supply, the dura and the spinal nerves, which exit from the spinal cord at each level. Figure 5.4 shows a magnetic resonance image (MRI) of the spinal cord, vertebral column and adjacent tissues.

There are many different tracts, ascending and descending, with differing functions; they are often not easily tested clinically. Three of the most important tracts clinically will be highlighted here. These are the spinothalamic tract, the posterior columns and the corticospinal tract as shown previously in Figures 1.4 and 1.5a.

The gray matter column is responsible for the interface between the peripheral nerves and the peripheral nerve input or output at each segmental level, as well as connections with either ascending or descending tracts. The gray matter is functionally divided into dorsal and ventral parts. The dorsal portion is where the sensory input from the peripheral nerves interfaces with the spinal cord either to perform local activity such as reflex withdrawal or to transmit sensory information via either the spinothalamic tract for pain and temperature or the posterior columns for vibration and proprioception.

The peripheral nerve fibers serving pain and temperature enter through the dorsal rami to terminate and synapse with interneurons and second-order neurons, which cross in the anterior commissure (anterior to the central canal) and ascend on the opposite side of the body from the sensation to the lateral thalamus. The spinothalamic tract is somatotopically organized such that the fibers from the sacrum and the legs are located laterally to the fibers serving the trunk and the arms. Third-order neurons transmit the sensory information from the lateral thalamus to the somatosensory cortex in the parietal lobe (Figure 2.9b).

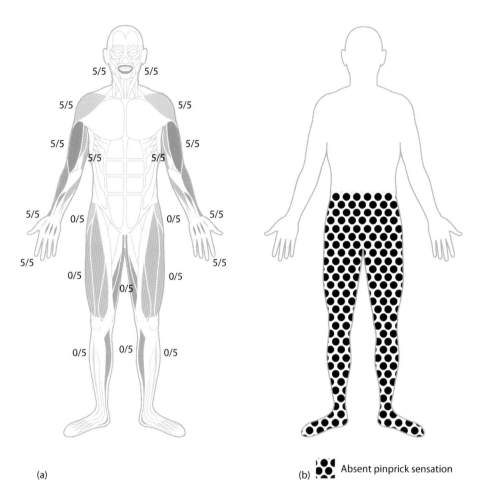

(a)

(b) Absent pinprick sensation

FIGURE 5.2: (a) Motor map and (b) sensory map for the chapter sample case.

The peripheral nerve fibers serving proprioception and vibration enter the spinal cord through the dorsal gray matter: these fibers do not synapse here but ascend to the lower medulla via a medially placed fasciculus for the legs, or via a laterally placed fasciculus for the arms, to terminate in the posterior column nuclei known as the nucleus gracilis for the leg fibers and the nucleus cuneatus for the arm fibers in the lower medulla. Second-order neurons synapsing from these nuclei then cross to the opposite side and ascend via the medial lemniscus to the thalamus. Third-order neurons originating in the thalamus then transmit this information to the somato-sensory area of the post-central gyrus of parietal lobe (Figure 2.9a).

The corticospinal tract is the major descending tract, which transmits information with respect to motor movement to the spinal cord. The corticospinal tract originates mainly in the frontal precentral gyrus with some contribution from the parietal cortex. These fibers descend in the internal capsule, through the cerebral peduncles in the brainstem and anterior white matter of the pons. In the lower medulla, the majority of these fibers cross in a structure

called the pyramidal decussation to descend in the lateral funiculus of the spinal cord as the lateral corticospinal tract, giving inputs to the lower motor anterior horn cells at each level. These lower motor neurons in turn send their axons to limb muscles to perform voluntary motor activity.

Corticoreticular tracts, which originate in the cortex, descend to reticular formation in the lower brainstem. These nuclei then give outputs which descend to the spinal cord in order to modulate muscle tone and reflexes. These outputs are called reticulospinal tracts.

Disruption of any of these tracts, plus the spinal gray matter, can therefore cause symptoms and signs of sensory and motor deficits below the level of the lesion, depending on which tracts are involved and at what level.

The blood supply of the human spinal cord, unlike the spinal cord of other animals, which tend to have segmental branches emanating from each spinal level, is relatively tenuous. The vertebral arteries join below the origin of the basilar artery to form the anterior spinal artery from above. The great radicular artery of Adamkiewicz joins the anterior spinal artery supplying the spinal cord at a variable level between T8 and L1 as a single large branch

Cervical spinal cord   Sacral spinal cord and cauda equina

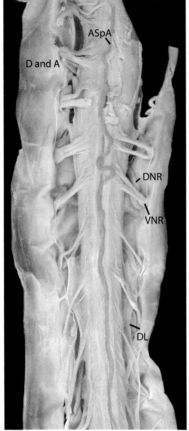
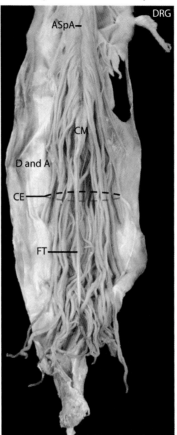

D and A – Dura and arachnoid (opened)   DRG – Dorsal root ganglion
ASpA – Anterior spinal artery   CM – Conus medullaris
DNR – Dorsal nerve roots   FT – Filum terminale
VNR – Ventral nerve root   CE – Cauda equina
DL – Denticulate ligament

FIGURE 5.3: Photographic views of the spinal cord: at the cervical level, including the dorsal and ventral roots; the anterior spinal artery is also seen at the lumbosacral level, including the conus medullaris; the spinal roots are found within the lumbar cistern, forming the cauda equina.

originating from the thoracic aorta. At any given level, the spinal cord is supplied by perforating branches of the relatively large anterior spinal artery whose territory is the anterior two-thirds of the cord, and by perforators of two smaller posterior spinal arteries, which have their origins at segmental levels and supply the posterior third of the spinal cord.

A word of caution about describing localization of spinal levels of dysfunction: often, the spinal cord level is used to describe the level of dysfunction based on clinical exam. However, as the spinal cord becomes more segmentally compact towards its termination at the L2 level, the spinal cord level and the vertebral level become increasingly discordant, the spinal cord level being several segments higher than the vertebral level. The usual neuroradiological practice is to describe the vertebral level and not the spinal cord level.

5.5 LOCALIZATION

Using the Extended Localization Matrix Worksheet (Table 5.1), the major findings to be entered are weakness, absent reflexes, sensory loss to pinprick and preserved vibration and proprioception in the lower limbs in a symmetrical fashion. Upper limb motor, reflex, sensory function, cranial nerves and mental status were all normal. The sensory deficit also involved the trunk below the level of the umbilicus. The plantar reflex was absent, which is of little localizing value.

Muscle disease and disease of the neuromuscular junction would be unlikely in this context as these two localizations do not cause sensory findings. A crush injury to the muscles in both legs could cause weakness, but there would be lots of pain.

An injury to the peripheral nerves might be another possibility. The problem with this is that Cletus' findings

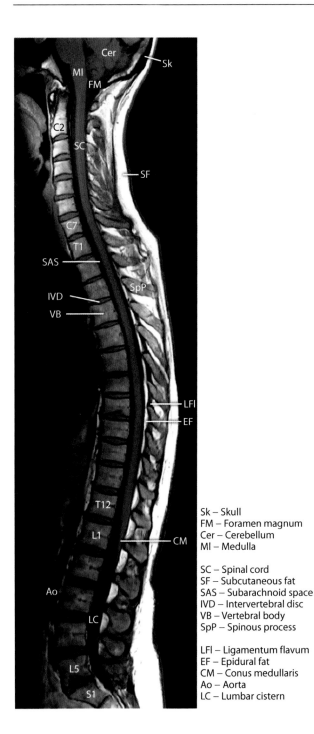

FIGURE 5.4: MRI of spinal cord showing the cord, the vertebra and soft tissue (fat). The C2 vertebra has a typical triangular configuration; other landmark vertebrae (T12, L1, and L5, S1) are also numbered.

Sk – Skull
FM – Foramen magnum
Cer – Cerebellum
Ml – Medulla

SC – Spinal cord
SF – Subcutaneous fat
SAS – Subarachnoid space
IVD – Intervertebral disc
VB – Vertebral body
SpP – Spinous process

LFl – Ligamentum flavum
EF – Epidural fat
CM – Conus medullaris
Ao – Aorta
LC – Lumbar cistern

are bilateral below the umbilicus. Injury to peripheral nerves would have to be widespread and symmetrical below a specific delineated level, a very unlikely possibility.

The presence of a motor and sensory level below the umbilicus would be consistent with a spinal cord injury about T10 (see Table 2.1), but what about the reflexes? Should not injury to the corticospinal tracts cause hyperreflexia?

In order to answer these questions, we have to refer back to the diagram of the spinal cord with the spinal nerves, gray matter and the three major tracts.

Figure 5.5 shows the spinothalamic tract (which carries pain and temperature), the posterior columns (which carry vibration and proprioception) and the corticospinal tracts, which carry motor information to the lower motor neuron of the spinal cord to control movement.

The clinical examination of Cletus indicated weakness of all muscles on both sides below the umbilicus. This finding is consistent with disruption of the corticospinal tracts at and below the T10 level; T10 motor nerves supply abdominal muscles around the level of the umbilicus. Lower limb reflexes are often absent in the situation of an acute spinal cord injury due the presence of a phenomenon called *spinal shock* in which the neurons of the spinal cord are diffusely dysfunctional or stunned for an average of 7–10 days after injury. Reflexes then usually become hyperactive. In the case of Cletus, there is a further complication. Interruption of the blood supply to the gray matter in the anterior portion of the spinal cord at the level of T10 as well as the area served by the anterior spinal artery above and below then leads to infarction of the gray matter, causing lower motor neuron weakness and therefore absent reflexes over the segments in which then gray matter has been infarcted.

The finding of loss of sensation to pinprick below the umbilicus needs to be localized. The sensory fibers for pain and temperature originate in the periphery and enter the dorsal aspect of the spinal cord in the substantial gelatinosa. There, the fibers synapse on to a second order neuron, which projects to the opposite side through the anterior funiculus of the spinal cord in front of the central canal to form the lateral spinothalamic tract on the opposite side. The lateral spinothalamic tract ascends to terminate in the ventroposterolateral nucleus of the thalamus. From there is a third order neuron, which connects the thalamus to the somatosensory cortex in the parietal lobe. To localize this finding, there would have to be widespread symmetric damage to all peripheral nerves below the umbilicus. More likely, the findings can be explained by transection of the spinothalamic tracts on both sides at the level of the lower thoracic spinal cord.

The last major finding is that the sensation for vibration and proprioception has been preserved. This suggests that the nerves and their spinal cord projections serving these modalities have been spared. This would suggest that the posterior columns of the spinal cord and their upward projections are intact.

Therefore, the localization for this combination of findings (including bilateral destruction of the corticospinal tracts and destruction of the spinothalamic tracts)

TABLE 5.1: Localization Worksheet Matrix

	Neuro Exam		MUS	NMJ	PN	SPC	BRST	WM/TH	OCC	PAR	TEMP	FR	BG	CBL
Mental status	N													
Cranial I, XII	N	N												
	Right	Left												
Tone	Abs below T10		⊘	⊘	⊘	○								
Power	Abs below T10		⊘	⊘	⊘	○								
Reflex	Abs below T10		⊘	⊘	⊘	○								
PP	Decr below T10		⊘	⊘	⊘	○								
Vib/prop	N	N	⊘	⊘	⊘	○								
Coord UL	N	N												
Coord LL	Abs	Abs	⊘	⊘	⊘	○								
Walk	U	U												
Toe/heel/invert/evert	U	U												
Tandem	U	U												
Romberg	U	U												

Abbreviations: Abs, absent; Decr, decreased; N, normal; U, untestable.

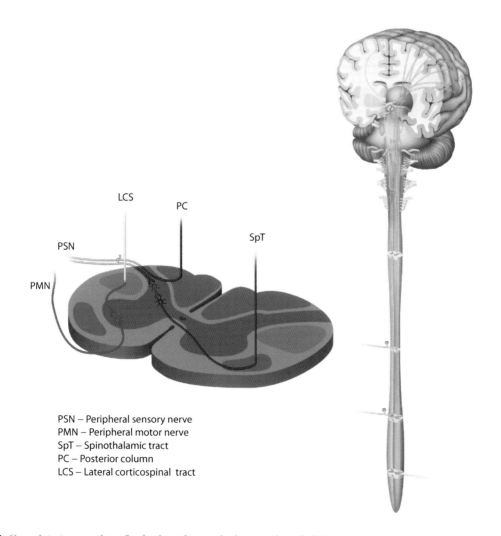

PSN – Peripheral sensory nerve
PMN – Peripheral motor nerve
SpT – Spinothalamic tract
PC – Posterior column
LCS – Lateral corticospinal tract

FIGURE 5.5: Complete transection of spinal cord: note the interruption of all the sensory pathways (on both sides) and the lateral corticospinal tracts bilaterally.

suggests that the damage has been confined to the anterior aspect of the spinal cord, with sparing of the posterior columns, the level of the cord lesion being at about T10.

A complete transection of the spinal cord at that level would result in total loss of function of the corticospinal and spinothalamic tracts and of the posterior columns on both sides. This is not the situation in the case of Cletus.

5.5.1 SPINAL CORD SYNDROMES

The major clinical findings in spinal cord injury relate to the relative degree of damage to the three tracts and the gray matter, as well as the level at which the damage occurs. Using the Expanded Localization Matrix, we see that, in any given patient with a spinal cord lesion, the probable major clinical findings are, variously, loss of sensation to pinprick and temperature; loss of sensation to vibration, proprioception and discriminative touch; weakness, reflex changes and changes in muscle tone. These findings need to be described in reference to a given dermatomal level as well as whether the findings are symmetric and replicate in both anterior and posterior dermatomes. This means that the sensory findings should correspond to equivalent levels when the patient is tested either supine or prone. Figure 14.4 provides a dermatomal map and standardized method for documenting spinal cord deficits.

Damage to the spinothalamic tract on one side will cause loss of sensation to pinprick and temperature below the lesion on the opposite side of the body. Damage to the posterior columns will cause loss of sensation for vibration, proprioception and discriminative touch on the same side below the lesion. Damage to the corticospinal tract will cause weakness, hyperreflexia and spasticity below the level of the lesion on the same side with the exception that hyperreflexia and spasticity may not be present for 7–10 days after an acute injury due to spinal shock.

Injury to just the gray matter at a given level will cause loss of all sensation and a lower motor neuron pattern of weakness with absent reflexes at the levels of damaged gray matter.

Several terms are used to describe spinal cord injury and level. *Paraplegia* refers to loss of function of the legs. Therefore, the lesion can be anywhere from a cord level of L2 to the cervical levels. Quadriplegia refers to the results of a spinal cord injury above C4, with paralysis of both arms and legs. *Pentaplegia* is a term sometimes used for high quadriplegics in which the roots of C2, 3, 4 are affected, resulting in paralysis of all four limbs and of the diaphragm. It is more precise to state the level of neurological injury, as these terms can be ambiguous. For instance, a complete C6 level cord injury results in lower limb paralysis and upper limb weakness below C6, in other words a partial quadriplegia.

Four major spinal cord syndromes are described to illustrate this anatomy in action.

5.5.1.1 COMPLETE TRANSECTION OF THE SPINAL CORD

This syndrome is the easiest to understand in that it consists of complete destruction of all spinal cord elements at a given level. Therefore, a complete spinal cord transection at T10 will cause paralysis, hyperreflexia, extensor plantar responses, spasticity, loss of control and sensation of bowel and bladder function, loss of sensation for pinprick, temperature, vibration and proprioception on both sides at and below that level of the spinal cord. All tracts are involved and all are damaged (Figure 5.5).

Depending on the cause, speed and intensity of the trauma that causes the transection, spinal shock may occur – typically with more acute trauma – during which time there is paralysis of the limbs below the level of the transection while spasticity and deep tendon reflexes can be decreased or absent. After recovery from spinal shock, hyperreflexia, increased tone and extensor plantar responses appear.

5.5.1.2 ANTERIOR SPINAL ARTERY SYNDROME

This is what Cletus has. While his aorta was being clamped, he had an interruption of the blood supply to his spinal cord in the anterior spinal artery territory due to occlusion of its major feeding artery in the lower thoracic region, the artery of Adamkiewicz. This caused an infarction to the anterior two-thirds of his cord, damaging the spinal cord gray matter, spinothalamic tracts and corticospinal tracts with sparing of the posterior columns. This explains why he could not move his legs or feel pinprick below the umbilicus but could feel the vibration sensation in his toes (Figure 5.6).

The extent of the spinal cord infarct can be over several segments such that the clinical level of deficit may not represent the actual entry level of the artery of Adamkiewicz.

5.5.1.3 BROWN–SEQUARD SYNDROME

This syndrome, named after a nineteenth-century French-American neurologist, occurs when there is damage to one side of the spinal cord with differential deficits resulting from the crossing of the spinothalamic tract in the spinal cord (Figure 5.7).

For example, if the right side of the spinal cord is destroyed, the corticospinal output to the right side will be damaged as well as the posterior columns on the same side. Therefore, the patient will lose motor power, and sensation for vibration, proprioception and discriminative touch on the right side below the level of damage. In contrast, the sensation for pinprick and temperature is lost on the left side as the spinothalamic tract serving the opposite side has already crossed the spinal cord below the level of the damage and therefore would be affected by a right-sided lesion.

The usual cause for this clinical situation is a tumor, usually metastatic, which grows from the lateral side of the vertebral column to compress the spinal cord and the structures within it.

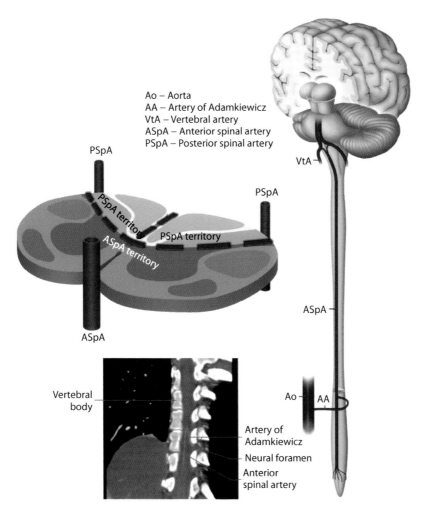

Ao – Aorta
AA – Artery of Adamkiewicz
VtA – Vertebral artery
ASpA – Anterior spinal artery
PSpA – Posterior spinal artery

FIGURE 5.6: Anterior spinal artery syndrome: note the interruption of both lateral spinothalamic tracts, sparing of the posterior columns and interruption of the lateral corticospinal tracts bilaterally.

5.5.1.4 SYRINGOMYELIA

This syndrome results from the expansion of the central canal of the spinal cord, a phenomenon known as a syrinx. The central canal is usually a small conduit for spinal fluid that extends down from the fourth ventricle in the medulla to provide cerebrospinal fluid (CSF) nutrition to the centre of the cord. Under circumstances of increased pressure in the spinal canal – such as from a blockage of upward CSF flow at the foramen magnum by a congenital descent of the cerebellar tonsils (a Chiari malformation) or by local trauma or tumors – the canal can undergo expansion and compress the adjacent structures in the spinal cord. The first structure that usually gets compressed is the anterior white matter commissure, which carries the crossing pain and temperature pathway from its second-order neuron cell body to connect to the spinothalamic tract in the anterolateral aspect of the spinal cord on the opposite side. The effect of compression of this commissure (carrying pain and temperature information from both sides, in the process of crossing each other) is to destroy the perception of pain and temperature on both sides on the body over the extent of the dilatation of the syrinx. The posterior columns and corticospinal tracts are usually not affected unless the syrinx becomes very large (Figure 5.8).

Usually syringomyelia occurs in the lower cervical and upper thoracic level; thus, the clinical effect is a suspended sensory level of decreased sensation to pain and temperature in a cape-like distribution or cuirass involving the upper trunk and the arms.

Syrinx formation at the level of the lower brainstem is known as syringobulbia. This dilated fluid cavity in the brainstem will interfere with cranial nerve, motor and sensory function depending on the nuclei or tracts which are compressed.

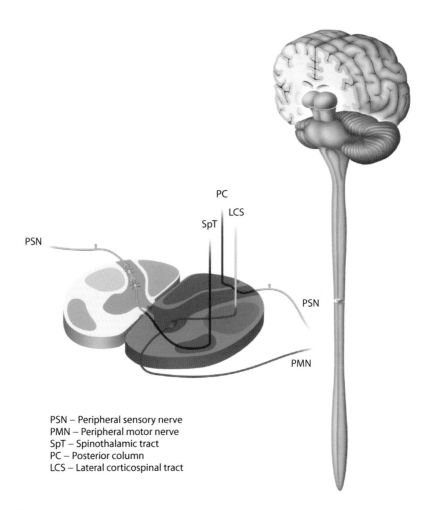

PSN – Peripheral sensory nerve
PMN – Peripheral motor nerve
SpT – Spinothalamic tract
PC – Posterior column
LCS – Lateral corticospinal tract

FIGURE 5.7: Brown–Sequard syndrome: a hemisection of the spinal cord: note the tracts interrupted and which side of the body would be affected.

5.6 ETIOLOGY – CLETUS' DISEASE PROCESS

The History Worksheet (Figure 5.1), Expanded Localization Matrix (Figure 5.5) and Etiology Matrix for this case tell us that this is an acute injury to the spinal cord following severe trauma as a result of the motor vehicle accident.

5.6.A PAROXYSMAL DISORDERS

There are several paroxysmal disorders of the spinal cord that cause abnormal movements secondary to uncontrolled discharges from anterior horn cells. A typical example is neuromyotonia, in which there is uncontrolled stiffening of the muscles of the arms and legs, resulting in so-called Stiff Person Syndrome. This is believed to be caused by an autoimmune attack on the spinal cord. These disorders tend to be acute in onset but are recurrent and eventually become chronic. There are several subtypes of these disorders characterized by specific antibodies against specific synaptic region proteins such as GAD (anti-glutamic acid decarboxylase); GlycRalpha1, a glycine receptor component; and Ampiphysin, a synaptic vesicle protein.

Spinal myoclonus consists of brief shock-like jerks usually in just a few muscle groups, typically due to segmental damage to the spinal cord.

The history and neurological findings in Cletus' case are in keeping with acute loss of function of two tracts of the spinal cord after an incident of trauma and surgery.

5.6.B TRAUMATIC SPINAL CORD INJURY

The commonest cause of acute complete spinal cord injury is trauma.

When the spinal cord is disrupted to the point that the various gray matter and white matter tracts are crushed or lose their anatomical proximity to their blood supply, complete destruction of the spinal cord occurs at that level. This leads to a bilateral loss of motor and sensory function below the level of the lesion as well as loss of bowel and bladder control. Spontaneous recovery is unlikely and the disability is profound and permanent.

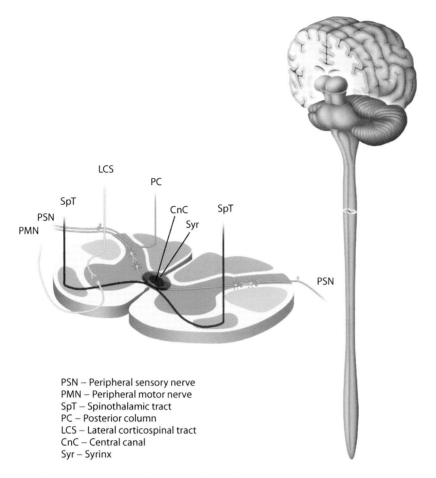

PSN – Peripheral sensory nerve
PMN – Peripheral motor nerve
SpT – Spinothalamic tract
PC – Posterior column
LCS – Lateral corticospinal tract
CnC – Central canal
Syr – Syrinx

FIGURE 5.8: Syringomyelia, an enlargement of the central canal: note the interruption of the crossing fibers carrying pain and temperature, involving both sides.

In the case of Cletus, there is preservation of function of one spinal cord tract bilaterally, the posterior columns. This could indicate a possible incomplete spinal cord transection due to trauma (Figure 5.5).

A more common and chronic form of spinal cord trauma is caused by degenerative changes, usually from osteoarthritis in the vertebral bodies and intervertebral disks. These degenerative processes lead to prolapse of intervertebral disks into the spinal canal and remodelling of the bone around joints; in turn, the bony protrusion(s) may impinge on the vertebral foramina (and thus one or more spinal nerves) and also the spinal canal, leading to compression of the spinal cord itself. This condition is known as *spinal stenosis*, which can occur at any level of the spinal axis. If the spinal stenosis is in the cervical or thoracic spine, it can lead to spinal cord compression or myelopathy with lower limb weakness and hyperreflexia. If this occurs in the lumbar spine below the termination of the spinal cord, it can cause compression of the nerves forming the cauda equina, leading to pain, sensory loss, decreased reflexes and weakness of the legs as well as difficulty with control of bladder and bowel continence.

5.6.C VASCULAR INJURY

In the localization exercise, we have already noted that Cletus' deficits fit best with an anterior spinal artery syndrome at or below the level of entry of the artery of Adamkiewicz. Occlusion of the artery of Adamkiewicz leads to infarction of the anterior two-thirds of spinal cord at several levels above and below the cord level of the artery. The infarction pattern will include the spinothalamic tracts, the gray matter and the lateral funiculi containing the corticospinal tracts.

If Cletus had been hypotensive and/or hypoxic at the scene of the accident, then there may be reason to suspect that his injury might have been due to hypoperfusion of the cord. Under such circumstances, there can be infarction in the watershed (or arterial border zone) region of the spinal cord. Arterial border zone areas are located where the regional blood supply comes from different arteries and the ischemic injury is in the border zone between those two arteries. In the spinal cord, such a border zone is usually located in the lower cervical/upper thoracic level at C7 – T1, where the border zone from the vertebral supply of the anterior spinal artery meets the supply from the anterior spinal artery segment which originated from the great radicular artery of Adamkiewicz.

Thus, with hypotension-induced border zone infarcts, the damage is usually confined to the low cervical or high thoracic cord above the entry point of the artery of Adamkiewicz. If the cord had been damaged due to Cletus' hypotensive state, however, he would not have been moving all limbs to pain at the scene of the accident.

Referring back to the conversation at the bedside between the physician and Cletus' mother gives us information that he had to have emergency vascular surgery to repair a partially ruptured thoracic aorta. This was caused by the deceleration trauma when the car hit the rock face and Cletus (unrestrained in the back seat) was subject to significant G forces, which pulled his thoracic aorta from its chest wall attachment. His severe systemic hypotension resulted from extensive bleeding from the torn, leaking aorta into the adjacent chest cavity.

During the emergency repair of Cletus' thoracic aorta, the surgeons were obliged to temporarily clamp his thoracic aorta above the artery of Adamkiewicz in order to put a patch on the ruptured aorta.

The effect of clamping the aorta above the artery of Adamkiewicz caused interruption of blood supply to the spinal cord at the level of entry of this artery. The blood supply to the anterior two-thirds of the spinal cord was interrupted, resulting in infarction of this region.

Other vascular causes could include direct trauma causing rupture or occlusion of the artery of Adamkiewicz.

Thus, in terms of localization and etiological diagnosis, Cletus most probably suffered an infarction to the anterior two-thirds of his spinal cord at the T10 level, involving several segments, resulting in loss of function of the corticospinal tracts and spinothalamic tracts on both sides.

The prognosis for this type of injury can be guardedly optimistic, in that part of the functional loss may be secondary to reversible ischemic damage rather than complete infarction. In addition, white matter injuries tend to recover better than gray matter injuries through the process of remyelination. The recovery, if it does occur, can be very slow – up 24 months. The outcome would have been significantly worse had he suffered a fracture dislocation of the spine at the same level, with complete cord transection.

Other vascular abnormalities of the spinal cord can exist such as arteriovenous malformations or cavernomas. These would, in general, have a more regional anatomical distribution based on the supplying artery and draining veins.

5.6.D DRUGS AND TOXINS

Various chronic toxic exposures can affect the spinal cord; examples include the heavy metals mercury, lead and arsenic. These disorders usually have a chronic course, with a combined picture of concurrent progressive cognitive and motor difficulties. A history of exposure is usually related to geographic factors of environmental pollution such as mercury poisoning in the Minamata region of Japan and in the aboriginal peoples of the Wabigoon-English River area in Northern Ontario, Canada.

5.6.E SPINAL CORD INFECTION

HIV infection can cause a vitamin B12-like myelopathy despite normal levels of vitamin B12. This is thought to be due to abnormal metabolism of vitamin B12 and its metabolites within the spinal cord.

Other infections of the spinal cord include agents such as West Nile virus, polio and Lyme disease.

West Nile is a viral infection transmitted from birds to humans by mosquitoes, causing a combined viral encephalitis and myelitis. Both gray matter and white matter elements of the spinal cord can be affected.

Poliomyelitis is a viral infection born through contaminated water, which causes necrosis of the anterior horn cells of the spinal cord, giving mainly asymmetrical segmental loss of motor function without sensory findings (see also Chapter 4). It has essentially been eliminated through mass population immunization programes.

Lyme disease is caused by a bacterial spirochete transmitted to humans from deer by a specific tick. This disorder has three distinct phases: an initial cutaneous expanding rash at the site of the bite, followed by a delayed phase of polyarthritis and then by involvement of the central nervous system (CNS).

Infectious myelitis due to herpes zoster can give a diffuse myelitis; this is often associated with disorders that interfere with cell-mediated immunity.

There have been outbreaks of flaccid paralysis associated with recent enterovirus D68 outbreaks mainly in children with lesions in the brainstem and spinal cord gray matter, similar to the neuroimaging seen in other forms of viral myelitis such as enterovirus 71 and poliomyelitis.

5.6.F METABOLIC INJURY

Metabolic disturbances, which may affect the spinal cord, include loss of the basic nutrients (oxygen and glucose); these disorders usually cause dysfunction in the brain long before the spinal cord.

Low levels of vitamin B12 cause degeneration of the posterior columns, corticospinal tracts and the large fiber peripheral nerves to cause both a neuropathy and myelopathy. This condition is called subacute combined degeneration; the term *combined* refers to the combined degeneration of both the posterior columns and the corticospinal tracts. Replacement of vitamin B12 and folate often results in prompt improvement of signs and symptoms. The clinical manifestations are progressive weakness and loss of sensation usually beginning in the lower limbs, with findings on exam of loss of sensation to vibration and proprioception, weakness and spasticity. Hyperreflexia is variable depending on the relative contribution of damage from the neuropathy versus the myelopathy.

5.6.G SPINAL CORD INFLAMMATION/ AUTOIMMUNE

Subacute causes include inflammatory conditions of the spinal cord such as multiple sclerosis, neuromyelitis optica (NMO) and transverse myelitis.

Multiple sclerosis causes spinal cord damage through recurrent attacks, resulting in demyelinating plaques in the cord white matter. Disability depends on the size, location and cumulative number of plaques.

NMO preferentially affects the optic nerves and spinal cord and has been increasingly recognized as a cause of autoimmune demyelination since the discovery of specific antibodies (anti-AQP4 IgG) against the protein aquaporin 4, located on the cell membranes of astrocytes. These proteins act as channels for the transport of water across the cell membranes.

Transverse myelitis is a disorder of subacute demyelination following primary infections – usually in the gastrointestinal tract or respiratory system – by various viruses or by mycoplasma. The mechanism is similar to the peripheral nerve demyelination associated with Guillain–Barré syndrome (see Chapter 4); a similar type of post-infectious demyelination can also affect the spinal cord. This is another example of molecular mimicry in that the immune system is primed to attack an outside invader such as a virus that happens to have an antigenic similarity with a tissue in the nervous system, in this instance the myelin of the spinal cord axon tracts.

Autoimmune or systemic vasculitis can affect the spinal cord. These disorders usually have significant systemic signs and symptoms and also affect the brain at the same time.

5.6.H NEOPLASTIC DISORDERS

Tumors of the spinal cord may be intrinsic to the cord (such as astrocytomas) or extra-axial (meningiomas and metastatic tumors). These usually present with subacute or chronic time courses, with weakness or sensory loss depending on which tracts or gray matter structures are damaged first.

5.6.H.1 Paraneoplastic

Paraneoplastic degeneration of the spinal cord can occur in association with a remote carcinoma usually of the lung, resulting in an immune-mediated necrotizing spinal cord degeneration.

5.6.I DEGENERATIVE DISORDERS

Chronic conditions affecting the spinal cord are usually the result of damage from pre-existing acute and subacute conditions. Chronic degenerative conditions are few in number, the most important example being amyotrophic lateral sclerosis (ALS). Degenerative diseases often overlap as late-onset genetic disorders.

ALS is a progressive degenerative disorder affecting both the upper and lower motor neurons in combination and at varying intensities. It is important to understand the term ALS. Amyotrophy (A) refers to muscle weakness not due to muscle disease. This represents the lower motor neuron component of the damage caused by ALS. Fasciculations are involuntary muscle movements that resemble a 'bag of worms' in multiple muscle groups. The presence of fasciculations is a cardinal sign that muscles have undergone denervation and reinervation secondary to progressive damage to lower motor neurons in the spinal cord or brainstem. The observation of fasciculations in the tongue is very significant in that it shows lower motor neuron damage located above the level of the cervical spine. The lateral sclerosis (LS) part of the name refers to the lateral columns or lateral funiculi in which the corticospinal tracts run. Therefore, the term *ALS* refers to both upper and lower motor neuron damage all in one acronym. The finding of an exaggerated jaw jerk is often present in ALS indicating bilateral upper motor neuron damage to corticobulbar tract affecting control of motor neurons in the trigeminal motor nucleus of the pons.

The cause of sporadic ALS is unknown, although there have been some cases due to genetic deficiencies of enzymes such as superoxide dismutase, and other mutations include TARDBP (TDP-43), FUS/TLS, VCP, Ubiquilin-2 and C9ORF72 gene. There are also cases resulting from a deficiency in an androgen receptor leading to both a predominantly bulbar pattern of ALS and to endocrine abnormalities (testicular atrophy, impotence, erectile dysfunction – Kennedy's disease). So far, there is no specific effective treatment for ALS, and the mean time from diagnosis to death is 48 months. Most treatment offers merely physical and psychological support. Important discussions concerning tracheostomy, artificial ventilation and end-of-life issues need to be addressed promptly after diagnosis.

5.6.J GENETIC DISORDERS

There are many genetic disorders that affect the spinal cord; a good example is spinal muscular atrophy, which is a specific genetic neurodegenerative disorder affecting the anterior horn cells. Other examples are a variety of genetic variations of ALS, including those noted in the previous section – among others – as well as the various types of spinocerebellar degeneration (SCA). There are now over 20 different types of SCA with varying involvements of the spinal cord.

5.7 CASE SUMMARY

In summary, Cletus has sustained severe trauma to his chest resulting from a rapid deceleration injury. This led to a partial avulsion and rupture of his thoracic aorta. He was hypoxic and hypotensive at the scene of the accident but noted to be moving all four limbs.

He was transported to the university hospital, where he was found to have absent pulses in the lower limbs, a widened thoracic aorta on chest x-ray and extravasation of contrast from the thoracic aorta on a contrast enhanced computed tomography scan of the chest.

He underwent an emergency procedure during which the surgeon had to clamp his lower thoracic aorta above the artery of Adamkiewicz in order to place a graft at the rupture site of the aorta.

Following recovery from anesthesia, he was found to have a spinal cord injury involving the corticospinal and spinothalamic tracts below T10, with sparing of the posterior columns.

This picture is consistent with an anterior spinal artery syndrome in which the blood supply to the anterior two thirds of the spinal cord was interrupted, resulting in infarction of the spinal gray matter, spinothalamic tracts and corticospinal tracts at that level.

After examination by the neurologist, Cletus underwent an MRI of his spine.

The MRI illustrated in Figure 5.9 shows that the anterior aspect of the spinal cord has been infarcted as evidenced by the increased signal on the sagittal and axial T2 images.

The MRI does not show any significant disruption of the vertebral column with subsequent compression of the spinal cord. The injury to the spinal cord therefore has to be due to an indirect cause.

The MRI therefore confirmed the localization and etiological diagnosis of the neurologist.

5.7.1 CASE EVOLUTION

Cletus recovered quickly from the effects of the chest surgery and was anxious to get back home to the bush; he was impatient with the medical, nursing and allied health staff during his stay in the acute care hospital and in the rehabilitation facility.

He and his family had difficulty understanding why he could not merely have surgery to his back and be permanently 'fixed'.

Cletus underwent spinal rehabilitation that focused on re-establishing his mobility with artificial devices such as wheelchairs, braces and crutches. He also underwent training to improve his bowel and bladder function. He received a vocational assessment and had significant psychosocial support during his recovery (see also Chapter 14).

The health care team worked hard to engage Cletus, his family and local social services to help him get back to his home and community and start a new life.

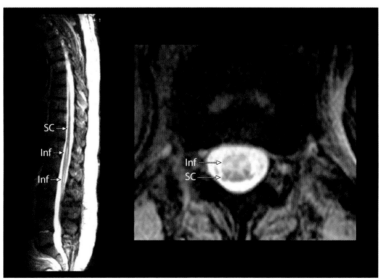

SC – Spinal cord (normal)   Inf – Infarct

FIGURE 5.9: MRI of spinal cord infarction.

5.8 E-CASES

5.8.1 CASE 5E-1: JEAN-PIERRE, AGE 35

Background: Jean-Pierre's job involves restocking shelves in the store, including moving heavy items to the top shelves with the help of mechanical lifting devices. He recently travelled with his wife to the Caribbean region, where he developed diarrhea for one to two days. Otherwise, his health has been good; he is a non-smoker and takes alcohol on occasion.

Chief complaint: Back pain; numbness and weakness of the legs; both present for three to four days

History:

- Pain is constant and is located in the lumbar region; does not radiate into the legs.

- *Both legs feel numb and heavy; Jean-Pierre can walk with difficulty and unsteadily; in going upstairs, he has to pull himself upwards using his arms.*
- *Has also noted some difficulty controlling his urge to urinate as well as with initiating erections.*

Examination:

- Normal mental status, cranial nerve examination and motor examination of the upper limbs
- *Grade 3/5 weakness of both legs, proximally and distally* (Figure 5.10a)
- *Reflexes:* Figure 5.10b
- *Decreased sensation (<50% of that over the neck/arms) in the lower abdomen and in the legs from the inguinal region down*

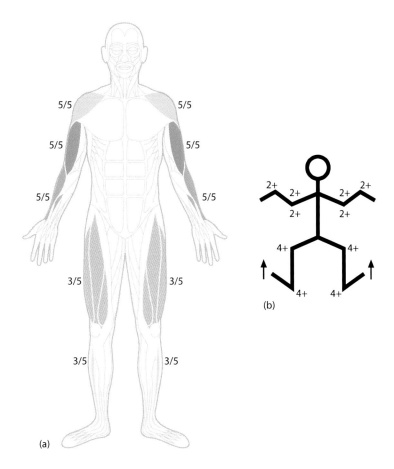

FIGURE 5.10: (a) Motor map and (b) reflex map for Jean-Pierre.

5.8.2 CASE 5E-2: ANNE, AGE 18

Background: This young woman was involved in a head-on collision; she was in the passenger seat and was wearing her seat belt, but unfortunately, she is now a paraplegic. Prior to the accident, she was in good health, a non-smoker and a weekend party drinker. She has been told that there is no chance for any functional recovery but refuses to accept this prognosis as she feels she has some sensation in her legs.

Chief complaint: Complete paralysis of the legs following an automobile accident six months ago

History:

- Head-on collision with a drunk driver; fracture *dislocation of her spine at T10*

- *In addition to leg paralysis, Anne has loss of most sensation in her legs and in the trunk below the umbilicus*

Examination:

- Depressed and teary
- Mental status, cranial nerve and upper extremity examinations are normal
- *Complete lower extremity paralysis (0/5) in all muscle groups; marked hyperreflexia* (Figure 5.11a)
- *Absent pinprick and temperature sensations below the umbilicus bilaterally*
- *Some retention of vibration sensation (128 Hz) in both knee-caps; absent in the ankles and feet* (Figure 5.11b)

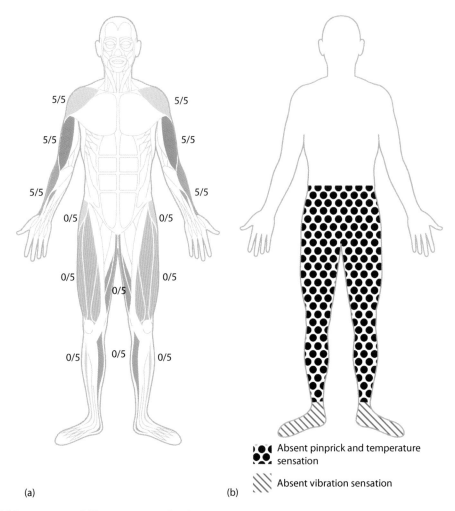

(a)   (b)

Absent pinprick and temperature sensation

Absent vibration sensation

FIGURE 5.11: (a) Motor map and (b) sensory map for Anne.

5.8.3 CASE 5E-3: INDIRA, AGE 68

Background: Indira has been living in North America for the last 20 years and has not travelled recently. She lives with her husband and they have three grown children. She does not use tobacco nor drink alcohol.

Chief complaints: Headache and fever for four weeks; back pain and leg weakness for 10 days

History:

- *Low-grade fever (38°C) and persistent headache* beginning about four weeks ago
- 10 days ago, Indira developed increasingly severe pain in the lower cervical and upper thoracic regions, accompanied by insidiously progressive weakness and stiffness of both legs
- At first, the weakness was evident only when walking upstairs, but today, her legs are so weak that she is unable to walk without assistance

- No history of recent travel; never diagnosed with tuberculosis but had received the Bacillus Calmette–Guérin vaccine many years ago in her native country
- 5 kg weight loss in the past month

Examination:

- Underweight (55 kg); normal chest and abdominal examinations; *appears chronically ill; temperature 38°C*
- Normal mental status, cranial nerve and upper limb motor examinations
- *Grade 3/5 weakness in the lower extremities, proximally and distally, worse on the left side* (Figure 5.12a)
- Generalized hyperreflexia in the legs; bilateral ankle clonus
- *Decreased pinprick, temperature and vibration sensation below the nipple line bilaterally* (Figure 5.12b)

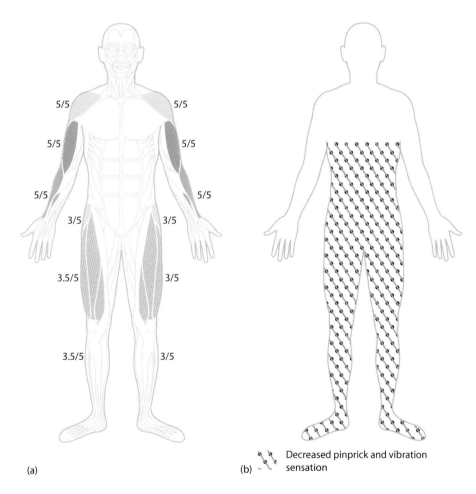

(a)

(b)

Decreased pinprick and vibration sensation

FIGURE 5.12: (a) Motor map and (b) sensory map for Indira.

5.8.4 CASE 5E-4: JOHN, AGE 80

Background: John, a former construction worker now aged 80, used to be physically active and has recently had to give up playing tennis and golf because of his physical challenges. Otherwise, he has been quite well; he quit smoking 15 years ago and enjoys a few beers with his companions at the Legion on the weekends.

Chief complaints: 'Tiredness' of the legs and difficulty voiding over the last six months

History:

- *Insidious onset of heaviness and weakness in his legs over a six-month period*
- *Increasing problems initiating voiding; stream of urine is weak*

- Dull lower back pain, worse with physical activity or if he sits still for long periods, e.g. driving his car on trips out of town
- No calf cramps when walking briskly; no chest pain, shortness of breath or palpitations

Examination:

- Normal mental status, cranial nerve examination and examination of upper limbs
- *Grade 4/5 weakness of both legs, proximally and distally* (Figure 5.13a)
- *Reflexes*: Figure 5.13b
- *Decreased pinprick and vibration sensation (256 Hz) at the ankles bilaterally*
- Rectal examination showed slightly enlarged prostate; abdominal exam revealed suprapubic fullness

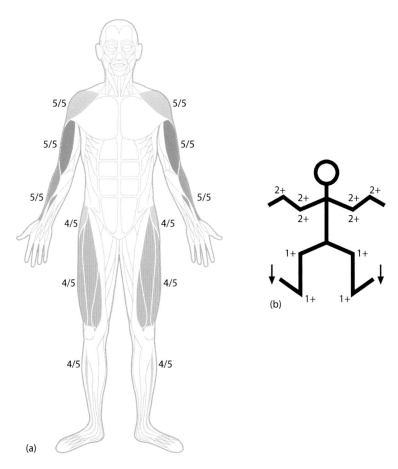

FIGURE 5.13: (a) Motor map and (b) reflex map for John.

5.9 SUMMARY OF KEY ANATOMICAL INFORMATION

- The spinal cord, an elongated portion of the CNS within the vertebral column, links the CNS with both the sensory (afferent) and motor (efferent) functions of the peripheral nervous system.
- Each spinal cord segment subserves a distinct portion of the body for the sensory systems (called a dermatome) and a group of muscles on the motor side (called a myotome).
- Reflex responses are functionally localized within the spinal cord, including that to pain (multisynaptic) and the response of a muscle to the elongation of its tendon – the myotatic (monosynaptic) reflex.
- The gray matter, located within the spinal cord, is functionally organized with the sensory aspects dorsally and motor aspects ventrally, surrounded by ascending and descending tracts (pathways).
- The major *sensory* tracts are the anterolateral (crossed, conveying pain and temperature) and the posterior (ipsilateral, conveying proprioception, vibration and discriminative touch) columns.
- The major *motor* tract is the lateral corticospinal tract carrying the voluntary motor commands directly from the cerebral cortex (upper motor neuron) to the motor neurons of the spinal cord (lower motor neuron).
- The autonomic nervous system innervation to smooth muscle and glands is provided by the sympathetic outflow from the lateral horn (T1–T12) and the parasympathetic outflow from the sacral segments (to bladder and bowel).
- Blood supply to the spinal cord (critical for all nervous tissue) is mainly via the anterior spinal artery (from the vertebral) to the anterior 2/3, and the posterior spinal arteries to the posterior 1/3, supplying both the gray matter and the tracts in each region.
- The anterior spinal artery is supplemented by a small branch of the aorta for its lower portions, known as the great radicular artery of Adamkiewicz.
- Knowledge of the location of the major tracts within the spinal cord and their blood supply is essential for clinical diagnosis of the localization of the level of the injury or disease.

5.10 CHAPTER-RELATED QUESTIONS

1. Describe the pathways of the two major ascending sensory tracts from their origin in the spinal cord and their termination in the cortex.
2. What is sacral sparing and how is it explained anatomically in the spinal cord?
3. What are the main clinical features that distinguish a spinal cord lesion from a peripheral nerve lesion?
4. List the blood supply to the spinal cord and the clinical consequences of vascular occlusion for each of the supplying arteries.
5. A 16-year-old girl is the passenger in a motor vehicle crash, which causes a fracture dislocation of the thoracic spine and a complete transection of the spinal cord at the level of T10. List her clinical deficits.
6. What is a cauda equina syndrome and what are the clinical features and risk factors?

SUGGESTED READING

Bitar, R., Leung, G., Perng, R., Tadros, S., Moody, A.R., Sarrazin, J., McGregor, C., Christakis, M., Symons, S., Nelson, A., Roberts, T.P. MR pulse sequences: What every radiologist wants to know but is afraid to ask. *Radiographics* 2006. 26(2): 513–537.

Bottiglieri, T. Folate, vitamin B12 and neuropsychiatric disorders. *Nutr Rev* 1996, 54: 382–390. Review of B12, folate and metabolic effects of deficiencies on the CNS.

Cheshire, W.P., Santos, C.C., Massey, E.W., Howard Jr., J.F. Spinal cord infarction: Etiology and outcome. *Neurology* 1996. 47: 321–330, 1996. Comprehensive collection of cases and review of spinal cord infarction.

Dharmarajan, T.S., Lakshmi Narayanan, S., Poduval R.D. Life threatening vitamin B12 deficiency: Will timely screening make a difference? *World J Gastroenterol* 2000. 6(3): 456–457.

Fraser, S. Cauda equina syndrome: A literature review of its definition and clinical presentation. *Arch Phys Med Rehabil* 2009. 90(11): 1964–1968.

Frohman, E.M., Wingerchuk, D.M. Clinical practice: Transverse myelitis. *N Engl J Med* 2010. 363(6): 564–572.

Kalsi-Ryan, S., Karadimas, S.K., Fehlings, M.G. Cervical spondylotic myelopathy: The clinical phenomenon and the current pathobiology of an increasingly prevalent and devastating disorder. *Neuroscientist* 2013. 19(4): 409–421.

Kasdan, R.B., Howard, J.L. Neuroimaging of spinal diseases: A pictorial review. *Semin Neurol* 2008. 28(4): 570–589.

Kirshblum, S.C., Burns, S.P., Biering-Sorensen, F., Donovan, W., Graves, D.E., Jha, A., Johansen, M., Jones, L., Krassioukov, A., Mulcahey, M.J., Schmidt-Read, M., Waring, W; American Spinal Injury Association (ASIA) International Standards Committee. International standards for neurological classification of spinal cord injury: The new worksheet. Available from: http://asia-spinalinjury.org/wp-content/uploads/2016/02/International_Stds_Diagram_Worksheet.pdf. Updated July 23, 2013.

Kumar, N. Neurologic presentations of nutritional deficiencies. *Neurol Clin* 2010. 28(1): 107–170.

National Spinal Cord Injury Statistical Center. *Spinal Cord Injury Facts and Figures at a Glance*. Birmingham, AL: University of Alabama at Birmingham, 2013.

Pagon, R.A., Adam, M.P., Ardinger, H.H., Wallace, S.E., Amemiya, A., Bean, L.J.H., Bird, T.D., Ledbetter, N., Mefford, H.C., Smith, R.J.H., Stephens, K., Eds. *GeneReviews®* [Internet]. Seattle, WA: University of Washington, Seattle, 2015. Hereditary Ataxia Overview.

Rosenblum, M., Goudreau, G., Carpenter, S., Beaulieu, N., Jolicoeur, P. Dissociation of AIDS-related vacuolar myelopathy and productive HIV-1 infection of the spinal cord. *Neurology* 1989. 39: 892–896. Study showing that HIV associated disease of the spinal cord is due to a metabolic disorder rather than primary infection of the spinal cord.

Schoenfeld, A.J., Bader, J.O. Cauda equina syndrome: An analysis of incidence rates and risk factors among a closed North American military population. *Clin Neurol Neurosurg* 2012. 114(7): 947–950.

Schwendimann R.N. Metabolic, nutritional, and toxic myelopathies. *Neurol Clin* 2013. 31 (1): 207–218.

Scott, T.F., Frohman, E.M., De Seze, J. et al. Evidence-based guideline: Clinical evaluation and treatment of transverse myelitis: Report of the Therapeutics and Technology Assessment Subcommittee of the American Academy of Neurology. *Neurology* 2011. 77(24): 2128–2134.

Transverse Myelitis Consortium Working Group. Proposed diagnostic criteria and nosology of acute transverse myelitis. *Neurology* 2002. 59: 499–505.

Wingerchuk, D.M., Banwell, B., Bennett, J.L., Cabre, P., Carroll, W., Chitnis, T., de seze, J., Fujihara, K., Greenberg, B., Jacob, A., Jarius, S., Lana-Peixoto, M., Levy, M., Simon, J.H., Tenembaum, S., Traboulsee, A.L., Waters, P., Wellik, K.E., Weinshenker, B.G.; International Panel for NMO Diagnosis. International consensus diagnostic criteria for neuromyelitis optica spectrum disorders. *Neurology* 85(2): 177–189.

Chapter 6

Ernesto

Objectives

- To demonstrate how disturbances of hearing (audition) and balance (vestibular function) can affect important aspects of everyday life
- To further the understanding of the brainstem and the cranial nerves and disease that may occur in that region
- To underscore the point that many people do not have immediate access to the latest technological advances for their medical care

6.1 ERNESTO

'Uncle' Ernesto is the 'oddball' of the family. At times almost non-communicative and at other times very friendly to all, he is treated by family members with respect. They have learned, however, never to call him 'Ernie' to his face! Although intellectual and 'philosophical' issues are referred to him and often resolved, he is considered 'a man of the world'. He is also the family guru for any technological questions and can usually fix problems with dysfunctional computers and has all the latest apps on his smart phone.

Ernesto, who recently turned fifty, has suddenly developed a love of pop music and, having purchased a commercial listening device, now spends many hours a day with his ears plugged in to the latest so-called hits.

Recently, Ernie has mentioned to anyone in earshot that he has a 'buzz' in his left ear that comes and goes; this type of noise, called tinnitus, is sometimes a ringing sound and at other times a 'hissing sound'. He finds this noise very distressing, seriously interfering with his ability to concentrate and occasionally interfering with his ability to fall asleep. No one, however, except for his six-year-old niece Julia, has had the audacity to suggest that he should perhaps turn down the volume of his music; this advice is, of course, deftly ignored.

After a few months, everyone, even Julia, begins to notice that Uncle Ernie is a little hard of hearing. He starts missing jokes (regardless of whether he laughs or not) and he consistently changes the telephone headset from his preferred left ear to his right ear. A couple of times recently, he has noticed that he is a little unsteady when walking to the washroom in his house (he lives alone). Now his friends are telling him to get a hearing aid. Others are telling him to get the wax out of his ears.

A little scared and very reluctantly, he decides to talk to 'the doc' and finally calls to make an appointment scheduled for a month later. By the time he arrives in the physician's office, he is a changed man, quite depressed and very worried.

The family physician notes the history, including the absence of any complaint of headache. A thorough physical examination reveals a 50-year-old male who is somewhat overweight (178 lb), with normal blood pressure (135/85), a normal chest and abdominal exam and no abnormality detected on a seven-minute neurological exam except for decreased hearing (using a tuning fork) in the left ear; normal eardrums are clearly visualized on both sides. The fundi are easily visualized and are normal.

The physician arranges for a hearing test at a local hearing clinic that fits hearing aids; Ernesto returns a few weeks later to be informed that there is audiologic evidence of a marked hearing loss in his left ear across all parts of the sound frequency range. This time, the physician arranges an urgent consultation with a neurologist from a large nearby city; she visits the local hospital one day a month.

Neurological examination: About three weeks afterwards, Ernesto arrives (on foot) at the local hospital for his appointment with the neurologist. He presents as a well-groomed middle-aged man who is very tense but fully cooperative.

Cranial nerve examination: CN I is not tested. Eye movements in all directions are intact and no nystagmus is seen on lateral gaze. The pupils are equal in size and react to light symmetrically. Visual fields on confrontation are intact. The corneal reflex is tested and found to be intact on both sides; when asked, he says that he can feel the cotton wisp touching his cornea on both sides. Sensation of the face to touch and pinprick is intact in all three divisions of the trigeminal nerve and equal on both sides; the jaw reflex is normal. Although the facial appearance looks symmetrical (Ernesto is a bearded male), testing reveals a reduction in the degree of wrinkling of the left forehead in comparison to the right, slight weakness on forced closure of the left eye and weakness of the left lower face when the patient is asked to smile or show his teeth. The gag reflex is difficult to elicit but seems intact; voice production also seems unaffected. Shoulder elevation is strong and symmetrical, and the tongue is normal in appearance with equal movements bilaterally.

Examination with the otoscope shows that the external ear is clear of wax and both eardrums are easily visualized; both have a normal appearance, with a cone of light reflecting in the lower quadrant of the drum. The hearing deficit of the left ear is confirmed using a vibrating tuning fork (at 256 Hz); even Ernesto notes that the sound of the tuning fork is heard longer and better on the right side. The neurologist's findings are consistent with nerve deafness due to loss of hair cells in the cochlea or a lesion affecting the VIIIth nerve, known as a sensorineural deficit; this conclusion cannot be considered reliable without a confirmatory audiogram performed by a certified audiologist.

On that particular day, the neurologist is working with a medical student who asks about the Weber and Rinne tests. Sound is transmitted most effectively through the tympanic membrane and middle ear structures to the cochlea, whose hair cells are activated. Any defect in the transmission of sound to the cochlea will cause a conductive hearing loss, whereas any defect of the cochlea or the auditory nerve will cause a sensori-neural hearing loss. One initially determines and compares the patient's hearing level between the two ears with a vibrating tuning fork by gradually moving the tuning fork closer to the patient's ear and asking the patient to respond when the sound is first heard. The distance of the response is noted and should normally be identical on both sides. A significant reduction in the distance from the ear of the response suggests the presence of decreased hearing on that side. The Weber test is next performed by holding the vibrating tuning fork in the midline of the skull; sound should be heard equally in both ears. If the sound is louder in the affected ear, then the hearing loss is conductive since the transmitting defect of the middle ear has been bypassed. If, on the other hand, the sound is louder in the better ear, then the hearing loss is sensori-neural. The Rinne test is next performed to confirm the presence or absence of a conductive problem. The perception of sound by air should normally be heard after the sound heard while holding the tuning fork over the mastoid is no longer perceived.

Sensation to touch and pinprick is intact and symmetrical in the upper limbs. Muscle bulk and tone are both normal; muscle power at the shoulders, elbows and hands are graded as 5 and symmetrical (see Table 2.2); biceps, triceps and brachioradialis reflexes are tested as 2+ (normal) on both sides (see Table 2.3). The findings on lower limb testing are also unremarkable; there is no ankle clonus, and the plantar reflex is downgoing on both sides.

Coordination in the upper and lower limbs is intact, as is Ernesto's gait. However, he is not able to accomplish tandem walking and is unsteady, with a tendency to stagger to the left. When asked to close his eyes and stand with his feet together, his balance does not get worse; i.e. the Romberg sign is not present.

Mental status examination is difficult to conduct because of the frequent need to repeat questions, but generally appears normal.

The fundi are examined with the room darkened. The discs on both sides demonstrate sharp margins and are assessed as normal. Finally, blood pressure is taken and recorded as 145/90. His weight is recorded as 175 lb. Before asking the patient to dress, the neurologist does a focused examination of the skin; no brown patches or small subcutaneous lumps are noted.

The neurologist, having decided upon the localization and probable etiology, explains to Ernesto the nature of his illness and the likely diagnosis, as well as the next steps that need to taken. These include a detailed documentation of his hearing loss in a properly equipped laboratory and an imaging study of his brain! All this will have to be done in the 'big' city, requiring detailed scheduling.

6.2 CLINICAL DATA EXTRACTION

At this point, the student should proceed to the Web site and complete the history, physical examination, localization and etiology worksheets, utilizing the information just provided concerning Ernesto's problem.

6.3 REVIEW OF THE MAIN CLINICAL POINTS

The history includes the following:

- Diminished hearing in his left ear, starting a few months ago and increasing over this time
- Intermittent 'ringing' or hissing sound (tinnitus) in the left ear
- Some unsteadiness when walking, otherwise termed *dysequilibrium*
- No complaint of headache

The neurological examination demonstrates the following:

- A hearing loss in the left ear, of a sensorineural type
- Facial weakness on the left side affecting muscles of both the upper and lower face
- Mild ataxia of gait

6.4 RELEVANT NEUROANATOMY

The history and neurological examination are consistent with a disease process affecting both divisions of CN VIII, the vestibulo-cochlear nerve, but primarily the auditory portion, with the additional involvement of CN VII. Because of the involvement of two cranial nerves, the locus of the disease is likely to be at the level of the brainstem.

The nerves supplying the head and neck, the cranial nerves, have their nuclei in the brainstem (as was introduced in Chapter 1). CN III and IV are found in the midbrain; CN V, VI, VII and VIII in the pons; and the remainder in the medulla (see Figure 1.6b). The three major pathways, two ascending somatosensory and one descending motor, travel through the brainstem (Figures 1.4a and b and 1.5a). In addition, the reticular formation, with its descending and ascending influences, is an important component (Figure 1.6c). Finally, the brainstem is located in the base of the skull, and these cranial nerves have to travel via various foramina and canals in order to reach their destinations peripherally.

6.4.1 FACIAL NERVE (CN VII)

The nucleus of CN VII is located in the lower pons, in the lateral aspect (see Figure 1.6c), supplying all the muscles of facial expression. The facial nerve has an unusual course within the brainstem, moving inwards and upwards and then coursing over the nucleus of CN VI before descending to exit at the cerebello-pontine (C-P) angle, adjacent to CN VIII. Both nerves are found initially in the internal auditory canal. The facial nerve then courses through the petrous temporal bone, exiting the skull at the stylomastoid foramen (just inferior to the ear canal) and distributing its branches (within the parotid gland) to supply the various muscles of facial expression.

The organization of the facial nucleus is unique in that there are two portions, a part supplying the muscles of the upper face (forehead wrinkling and eye closure) and a part supplying the muscles of the lower face (controlling the movements around the mouth and the lips). A lesion affecting the facial nucleus or the exiting fibers of CN VII within the skull will affect both sets of muscles, i.e. a lower motor neuron lesion. The portion of the nucleus controlling the upper facial muscles receives its commands from both motor cortices, whereas the portion of the nucleus controlling the lower facial muscles only receives commands from the contralateral motor cortex. Therefore, when a patient has weakness confined to the lower face one looks for a lesion in the contralateral motor cortex or in the descending fibers from the cortex to the pons, i.e. an upper motor neuron lesion.

Additional functional parts of the facial nerve are afferent (sensory) fibers carrying taste from the anterior two-thirds of the tongue, as well as efferent parasympathetic fibers to the submandibular and sublingual salivary glands and to the lacrimal gland; both of these constitute a separate branch of CN VII (the nervus intermedius).

6.4.2 VESTIBULO-COCHLEAR NERVE (CN VIII)

There are two parts to CN VIII, the auditory component and the vestibular component; the nerve itself is large and divisible into these two major portions.

6.4.2.1 AUDITORY COMPONENT

The auditory part of the nerve originates in the region of the hair cells of the cochlea; the nerve's cell bodies, which form the peripheral (sensory) ganglion, the spiral ganglion, are also found within the cochlea. The nerve then courses in the internal auditory canal, a canal within the petrous portion of the temporal bone of the skull; after leaving the canal, the nerve goes through the subarachnoid space. It enters the brainstem at the C-P angle, very close to CN VII. Up to this point, the nerve is a peripheral nerve with the myelin of the peripheral nervous system (PNS) and its associated Schwann cells.

6.4.2.1.1 Central Auditory Pathway

There are several auditory nuclei, the first of which are found along the nerve itself as it enters the brainstem, with others found at the level of entry, in the lowermost pons and uppermost medulla (Figure 6.1). At this point, the sound is analyzed as to its laterality. After multiple synapses, some of the auditory fibers cross the midline (within the trapezoid body), while others remain on the same side. Sound data are then transmitted upwards, towards the cortex, on both sides, in the lateral lemnisci (singular, lemniscus).

Hearing is unique amongst the sensory systems in that the projection is bilateral, although the crossed fibers are more numerous. There is an additional unique feature of the auditory system whereby some neurons in the pontine auditory nuclei project from the central nervous system (CNS) back to the sensory hair cells of the cochlea; the function of these axons is thought to be the modification of thresholds in the hair cells.

The upwardly projecting fibers synapse in the midbrain (the inferior colliculi, at the lower midbrain level), where some auditory processing occurs. The auditory pathway then travels upwards to a specific relay nucleus of the thalamus, the medial geniculate nucleus, after which it arrives at the cortex along the transverse gyri of Heschl, located along the superior part of the temporal lobe, within the lateral fissure (see Figures 1.10, 6.1 and 13.3). Sound frequency (pitch) is preserved throughout the pathway up to and including the primary auditory cortex, where there is tonotopic localization. Adjacent areas of the cortex are

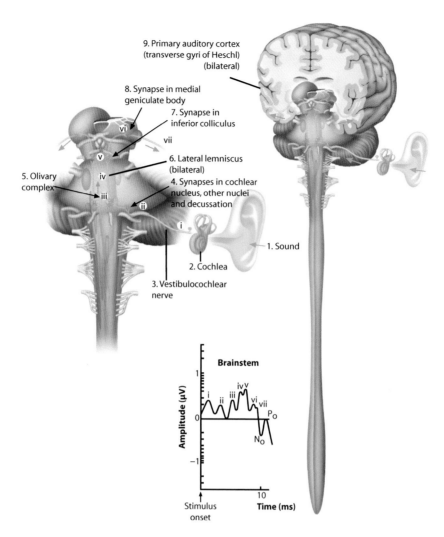

FIGURE 6.1: The pathway for audition (hearing). Sound waves enter the ear (1) and, after impinging on the tympanic membrane and activating the middle ear ossicles, are converted to neural signals in the cochlea (2). These impulses are carried via CN VIII, the vestibulo-cochlear nerve (3), to the cochlear nuclei (4) in the lower pons. Some of the fibers relay in the (superior) olivary complex (5) on the same side or on the opposite side after decussating in the trapezoid body. The auditory fibers ascend through the brainstem in the lateral lemniscus on both sides (6) and relay in the inferior colliculus in the lower midbrain (7). The next relay is the medial geniculate body (MGB) of the thalamus (8) and then to the primary auditory cortex (transverse gyri of Heschl) on the superior surface of the temporal lobe on both sides (9). The brainstem evoked potential (BSEP) is included below, showing the waves generated in each part of the brainstem pathway (labelled i to vii). Note: The auditory pathway is animated on the text Web site.

association areas for sound interpretation, including the nearby Wernicke's area (in the dominant hemisphere, as discussed in Chapter 1).

Deficits of hearing caused by failure of the sound waves to transmit, known as *conductive* hearing loss, can occur because of blockage in the external ear (e.g. ear wax), damage or disease of the ear drum itself, disease or fluid within the middle ear or damage to the ossicles of the middle ear. The other type of hearing loss is a *sensori-neural* deficit, caused by a lesion of the cochlea (loss of hair cells), disruption of transmission of the impulses peripherally (by the cochlear division of the VIIIth nerve within the skull) or lesions of the cochlear nuclei within the brainstem (at the level of the lowermost pons).

6.4.2.1.2 Testing

Hearing should be tested by a certified audiologist working with an ear, nose and throat specialist. The testing includes the transmission of sound (ear canal, tympanic membrane, middle ear ossicles) and testing of hearing of sounds of different pitch. The test result for hearing is displayed as an audiogram (Figure 6.2a, normal, b, abnormal). Particular attention is paid to the discrimination of speech sounds (500–3000 Hz, tested at 55 decibels).

A more sophisticated (and complicated) testing of the auditory system is based upon the processing of information by the auditory relay nuclei and associated CNS pathways, known as a *brainstem auditory evoked potential* test (BAEP, also BSEP and BAER). The normal evoked

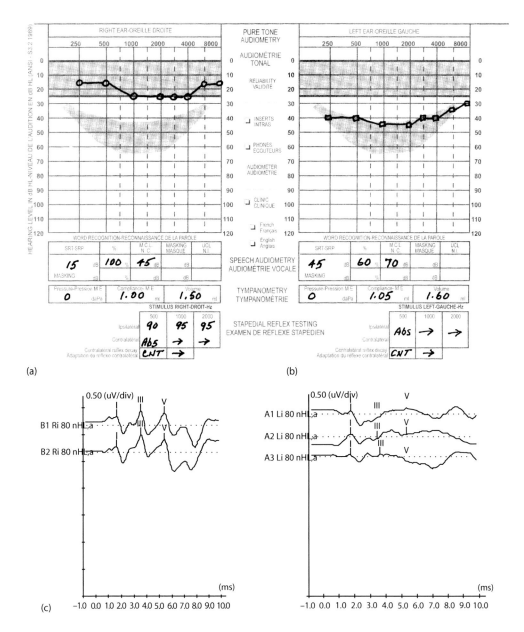

(a)

(b)

(c)

FIGURE 6.2: An audiogram for a patient such as the one presented in this chapter, with the normal right ear (a) on the left side of the page (the right side of the patient) and the hearing deficit (as described in the text) in the left ear (b) on the right side of the page. (Courtesy of Dr. J. Marsan.) (c) The BSEP for the same patient is also presented showing the normal response for the right ear and the abnormal response in the affected left ear, including both delayed onset as well as the poor configuration of the waves; note the expanded timeline for that ear. (Courtesy of Dr. J. Marsan.)

brainstem potential is shown in the lower portion of Figure 6.1, and each wave is paired with the corresponding neuroanatomical component.

The stimulus is a clicking noise (which consists of all frequencies) of defined intensity presented through an earphone; recording of the response is achieved by electrodes attached to the mastoid processes and the scalp (and averaged by an online computer after hundreds of repetitions). The response consists of a number of well-defined waveforms ascribed to the auditory pathway and

abbreviated with the acronym COLI: the auditory nerve (C), the olivary complex (O) in the lower pons, the lateral lemniscus pathway (L) and the inferior colliculus (I); additional recordings from the auditory cortex can also be obtained. This testing would be done in order to better define the nature of the lesion of the auditory system. The most common changes of auditory evoked responses are abnormalities of the configuration of the waves or a delay between the different waves, when comparing one ear to the contralateral ear (as shown in Figure 6.2c).

6.4.2.1.3 Tinnitus

Tinnitus is a sound heard by a person in the absence of an external source for the sound. It is described as a meaningless noise, such as a ringing, buzzing, hissing, clicking, roaring or chirping sound and can be reported in one or both ears; other people cannot hear it. It may be continuous or intermittent and often depends upon the ambient noise level. It is often associated with a hearing loss. Tinnitus can be quite distressing for a person, causing difficulty with concentration and reading and perhaps other tasks; interference with sleep is quite common. Some will react with emotional distress, which may require active treatment. There is some experimental evidence that tinnitus is a central phenomenon due to a reorganization of the primary auditory cortex, associated with diminished intracortical inhibition possibly related with aging.

6.4.2.2 VESTIBULAR COMPONENT

The vestibular portion of CN VIII originates in other parts of the inner ear, the three semicircular canals (for detection of angular acceleration) and the utricle and saccule (for detection of linear acceleration and the head's static orientation in space, and is hence important for balance and equilibrium). The sensory ganglion (Scarpa's ganglion) is located in the internal auditory canal and the central processes of these neurons, as peripheral myelinated nerves (with their associated Schwann cells), are also found within the internal auditory canal. The nerve enters the brainstem with its auditory portion at the C-P angle.

6.4.2.2.1 Central Vestibular Pathway

The information from the receptors is conveyed by the sensory neurons via the vestibulo-cochlear nerve, CN VIII, to the vestibular nuclei in the brainstem. There are four nuclei, located in the uppermost medulla and the lowermost pons, where these fibers synapse (see Figure 1.6c). Information from these nuclei is conveyed to the nuclei controlling the extra-ocular muscles (visuomotor nuclei – CN III, IV and VI), forming the basis for the necessary reflex adjustments of eye movements in response to acceleration and to changes in body position in relation to gravity. This reflex, consisting of conjugate movement of the eyes in the direction opposite to the direction of the head movement in progress, is intended to compensate for the head movement in order to keep the eyes fixed on an object and is called the *vestibular-ocular reflex* (VOR). If this reflex is not functioning, every head movement would cause a corresponding eye movement and the entire visual world would move as well (this is seen in people with bilateral vestibular dysfunction and is quite disorienting).

Vestibular data are also provided to the reticular formation of the brainstem as well as to the cerebellum so that the movements of the head by the neck muscles and other trunk and limb postural adjustments can be coordinated with those of the eyes; all this is handled by the so-called Nonvoluntary (postural) motor system (discussed in Section 2.4.4). There is also a descending pathway from one of these vestibular nuclei (the lateral) to the spinal cord for postural (non-voluntary) adjustments to the effects of gravity and alterations in acceleration (as in keeping one's balance hanging onto the strap when the bus starts moving).

Vestibular information is delivered, via the ventral posterior nucleus of the thalamus, to several cortical areas. These include area 3 (of the sensory postcentral gyrus), the posterior parietal lobe and the retroinsular cortex.

Disease can affect the vestibular apparatus itself, the vestibular portion of the VIIIth nerve (within the skull), the vestibular nuclei or the connections to the cerebellum or to the visuomotor nuclei.

6.4.2.2.2 Testing

Testing of vestibular function is most often done by checking the VOR and can be done at the bedside using the head-impulse test; alternatively, testing can be performed in a controlled environment using caloric testing. Both tests rely on causing motion of endolymph within the semicircular canals, which produces a deflection of the hair cells within the ampulla of the canals, resulting in either stimulation or inhibition of vestibular nerve impulses depending on the direction of deflection.

The bedside head impulse test achieves this in a physiologic manner using head movements. The patient is asked to fixate on a target in the distance while the examiner grasps the head and rapidly moves it in the plane of one of the semi-circular canals (most often the lateral canal which is tested with a lateral head turn). The endolymph (having its own inertia) doesn't move along with the head and instead causes a deflection of the hair cells within the canal. This stimulates the ipsilateral vestibular nerve and, through the central connections described earlier, results in compensatory eye movement in the opposite direction, thus keeping the eyes stable on the target. However, if the VOR is defective, the eyes will move with the head and the patient will need to make a 'refixation saccade' with the eyes to keep them on the target – this is the sign of an abnormal VOR. Each individual semicircular canal can be tested by moving the head in the plane of orientation of that canal. More recently, video recording of these eye movements in a laboratory environment can provide quantitative measurements of the gain of the VOR (ratio of eye movements to head movement) for each semicircular canal.

In the more traditional caloric test, the movement of endolymph is achieved through heating or cooling one part of the endolymph, which creates a convection current within the semicircular canal and, thus, deflection of the hair cells. This is achieved by introducing water (either warm or cold) into the external ear canal and observing the reflex eye movements (Figure 6.3). The temperature of the water is transmitted across the tympanic membrane to the middle

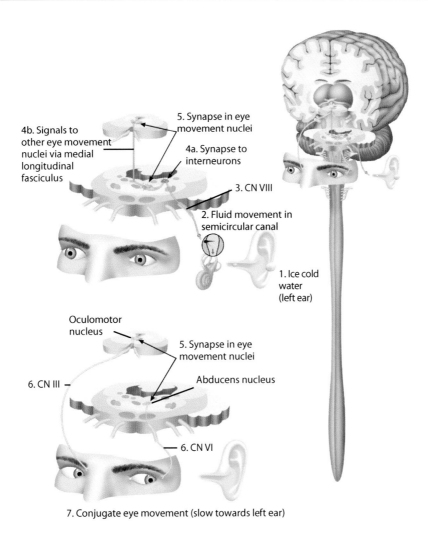

4b. Signals to other eye movement nuclei via medial longitudinal fasciculus

5. Synapse in eye movement nuclei

4a. Synapse to interneurons

3. CN VIII

2. Fluid movement in semicircular canal

1. Ice cold water (left ear)

Oculomotor nucleus

5. Synapse in eye movement nuclei

6. CN III

Abducens nucleus

6. CN VI

7. Conjugate eye movement (slow towards left ear)

FIGURE 6.3: The vestibular-ocular reflex (VOR). Upper illustration: Ice-cold water is introduced into the external ear canal of the left ear (1), which stimulates fluid movement in the (horizontal) semicircular canal on that side (2). The impulses are carried in the vestibulo-cochlear nerve (3) and after multiple synaptic connections in the vestibular-related nuclei (4a) are relayed to the visuomotor nuclei CN VI and also to CN III on the opposite side via the medial longitudinal fasciculus, the MLF (4b), synapsing in these nuclei (5). Lower illustration: Stimulation of the abducens nucleus on the same side (5) activates the lateral rectus muscle via CN VI (6) while stimulation of the oculomotor nucleus on the opposite side (5) activates the medial rectus muscle via CN III (6), causing the eyes to move slowly to the same side (to the left) as a conjugate movement (7). The rapid 'corrective' return—the direction of the named nystagmus—is to the opposite side and is not shown; this corrective return would not occur in a comatose patient. Note: The VOR is animated on the text Web site.

and inner ear and, as the lateral semicircular canal is closest to the middle ear, preferentially to the lateral canal. The head of a supine patient is elevated 30 degrees to bring the lateral canal into a vertical position, thereby allowing convection currents to be induced by gravity. In this position, if cold water is used, the fluid in the lateral part of the semicircular canal will cool, condense and begin to sink, thus producing the same message as if the head were rotating in the opposite direction. This induced fluid movement evokes a conjugate movement of the eyes towards the cooled ear; the converse would occur (i.e. the fluid would begin to rise and the eyes would deviate contralaterally) if warm water were used. If the test is done on an awake person, there is a natural tendency to keep the eyes in central gaze such that

compensatory saccades occur in the opposite direction, creating an oscillating eye movement. This combination of slow conjugate eye drift in one direction with a compensatory fast saccade in the other direction is called jerk nystagmus, and it is named for the direction of the *fast phase* (e.g. fast phases to the right is named right beating jerk nystagmus). It is important to note that a conscious person undergoing the test may experience a rotatory ('dizzy') sensation and this can be very disconcerting. It is also very important to note that the response to the test depends on the integrity of the vestibular part of the VIIIth nerve, the integrity of the ponto-medullary region of the brainstem and that of CN III and VI.

As illustrated in Figure 6.3, ice-cold water is put into the left ear. The stimulus (which in this case is inhibitory) is conveyed via the vestibular division of CN VIII to the vestibular nuclei in the lowermost pons (and uppermost medulla). The basis for the eye movements is connections from the vestibular nuclei to two of the extraocular motor nuclei, CN III and CN VI. Two eye muscles are needed for the eyes to move together in the horizontal plane, the medial rectus of one eye (controlled by CN III – at the level of the midbrain) and the lateral rectus of the other eye (controlled by CN VI – at the level of the lower pons). The vestibular nucleus sends inhibitory connections to neurons in CN VI on the same side and excitatory connections to CN VI on the opposite side. A subpopulation of the CN VI neurons has axons that cross immediately and ascend in a specific pathway through the brainstem (the medial longitudinal fasciculus [MLF]) to medial rectus subnucleus of CN III in the midbrain. Cold water causes inhibition of the left vestibular nucleus and thus relative activation of the left CN VI and right CN III. This causes the slow phase of the movement of the eyes to the same side, the left; the rapid saccadic return of the eyes is in the opposite direction, to the right, and is the basis for the classification of the nystagmus. The mnemonic used for this test is COWS: cold opposite and warm same, referring to the fast component of the nystagmus. (This reflex and the accompanying nystagmus are shown on the text Web site.)

The same tests can be applied in the comatose patient to evaluate whether the VOR (and thus CN III, VI and VIII and their central connections as previously described) is intact. In this scenario, the patient is not conscious to perform refixation saccades, so the interpretation is less exact. If the VOR is intact, a head movement should produce eye movements in the opposite direction (termed a *dolls-eye* or oculocephalic reflex). Similarly, cold-water caloric irrigation will produce an ipsilateral eye deviation without corresponding nystagmus.

6.4.2.2.3 Dizziness

Dizziness is a term used by patients to describe various abnormal sensations, including spinning, light headedness, feeling 'faint', unsteady or 'off balance' and spatial disorientation to name only a few. Vertigo bears defining as the medical term for a false sensation of motion, whether of the self (internal vertigo) or the environment (external vertigo), and can be spinning, rocking or swaying. It was long believed that the character of dizziness defined the type of problem the patient was having; i.e. if a patient said he or she was 'lightheaded', then presyncope was the cause. However, it turns out it is quite difficult for patients to describe their vestibular sensations; what is more consistent is the timing and trigger(s) of the sensation. For instance, if a patient describes 'light-headedness' triggered by rolling over in bed or lying down and lasting seconds to minutes, it is most likely due to benign paroxysmal positional vertigo (BPPV) despite the patient's lack of use of the word 'vertigo'. This isn't to say that patients' descriptions are not important, only that they might carry less weight in

the diagnostic process. This approach, based on the timing and triggers of dizziness, is identical to what is done in other areas of neurologic diagnosis. Diagnostic possibilities are then further refined through querying associated symptoms, performing a bedside examination and interpreting subsequent laboratory tests or imaging if required. While dizziness would also seem to imply vestibular or balance pathology, it is one of the most non-specific symptoms encountered in clinical medicine, with over half of people with dizziness being diagnosed with a non-vestibular cause. The other causes are quite variable: stroke, anxiety, electrolyte imbalances, cardiac arrhythmias and vestibular neuritis to name only a few. No single cause accounts for more than 10% of cases, making a symptom of dizziness one that requires careful thought and investigation.

Some individual causes of dizziness and vertigo are worth discussing due to their frequent occurrence and how they demonstrate the physiology of the vestibular system. BPPV is the most common vestibular disorder and is caused by otoconia: calcium carbonate crystals that are usually stuck onto the saccule and utricle and become dislodged and trapped in the semicircular canals. These crystals have more inertia than the endolymph, and thus, when normal endolymph slows after a head movement, the crystals continue to move. This creates a false neural signal of head rotation in the plane of the semicircular canal, with associated vertigo and nystagmus. The slow component caused by the motion of the otoconia is then compensated by the fast phase, which is in the direction of the false head movement. The symptoms last only until the otoconia settle and stop moving, usually 30–60 seconds. It is treated by repositioning manoeuvres that place otoconia back in the vestibule (a reservoir of endolymph to which all the semicircular canals connect and where the utricle and saccule are located). *Vestibular neuritis* is a presumed inflammation of the vestibular nerve causing reduced neural impulses from the affected nerve. The vestibular nerves are paired and any reduction in activity of one side is interpreted by the brain as activation of the other, so vestibular neuritis tricks the brain into thinking the head is constantly rotating with resultant vertigo and nystagmus (fast phase away from the lesion). Over days to weeks, the cerebellum can compensate and rebalance the inputs so that vertigo and nystagmus disappear even if there is some residual vestibular damage. Finally, *presbyequilibrium* is a common disorder in the elderly causing poor balance especially when ambulating. It is due to a combination of age-related degeneration in the somatosensory, visual and vestibular systems and their central integration and highlights the multisensory nature of human balance.

6.4.2.2.4 Sensory and Motor Pathways

The somatosensory pathway for discriminative touch and proprioceptive sensation in the spinal cord, the posterior column, crosses in the lower brainstem and becomes the medial

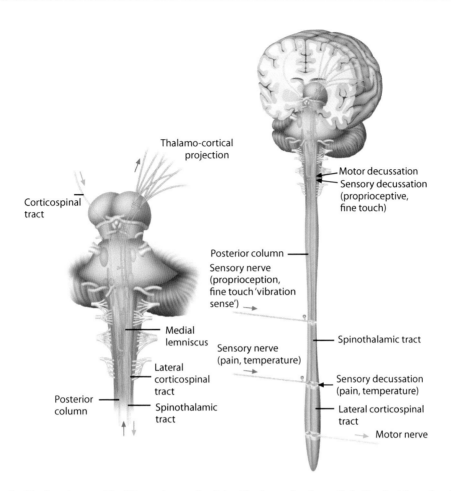

Thalamo-cortical projection

Corticospinal tract

Motor decussation

Sensory decussation (proprioceptive, fine touch)

Posterior column

Sensory nerve (proprioception, fine touch 'vibration sense')

Medial lemniscus

Spinothalamic tract

Lateral corticospinal tract

Sensory nerve (pain, temperature)

Posterior column

Spinothalamic tract

Sensory decussation (pain, temperature)

Lateral corticospinal tract

Motor nerve

FIGURE 6.4: Longitudinal pathways. The illustration on the right side shows the three clinically significant longitudinal pathways: the sensory posterior (dorsal) column-medial lemniscus (for discriminative fine touch, proprioception and vibration), the sensory spinothalamic tract (for pain and temperature) and the motor corticospinal tract, including the peripheral nerves, the decussations of these tracts and the cortical projections. The illustration on the left side shows the details of these pathways through the upper spinal cord and the brainstem.

lemniscus (Figure 6.4). It ascends through the medulla in the midline area, moving more laterally as it ascends through the pons and into the midbrain. The pathway for pain and temperature, the spinothalamic tract, crosses in the spinal cord and ascends; in the brainstem, it is situated more laterally than the medial lemniscus. Both tracts merge at the level of the uppermost pons and are found together in the midbrain. Lesions affecting the brainstem would result in a contralateral loss of either or both sensory modalities, depending upon the location and extent of the lesion.

The voluntary motor pathway, the corticospinal tract, is found in the anterior region of the brainstem, occupying the cerebral peduncles in the midbrain, lying in the anterior bulge of the pons, and found within the pyramids in the medulla anteriorly (hence its name 'the pyramidal tract'; see Chapter 7). As will be seen in Chapter 7, a lesion of this pathway, if accompanied by disruption of the cortico-reticulospinal and corticorubrospinal pathways, will reveal on neurological testing a contralateral weakness (paresis) or paralysis and an increase in reflex responsiveness; an

extensor plantar response would indicate involvement of the corticospinal tract at any point along its pathway.

6.5 LOCALIZATION

Based on the neuroanatomical information just presented, the history and neurological examination are consistent with the hypothesis that there is a disease process affecting both divisions of CN VIII as well as CN VII. The major question is whether the process is occurring outside the brainstem (extra-axial) or within the brainstem (intra-axial).

The hearing loss in our case is restricted to one ear (the left one). As the central auditory pathway is bilateral within the brainstem, loss of hearing due to a lesion within the CNS should cause a hearing deficit in both ears, although not equally. This shifts the locus of the lesion in this case outside the brainstem itself (extra-axial). In addition, if the lesion were within the brainstem, one would anticipate some involvement of the long tracts, sensory or motor, as these tracts course through the brainstem; one might also see evidence of dysfunction in other cranial nerves such as V and VI. No somatosensory deficits were noted in our patient and there were no changes in motor strength or in tendon reflex responsiveness, thereby pointing again to a lesion outside the brainstem, but within the skull.

The fact that the facial weakness involves both the upper and lower face is very consistent with involvement of the VIIth nerve nucleus within the lower pons or the nerve itself after it exits the brainstem, i.e. a lower motor neuron lesion. If the lesion were to be within the brain, above the level of the pons, affecting the corticobulbar fibers to the nucleus of CN VII on one side, there would be a deficit of only the lower face (on the opposite side). Again, since there is no involvement of the long tracts (sensory or motor), the locus is again pointing to a lesion of the facial nerve itself, outside the brainstem.

It should be noted that the neurologist also tested the corneal reflex on the left and found it to be normal. This result certainly indicates that the sensory component of CN V has not (as yet) been affected by the disease process. Involvement of the upper sensory division of this nerve would lead to a decreased blink response in both eyelids when the testing is done on the side of the lesion (see Figure 2.4), but the response would be normal when tested on the other side. A decrease in the response on one side could also be due to the involvement of CN VII, as the palpebral (eyelid) muscles are supplied by CN VII. In Ernesto's case, the pathology in CN VII was too mild to interfere with the corneal reflex; with more severe facial nerve pathology, the corneal reflex would be diminished on the side of the weakness, normal on the opposite side, and the patient would report that the sensation of corneal stimulation was the same in both eyes.

The unsteadiness of walking that was seen when the patient was asked to tandem walk or change direction could be caused by a disease affecting the VIIIth nerve (sensory) within the skull (before it enters the brainstem), by disease of the brainstem affecting the vestibular nuclei, or could also reflect cerebellar pathology. The vestibular nuclei project to the central region of the cerebellum (the vermis) to assist in the coordination of gait. Other cerebellar coordination tasks – regulated by the lateral regions of the cerebellum (the cerebellar hemispheres, discussed in Chapter 7) – were tested and found to be within normal limits. The problem with gait could therefore be due to the involvement of the nerve itself peripherally or to a lesion within the CNS.

In a way, Ernesto's tendency to veer to one side while walking can be likened to what occurs when a two-engine airplane has one engine that is malfunctioning and has less power: the airplane will veer to one side unless the pilot makes constant course changes. A dysfunction in one vestibular nerve or its central connections will result in a similar difficulty maintaining a steady course.

As both CN VII and CN VIII are found in the internal auditory canal and both enter/exit the brainstem at the C-P angle, the likelihood is that there is a disease process occurring in either of these locations affecting both cranial nerves.

6.6 ETIOLOGY: ERNESTO'S DISEASE PROCESS

Ernesto's symptoms have been noted for more than three months, therefore falling into the category of a chronic disease process (see Figure 4.8). In fact, his symptoms have become worse in this time period, indicating the progressive nature of the disease process.

Hearing loss: In considering a hearing loss, there are certain parameters that are specific for the auditory system. One of the most common occurrences is an accumulation of earwax in the external ear canal, more often affecting men; this was not the case with Ernesto. Hearing loss for the higher frequencies is quite common as people age, but this usually affects both ears (to some degree) and is most commonly found after the age of 60. Another common reason for hearing loss is exposure to excessively loud sound, leading to a destruction of the hair cells of the cochlea, usually affecting both ears; Ernesto is not involved in an industrial job or a hobby with chronic exposure to loud noise, although he has recently been listening to music through headphones at high volume. It is not clear whether his hearing problem started before or after this high volume sound exposure, but one would expect both ears to be affected if loud music was the cause.

6.6.A PAROXYSMAL DISORDER

A paroxysmal disorder is an acute phenomenon or an acute event superimposed on a chronic condition. The history in this case does not coincide with such a disease process.

6.6.B TRAUMATIC INJURY

There has been no accident or trauma to the head in this case.

6.6.C VASCULAR DISORDER

Again, vascular events fall under the category of acute, and our patient's history is not compatible with a vascular event, whether related to acute thrombotic or embolic occlusion and resulting infarction, to vasculitis, or to hemorrhage.

6.6.D TOXIC INJURY (INCLUDING DRUGS)

Ernesto has not taken any medication that could lead to deafness (e.g. the antibiotics streptomycin and gentamycin;

quinine; chronic abuse of non-steroidal anti-inflammatory drugs). Aspirin overdose can cause bilateral tinnitus, but nothing in the history suggests that he has been taking painkillers for arthritis or any other pain condition. Chronic lead poisoning has been associated with progressive deafness (e.g. postulated for the composer Ludwig van Beethoven), but Ernesto is not making his own wine nor does he store wine in lead-lined containers (such as crystal) or use pottery dishes with lead glazes (one needs to ask these questions specifically). Besides, a toxic type of process would be expected to affect both ears equally.

6.6.E INFECTIOUS DISEASES

Although chronic infectious processes are also unlikely as causes of unilateral sensorineural hearing loss and ipsilateral facial weakness, for purposes of completeness, a few acute and subacute-onset infectious causes of unilateral deafness are worthy of mention.

Herpes zoster virus can involve the VIIIth nerve and its root entry zone in the brainstem, causing hearing loss and vertigo of central origin. Unilateral deafness is also a common complication of bacterial meningitis, which can often affect cranial nerves VII and VIII.

An increasingly common infectious cause of somewhat slower-onset unilateral hearing loss and facial weakness is Lyme disease, caused by the spirochete *Borrelia burgdorferi* and transmitted to humans by a tick vector. The onset of cranial nerve deficits in Lyme disease is typically considerably more rapid than we have seen in Ernesto's story; nevertheless, in endemic regions, this diagnosis would have to be considered and excluded by specific testing.

6.6.F METABOLIC DISEASES

Although metabolic diseases (e.g. diabetes or other endocrine disorders) are, by their nature, chronic for the large part, nothing in the history indicates that Ernesto is suffering from a chronic systemic disease. He has not complained of fatigue or shortness of breath while walking or cycling. Besides, again, a metabolic type of disease process would be expected to affect both ears equally.

6.6.G INFLAMMATORY/AUTOIMMUNE

Chronic inflammatory/autoimmune diseases of the CNS seldom produce unilateral sensorineural hearing loss. An important exception is cholesteatoma, a chronic granulomatous disease complicating a longstanding otitis media, primarily in children. In addition, mention should be made of sarcoidosis, a chronic granulomatous process of unknown etiology that may produce gradually evolving multiple cranial nerve pareses, sometimes unilaterally.

6.6.H NEOPLASTIC

One must consider the neoplastic disease category in any chronic disease. In the present case, we would be considering a tumor in the region of the petrous temporal bone of the skull or, alternatively, affecting the nerves at the C-P angle. There is a particular tumor of the VIIIth nerve, known as a *vestibular schwannoma* (commonly misnamed an acoustic neuroma), which would take into account all the neurological findings and be consistent with the history of Ernesto's case. The growth of this Schwann cell tumor occurs at the C-P angle and/or within the internal auditory canal, causing pressure on both divisions of the VIIIth nerve and eventually also on the VIIth nerve. Although benign, continued growth of this tumor will lead to involvement of CN V, as well as brainstem compression at that level.

A C-P angle meningioma, although somewhat less common, would also be possible. A brainstem glioma is also to be considered, but these are always associated with long tract sensory and motor deficits. CNS lymphoma may also involve cranial nerves but typically affects many nerves simultaneously. A condition known as metastatic carcinomatosis of the meninges can cause multiple cranial nerve palsies including CN VII and VIII. A solitary metastasis is unlikely in this case, as there is no known primary.

6.6.I DEGENERATIVE DISORDERS

Menière's disease, a degenerative disease of the vestibulocochlear apparatus, is also a likely possibility in this case. This disease is characterized by deafness (particularly in the low frequency range) and tinnitus, usually involving only one ear. It affects both men and women, with a disease onset about the age of 50. The hallmark of this disease is the occurrence of *acute* attacks of dizziness, sometimes described as 'explosive vertigo', lasting minutes to hours, accompanied by nausea and sometimes leading to vomiting. Once the unilateral degenerative process is complete and the sensorineural hearing loss in the involved ear is profound, the vertiginous attacks stop occurring. In a severe attack, the person may fall and there is an intense sensation of rotation of the environment. Our patient Ernesto has not reported any such episodes. In addition, Menière's disease is not accompanied by unilateral lower motor neuron facial weakness.

6.6.J GENETIC DISORDERS

Tumors involving peripheral nerves, including cranial nerves, are part of the genetic disease neurofibromatosis, especially the type 2 variety. The disease may be unilateral or can occur bilaterally. In this disorder, there are typically also congenital brownish pigmentary skin patches

Disease type	Secs	Mins	Hrs	Days	Weeks	Months	Years
Paroxysmal (seizures, faints, migraine)							
Traumatic							
Vascular (ischemic stroke, bleed)							
Toxic						⊘	⊘	
Infectious					⊘	⊘	⊘
Metabolic					⊘	⊘	⊘
Inflammatory/autoimmune					⊘	⊘	⊘
Neoplastic					O	O	O
Degenerative						⊘	⊘
Genetic					⊘	⊘	⊘
			<-- Acute -->		<-- SubAcute -->		<-- Chronic -->

FIGURE 6.5: Completed etiology matrix for Ernesto.

referred to as 'café-au-lait' spots, as well as small, rubbery neurofibromas underneath the skin. (Remember that the neurologist did do an examination of the skin at the end of the physical.)

In addition, there are many familial forms of deafness, producing either deafness alone or deafness along with other disease processes (retinitis pigmentosa; peripheral polyneuropathies such as Charcot–Marie Tooth disease and Fabry's disease; and cardiac conduction deficits, among many). In the vast majority of genetic forms of deafness, the hearing loss is sensorineural and bilateral. An exception is otosclerosis, often a familial disorder of unknown etiology in which there is progressive sclerosis of the middle ear ossicles leading to conductive hearing loss. In many cases of this disorder, hearing loss appears first in one ear before extending to the other; associated facial weakness does not occur.

Figure 6.5 summarizes the discussion concerning the most probable etiology for Ernesto progressive left-sided hearing loss and facial weakness.

6.7 CASE SUMMARY AND SUPPLEMENTARY INFORMATION

At this point, with the knowledge of the relevant anatomy, the localization of Ernesto's disease is the left vestibulo-cochlear nerve, likely within the internal auditory canal, as well as the left facial nerve (Table 6.1, Expanded localization matrix). The most likely etiological diagnosis is a tumor affecting the left CN VIII, a *vestibular schwannoma*, most often called an acoustic neuroma.

6.7.1 EVOLUTION OF THE CASE

After the expenditure of much energy and time by others, arrangements are finally completed for the additional testing to be done three weeks following the neurological examination.

Ernesto's symptoms continue to worsen. He is now almost totally deaf in one ear, has definite one-sided facial weakness and is having more unsteadiness of gait. The tinnitus is now less of a problem, for some unknown reason.

6.7.2 INVESTIGATIONS

Ernesto's left-sided hearing deficit is found to be significant and is recorded as a moderate to severe sensori-neural hearing loss, with loss of speech discrimination (shown in Figure 6.2b for a similar case). His hearing is completely normal in the non-affected (right) side.

The results of the testing of auditory evoked response on the right side show normal wave configuration and wave I to wave V latencies. On the left side, there is a loss of configuration of wave III and wave V suggesting the presence of a central retro-cochlear disorder along the auditory pathways (see Figure 6.2c).

Caloric testing is also done, much to Ernesto's dismay. The left (abnormal) ear is tested first, and there is no reflex response of the eyes (no nystagmus) and hence no subjective response either, indicating a loss of the function of the vestibular portion of CN VIII on that side. The involvement of the vestibular nerve to this degree explains some of the patient's difficulties with gait. Testing of the non-affected right ear, however, elicits the normal response of nystagmus, along with extreme dizziness, accompanied

TABLE 6.1: Completed Localization Matrix

	Neuro Exam		MUS	NMJ	PN	SPC	BRST	WM/TH	OCC	PAR	TEMP	FR	BG	CBL
Attention	N													
Memory	N													
Executive	N													
Language	N													
Visuospatial	N													
CRN I	N	N												
CRN II	N	N												
CRN III, IV, VI	N													
CN V	N	N												
CRN VII	Wk U/L	N			O		⊘	⊘		⊘		⊘	⊘	⊘
CRN VIII	Decr	N			O		O	O		⊘	O	⊘	⊘	⊘
CRN IX, X	N	N												
CRN XI	N	N												
CRN XII	N	N												
Tone	N	N												
Power	N	N												
Reflex	N	N												
Involuntary Mvt	N	N												
PP	N	N												
Vib/prop	N	N												
Cortical sensation	N	N												
Coord UL	N	N												
Coord LL	N	N												
Walk	N	N												
Toe/heel/invert/evert	N	N												
Tandem	Decr	Decr			O		O	O	⊘	⊘			⊘	O
Romberg	N	N												

Abbreviations: N, normal; Decr, decreased; Wk U/L, weakness upper and lower face.

by nausea and vomiting. (Thankfully, Ernesto had followed the instructions and had not taken any food after midnight.) It takes almost four hours before he is able to get up and walk about without feeling dizzy or almost falling over.

Later that day, Ernesto is taken to neuroradiology. A gadolinium-enhanced magnetic resonance image (MRI) reveals the presence of a significant tumor of CN VIII located in the internal auditory canal and extending to the C-P angle (Figure 6.6). The tumor measures 2.5 cm in greatest diameter and is beginning to compress the brainstem in that area. (The MRI for cases involving a possible C-P angle tumor can also be viewed with most newer machines using finer 'cuts' of the brainstem and the skull in that region, without enhancement.) Being strapped inside this space-age machine, along with experiencing all the clunking and other noises, is a 'great' experience for Ernesto, one that he loves to talk about whenever asked.

The results are relayed to the neurologist, who wisely decides to fax her report followed by a telephone call to the family practitioner. The neurologist explains the diagnosis of *vestibular schwannoma* (an acoustic neuroma, as discussed below) as the most likely cause of Ernesto's problems and the likelihood of requiring neurosurgery.

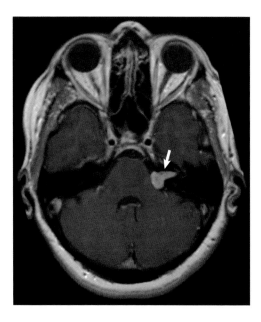

FIGURE 6.6: Radiograph: The contrast-enhanced radiograph of the brain shows a typical-shaped lesion on the left side (of the patient – the right side of the radiograph), with a narrow 'stem' laterally and an expansion medially impinging on the brainstem at the cerebello-pontine angle, compatible with a vestibular schwannoma. (Courtesy of Dr. M. Kingstone.)

6.7.3 TREATMENT, MANAGEMENT AND OUTCOME

Ernesto is contacted and brought to the physician's office for a special appointment to be informed of the diagnosis. Notwithstanding his initial reluctance to consider surgery 'to his brain', he quickly realizes, after explanations utilizing some anatomical diagrams, the actual location of his tumor and what might happen if it continues to grow. Ernesto agrees to have a consultation with a neurosurgeon, back in the big city. A few phone calls more and the appointment is confirmed for the following month. It is now eight to nine months since the onset of symptoms and three to four months since Ernesto first saw his family physician.

At the time of the consultation, the neurosurgeon explains the two main treatment options: fractionated stereotactic radiotherapy to the tumor bed or surgical removal. Ernesto opts for the latter and the surgery is done four weeks later; the tumor is located as expected. This is very delicate and difficult surgery as the neurosurgeon must identify the various cranial nerves and other structures, and attempt to remove the tumor without damaging the acoustic division. As it is now almost one year since he started complaining of his hearing loss, the tumor is now quite large, making the surgery more difficult. The surgery lasts about seven hours and the tumor is finally removed. The post-operative recovery is without complications. The pathology confirms the putative diagnosis of a benign vestibular schwannoma. Ten days after the surgery, Ernesto is able to go home, where he receives supportive services while his wound heals and until he is able to look after himself independently.

Ernesto is seen in follow-up by his family physician about one month later. He is in a much better frame of mind and almost back to his usual outgoing self. There is some residual facial weakness noted on examination, but most people would not notice this (especially after his beard grows back). There is no longer any complaint of tinnitus, but unfortunately, there is a complete loss of hearing in his left ear. Ernesto slowly learns to live with this deficit and everyone around him learns to speak 'a little louder' or one at a time or to his intact side.

A hearing aid is not usually recommended in such cases, as there has been a loss of the nerve itself, a sensorineural deficit, and hearing aids can be useful only if there is partially reduced hearing that can respond to sound amplification. Hearing rehabilitation can be achieved by using the *C*ontralateral *R*outing *of S*ignals (CROS) type of hearing aid that transmits sound wirelessly from the poorer ear to the ear with better hearing. However, newer devices are being tried, one of which is a bone-anchored hearing aid, which could transfer sounds via bone conduction from the deaf side to the non-affected side. Ernesto is seriously considering getting one of these devices.

6.7.4 PATHOLOGY

The pathological examination of the tumor revealed a *vestibular schwannoma* (Figure 6.7). These tumors in fact usually originate on the vestibular portion of the VIIIth nerve, and the cells giving rise to this tumor are the Schwann cells, the myelin-forming cells of peripheral nerves. The term *acoustic neuroma* has been used for these tumors but is not appropriate, as neuromas are of connective tissue origin. Another erroneous term that has been used for these tumors is *neurolemmoma*. The correct nomenclature is *vestibular schwannoma*.

Growth of these tumors is slow and leads to pressure on the acoustic division, causing hearing loss and tinnitus. Continued growth of the tumor leads to pressure on CN VII at the C-P angle or within the internal auditory canal, thereby causing the facial motor weakness, of both upper and lower facial muscles, i.e. lower motor neuron involvement. On occasion, the tumor may become so large that it will compress the brainstem sufficiently to compress the fourth ventricle, impede cerebrospinal fluid circulation and cause hydrocephalus with raised intracranial pressure (see Chapter 9).

It should be noted that since vestibular schwannomas are benign and often very slow growing, small tumors that are not causing significant symptoms can be monitored for growth; sometimes, the decision is made to leave them untouched, without surgical intervention, particularly in older patients.

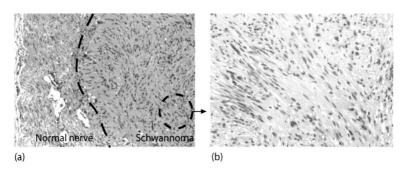

(a)   (b)

FIGURE 6.7: Stained sections from a vestibular schwannoma: (a) The normal portion of the nerve with a more linear appearance is shown on the left; the pattern of cells of the schwannoma on the right side is more random. (b) The tumor is shown at higher magnification. Note that the cells in both areas, normal and tumor, are elongated (spindle-shaped). (Courtesy of Dr. J. Woulfe.)

6.8 E-CASES

6.8.1 CASE 6E-1: ERNO, AGE 15

Background: Erno is a typical cell-phone teenager who is constantly occupied with social media friends and sports activities at school and is also an avid chess player. He is now having neurological problems.

Chief complaint: Unsteady gait for the past four days
History:

- Acute illness two weeks ago consisting of recurrent vomiting and diarrhea; recovered after a few days.
- Four days ago, Erno awoke to find that he was *unable to look to the right*; when trying to walk, he experienced a feeling that his body was turning towards the left.
- Increasing dizziness and unsteadiness accompanied by intense nausea.
- Two younger siblings, both well; 36-year-old father has multiple sclerosis, onset age is 31.

Examination:

- Alert, oriented, cooperative
- Head and eyes deviated slightly to the left
- Normal visual acuity in both eyes; normal fundi
- *Unable to either adduct his left eye or abduct his right eye*; vertical gaze normal (see Figure 6.8)
- On gaze left, nystagmus of the left eye with fast phase to the left
- With passive head rotation, *his eyes turn reflexively to the left but not to the right*

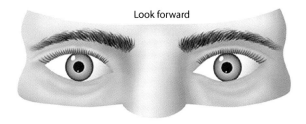

Look forward

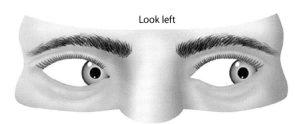

Look left

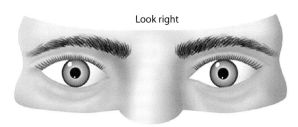

Look right

FIGURE 6.8: Eye movement findings in case 6e-1 (Erno).

- Remainder of the cranial nerve examination is normal; normal tympanic membranes and hearing acuity bilaterally
- Normal muscle tone, bulk and power in all four limbs; normal finger-nose and heel-shin testing; normal tendon reflexes and plantar responses
- Attempt to walk resulted in horizontal nystagmus, *a tendency to lurch to the left* and vomiting

6.8.2 CASE 6E-2: ESTHER, AGE 80

Background: This well-dressed elderly woman presents initially as well composed but immediately reveals that she is in a state of acute distress, with head pain and having hardly slept at all in three nights.

Chief complaint: Severe pain in the right forehead and scalp of three days duration

History:

- Three days ago, overnight appearance of burning pain in the right forehead and scalp, at times with shooting pain into the right eye

- No associated nausea, visual disturbance or vertigo, but Esther has a poor appetite and feels tired
- No significant previous illnesses; not hypertensive

Examination:

- Pulse 86/min; blood pressure 140/85; respirations 18/min; temperature 38°C
- Four slightly raised, pinkish-red 2-mm-diameter skin lesions in the right forehead area; right eyelid is erythematous and slightly oedematous (see Figure 6.9 for similar picture)
- With one exception, a normal cranial nerve and motor system examination

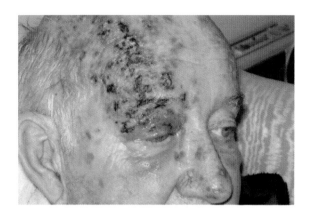

FIGURE 6.9: Typical findings in a patient with this disorder (case 6e-2, Esther).

- *Sharply demarcated region of hyperesthesia to pinprick and dysesthesia to touch and temperature in the skin of the right forehead extending inferiorly to the upper eyelid and right side of the nose and posteriorly 2/3 of the way through the right scalp*
- Prominent, tortuous but non-tender superficial temporal arteries

6.8.3 CASE 6E-3: CHRISSY, AGE 4

Background: Your patient Chrissy is brought to your office by her young mother accompanied by her parents, all very anxious. Chrissy herself behaves (and is dressed) like a little lady and does not seem at all perturbed by this medical appointment.

Chief complaint: Deviation of the tongue to the right and unilateral atrophy, duration unclear

History:

- Tongue deviation noted by a physician in a walk-in clinic, where the patient had been taken for assessment of fever and sore throat; *the right side of the tongue appeared to be shrunken in size and to be fasciculating* (spontaneous multifocal twitching or writhing movements)
- Nystagmus of both eyes has been present, and stable, since at least age 2; assessed by ophthalmology at the time and felt to be 'congenital'
- Periodic, sudden pain in the back of the head, triggered by coughing or sneezing

Examination:

- *Obvious head tilt to the left*; a review of family photographs shows that the tilt was not present prior to age 2.

- *Coarse nystagmus with horizontal gaze*, worse with gaze right; vertical gaze is normal.
- *Right half of the tongue is atrophic and fasciculating*; on protrusion, the tongue curves to the right.
- Otherwise, the cranial nerve examination, including the fundi, is normal.
- Normal muscle strength/tone in all four limbs; normal upper and lower limb coordination.
- *Tendon reflexes*: see reflex map (Figure 6.10).

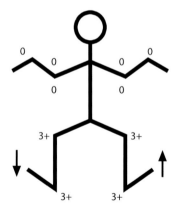

FIGURE 6.10: Reflex map for case 6e-3 (Chrissy).

6.8.4 CASE 6E-4: BRENDA, AGE 9

Background: During one of the busiest evening shifts in the emergency department of your local hospital, you are asked to see a child accompanied by a very worried mother. Your patient Brenda is not in acute distress but clearly very subdued, although very cooperative and communicative.

Chief complaint: Progressive loss of function of the left side over the past three months

History:

- Beginning three months ago, increasing difficulty using her left hand in bilateral hand activities such as tying shoelaces and doing up coat buttons

- Tendency to trip over her left foot on stairs or on uneven ground; left leg increasingly stiff and awkward
- Slurring of speech evident since age 2, worse over the past few months
- Mother has deferred seeking assessment of Brenda's symptoms because her husband is recovering from major surgery

Examination:

- *Nystagmus on horizontal and vertical gaze*, equal in amplitude in all directions; no restriction of gaze
- *Bilateral weakness of the lower face*; trouble pursing her lips (Figure 6.11a)
- Impaired ability to move the tongue from side to side
- *Extremely brisk jaw jerk*
- Otherwise, cranial nerve examination unremarkable with normal fundi
- Weakness of left hand grip, foot dorsiflexion and ankle eversion (Figure 6.11)
- *Increased muscle tone in the left ankle flexors*; otherwise muscle tone uniformly *low*
- Rapid alternating hand movements are *slow bilaterally*, worse on the left side
- Absent independent finger and toe movements, as well as toe tapping, on the left side
- *Dysmetria and intention tremor in the right arm*
- *Broad-based, unsteady gait*; walks on the toes of her left foot
- *Tendon reflexes:* see reflex map (Figure 6.11b)

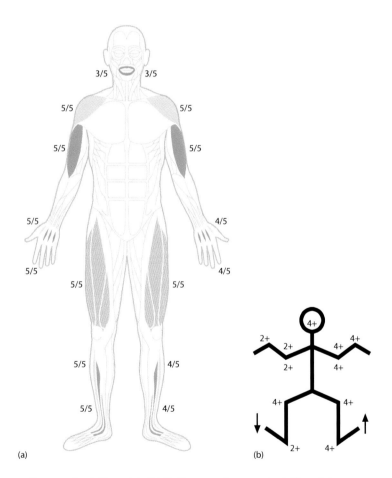

(a) (b)

FIGURE 6.11: (a) Motor map for case 6e-4 (Brenda). (b) Reflex map for case 6e-4 (Brenda).

6.8.5 CASE 6E-5: ALAIN, AGE 6

Background: A little boy is rushed in to the emergency room of the children's hospital just at the beginning of the morning shift, accompanied by his parents in a state of panic. Alain, a previously healthy active child, is now lying on a stretcher in a seemingly non-responsive state.

Chief complaint: Impaired upper limb coordination and unsteady gait of two days' duration

History:

- Upper respiratory tract infection two weeks ago followed three days later by fever (38.5°C) and left ear pain; treated successfully for left otitis media
- Lethargy, sleepiness, dizziness and anorexia beginning one week ago
- Beginning two days ago, *difficulty feeding himself*; kept missing his mouth
- On the day of assessment, a *staggering gait* with frequent falls; also appeared *not to recognize his parents*

Examination:

- Drowsy and lethargic; *disoriented for time and place*
- Normal vital signs while awake; once asleep, however, Alain became *hypotensive and hypopneic* (respiratory rate decreased from 22/min while awake to 8/min when asleep)
- *Dramatic horizontal and vertical nystagmus* but full range of eye movement
- Moderate weakness of eye closure and of perioral movement bilaterally
- Slurred speech with irregular rate of word production (*scanning speech*); absent jaw jerk
- Otherwise normal cranial nerve examination including fundi
- Generalized low limb muscle tone but normal strength
- Impaired alternating forearm movements; *marked dysmetria with past-pointing on finger-nose testing*
- *Severe truncal ataxia*; unable to sit without support
- Normal tendon reflexes and plantar responses

6.9 SUMMARY OF KEY ANATOMICAL INFORMATION

Facial nerve:
- The *facial nerve (cranial nerve, CN VII)* consists of *motor fibers* to all of the muscles of facial expression; *parasympathetic fibers* to the lacrimal, sublingual salivary and submandibular salivary glands; and *sensory taste* fibers from the anterior 2/3 of the tongue. The motor fibers are contained in the main trunk of the facial nerve while the parasympathetic and two types of sensory fibers are contained in a smaller adjacent branch, the *nervus intermedius*, located in the internal auditory meatus between the rest of CN VII and CN VIII.
- Facial nerve motor fibers originate in the *facial nerve nucleus*, located in the *lower lateral pontine tegmentum* (Figure 1.6b). Before exiting the brainstem, these fibers move superomedially around the nucleus of CN VI (abducens) before looping back inferiorly and laterally to exit the brainstem at the pontomedullary junction, just medial to CN VIII.

- The facial nerve nucleus has two components or subnuclei: one sending motor fibers to the *upper facial muscles* (forehead, eyelid regions), the other projecting to the *lower facial muscles* around the mouth. Upper motor neuron projections from the primary motor cortex for the *upper* facial muscles come from both motor cortices, while those projections destined for the *lower* facial muscles originate exclusively from the *contralateral* motor cortex.

Vestibulo-cochlear nerve:
- The *vestibulo-cochlear nerve (CN VIII)* contains special sensory fibers from two sources: *auditory* nerve fibers from the *cochlear labyrinth* and *vestibular* nerve fibers originating in the *vestibular labyrinth* (the *three semicircular canals* for the detection of angular acceleration of the head and body; the *utricle and saccule* for the analysis of the position of the head with respect to the ground, or source of gravitational pull).
- *Auditory nerve* fibers enter the brainstem at the pontomedullary junction, where they synapse with a group of ipsilateral and contralateral *auditory subnuclei* (Figure 6.1). From there, the auditory pathways derived from the inner ear in question ascend

bilaterally (contralateral > ipsilateral) forming the *lateral lemnisci* to the *inferior colliculi* in the lower midbrain; they then move onwards to the *medial geniculate nuclei* of the thalamus before terminating in the *primary auditory cortex (Heschl's gyrus)* in the floor of the lateral fissure.

- *Vestibular nerve* fibers enter the brainstem adjacent to the auditory fibers and synapse in one of four ipsilateral *vestibular nuclei* located in the lateral pons and medulla (Figure 6.3). From these nuclei, there are direct connections with the nuclei controlling the *extraocular muscles* (CN III, CN IV, CN VI) to mediate reflexive horizontal and vertical eye movements that serve to keep visual attention fixed on important targets while the head or body is tilting or rotating in space. There are also secondary *vestibulospinal connections* that mediate postural adjustments of the trunk and limbs while the body (or the 'platform' on which one is standing or sitting) is in motion.

Somatosensory:

- Sensory pathways originating in the trunk and limbs ascend through the entire brainstem en route to the *ventral posterolateral nucleus of the thalamus*. Discriminative touch and proprioceptive information ascend in the posterior columns to the lower medulla and, after synapsing in the *posterior column nuclei* (the gracilis nucleus medially and the cuneate nucleus laterally), cross the midline to form the *medial lemniscus* in the contralateral medulla. While ascending through the pontine and midbrain tegmentum, the medial lemnsicus moves gradually laterally before entering the thalamus (Figure 2.6a).

- Pain, temperature and crude touch pathways have already synapsed in the spinal cord and crossed near the point of entry and ascend as the lateral spinothalamic tract (Figure 2.6b; also called the anterolateral system). The tract ascends through the lateral medulla and continues through the lateral pons and midbrain to mingle with the ascending posterior column pathway (Figure 6.4).

Motor:

- *Descending motor pathways* (Figure 6.4) intended for the control of distal limb movements pass via the cerebral peduncles and pons proper to the medullary pyramids, where they decussate to form the *lateral corticospinal tracts* in the spinal cord (Figure 2.5 and Figure 6.4; see also Chapter 7 for more details).

6.10 CHAPTER-RELATED QUESTIONS

1. A 45-year-old man presents with a two-week history of increasingly severe right-sided facial weakness. There is no disturbance in speech. Physical examination reveals a moderately severe right-sided facial paralysis affecting the muscles involved in wrinkling of the forehead, eye closure and mouth movement. He also has absence of taste sensation in the anterior 2/3 of his tongue.

   Based on the clinical findings, which of the following sites is the most likely location of the disease process?

   a. The left precentral gyrus

   b. The right pontine tegmentum

   c. The proximal right facial nerve

   d. The distal right facial nerve near the stylomastoid foramen

2. Severe damage to the auditory component of the right vestibular-cochlear nerve results in marked hearing impairment in the right ear. What kind of hearing deficit would result from damage to the right medial geniculate body in the thalamus? Explain the reason(s) for your answer.

3. Imagine that you encounter a patient who has gradually lost the ability to understand spoken language and who behaves as though he were deaf. His ability to read is unimpaired. Physical examination reveals that, while he cannot understand a word of what you are saying to him, he responds normally to sound and has normal hearing thresholds. When tested using written instructions, he is unable to distinguish a high musical note from a low one.

   a. What is the most probable location of the disease process?

   b. If he could correctly identify pitch but not understand spoken language, what then would be the most likely location of the disease?

4. Based on information provided in Chapters 2 and 6, and working from first principles, reconstruct the resting position of the eye in a patient with a severe lesion in, and complete loss of

function of the right oculomotor nerve (CNIII). Explain your reasoning.

What will happen to the pupil of the involved eye?

5. Imagine that you are sitting in one of the chairs of a high-speed rotation ride at the Midway of an amusement park. Strapped into your seat and facing outwards, you are being rapidly spun in a circle in a clockwise direction. Assuming that you keep your eyes open and you attempt to look at your surroundings, describe what will happen to your eye movements.

6. a. Describe the sensory findings below the neck in a patient with a demyelinating lesion in the medial portion of the left side of the upper medulla.

 b. Same as for 6a, but with a lesion of the lateral portion of the upper medulla.

 c. What would be the sensory findings below the neck in the case of a similar lesion located instead in the lateral portion of the left upper pontine tegmentum?

7. a. What would be the motor system findings in a patient with a disruption of the corticospinal and corticobulbar pathways in the right cerebral peduncle in the midbrain?

 b. What would you expect to find if the lesion instead involved the corticospinal pathways in the right lower pons, as well as the right facial nerve nucleus?

SUGGESTED READING

Jackler, R.K., Driscoll, C.L.W., Eds. *Tumours of the Ear and Temporal Bone.* Philadelphia: Lippincott, Williams and Wilkins, 2000.

Musiek, F.E., Baron, J.A. *The Auditory System: Anatomy, Physiology and Clinical Correlates.* Boston, MA: Allyn & Bacon.

Roland, P., Marple, B.F., Samy, R.N. *Diagnosis and Management of Acoustic Neuroma. A self instructional package*, 2nd ed. Alexandria, VA: American Academy of Otolaryngology, 2003.

See also Annotated Bibliography.

Chapter 7

Bernie

Objectives

- Learn the main anatomical components of the central nervous system involved in the initiation, cessation and coordination of movement
- Understand the ways in which the cerebral cortex, basal ganglia, cerebellum, thalamus, brainstem and spinal cord collaborate in the control of posture, locomotion and complex hand movements
- Understand the symptoms and signs resulting from a disease process affecting each of the three principal central components of the motor control system: the direct and indirect descending motor pathways; the basal ganglia and their connections; the cerebellum and its connections
- Review the principal neurotransmitter molecules involved in the central movement control circuits

7.1 BERNIE

Bernadette, aged 32, and her partner Ben, a chef, own a French restaurant in Fort Lauderdale, Florida. Affectionately known to the locals as Benny and Bernie, Ben and Bernadette have become highly regarded fixtures in the dining establishment, offering an eclectic mixture of traditional French dishes and modern Continental cuisine. Bernie's medical history opens with the visit of her younger sister, whom she has not seen for three years.

After the initial excitement over the long-awaited reunion had dissipated, Bernie's sister became concerned about Bernie's changed appearance. Always an exuberant, vivacious (not to say hyperactive) personality, Bernie had become slow-moving, almost lethargic. Her face, normally a kaleidoscope of changing emotions, seemed devoid of expression, almost a mask. At times, particularly later in the day, Bernie had trouble keeping her head erect, tending to slouch forward at her desk or table, holding her head with her left hand.

At first, her sister told herself that Bernie was just very tired, possibly depressed. After all, the restaurant business is unrelenting in its demands on the owner's time and energy, with few possibilities for time off. As the days passed, however, the sister began to notice several other alarming symptoms in Bernie.

The most obvious problems occurred when Bernie walked from station to station in the restaurant kitchen. She had a tendency to keep her right arm held rigidly in front of her body rather than by her side, her elbow and fingers partially flexed. In addition, her right foot had a pronounced tendency to turn inwards at the ankle as she walked. Toward the evening, she typically walked slightly on her toes, unsteadily, and sometimes stumbled when walking upstairs. At the best of times, she walked at a slow pace with short steps. Finally, Bernie's handwriting had deteriorated enormously: her formed letters were very small, sometimes rendering her writing illegible. In the past, Bernie had always been the one who wrote the daily specials on wall-boards in the restaurant; about six months ago, she had to delegate this responsibility to someone else.

When Bernie's sister at last mentioned her concerns, Bernie conceded that she was concerned about her state of health but had concluded that she was just over-tired and needed a rest. After some reflection, her sister replied: 'Bernie, this is more than just being tired. I just realized that you are starting to look a bit like Daddy did around the time you married Ben. Daddy's doctor thinks he has a mild form of Parkinson's disease. At any rate, Daddy has done very well on his medication. Perhaps you need the same thing!'

Bernie required some convincing, but eventually she made an appointment with her family physician who – of course – turns out to be *you*.

Upon reviewing Bernie's file, you note that she has previously been in excellent health: your only contact with her has been for prenatal care during her pregnancy, then for the delivery of her son. She has been too busy at work to come for regular check-ups.

Your assessment of Bernie takes place at the end of the afternoon, in your 'add-on' slot. Bernie comes with her sister, having worked all through the lunch sitting at the restaurant. The general physical examination is normal; all of the abnormal findings concern the nervous system.

Bernie sits on the examining table in a stooped position; she seems to have trouble holding her head and upper trunk erect enough to look at you face-on. You perform a brief mental status exam that is perfectly normal, although the volume of Bernie's voice appears decreased or muffled. In conversation, she has virtually no facial expression, even when making an amusing, self-deprecating remark. In addition, you are struck by the fact that she almost never blinks.

During your examination of her upper limbs, you find that she has consistently normal muscle strength, proximal and distal. Passive movement around the wrists, elbows and shoulders, however, reveals a marked increase in muscle tone throughout the range of movement: at the wrist, for example, there is equal resistance to passive stretch whether the wrist is being flexed or extended. At the elbows, there is also a jerky, ratchet-like quality to the resistance, more on the right side. The fingers of the right hand are maintained in flexion at the metacarpo-phalangeal joints, and extension of the inter-phalangeal joints. On both sides, there is marked slowing in rapid hand and finger movements. A handwriting sample reveals a strikingly small character size.

In the lower extremities, muscle strength is also normal, but muscle tone is increased at the knees and ankles. In particular, there is high tone in the invertors of both ankles, worse on the right. When she walks, she is stiff-legged and tends to go up on her toes, her feet pointing inwards, right more than left – a configuration otherwise known as an equinovarus posture. She also seems unsteady and takes tiny steps. At the same time, the flexed posture of her right arm becomes more striking; there is no arm swing on either side. When she changes direction while walking, her whole body turns stiffly, as a single unit – 'en bloc'. Tendon stretch reflexes are uniformly increased while the plantar responses are flexor bilaterally; there is no ankle clonus.

When you comment on your positive findings to your patient, her sister points out that Bernie's gait and sitting posture, although far from normal, were much better at breakfast time than at present.

7.2 SOLVING THE CLINICAL PROBLEM

Before proceeding, as in previous chapters, the student should develop hypotheses concerning the probable localization of Bernie's disease process within the nervous system and the nature of that process, utilizing the history, examination, localization and etiology worksheets on the Web site. Once you have completed this process, you are ready to proceed with the remainder of the chapter.

7.3 MAIN CLINICAL POINTS

Bernie's *chief complaints* are that *she has difficulty walking and climbing stairs, as well as progressively deteriorating hand-writing.*

The essential elements of Bernie's medical history and neurological examination are as follows:

History
- Insidious onset of symptoms over a period of at least six months, possibly as long as three years, in the early fourth decade of life.
- Difficulty in maintaining the normal position of the head and upper trunk.
- Increasing difficulty walking and climbing stairs; a tendency to trip over her own feet.
- A progressive deterioration in hand-writing with a tendency to form excessively small letters (also known as *micrographia*).
- A worsening of symptoms over the course of the day.
- A parent with 'mild' Parkinson's disease, responding well to medication.

Physical examination
- On examination, normal vital signs and systems overview; normal mental status.
- A flexed posture of the head and upper trunk.
- Low voice volume; muffled speech; lack of facial expression; paucity of spontaneous blinking.
- Normal muscle strength, proximal and distal, but a diffuse increase in muscle tone in response to passive stretch; at the elbows, there is a 'ratchet' quality to the tone during passive movement.
- Abnormal posture of the right arm at rest, with flexion of the elbow, flexion of the metacarpophalangeal joints and extension of the interphalangeal joints.
- A tendency for the feet to turn inwards at the ankles (exaggerated inversion), with toe-walking, right side worse than left.
- Abnormally small stride length; absent arm swing while walking; turning en bloc.
- Increased muscle tendon reflexes.

As with the previous cases, it is important to recognize some important 'negatives':

- Although Bernie is very slow moving in general and has reduced finger movement speed, she is not weak, either proximally or distally, and does not have muscle wasting.
- She does not have a Babinski sign in either foot (see Chapters 2 and 3).

In short, Bernie has a gradual-onset disorder of postural control and speed of movement, possibly hereditary. Except for her motor control problem, she seems to be functioning normally.

7.4 RELEVANT NEUROANATOMY AND NEUROCHEMISTRY

As always, our first question concerning Bernie must be *where* in the motor system is the problem?

Unlike Fifi (in Chapter 4), Bernie is not weak, has increased rather than decreased muscle tone in response to stretch and has exaggerated rather than absent muscle tendon reflexes. In consequence, we can immediately exclude the presence of a lesion involving lower motor neurons. Although she does not have Babinski signs, Bernie's clinical findings otherwise suggest the presence of a problem at the level of the upper motor neuron.

This being the case, we should examine in more detail the organization of the upper part of the motor system.

As a first step, building on material provided in Chapters 1, 2 and 4, let us look at the ways in which movement commands are transmitted from the motor areas of the cerebral cortex to the lower motor neurons (final common pathway) in the brainstem and spinal cord.

7.4.1 DESCENDING MOTOR PATHWAYS

As we noted in Chapter 2, there are two principal pathways for the downward transmission of motor messages, the so-called direct and indirect pathways, both illustrated in Figure 7.1. The two pathways work together in (a) carrying out a specific complex movement with hand or foot and (b) achieving the body positions necessary to permit the performance of such a movement.

The *direct pathway* for movement control appears to have evolved later and has been added to the indirect pathway to allow for voluntary precision movements, primarily of the hands and fingers. As might be expected, it is particularly well developed in primates. The neurons transmitting messages required, let us say, to employ the right thumb and index finger in holding a piece of chalk while writing a menu have their cell bodies in the deeper layers of the motor cortex anterior to the left central fissure. The axons originating in these neurons carry their action potentials all the way to the gray matter of the cervical spinal cord in one fell swoop: they pass through the cerebral hemispheric white matter and enter the internal capsule, a large white matter tract adjacent to the basal ganglia. (As was discussed in Chapter 1, the basal ganglia are large central hemispheric gray matter nuclei that are involved in the initiation of movement – see Section 7.4.2.) The motor axons in the internal capsule then descend

lateral to the thalamus, subsequently passing downward into the anterior midbrain via the cerebral peduncles. From there, the axons voyage through the ventral part of the pons into the ventral medulla, where they form part of a visible bulge in the ventral medulla referred to as the pyramid (due to its pyramidal shape in cross-section).

From the left pyramid, the motor axons destined for the right cervical cord region pass posteriorly and downward through the lower medulla, crossing the midline and entering the upper cervical spinal cord as part of a distinct bundle known as the (right) *lateral corticospinal tract*. Once they have arrived at the level of the spinal cord involved, let us say, in finger-thumb apposition (C8 and T1), the axons enter the right anterior horn of the gray matter and synapse either directly on the appropriate anterior horn cells (lower motor neurons) or indirectly via an interneuron.

Thus, when Bernie used to write out the menu-du-jour on the restaurant blackboard, the intricacies of letter and word formation were primarily transmitted to the final common pathway in the cervical cord by the 'direct motor pathway.' Since the pathway passes through the medullary pyramid (while the soon-to-be-described 'indirect motor pathway' does not), it has come to be known as the *pyramidal motor system*.

Clearly, in order to write out the lunch menu, Bernie needs the cooperation of much more than the muscles involved in manipulating a stick of chalk. She will need to be able to maintain a stable, standing position, with her head steady, her eyes facing the board. She must keep her shoulder partially abducted (and stable), her elbow partly flexed and her wrist in a neutral position. All of the body and limb positioning, with respect to the targeted blackboard, is the job of the 'indirect motor pathway', often referred to by clinicians as the 'extrapyramidal motor system', a term that is simplistic and somewhat misleading.

Neurons involved in carrying messages to the necessary trunk and postural limb muscles have their cell bodies in more widespread areas of the cerebral cortex: not just the pre-central (motor) gyrus, but also the so-called 'premotor' cortex (anterior to the primary motor cortex) and even the 'sensory' cortex just posterior to the central fissure. These axons also travel downward through the cerebral white matter and internal capsule to the midbrain, pons and medulla, where they terminate in a variety of intermediary areas, the most important of which are the red nucleus (midbrain) and the pontine and medullary reticular formation (central brainstem). These 'indirect' motor relays are therefore referred to, respectively, as the corticorubral and corticoreticular pathways.

In turn, motor axons from the intermediary relay centres pass inferiorly to the spinal cord to the cervical, thoracic and lumbosacral cord regions, as appropriate, to connect via interneurons with anterior horn cells projecting to proximal arm and leg muscles, as well as trunk musculature. The spinal cord tracts containing these

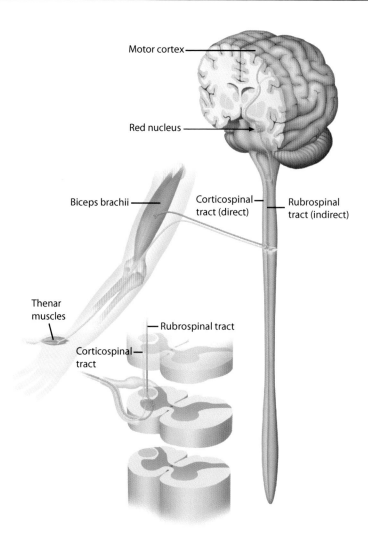

FIGURE 7.1: Direct and indirect descending motor pathways. The light green neurons constitute the direct pathway, and the dark green neurons, the indirect pathway.

motor axons are the rubrospinal tract (lateral column white matter) and the reticulospinal tracts (medullary and pontine – lateral and anterior columns respectively) – see representative examples in Figure 7.2.

All of these indirect pathway tracts, if disconnected from the sensorimotor cortex, exert differential tonic influences upon their anterior horn cell targets, the net result of which is an increase in tone in the trunk and proximal limb muscles. Lack of cortical modulation of the rubrospinal and medullary reticulospinal pathways results in increased tone primarily in the flexor muscles, especially of the upper limbs. Similar disinhibition of the pontine reticulospinal tract causes increased tone that predominates in the extensor muscles of all four limbs. It is for this reason that 'upper motor neuron' lesions are accompanied by increased muscle tone.

We will return to the effects of lesions in one or other of the 'direct' and 'indirect' motor control pathways when we consider in more detail the localization of Bernie's disease process. First, however, we must describe the components of the motor system involved in the initiation and cessation of movement (the basal ganglia connection), and in the precise coordination of movement, once initiated (the cerebellar connection).

7.4.2 BASAL GANGLIA: THE 'ACCELERATOR' AND THE 'BRAKE'

The location and anatomical description of the basal ganglia were outlined in Chapter 1. With the exception of the caudate nucleus, components of all the other basal ganglia participate in the overall apparatus of motor control. To these we must add two specific thalamic nuclei located in the anterolateral part of that structure: the ventral-lateral (VL) and ventral-anterior (VA) nuclei. Figure 7.3 shows the anatomical locations and relationships of the basal ganglia and thalamus.

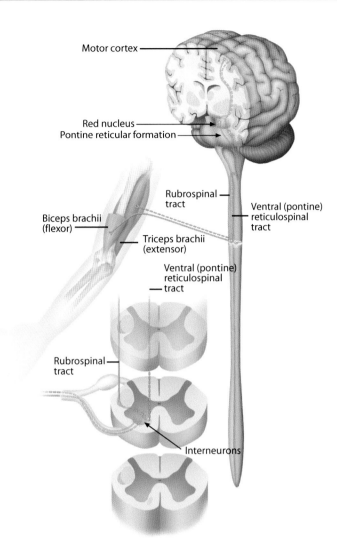

Motor cortex

Red nucleus
Pontine reticular formation

Rubrospinal tract

Ventral (pontine) reticulospinal tract

Biceps brachii (flexor)

Triceps brachii (extensor)

Ventral (pontine) reticulospinal tract

Rubrospinal tract

Interneurons

FIGURE 7.2: Indirect descending motor pathway: proximal flexor and extensor activities. The solid green neurons represent the pathway for the flexor muscle, and the dashed neurons, the pathway for the extensor muscle.

With respect to the initiation and cessation of movement circuitry, the key structures are, in what amounts to a functional sequence, the motor cortex, the putamen, the globus pallidus externa (GPe), the subthalamic nucleus (STN), the globus pallidus interna (GPi) and the VA/VL nuclei of the thalamus. The latter nuclei then complete an elaborate loop by projecting back to the motor cortex.

The anatomical circuitry and neurochemical sequences involved in the basal ganglia's role in movement control are complex. Since this is an introductory text, it is necessary, and useful, to over-simplify the situation sufficiently in order to communicate the basic principles required for an initial clinical approach to a patient with a movement disorder. From the practical point of view, therefore, we can divide the movement control circuitry of the basal ganglia into two components: a so-called 'direct' circuit responsible for the initiation of movement and an 'indirect' circuit whose function is the cessation of movement.

It is important for the student not to confuse the terms 'direct' and 'indirect' (in the context of basal ganglia circuitry) with the already described direct and indirect pathways conveying motor messages to the brainstem and spinal cord (see previous section).

The basic circuits for the direct and indirect basal ganglia connections are illustrated in Figure 7.4. The *direct* circuit is so named because there is a direct connection between the putamen and the GPi, the output from the GPi proceeding thence to the VA/VL thalamic nuclei. In contrast, the *indirect* circuit involves a diversionary loop from the putamen to GPe then STN, finally to GPi. In effect, the net role of the direct circuit is to *excite* the motor cortex, hence initiate movement, while the role of the indirect circuit is to *inhibit* the motor cortex.

In a sense, the direct circuit can be compared to the accelerator pedal in the automobile we have been intermittently employing to illuminate aspects of human

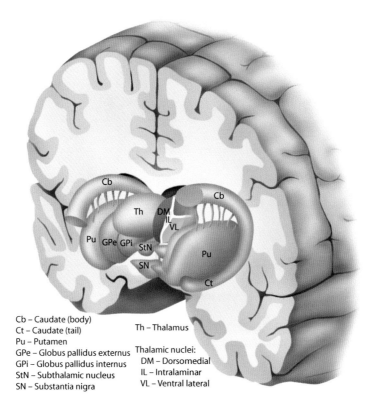

Cb – Caudate (body)
Ct – Caudate (tail)
Pu – Putamen
GPe – Globus pallidus externus
GPi – Globus pallidus internus
StN – Subthalamic nucleus
SN – Substantia nigra

Th – Thalamus

Thalamic nuclei:
  DM – Dorsomedial
  IL – Intralaminar
  VL – Ventral lateral

FIGURE 7.3: Basal ganglia: anatomical location and relationships. The left thalamus is sectioned in the coronal plane in order to illustrate some of its component nuclei.

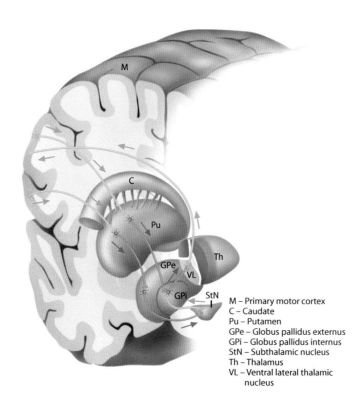

M – Primary motor cortex
C – Caudate
Pu – Putamen
GPe – Globus pallidus externus
GPi – Globus pallidus internus
StN – Subthalamic nucleus
Th – Thalamus
VL – Ventral lateral thalamic
       nucleus

FIGURE 7.4: Basal ganglia: motor circuitry, phase 1. For this figure, a green colour indicates an excitatory neuron (not a motor pathway neuron as is standard for this text); a red colour indicates an inhibitory neuron.

locomotion – push on the gas pedal and our car moves forward. The indirect circuit, on the other hand, is like the brake – press on it and our car stops moving. Using this analogy, and to avoid confusion, we will henceforth refer to the direct basal ganglia circuit as the *accelerator circuit*, and the indirect circuit as the *brake circuit*.

The normal function of the VA/VL thalamic nuclei is to excite the motor cortex while the normal function of GPi is to inhibit that excitation (see Figure 7.4). Thus, the 'normal' output from the last element in the chain within the basal ganglia would be to activate the brake; in order to initiate movement, the accelerator circuit inhibits GPi, *removes* the brake and permits increased activity in VA/VL. While this concept undoubtedly seems confusing, the student simply has to recall the basic mathematical principle that a 'minus' plus a 'minus' yields a 'plus': *inhibition of inhibition = excitation*.

For the brake circuit, in contrast, the output from the STN is excitatory: thus GPi's inhibitory function is increased, VA/VL activity is decreased and movement cannot occur. In effect, the *STN* represents a foot applied to the brake.

It is important to reiterate that the basal ganglia circuitry described in this section is a simplified version of a more complex picture that also involves cholinergic and gabaergic interneurons and inputs from other regions of the brain. Any reader whose interest has been piqued by the information in this section should read the excellent, recent review by Benarroch cited at the end of this chapter.

In summary, for Bernie to initiate a movement, whether a spontaneous smile or an upstroke with a piece of chalk, her putamen–GPi accelerator circuit must be functioning properly. In order to stop the chalk upstroke so that the succeeding down-stroke may take place, the GPe–STN–GPi brake circuit must also be doing its job.

Thus, for the 'soup-du-jour' to appear on the blackboard, both basal ganglia circuits must be in good working order. For the writing to be performed evenly and efficiently, and for it to be legible, however, a second set of motor control circuitry will be required: the cerebellar connection.

7.4.3 CEREBELLUM: THE 'STEERING WHEEL'

Put simply, the function of the cerebellum, as concerns movement control, is to compare information from cerebral cortex about movement commands just being transmitted with information from peripheral sensory receptors concerning the posture and speed of the part of the body being moved. This comparator function is a continuous process occurring, epoch by epoch, or millisecond by millisecond, throughout a given movement.

In other words, returning to the chalk-board analogy, the cerebellum at a given point in time is simultaneously receiving information from Bernie's left motor cortex concerning commands just being sent to her cervical cord (to allow for a further component of an upstroke movement) and information from muscle and joint sensors in her right arm and fingers concerning the part of the upstroke movement that took place a few milliseconds earlier. Her cerebellum, having analyzed these data, then sends a message back to the motor cortex that in essence is saying: 'fine, your previous command had this result, in comparison with what was intended – here is what should happen next to allow the action to continue in the manner desired'.

As was the case with the basal ganglia, the anatomy of the cerebellum is extremely complex. In order to address Bernie's clinical problem in a logical fashion, it is simply necessary for us to outline the broad aspects of the cerebellar connections and how they link with the rest of the apparatus controlling movement.

As for the basal ganglia, the location, anatomical components and attachments of the cerebellum were described in Chapter 1. With respect to its role in movement control, the main connections of the cerebellum are outlined in Figures 7.5 and 7.6.

Information from skin, joints and muscle spindle receptors is transmitted via the dorsal roots to the spinal cord by large myelinated axons. While some of these sensory data are then transmitted to the contralateral sensory cortex via posterior columns and the thalamus as conscious sensation, parallel sensory information not reaching conscious awareness is transmitted by *spinocerebellar tracts* to the cortical gray matter in the cerebellar hemisphere on the same side as the limb(s) concerned. Since, in the specific instance of Bernie writing on the blackboard, her right arm and hand are primarily doing the work, sensory information about right arm posture and position is being conveyed by the right spinocerebellar tracts to the right cerebellar hemisphere.

At the same time, axons originating in Bernie's left cerebral motor cortex (in parallel to corticospinal and corticoreticulospinal pathways destined for the right cervical spinal cord) pass downwards to the left side of the pons to synapse with a collection of relay neurons located in scattered nuclei in the base of the pons. Axons from these pontine neurons cross the midline and enter the *right* middle cerebellar peduncle, whence they travel to the same area of the right cerebellar hemispheric cortex as the sensory axons carrying information from the right arm.

Once these two streams of converging information have been compared and analyzed by the cerebellar cortex, appropriately consolidated information is then transmitted to neurons located in large nuclei in the central part of the cerebellar hemisphere (see Figure 7.5). For right arm movement, this information,

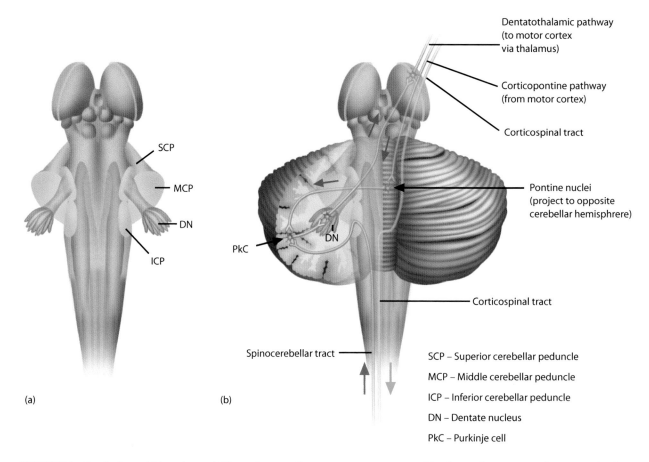

FIGURE 7.5: Cerebellum: (a) location and (b) attachments. Green represents neurons of the descending motor pathway; purple, those of the spinocerebellar (afferent) pathway; and blue, those involved in the dedicated cerebro-cerebellar circuitry.

originating in the largest neurons of the cerebellar cortex, the *Purkinje cells*, is transmitted to the right *dentate nucleus* in the central region of the cerebellar hemisphere. From there, information is relayed upward in axons located in the right superior cerebellar peduncle. In the midbrain, the axons cross the midline and enter the left red nucleus and either synapse with red nucleus neurons or pass through en route to the left thalamus, where they synapse with neurons in the VL thalamic nucleus. These latter neurons then project back to the left sensorimotor cortex to complete the feedback loop.

As we have already intimated, there are functional differences between parts of the cerebellum. The midline portion of the cerebellum (the vermis) and the para-midline regions of the cerebellar hemispheres are primarily concerned with integration of trunk postural control and leg movement, i.e. the facilitation of sitting, standing and walking. The lateral regions of the cerebellar hemispheres, by far the largest in terms of volume and surface area, are concerned with arm, hand and finger movements.

To return to Bernie and her menu board, the right cerebellar cortex is conveying information rostrally to the left cerebral motor cortex in a just-in-time fashion to help instruct the cortex how to direct the next part of the movement previously planned by the left supplementary motor area. In this way, for example, the chalk is held with consistent (rather than varying) pressure against the board and, in creating the letter D (for D'Hôte), is made to move upward just enough to make the letter the same height as all the other capital letters in the line. The chalk is then made to change direction, proceeding downward and to the right in a curving fashion to eventually arrive at the starting point of the letter D, no more and no less. One can thus imagine that, without the input of the cerebellum, the letter D might be too short or too high, the curve back to the origin poorly formed, and the line perhaps continued past the origin to produce a symbol unrecognizable as a 'D'.

Clearly, the process of successfully producing a letter D will also require constant visual feedback, information relayed from the visual cortex both to the motor cortex and to the cerebellum.

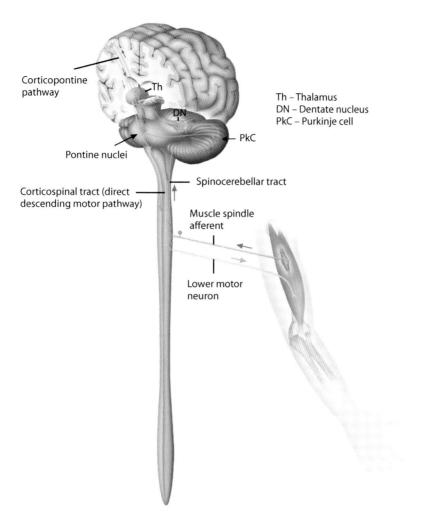

Corticopontine
pathway

Th

DN

Th – Thalamus
DN – Dentate nucleus
PkC – Purkinje cell

PkC

Pontine nuclei

Spinocerebellar tract

Corticospinal tract (direct
descending motor pathway)

Muscle spindle
afferent

Lower motor
neuron

FIGURE 7.6: Cerebellum: basic circuitry for motor function. The colour code is the same as for Figure 7.5.

Finally, to complete the automobile analogy, our car may be moving forward, at speed, but will drive into a ditch (or worse!) if, based on constant feedback as to the car's speed and direction, steering wheel adjustments are not constantly being made. Without the basal ganglia circuit, the 'automobile' will not leave its parking spot; without the cerebellar circuit, the 'automobile' will not stay on the road. Both components are required to get from point A to point B.

Although we have confined our discussion of the cerebellum to its role in movement control, it is important to recognize that the cerebellum also plays a role in the integration of higher brain functions such as emotional regulation, language, memory and learning.

7.4.4 NEUROTRANSMITTERS INVOLVED IN MOVEMENT CONTROL

The structure and function of neurotransmitter molecules will be considered in some depth in Chapter 11. At this stage in the text, it is simply necessary to note that a variety of transmitter molecules are involved in cell-to-cell signalling in the components of the motor control system discussed in this chapter.

The most important transmitters are glutamate (an excitatory compound) and gamma-aminobutyrate (also known as GABA, an inhibitory compound). Glutamate is the transmitter operating in the descending motor pathways (the upper motor neuron), as well as the excitatory components of the basal ganglia and cerebellar circuitry; the inhibitory components of the latter circuits are primarily mediated by GABA (see Figure 7.4).

Other transmitters implicated in basal ganglia circuitry are substance P, enkephalins and the monoamine transmitter dopamine; dopamine is the main neurotransmitter produced by neurons in the substantia nigra (SN) and will be discussed in more detail later in this chapter.

As is well known, the lower motor neuron transmitter at its synapses with muscle fibers is acetylcholine.

7.5 LOCALIZATION PROCESS

We have already concluded that, with hyperactive rather than absent muscle tendon reflexes and increased rather than decreased muscle tone, it is highly unlikely that Bernie has a disease process affecting peripheral nerves, neuromuscular junctions or muscles. Since her disorder includes diminished action in facial muscles, it is also improbable that she has a lesion confined to the spinal cord.

Bernie's symptoms and signs point to a disease process affecting the upper levels of the motor control system, but without any associated sensory or cognitive abnormalities. While she may have impaired abilities to walk, talk and write, she has lost none of her capacity to reason, to learn and to remember. This means that a diffuse disease process affecting cerebral cortex, central white matter, or entire central gray matter structures like the thalamus, is not likely.

Based on what we have already learned about the organization of the upper levels of the motor system, we have to consider three main possibilities:

1. The disease could be affecting the descending motor pathways, both direct (e.g. impaired finger movement control) and indirect (e.g. impaired postural control of body and limbs), either in the cerebral hemispheres or in the brainstem;
2. There could be a disorder involving the basal ganglia circuitry; or
3. A disorder affecting the cerebellum and its connections.

7.5.1 DESCENDING MOTOR PATHWAYS

Pure disruptions of the direct or 'pyramidal' motor control pathways are relatively rare as they require the presence of a lesion confined to the motor cortex (the precentral gyrus) or to the medullary pyramids. Lesions involving the cerebral white matter, internal capsules or cerebral peduncles tend to affect both the direct and indirect pathways as they travel intermixed at these levels. When an isolated dysfunction of the direct pathway does occur, it is characterized by a loss of independent finger movements and fine motor control in the hands (as well as an equivalent loss of rapid movements in the feet and toes) and an extensor plantar response (positive Babinski reflex). There is *no* significant disturbance in muscle tone. Thus, with an isolated direct pathway lesion, one could still pick up an apple and eat it but would have difficulty paring the skin off the fruit.

In contrast, the effects of an indirect 'extrapyramidal' motor pathway lesion are far more dramatic. Disruption of the corticoreticular and corticorubral connections (as, for example, with a lesion of the posterior limb of the internal capsule) results in an inability to move the hands and feet in a purposeful fashion, an increase in muscle tone in the limbs, hyperactive muscle tendon reflexes and extensor plantar responses. If the lesion is unilateral, the control of the posture of the corresponding arm and leg may be so compromised that the patient may be unable to bear weight on the leg or use the arm to reach for an object. When the lesion is bilateral, the patient cannot walk or even sit without support and requires a wheelchair.

In practice, most disorders affecting the descending motor pathways disrupt *both* the direct and indirect pathways, leading to a combination of the deficits outlined earlier. A lesion of the left internal capsule, for example, will result in a loss of fine finger movement in the right hand, loss of a pincer or even a crude grasp, difficulty extending the fingers and wrist and inability to dorsiflex the right foot and toes. Muscle tone in response to stretch on the right side will be increased, particularly in the finger flexors, wrist flexors and pronators, elbow flexors, knee flexors, plantar flexors and foot invertors. This combination of weakness and increased muscle tone results in a characteristic 'hemiparetic' posture, with fisted hand, drop-wrist posture, flexed elbow and tendency to walk on the toes with the foot inverted (an *equinovarus* posture). An example is shown in Figure 7.7.

With this description in mind, let us examine Bernie's clinical findings for similarities and differences. On the plus side, Bernie had a flexed posture of the forearm on the right, an equinovarus posture of the right foot, increased muscle tone in the arms and legs and abnormally brisk muscle tendon reflexes. That said, however, there are many important differences. In particular, formal muscle strength testing revealed *no* weakness, not even in intrinsic hand muscles. Rapid hand and finger movements are not abolished: they are just very slowly performed. The posture of her right hand was not characteristic of an indirect motor pathway deficit in that her interphalangeal joints were maintained in extension rather than flexion. Finally, Bernie did not have Babinski reflexes, a sine qua non of lesions involving the descending motor pathways.

Thus, the result of our comparison suggests that it is unlikely that Bernie has a pathological process primarily affecting the descending motor pathways.

7.5.2 BASAL GANGLIA CIRCUITRY

If we review the information already given about the motor control system and compare this with what we found on Bernie's neurological examination, we find a number of similarities. A major difficulty for Bernie was the spontaneous and voluntary *initiation* of movement: she did not blink; her face was expressionless; fine finger movements

FIGURE 7.7: Patient with left spastic hemiparesis following cerebral vessel thrombosis in the context of a relapse of inflammatory bowel disease. There is weakness of voluntary facial expression on the left, flexion of the left arm at the elbow and fisting of the left hand. (Patient's release and with permission; courtesy of Dr. P. Humphreys.)

were slow and her handwriting abnormally small; she was able to walk, but awkwardly and at a slow rate.

There were also several things, previously mentioned, that we did *not* find in our examination of Bernie, and that also support the possibility of a basal ganglia disturbance: the lack of muscle weakness per se, and the flexor, rather than extensor plantar responses.

Bernie also had two other findings consistent with basal ganglia dysfunction that we have not yet addressed in our consideration of this component of the motor control system: she had difficulty maintaining an upright posture of her head and upper trunk; there was a diffuse

increase in muscle tone affecting both agonist and antagonist muscles, sometimes with a 'ratchet'-like quality.

That the basal ganglia play a role in postural control is based largely on evidence from clinico-pathological correlations. Individuals with acquired bilateral injury to the basal ganglia (for example, children with dystonic cerebral palsy following perinatal anoxic damage to the putamen; adults with carbon monoxide poisoning and injury to the globus pallidus) characteristically have difficulty maintaining head control and a normal sitting/standing posture. This postural control mechanism is presumed to be mediated via the sensorimotor cortex and indirect motor pathway projections to the brainstem reticular formation.

Similarly, basal ganglia pathology, whether unilateral or bilateral, is characteristically accompanied by an abnormal tone pattern quite distinct from that resulting from lesions in the indirect descending motor pathways. Instead of tone being primarily increased in proximal and distal flexor muscles (see previous section), muscle tone in basal ganglia pathology typically involves opposing flexor and extensor muscles simultaneously, i.e. both agonist and antagonist muscles. If one were to passively flex and extend the elbow joint(s) in someone with an indirect motor pathway lesion (e.g. internal capsule infarct), there would be a marked resistance from the biceps brachialis but little resistance from the triceps. In addition, in this scenario, continued stretching of the biceps would typically result in an abrupt reduction in resistance as the muscle lengthened: the so-called 'clasp-knife' phenomenon. In contrast, a similar manoeuvre in an individual with a lesion in the basal ganglia (e.g. globus pallidus) would result in an equal degree of difficulty extending *and* flexing the elbow, without any clasp-knife phenomenon towards the end of the range of movement. We have clearly documented this latter phenomenon, also known as 'lead-pipe rigidity', in our examination of Bernie.

In certain specific basal ganglia disorders (to be described later), the rigid tone may be accompanied by a rhythmic series of brief tone reductions throughout the range of movement. This rhythmic tone fluctuation, present in muscles acting around Bernie's elbow joints, is known as 'cog-wheel' rigidity. Some individuals with cog-wheel rigidity (although not Bernie) also have a coarse *tremor* of the hands and fingers when the hands are not in use; this *resting tremor* is often referred to as a 'pill-rolling' tremor because it appears as if the person is rolling a small object between the thumb and fingers. In essence, the pill-rolling tremor is a low-frequency oscillation between wrist pronator and supinator muscle groups.

Finally, for reasons that are unclear but presumably relate to relative brain immaturity, children and young adults with basal ganglia dysfunction often demonstrate twisted trunk and limb postures. The patient's head may be uncontrollably turned to one side (torticollis); one arm may be held in extension across the front of the body, the wrist flexed and the fingers extended; one leg may be flexed at the hip joint while the other is extended. These various twisted postures are collectively referred to as *dystonia*.

On the whole, therefore, Bernie has demonstrated many characteristic features of a disorder of basal ganglia circuitry.

7.5.3 CEREBELLUM AND ITS CONNECTIONS

Since, as we have seen, the cerebellum is primarily involved (with respect to movement control) in the ongoing smooth coordination of a specific movement throughout its performance, a disturbance in cerebellar function would be expected to result in poorly coordinated movement. Such is, in fact, the case. Movements in a person with cerebellar dysfunction, whether he or she is walking across a room or writing on a blackboard, are not so much slow as they are erratic.

We have already noted (Cerebellum: The 'Steering Wheel' section) that the midline and paramedian portions of the cerebellum are largely concerned with control of trunk movements. Thus, it would not be surprising to learn that patients with midline cerebellar dysfunction tend to stagger when walking and to veer sideways. With severe disruption of function, they may be unable to maintain a standing or sitting posture without lurching over, a phenomenon known as *truncal ataxia*. It is often said that patients with midline (vermis) cerebellar lesions walk as if they are drunk: this is no coincidence as midline cerebellar function is easily suppressed by ethanol. Indeed, the vermis is selectively vulnerable to permanent damage from chronic ethanol abuse.

On the other hand, impaired function in the cerebellar hemispheres results in incoordination of limb movements, both of arms and legs. In reaching for an object, for example, the patient is unable to gauge exactly how far to extend the arm and may either undershoot or overshoot the object (a phenomenon referred to as *dysmetria*). At the same time, given that antagonistic muscles cannot be respectively contracted and relaxed in a smooth, graded fashion, the antagonistic muscles tend to shorten or lengthen in an irregular, step-wise fashion – this results in a coarse, oscillating tremor during the intended activity, known as an *intention tremor*.

As was the case with basal ganglia dysfunction, muscle strength is fairly normal in cerebellar disorders. When someone with cerebellar dysfunction is requested to contract a muscle as strongly as possible (for example, to resist a powerful pull against the muscle), there will be difficulty contracting the muscle in a consistent fashion such that the tension may appear to vary from moment to moment. Likewise, there may be difficulty reducing the amount of muscle contraction in a regular, timely fashion

(as, for example, when the pulling force against the limb has been suddenly reduced), resulting in the limb overshooting – the *rebound* phenomenon.

As might be anticipated, handwriting in an individual with cerebellar dysfunction is irregular in size and spacing of letters. Rather than being uniformly small, the letters tend to be large and poorly formed, often unrecognizable.

Rather than being increased, muscle tone in cerebellar disorders is typically low, and muscle tendon reflexes are somewhat reduced, rather than hyperactive. In addition, once a patellar reflex has been elicited in someone with a cerebellar disorder, there is a transient, low-amplitude anterior–posterior oscillation of the leg below the knee for one to two seconds. This phenomenon is known as a *pendular reflex*; in essence, the reflex is not properly damped following its appearance.

Taken individually and together, the characteristic motor abnormalities in an individual with cerebellar pathology appear completely different from what we have noted on Bernie's neurological examination.

Thus, of the three main components of the upper echelon of the motor control system, the *basal ganglia circuit* seems the likely target of Bernie's disease process. Our discussion of the various localization possibilities for Bernie's disease process is summarized in a completed *Localization Matrix* (Table 7.1).

7.6 ETIOLOGY – BERNIE'S DISEASE PROCESS

Given that Bernie has had her symptoms for months, possibly years, a review of Figure 4.8 would lead one to conclude that the timeframe is too long for the paroxysmal, traumatic and vascular disease categories, but that all of the other categories are theoretically possible. As in other chapters, we will briefly consider each disease category, focusing most of our attention on the degenerative and genetic categories as these appear by far the most likely.

7.6.A PAROXYSMAL

Paroxysmal disorders of the basal ganglia are rare but well described. The most important example is paroxysmal kinesigenic dystonia, an autosomal dominant hereditary disorder. Affected individuals, appearing otherwise normal, suddenly develop unilateral or bilateral limb dystonic posturing while in the process of standing up or starting to walk or run. As a result, the patient is unable to continue with the activity and may fall to the ground. Even though this is not an epileptic disorder, the symptoms usually respond well to antiepileptic drugs such as carbamazepine.

TABLE 7.1: Expanded Localization Matrix

	Right	Left	MUS	NMJ	PN	SPC	BRST	WM/TH	OCC	PAR	TEMP	FR	BG	CBL
Attention	N													
Memory	N													
Executive	N													
Language	N													
Visuospatial	N													
	Right	Left												
CRN I	N	N												
CRN II	N	N												
CRN III, IV, VI	N													
CN V	N	N												
CRN VII	Decr expr	Decr expr	⊘	⊘	⊘		⊘	⊘				⊘	○	
CRN IX, X	Decr voice	Decr voice	⊘	⊘	⊘		⊘	⊘				⊘	○	⊘
CRN XI														
CRN XII														
Tone	↑	↑				⊘	⊘	⊘				⊘	○	⊘
Power	N	N				⊘	⊘	⊘				⊘	○	⊘
Reflex	↑	↑				⊘	⊘	⊘				⊘	○	⊘
Involuntary Mvt														
PP														
Vib/prop														
Cortical sensation														
Coord UL	Sm print	Sm print				⊘	⊘	⊘				⊘	○	⊘
Coord LL														
Walk	Sm steps	Sm steps				⊘	⊘	⊘				⊘	○	⊘
Toe/heel/invert/evert														
Tandem														
Romberg														

Abbreviations: Decr expr, decreased expression; Decr voice, decreased voice intensity; N, normal; Sm print, small print (handwriting); Sm steps, small steps.

7.6.B TRAUMATIC

Although rather unusual, one of the lentiform nuclei may undergo hemorrhagic contusion with a closed head injury. Initially, the patient is typically in deep coma with a flaccid hemiparesis; the latter eventually evolves into a severe dystonic hemiparesis. Clearly, neither the history nor the timeframe supports the idea of a traumatic etiology.

7.6.C VASCULAR

Both ischemic and hemorrhagic strokes may selectively damage the basal ganglia, usually the caudate head, putamen and globus pallidus in combination, typically in one cerebral hemisphere. If the patient survives the event – less likely in the case of a basal ganglia hemorrhage – there will again be a dense dystonic hemiparesis. Isolated infarction or hemorrhagic injury to the STN in one hemisphere leads to a characteristic movement disorder in which there are wild, flinging movements of the contralateral arm and leg, a phenomenon referred to as *hemiballismus*. Vascular disorders of the cerebral hemispheres will be considered in detail in the next chapter.

7.6.D TOXIC

Both basal ganglia may be selectively damaged by a number of neurotoxic agents of which the most common examples are carbon monoxide, manganese and methylmercury. In the case of the former toxin, the onset of symptoms is acute; they appear shortly after the patient awakes from an initial comatose state. With methylmercury poisoning – typically following the protracted consumption of contaminated fish – the symptoms evolve over weeks to months.

7.6.E INFECTIOUS

Some neurotropic viruses (e.g. herpes simplex, influenza A, Japanese encephalitis virus) may primarily affect the basal ganglia and produce either a hyperkinetic movement disorder such as choreoathetosis or an isolated hypokinetic syndrome, typically with hypomimia and difficulty chewing and swallowing. Such localized encephalitides, however, are invariably of acute (days) or subacute (weeks) length and do not gradually evolve over many months.

Bacterial meningitis – especially tuberculous meningitis – may produce thromboses of small vessels entering the brain from the Circle of Willis (see Chapter 8). The result is a putaminal infarction and a characteristic form of dystonic hemiparesis in which the involved arm is extended at the elbow, hand fisted, and partly abducted or flexed at the shoulder, as if the person were trying to push open an imaginary door.

7.6.F METABOLIC

Most metabolic diseases affect the cerebral hemispheres as a whole, rather than specifically targeting the basal ganglia. A characteristic gliosis with large protoplasmic astrocytes develops in the basal ganglia, as elsewhere in the brain, in some patients with chronic hepatic failure; this phenomenon may help explain flapping tremor (or asterixis) and cog-wheel rigidity sometimes seen in such patients. Although this suggests some similarity with Bernie's clinical picture, the presence of hepatic failure would be known well in advance of the development of neurologic symptoms.

Selective metabolic injury to the basal ganglia is well known to occur in neonates. Short-duration systemic anoxia (such as may occur in the context of uterine rupture during labour or with premature placental separation) may result in selective injury to the posterior putamen, lateral thalamic nuclei and motor cortex. Injury to the globus pallidus, on the other hand, occurs in neonates with very high circulating blood levels of unconjugated bilirubin.

7.6.G INFLAMMATORY/AUTOIMMUNE

Some post-infectious autoimmune disorders may produce basal ganglia dysfunction, usually in a subacute fashion over several weeks. An excellent example is post-streptococcal choreoathetosis, or *Sydenham's chorea*. *Chorea* refers to an uncontrollable, rapid, dance-like movement of the fingers, hands, toes, feet or even the face and the vocal apparatus. *Athetosis* refers to a slower, more proximal, twisting movement, usually of the upper limbs, when an attempt is made to reach for an object. Chorea, typically unilateral, is also a common manifestation of cerebral involvement in the collagen-vascular disease *systemic lupus erythematosus*. Quite apart from the fact that the time course of Bernie's neurological disorder is much more protracted than for Sydenham's chorea or lupus, her movement disorder is a paucity or absence of movement, rather than an excess.

7.6.H NEOPLASTIC

Primary central nervous system tumors (gliomas, germinomas, meningiomas) and metastatic malignancies may selectively involve the basal ganglia. A neoplastic disease is highly unlikely in Bernie's case, however, for the simple reason that the disease process seems to involve the basal ganglia bilaterally. A tumor would likely affect one cerebral hemisphere, but not both sets of basal ganglia simultaneously without involving other structures such as the internal capsules and thalami.

7.6.I/J DEGENERATIVE/GENETIC

There is a wide variety of degenerative disorders affecting the basal ganglia; many, but not all, are of genetic origin. Before considering the most important of these disorders – and the most probable in Bernie's case – it would be helpful at this point to consider in more detail the specific basal ganglia syndrome Bernie has developed.

7.6.I/J.1 THE PARKINSONIAN SYNDROME

That Bernie might have a type of parkinsonian syndrome (often abbreviated as 'parkinsonism') was recognized by her sister, who had already witnessed similar symptoms appearing at a somewhat later age in their father. Parkinsonism refers to a specific collection of symptoms and signs that result from a common sporadic degenerative disease of a component of the basal ganglia, *Parkinson's disease*. This disease, first described in the early nineteenth century by James Parkinson, is one of the most common causes of progressive, severe motor disability in the sixth to eighth decades of life.

The principal features of idiopathic Parkinson's disease include a resting *t*remor of the pill-rolling type, increased muscle tone of a *r*igid type, slowness of movement (bradykinesia) or complete inability to initiate movement (*a*kinesia) and *p*ostural control problems. These four main features are most readily recalled by the use of the acronym t-r-a-p, standing respectively for tremor, rigidity, akinesia and posture. Affected individuals typically have an expressionless face with a fixed stare and reduced blinking; drooling is common (Figure 7.8a). Movements are performed slowly and with effort; handwriting is micrographic (Figure 7.8b). In standing, parkinsonian individuals tend to be stooped forward, with neck and elbows semiflexed (Figure 7.8c). The gait is slow and shuffling, with reduced arm swing and a continuous tremor of the fingers. Postural control is poor and balance easily lost, particularly if the ground is uneven. The rigid tone is obviously more apparent on physical examination, with cog-wheeling at the elbows particularly prominent.

Later in the illness, many affected persons may develop features of a progressive dementia (see Chapter 12).

(a)

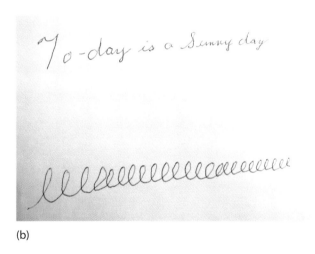

(b)

(c)

FIGURE 7.8: Clinical manifestations of Parkinson's disease. (a) Mask-like facies. (b) Micrographia. (c) Lateral view of patient showing typical trunk posture. (Patient's release and with permission; courtesy of Dr. D. Grimes.)

Postural hypotension is also a common problem, as are other autonomic instabilities.

The cardinal pathological abnormality in idiopathic Parkinson's disease is degeneration of the large, pigmented neurons in the SN, the most inferiorly placed component of the basal ganglia located in the midbrain posterior to the cerebral peduncles (Figures 7.3, 7.9a and 7.10b). Some degenerating neurons have prominent intra-nuclear inclusions, so-called Lewy bodies (Figure 7.9b); these are accumulations of a compound known as α-*synuclein* or *ubiquitin*. Neuronal degeneration and α-synuclein aggregation also occur in a number of other brainstem nuclei as well as the hypothalamus, amygdala (see Chapter 13) and the cholinergic basal forebrain nucleus of Meynert (see Chapter 12).

The cause of Parkinson's disease is incompletely established and probably varies among individual patients. Known predisposing factors, probably acting in varying combinations, include a genetic predisposition, pesticide exposure, head injury earlier in life, use of beta-blocking drugs, drinking well-water and agricultural occupations (the latter two often seen in the same individuals). In contrast, there are a number of factors that seem to protect against the development of Parkinson's disease; these include the use of calcium channel blocking drugs and non-steroidal anti-inflammatory agents, regular coffee drinking and less desirable practices such as tobacco smoking and ethanol abuse.

While Bernie obviously has many features of sporadic idiopathic Parkinson's disease, at 32, she is well

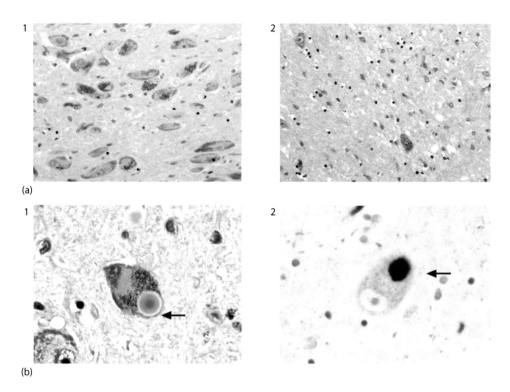

FIGURE 7.9: Neuropathological features of Parkinson's disease. (Courtesy of Dr. J. Woulfe.) (a) Panel 1 – photomicrograph of normal SN neurons. Panel 2 – SN neurons in Parkinson's disease showing depigmentation and neuronal loss. (b) Panel 1 – high power photomicrograph showing typical Lewy body (arrow). Panel 2 – Lewy body, ubiquitin stain (a different neuron from the one shown in Panel 1).

below the typical age range for the disorder. In addition, her father also has parkinsonism, apparently of a mild, less aggressive form than would normally be seen in someone who is presumably around 60 years of age.

As we have just suggested, familial forms of parkinsonism do exist, typically beginning in the third to fifth decades of life, and are pathologically distinct from sporadic Parkinson's disease; the clinical picture is often accompanied by some elements of *generalized dystonia*. Dementia and severe postural instability do not typically develop in early-onset, familial parkinsonism. A large number of genes have been implicated in the production of familial parkinsonism, the most common mutations being in the *parkin* gene at 6q25.2. *Parkin* mutations do not usually produce significant nigral neuronal degeneration unless homozygous and thus appear as autosomal recessive disorders, sometimes as early as age 10. There are also a number of parkinsonism-associated genes that are typically autosomal dominant in inheritance pattern, the most important of which are *SNCA* and *LRRK2*.

Not all parkinsonian syndromes are neurodegenerative in nature, however. One of the most striking exceptions is *dopa-responsive dystonia*, usually an autosomal dominant disorder with variable expressivity whose mechanism is a defect in dopamine synthesis. In its most aggressive form, the disorder becomes manifest around

the middle of the first decade of life with progressive limb dystonia, legs more than arms, and impaired control of head and trunk posture. Milder forms of the disease produce a mixture of dystonic and parkinsonian features beginning in early- to mid-adulthood. For both the early and later forms of the disorder, there is often a characteristic diurnal variation in symptom severity, with symptoms becoming progressively more severe over the course of the day, and improved after a nap or rest. You will recall that Bernie had a significant diurnal variation in symptom severity, as witnessed by her sister.

From what we have learned in this section, it appears most likely that Bernie has either a form of familial, early-onset parkinsonism/dystonia or a later-onset form of dopa-responsive dystonia. Before considering other potential diagnoses, let alone what might be done to help Bernie, however, we must focus on the SN and its role in parkinsonian-type movement disorders. As well, as suggested by the self-explanatory term dopa-responsive dystonia, we must consider the role played by the monoamine transmitter dopamine in the control of posture and movement.

7.6.1/J.2 The SN

In considering the contribution of the basal ganglia to the control of movement, we have presented, for the sake of clarity and simplicity, only the basic elements of basal ganglia circuitry: the accelerator and brake mechanisms. We must now review the subsidiary and crucial role played by the SN.

There are reciprocal connections between the putamen and SN that complement the previously described circuits involving putamen, globus pallidus and STN. With respect to the observed deficits in the parkinsonian syndrome, the key component in the 'dialogue' between putamen and SN is the projection returning from SN to putamen: the *nigrostriatal pathway* (see Figure 7.10a).

As we noted earlier, the SN contains large pigmented neurons whose axons project to the corpus striatum (caudate, putamen, nucleus accumbens) as well as to the basal forebrain and frontal lobe cortex (see Chapter 13). Axons destined for the corpus striatum (in the case of movement control, the putamen) originate in the compact portion of the SN (pars compacta), while those destined for the basal forebrain originate in the adjacent 'loose' part, the pars reticulata. The main neurotransmitter elaborated by the synaptic terminals of the SN neurons, as you may already have guessed, is *dopamine*.

In the putamen, the dopaminergic terminals impinge on neurons involved in *both* the accelerator and brake circuits (Figure 7.10a). Even though the SN axons project the same chemical 'message' to both pathways, the effects are *opposite*. The explanation for this apparent contradiction can be found in the dopamine receptor subtypes present on the respective putaminal neurons for the two pathways. Putaminal neurons participating in the accelerator circuit (direct to GPi) elaborate a class of dopamine receptors that respond by exciting the cell (*D1 receptors*). The end result is a facilitation of the accelerator circuit and initiation of movement. On the other hand, the putaminal neurons participating in the brake circuit (via the STN) contain receptors that respond by inhibiting the cell (*D2 receptors*). In this case, the result is an inhibition of the brake circuit or, again, a facilitation of movement. Thus, through both accelerator and brake circuits, the main contribution of the nigrostriatal dopaminergic projection is, normally, to facilitate the initiation of movement.

You will immediately recognize, therefore, that a degeneration of pars compacta neurons, or a metabolic 'failure' of the nigrostriatal connection, will result in the opposite: a profound inability to initiate movement, one of the cardinal features of parkinsonism.

For sporadic Parkinson's disease and for early-onset familial parkinsonism/dystonia, as we have seen, the pathological substrate is progressive degeneration of dopaminergic neurons in the pars compacta of the SN. In the case of dopa-responsive dystonia, the nigral neurons remain intact but are depigmented; they are lacking an enzyme, usually GTP cyclohydrolase 1 (*GCH1*), whose function is to facilitate the first step in the synthesis of tetrahydrobiopterin (BH4), a key factor in the synthesis of dopamine (see Figure 7.11). BH4 is a necessary cofactor for conversion of the essential amino acid phenylalanine to tyrosine and, in turn, the conversion of tyrosine to dihydroxyphenylalanine (dopa), the immediate precursor of dopamine. Thus, a deficiency in the production of BH4 will result in a deceleration in the production of dopa/dopamine, particularly in situations where the nigrostriatal pathway is active for long periods. This inability of dopamine synthesis to keep up with 'demand' may explain why patients with dopa-responsive dystonia become more incapacitated as the day goes on, and partially recover overnight.

7.6.1/J.3 OTHER CAUSES OF PROGRESSIVE DYSTONIA

Since Bernie's clinical picture includes dystonic elements, we would also have to briefly consider other causes of progressive dystonia that are not typically associated with the parkinsonian characteristics of tremor, rigidity, akinesia and postural instability. There are, for example, a variety of hereditary dystonias, of which the most common is *idiopathic torsion dystonia*, or hereditary dystonia type 1, due to a mutation in the torsin gene. *Wilson's disease*, an autosomal recessive disorder of copper transport, produces a progressive dystonia, dysarthria and dysphagia typically beginning in the teenage years and is accompanied by visible evidence on magnetic resonance imaging (MRI) of degeneration in the corpus striatum. Rigidity and dystonia can also be seen in the early-onset forms of *Huntington's chorea*, an autosomal dominant neurodegenerative disorder of the corpus striatum due to a mutation in the huntingtin gene. At Bernie's age, however, chorea would be the predominant symptom. Finally, but not inclusively, progressive rigidity and dystonia can be seen in *familial strionigral degeneration*, an autosomal dominant disorder usually becoming manifest in the second to fourth decades.

The reasoning process behind the previous discussion of the most likely disease processes that can account for Bernie's clinical picture is summarized in a completed *Etiology Matrix* (Figure 7.12). As in other chapters, only those disease categories associated with clinical histories extending over the time period described in Bernie's story (months to years) are considered in this table.

7.7 SUMMARY AND SUPPLEMENTARY INFORMATION

To recapitulate, Bernie has a slowly progressive disease characterized by increasing difficulty maintaining

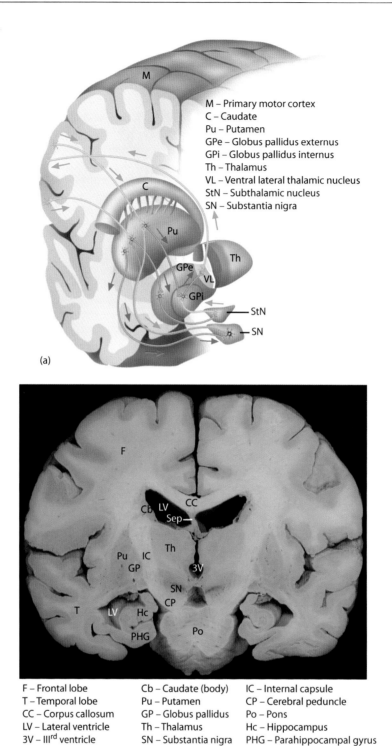

M – Primary motor cortex
C – Caudate
Pu – Putamen
GPe – Globus pallidus externus
GPi – Globus pallidus internus
Th – Thalamus
VL – Ventral lateral thalamic nucleus
StN – Subthalamic nucleus
SN – Substantia nigra

(a)

F – Frontal lobe   Cb – Caudate (body)   IC – Internal capsule
T – Temporal lobe   Pu – Putamen   CP – Cerebral peduncle
CC – Corpus callosum   GP – Globus pallidus   Po – Pons
LV – Lateral ventricle   Th – Thalamus   Hc – Hippocampus
3V – IIIrd ventricle   SN – Substantia nigra   PHG – Parahippocampal gyrus
Sep – Septum pellucidum

(b)

FIGURE 7.10: (a) Basal ganglia: motor circuitry, phase 2. For reasons explained in the text, nigrostriatal neurons have both a red and a green colour. (b) Coronal section of cerebrum showing the anatomical location of the SN and other central gray matter structures.

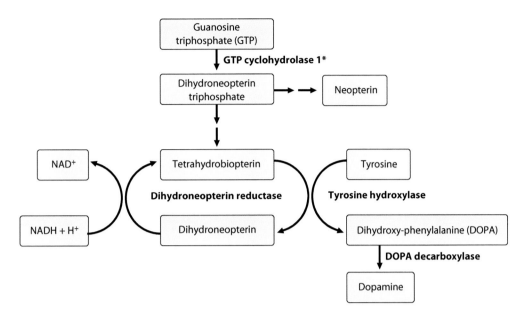

FIGURE 7.11: Dopamine synthesis pathway; * indicates the enzyme that is deficient in Bernie.

Disease Type	Secs	Mins	Hrs	Days	Weeks	Months	Years
Paroxysmal (seizures, faints, migraine)							
Traumatic							
Vascular (ischemic stroke, bleed)							
Toxic							⊘	
Infectious							⊘	⊘
Metabolic							⊘	⊘
Inflammatory/ autoimmune							⊘	⊘
Neoplastic							⊘	⊘
Degenerative							O	O
Genetic							O	O
	<-- Acute -->							
				<-- SubAcute -->			
						<-- Chronic -->	

FIGURE 7.12: Etiology matrix.

postural control, walking and writing. The main clinical findings of a mask-like facies, muscle tone rigidity, cogwheeling and abnormal twisted limb postures suggest a parkinsonism/dystonia disorder; this, in turn, places the likely focus of disease in the basal ganglia, in particular the SN. While idiopathic Parkinson's disease is the most common degenerative disease affecting these structures, Bernie's relatively young age and positive family history suggest that she has either familial parkinsonism/dystonia syndrome or dopa-responsive dystonia.

7.7.1 APPROACH TO THE INVESTIGATION OF PATIENTS WITH PARKINSONISM/ DYSTONIA

The initial work-up should include a cranial imaging study, preferably MRI. While no abnormality would be expected in sporadic Parkinson's disease, hereditary early-onset parkinsonism, dopa-responsive dystonia, or idiopathic torsion dystonia, evidence of basal ganglia atrophy and/or signal abnormalities may be found in disorders such as Wilson's or Huntington's disease. Wilson's disease, if suspected, can be confirmed by the finding of a low serum level of the copper transport protein ceruloplasmin.

In both familial and sporadic parkinsonian disorders, whether or not accompanied by nigral degeneration, there is a marked decrease in output of dopamine from the SN. This phenomenon can be documented by the finding of an abnormally low level of homovanillic acid (HVA), the main metabolite of dopamine, in the cerebrospinal fluid (CSF). CSF studies can distinguish between juvenile parkinsonism and autosomal dominant dopa-responsive dystonia/parkinsonism in that, while HVA levels will be low in both disorders, neopterin and tetrahydrobiopterin levels are also low in the latter disorder (see Figure 7.11) and normal in the former.

For Bernie, the most specific investigative tool is leukocyte DNA analysis for possible mutations in relevant genes. Thus, for early-onset autosomal recessive parkinsonism/dystonia, characterization of the parkin gene would likely detect a mutation. Likewise, for dopa-responsive dystonia, one would search for mutations in the GTP cyclohydrolase 1 gene; for idiopathic torsion dystonia, defects in the torsin gene; and so on.

If, after extensive investigation, a specific cause for the parkinsonian syndrome had not been found, it is possible to document the presence or absence of nigral degeneration with positron emission tomography using a labelled isotope of dopa, ^{11}F-fluorodopa. Fluorodopa, when injected into a patient, is taken up by nigral axonal terminals in the corpus striatum; the degree of uptake can be quantified by tomographic scanning of the corpus striatum. In both sporadic Parkinson's disease and familial early-onset parkinsonism/dystonia, fluorodopa uptake is markedly diminished in comparison with normal controls. In dopa-responsive dystonia, the number of dopaminergic terminals is not diminished and fluorodopa uptake is normal.

7.7.2 BERNIE'S INVESTIGATIONS

Bernie was referred to a local neurologist, who, after arranging for a series of investigations, gave her a trial of levodopa/carbidopa 100/10, ½ tablet three times daily (for explanation, see next section). Within 48 hours, there was a dramatic improvement in Bernie's gait, upper trunk posture and hand-writing. This result eliminated the possibility of idiopathic torsion dystonia as it does not respond to dopamine precursor medications.

An MRI study of the head was normal, as was a serum ceruloplasmin level.

A lumbar puncture (done with Bernie temporarily off levodopa/carbidopa) revealed low levels of HVA, neopterin and tetrahydrobiopterin. These results strongly suggested the diagnosis of dopa-responsive dystonia; the diagnosis was confirmed by the finding of a mutation in the *GCH1* gene. A subsequent extended investigation of the family revealed that Bernie's father had the same mutation while her sister's study was normal.

7.7.3 TREATMENT/OUTCOME

After her lumbar puncture, Bernie was restarted on the same small dose of levodopa/carbidopa. She continued to improve to such an extent that after three months on medication, her neurological examination was completely normal. Bernie has had no side effects of the medication and, a year later, has delivered a second child – a daughter who is eventually found to have her mother's GCH1 mutation and who will be treated promptly as soon as she becomes symptomatic.

Why the use of the levodopa/carbidopa combination?

Orally administered dopamine does not cross the blood–brain barrier in significant amounts. Its immediate precursor dopa does cross the barrier, particularly if combined with an agent that inhibits the peripheral degrader of dopa, dopa decarboxylase. Such combinations (known as levodopa/carbidopa) readily enter the brain, where the dopa is taken up by dopaminergic nerve terminals. The end result is a sustained, dramatic improvement in parkinsonian symptoms in both familial early-onset parkinsonism/dystonia and dopa-responsive dystonia, as well as in the early stages of sporadic Parkinson's disease.

When the parkinsonian disorder is due to nigral neuronal degeneration (as in classic Parkinson's disease), the number of nigrostriatal nerve terminals eventually becomes so small that dopa/carbidopa combinations are no longer effective. Up-regulation of dopamine receptor sensitivity also results, with exposure to dopa/carbidopa, in a variety of troublesome involuntary movements, referred to as *dyskinesias*. Loss of clinical drug response and dyskinesias are either avoided or delayed if one uses instead a dopamine agonist drug such as bromocriptine, pramipexole or ropinirole.

7.8 E-CASES

7.8.1 CASE 7E-1: FATIMA, AGE 69

Background: This well-dressed woman, a retired school teacher, presents at your office accompanied by her husband. They had been planning to spend the winter in Florida but decided to look into her condition first – a tremor affecting her hands.

Chief complaint: Increasingly severe hand tremor present for one year

History:

- Hand tremor *at rest*, right worse than left; disappears during hand activities
- Increasing stiffness of the right arm while brushing hair or teeth; handwriting small and difficult to interpret
- Increasing tendency to drool
- A tendency to lose her balance when changing direction while walking; trips over her feet while walking up an incline
- No family history of tremor or progressive gait deterioration.

Examination:

- Moderately obese; blood pressure = 160/90; otherwise, vital signs are normal.
- Normal mental status examination except for relative difficulty in remembering number sequences both forwards and backwards.
- *Almost no spontaneous facial expression; rarely blinks*; saliva sometimes escapes from the side of her mouth.
- *3–4 Hz resting hand tremor*, worse on the right.
- Normal muscle strength in all four limbs.
- Increased tone with passive stretch of both flexor and extensor muscles at the right elbow and wrist, with some *cog-wheeling.*
- Slow finger movement speed; normal finger-nose testing.
- *Walks with small, shuffling steps*; unable to maintain stance if she is pulled off balance.

7.8.2 CASE 7E-2: TADEUSZ, AGE 49

Background: This master carpenter is coming for his annual check-up, which is a routine part of your family practice unit. You had seen him six months previously for the assessment of low back pain and for other neuromuscular problems and had ordered some 'tests'.

Chief complaint: Muscle cramps and weakness in the left leg for about two years

History:

- Low back pain provoked by heavy lifting for 10 years.
- Cramps in the left calf and foot accompanied by gradual development of a foot-drop gait over a two-year period.
- Six months ago, your examination had shown *weakness and wasting of muscles in the left anterior tibial compartment* (foot dorsiflexors and evertors); MRI showed degenerative disc changes at L4–L5, L5–S1; referred to a neurosurgeon and put on waiting list for surgery – still there!
- In the past six months, *increasing difficulty walking upstairs.*
- Decreased grip strength in the right hand.
- Irregular twitches have developed in the muscles of both thighs.

Examination:

- Occasional spontaneous muscle twitches (fasciculations) in the *tongue*
- *Fasciculations in the muscles of the right forearm and both quadriceps*
- Obvious muscle wasting in the left anterior tibial area, the *left calf and in the intrinsic muscles of the right hand*
- *Bilateral spasticity in the hamstring muscles*
- Mild-moderate weakness of the right hand interossei, finger flexors/extensors, wrist flexors/extensors; both *deltoids*; left tibialis anterior, foot evertors and toe extensors; *right tibialis anterior* (see Figure 7.13a)
- Reflexes: Figure 7.13b
- Normal sensory examination

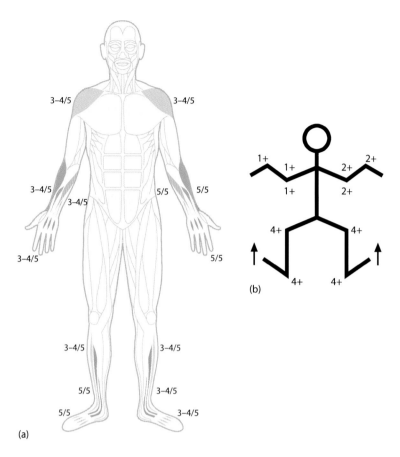

FIGURE 7.13: Motor examination (a) and reflex (b) maps of case 7e-2 (Tadeusz).

7.8.3 CASE 7E-3: KHEDER, AGE 8

Background: Your normally calm afternoon pediatric clinic is disrupted by the noisy entrance of a distraught mother and her son, one of your regular patients.

Chief complaint: Restlessness and personality change present for the past six weeks

History:

- Restless, fidgety, distractible and irritable for the past six weeks; previously calm and introspective.
- Inability to concentrate on school-work; he failed a math test for the first time in his life.
- *Uncharacteristically clumsy*, injuring himself by collisions with door jambs or by falling upstairs.
- *Preoccupied with avoiding 'germs'*; Kheder washes his hands many times a day and avoids touching family members.
- Episode of streptococcal pharyngitis two months ago, treated with antibiotics.

Examination:

- Normal vital signs; substernal mid-systolic murmur
- Cooperative but easily distracted
- *Intermittent slurring of speech*; irregular rhythm of word production
- *Frequent involuntary movements*: sudden, transient, irregular movements of fingers, toes; occasional dramatic proximal limb movements that he tries to mask by converting them to something purposeful; *frequent, involuntary facial grimaces*
- Normal formal muscle strength testing but difficulty *sustaining hand grip*, with intermittent relaxation/reapplication
- With arms outstretched, wrists are flexed and fingers extended
- Mild generalized hypotonia; normal tendon reflexes and plantar responses

7.8.4 CASE 7E-4: DEEPA, AGE 1

Background: Young parents have brought their daughter in for assessment of her development; they have recently moved into your town and are clearly extremely anxious.

Chief complaint: Severe gross motor developmental delay

History:

- *Deepa is bright, sociable and vocal but is unable to sit without support, to crawl or to pull to a standing position*
- Unable to reach for and to grasp an object because of *involuntary proximal arm movements*
- Able to swallow infant formula and pureed foods but chokes on chopped food
- Delivered at term by emergency caesasian section because of *umbilical cord prolapse* before the advancing head and resulting severe fetal bradycardia
- *Apgar scores* at 1, 5 and 10 minutes were *1/10, 3/10* and *6/10*, respectively; required vigorous resuscitation with intubation and temporary mechanical ventilation
- *Unable to suck or swallow for the first 3 weeks of life*, obliging the use of a nasogastric tube; thereafter was unable to breast-feed and required bottle-feeding with a large-holed nipple

Examination:

- Alert, sociable infant with good eye contact; borderline small head circumference (44 cm)

- *Mouth tends to remain open most of the time, with periodic, stereotypic tongue movements*
- Unable to sit without support; periodic loss of head control
- *Erratic, swinging arm movements when Deepa attempts to reach for objects*; unable to bring an object placed in her hand toward her mouth
- Generally low muscle tone when Deepa is tested in a supine position; *high tone in the limb extensor muscles when she is held upright*
- When her head is turned to one side, Deepa adopts an invariable, stereotyped posture of the limbs: *extension of the arm/leg on the side toward which her head is turned, and flexion of the contralateral arm/leg* (see Figure 7.14)
- *Reflexes*: see Figure 7.15

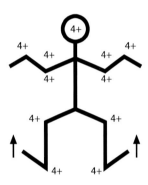

FIGURE 7.15: Reflex map (Deepa).

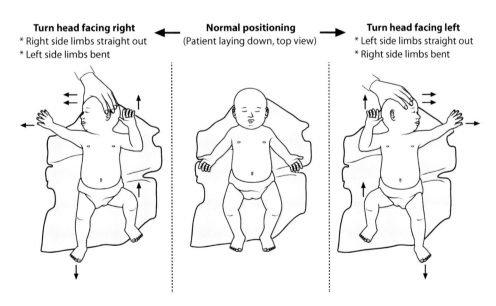

Turn head facing right
* Right side limbs straight out
* Left side limbs bent

Normal positioning
(Patient laying down, top view)

Turn head facing left
* Left side limbs straight out
* Right side limbs bent

FIGURE 7.14: Drawing demonstrating the clinical finding of an obligatory asymmetric tonic neck reflex (Deepa).

7.8.5 CASE 7E-5: MAURICE, AGE 13

Background: One of your teenaged patients is coming in for his annual assessment. His parents are increasingly concerned about his motor coordination. His visit the previous year was cancelled because of a death in the family.

Chief complaint: Increasing generalized clumsiness over the past two years

History:

- Progressively decreasing maximum walking speed; *a tendency to lose his balance and fall on uneven ground*
- Has *lost* the ability to ride his bicycle
- Does well academically but there has been a *marked deterioration in hand-writing*, forcing him to revert to printing
- Increasing *slurring of speech*
- Normal developmental history; parents are both French-Canadian, but unrelated

Examination:

- High-arched feet; toes are hyper-extended at the proximal interphalangeal joints and flexed at the distal interphalangeal joints ('hammer-toe deformity' – see Figure 7.16); mild scoliosis
- Slurred speech with irregular speed of word production
- *Bilateral horizontal nystagmus*
- Has difficulty moving his tongue from side to side
- Normal muscle strength in limbs but low tone
- *Dysmetria and past-pointing on the finger-nose test*

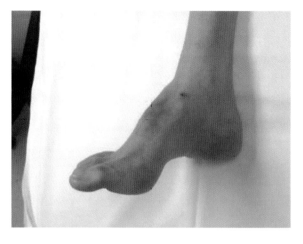

FIGURE 7.16: Photograph of the characteristic appearance of the feet in a patient with hammer-toe deformities (Maurice).

- *Reflexes*: see Figure 7.17
- Broad-based, ataxic gait
- *Impaired proprioceptive sensation in the toes of both feet*

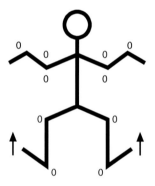

FIGURE 7.17: Reflex map for case 7e-5 (Maurice).

7.8.6 CASE 7E-6: MORDECAI, AGE 18

Background: You have agreed to see this young man for a second opinion because of concerns regarding his unusual gait. The cause of this problem that has been proposed by the previous physician has provoked a vehement rejection of the diagnosis by the parents and their subsequent decision to change physicians.

Chief complaint: Increasingly abnormal limb postures over the past six years

History:

- At age 12, Mordecai gradually developed an extremely *dorsiflexed posture of his left foot*, leading to difficulty walking and loss of the ability to wear hockey skates.
- At around age 14, gradual onset of 'stiffness' in the right hip, accompanied by an *inability to flex his hip during the gait cycle*; instead, he had to circumduct his leg at the hip in order to bring the leg forward.
- The gait problem would gradually *improve if Mordecai started running*; his ability to ride a bicycle was *unimpaired* – in consequence, many physicians concluded that he was 'faking'.
- At age 16, Mordecai developed 'writer's cramp' and had to begin writing with his left hand.
- His father also has writer's cramp; father's sister has had progressive generalized dystonia since age 18, etiology unknown but under investigation.

Examination:

- Normal general physical examination and vital signs

- Mordecai sits on the examination table with his spine hyper-extended, his left hip flexed, his left knee extended and his foot dorsiflexed; he is able to *eliminate this postural abnormality if requested* but it returns as soon as he is distracted.
- Normal cranial nerve examination
- Normal muscle strength and bulk in all four limbs
- *Increased muscle tone in the dorsiflexor muscles of the left foot, the left toe extensors and the right hip extensors*

- Normal rapid finger movements, toe tapping, tendon reflexes and plantar responses
- When walking, his left foot is dorsiflexed and his right leg circumducts; at the same time, *he fists his left hand and does not swing his left arm*; he also has a cane in his right hand to prevent falling
- Mordecai requires 2–3 seconds to change from walking to running, discards his cane and then has fisting of both hands; he has trouble changing from running to walking without falling

7.9 SUMMARY OF KEY NEUROANATOMICAL AND NEUROPHYSIOLOGICAL INFORMATION

There are two separate final common upper motor neuron pathways in the brain and spinal cord: the *direct* and *indirect* pathways (Figure 7.1).

- The *direct pathway* consists of motor axons originating in the precentral gyrus and descending 'directly' through the internal capsule, through the brainstem (via the cerebral peduncle [midbrain], ventral pons, pyramid [medulla]) to the appropriate cranial nerve motor nucleus, then *crossing*, and becoming the lateral corticospinal tract to the motor neurons in the anterior horns (lateral regions of the spinal cord). This pathway is primarily devoted to the control of intricate movements of the *face, hands, fingers* and *toes.*
- The *indirect pathway* originates more widely in the motor and sensory cortices and passes downwards via a *polysynaptic* pathway that includes relay nuclei in the midbrain, pons and medulla (primarily the *red nucleus* and the *pontine and medullary reticular formation*) (Figure 7.2). The last components of this pathway terminate in the medial regions of the anterior horns and are devoted to the control of *trunk and proximal limb posture and movement.*

Motor modulation

i. Basal ganglia

The *initiation and cessation of movements* – as well as the organization and retention of learned *complex motor programs* – requires the participation of a chain of neurons that originates in the motor cortex and passes through the basal ganglia and lateral thalamus en route back to the motor cortex. The main participants in this circuit are the *motor cortex, posterior putamen, GPe, STN, GPi and the VL and VA thalamic nuclei* (Figure 7.3).

Within this basal ganglia circuit are two main components: the *direct circuit* (putamen–GPi–thalamus) and the *indirect circuit* (putamen–GPe–STN–GPi–thalamus). The former circuit *facilitates* the production of movement via the motor cortex while the latter circuit *inhibits* movement (Figure 7.4).

ii. SN

An important subsidiary component of the basal ganglia circuitry involves reciprocal connections between the putamen (with respect to motor control) and the *SN* in the midbrain. The latter structure makes an important contribution to the initiation of movement by projections to both the direct and indirect basal ganglia circuits (Figure 7.10). *Dopamine-producing neurons* in the SN can facilitate movement either by exciting direct pathway neurons in the putamen (via *type 1 or D1 dopamine receptors*) or by inhibiting indirect pathway putaminal neurons via *type 2 (D2) receptors*. Destruction or impaired function in the nigrostriatal pathway results in the *parkinsonian syndrome.*

iii. Cerebellum

The *coordination* of the various muscles participating in a given movement sequence is the responsibility of a different neuronal chain that passes down the neuraxis via the pons to the cerebellum, and then back upward to the motor cortex, which originated the command via VL. The main participants in this circuit are the *motor cortex, pontocerebellar relay nuclei, cerebellar cortex* (where the cerebellum receives comparison data indirectly from peripheral sensory receptors), *the cerebellar central nuclei (particularly the dentate nucleus) and the VL thalamic nucleus* (Figure 7.6). In effect, the cerebellum compares efferent motor data from the motor cortex with afferent sensory data from the periphery and sends instructions back to the motor cortex as to how the ongoing movement should be modified in order to completely achieve the desired movement.

7.10 CHAPTER-RELATED QUESTIONS

1. Which of the two descending motor pathways would be *predominantly* involved when you:
   a. Shrug your shoulders?
   b. Play the guitar?
   c. Lick your lips?
   d. Kick a soccer ball?
   e. Tap your index finger on the table while you are trying to think of the answers to these questions?

2. In this chapter, you learned that the direct descending pathway is compacted together as a distinctive bundle in the medullary pyramid, just rostral to the point where its constituent motor axons cross the midline to (predominantly) form the contralateral lateral corticospinal tract. Although pathological processes confined to one medullary pyramid are rare, they do occur. Based on what you have been told about the direct motor pathway, what kind of neurological deficits would you expect to see in such a patient?

3. Outline the motor deficits you would expect to see in a patient with an old traumatic injury to the direct and indirect descending motor pathways
   a. In the right cerebral hemispheric white matter?
   b. In the right half of the upper cervical spinal cord (at C2)?

4. Based on what you have learned about basal ganglia circuitry, what type of movement control disorder would be present in someone with an isolated lesion involving the
   a. Left STN?
   b. Right SN, pars compacta?

5. Malignant tumors originating in the cerebellar vermis are unfortunately not unusual in young children. If such a tumor were to be largely confined to the vermis but had caused partial destruction of this structure, what pattern of neurological deficit would you expect to see in such a child?
   What about the deficits if there were a malignant lesion in the right cerebellar hemisphere?

6. You have probably had the experience of being awoken from a deep sleep in the middle of the night by an urgently ringing telephone – only to discover that you must have fallen asleep on your arm and it is numb and tingling. When you attempt to reach for the phone, you discover that your arm won't do what you want and your hand misses the phone by a wide margin. Based on what you have learned in this chapter, can you think of a neurophysiological explanation for this phenomenon?

7. One of the limitations of some of the earliest antipsychotic medications (e.g. chlorpromazine, stelazine, perchlorphenazine) was the gradual development of parkinsonian symptoms: mask-like facies, drooling, stooped posture, shuffling gait. Within a short time after withdrawal of the medication, the parkinsonian features disappear. What is the most likely mechanism of this untoward side effect?

REFERENCE

Benarroch, E.E. Intrinsic circuits of the striatum: Complexity and clinical correlations. *Neurology* 2016. 86: 1531–1542.

SUGGESTED READING

Clarke, C.E. Parkinson's disease. *BMJ* 2007. 335: 441–445.
Hermanowicz, N. Drug therapy for Parkinson's disease. *Semin Neurol* 2007. 27: 97–105.
Kalia, L.V., Lang, A.E. Parkinson's disease. *Lancet* 2015. 386(9996): 896–912.
Segawa, M., Nomura, Y., Nishiyama, N. Autosomal dominant guanosine triphosphate cyclohydrolase 1 deficiency (Segawa disease). *Ann Neurol* 2003. 54(Suppl 6): S32–S45.
Tuite, P.J., Krawczewski, K. Parkinsonism: A review-of-systems approach to diagnosis. *Semin Neurol* 2007. 27: 113–122.

WEB SITES

National Institute for Neurological Disorders and Stroke: www.ninds.nih.gov/disorders/parkinsons_disease/detail_parkinsons_disease.htm
www.nlm.nih.gov/medlineplus/movementdisorders.html

Chapter 8

Etienne

Objectives

- Learn the anatomy of the vascular supply of the brain
- Learn the mechanisms and process of brain cell death from ischemic and hemorrhagic damage
- Learn the effects of vascular damage to affected regions of the brain and their clinical manifestations
- Learn the major stroke syndromes and their clinical manifestations
- Become sensitized to the medical, physical and psychosocial implications of damage caused by cerebrovascular disease

8.1 ETIENNE

Etienne is a used-car salesman with little regard for his lifestyle. He works long hours, frequently eating greasy take-out food at work.

At age 57, his weight is 240 pounds; at 5'6", this gives him a body mass index of 39 kg/m^2 – normal range is 18.5–25 kg/m^2. His doctor has told him on several occasions that his weight, blood pressure and cholesterol were too high and that he should quit smoking. His wife continually complains of his snoring; as well, he frequently has spells during which he would temporarily stop breathing when asleep. His father had died of a stroke at age 55.

About 10 days ago, while at work, he suddenly experienced an episode lasting five minutes, during which he had slurring of his speech, with some left face and arm weakness. He shook it off as just being tired and overworked.

Recently, he has been under considerable stress because of his manager's complaints about his poor sales over the last month. Today, while he was having a heated exchange with his manager, he developed sudden weakness of his left arm, face and leg to the point that caused him to collapse to the floor. His boss called 911; the emergency services arrived within 15 minutes and called the Regional Stroke Center, declaring a potential *Stroke Code*; this made Etienne a possible candidate for intravenous (IV) *tissue plasminogen activating factor (tPA)*.

His examination on arrival in the emergency department showed a blood pressure of 190/100 and a heart rate of 84, which was regular.

He was alert and able to speak and comprehend fully. Both eyes were deviated to the right but could be moved across the midline to the left side with rapid movement of the head to the left. The left lower face was weak (Figure 8.1a), and there was decreased movement of the left side of the palate. He responded to visual threat on the right side but not on the left.

His motor exam showed profound weakness in the upper and lower limbs (Figure 8.1a), with hyperreflexia in the biceps, triceps, brachioradialis, knee and ankle jerks on the left; all reflexes on the right were graded as normal (Figure 8.1a and b). The plantar responses were tested: on the left, the reflex response was abnormal showing an extensor plantar response (positive Babinski's sign); on the right, the response was normal showing a flexor plantar response (negative Babinski's sign). A sensory exam showed that he could feel pinprick sensation on the right over the face, trunk, arm and leg, but not on the left (Figure 8.1c). Coordination testing and gait testing could not be performed due to the profound left-sided weakness.

The neurologist on-call performed a 'Stroke Code' assessment and determined that his National Institutes of Health score was 20 (see reference).

8.2 CLINICAL DATA EXTRACTION

Before proceeding with this chapter, you should first carefully review Etienne's story in order to extract key information concerning the history of the illness and the physical examination. On the Web site accompanying this text, you will find a sequence of tools to assist you in this process.

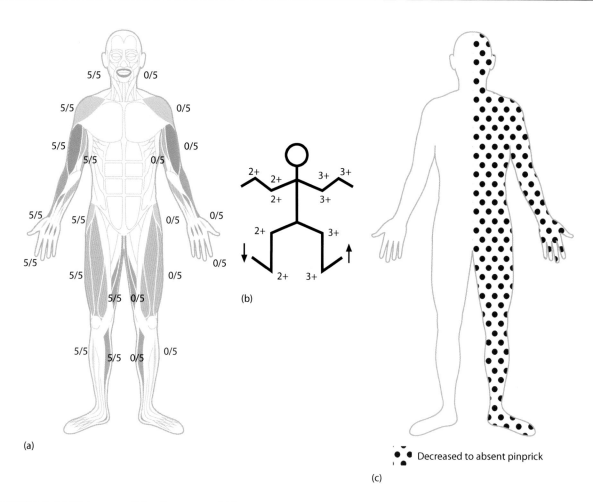

FIGURE 8.1: (a) Motor map, (b) reflex map and (c) sensory map.

8.3 MAIN CLINICAL POINTS

Etienne's principal clinical findings are listed in the following; the evolution of his symptoms is portrayed in the History Worksheet (Figure 8.2).

- One episode of transient left-sided weakness one month before the major event
- Significant risk factors for vascular disease, including obesity, smoking, male gender, middle age, hypertension, hyperlipidemia, probable obstructive sleep apnea and a family history of vascular disease
- Sudden onset of dense weakness of the left lower face, arm and leg; hyperreflexia; Babinski sign with normal speech
- Conjugate eye deviation to the right side
- Lack of visual attention to the left visual field and loss of sensation on the left side of the body
- Hypertension on arrival in the emergency room (ER)

8.4 RELEVANT NEUROANATOMY AND PHYSIOLOGY

As the clinical history suggests, Etienne's predisposing features and clinical deficits suggest a disorder of the cerebral circulation. Subsequent discussion with respect to ischemic cerebrovascular disease must be based on an understanding of the mechanisms of cell death due to ischemia and a detailed knowledge of the blood supply to the brain. The mechanisms and effects of disturbances of that blood supply on different areas of the brain and their clinical effects will be examined.

The sequence of events that occurs when the blood supply is interrupted to a given area of the brain and spinal cord needs to be understood in order to facilitate the process of rapid diagnosis and treatment.

8.4.1 CEREBRAL CIRCULATION

The cerebral circulation is unique in several ways. The major cerebral arteries and their branches serve specific areas of the brain, as indicated in Table 8.1.

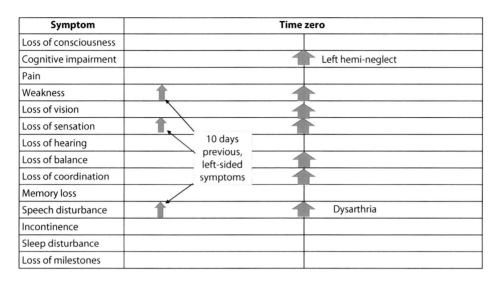

FIGURE 8.2: History worksheet for the chapter case.

TABLE 8.1: Summary of the Territories of Cerebral Artery Supply Showing the Areas of Overlap (Occurring Mainly in the Deep Structures) and Areas of White Matter and Cortex Lying between Two Main Arterial Supplies

Artery	Lobe	Function
Anterior cerebral (ACA)	Frontal, parasagittal	Contralateral motor leg, bladder control, executive function
Middle cerebral (MCA)	Anterior, lateral frontotemporal, parietal, anterior thalamus, thalamus, internal capsule	Contralateral motor, sensory of face, arm, language, memory
Posterior cerebral (PCA)	Parietal, occipital, posterior thalamus, inferior and mesial temporal	Visual fields, contralateral sensory, memory, visual recognition
Basilar	Brainstem, cerebellum, thalamus	CN III–XII, sensory, motor, balance, coordination
Vertebral	Brainstem, spinal cord	CN III–XII, sensory, motor, balance, coordination
Posterior inferior cerebellar artery (PICA)	Inferior surface of cerebellum and lateral medulla	Coordination, Horner's syndrome, contralateral sensation

Unfortunately, there is little collateral flow between the various cerebral arteries *after* their major branching points. The consequence of this pattern of vascular anatomy is that occlusion of a major cerebral artery or one of its branches will lead to ischemic damage in the area served by that artery. This is the bad news. The good news is that the cerebral circulation has been designed with redundancy between the major cerebral arteries. This is achieved through the supplementary arterial branches known as the communicating arteries. For instance, the posterior communicating (PCom) arteries join the posterior cerebral arteries (PCAs) on each side to the internal carotid arteries on the same side. In addition, the single anterior communicating (ACom) artery joins the two anterior cerebral arteries together.

This arterial network, known as the circle of Willis, allows for collateral flow from one carotid to the opposite hemisphere and supply from the basilar artery to the cerebral hemispheres through the PCom arteries. This system

of redundancy – similar to an aircraft fuel system – allows for blockage of one or more of the major cerebral arteries, with flow being maintained through the circle of Willis. Figures 8.3 through 8.6 show the relationships of the major cerebral arteries and the territories of the brain that are supplied by these arteries.

There is another aspect of the cerebral circulation that is unique in comparison with other body organs: cerebral arteries are capable of regulating blood flow without any influence from their own nerve supply (the vasa nervorum). If blood pressure drops significantly, cerebral arteries will automatically dilate in order to maintain adequate cerebral blood flow. Conversely, if systemic blood pressure abruptly becomes dangerously high, cerebral arteries will automatically constrict in order to prevent excessive brain perfusion and a possible hemorrhage. This process is known as *autoregulation*. While there is a limit to autoregulatory capacity in the brain, in most instances,

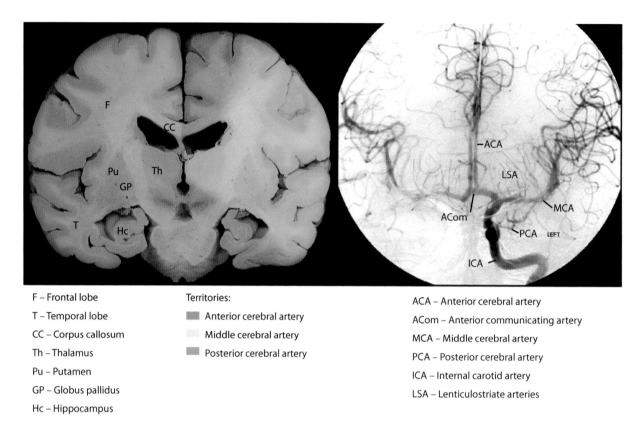

F – Frontal lobe

T – Temporal lobe

CC – Corpus callosum

Th – Thalamus

Pu – Putamen

GP – Globus pallidus

Hc – Hippocampus

Territories:

▓ Anterior cerebral artery

░ Middle cerebral artery

▓ Posterior cerebral artery

ACA – Anterior cerebral artery

ACom – Anterior communicating artery

MCA – Middle cerebral artery

PCA – Posterior cerebral artery

ICA – Internal carotid artery

LSA – Lenticulostriate arteries

FIGURE 8.3: Blood supply to the brain – coronal view. Blood supply of the brain: the basic pattern of the areas of the cerebral hemispheres of the brain which are served by the ACA, MCA and PCA arteries in a coronal brain section and an accompanying left carotid angiogram viewed in the same plane. The MCA serves a large area of frontal, temporal and parietal cortex as well as white matter and the lateral thalamus. The ACA serves the medial part of the frontal and parietal lobes including the cingulate gyrus. The PCA serves the medial temporal lobe, the posterior aspect of the parietal lobe, the occipital lobes and medial thalamus.

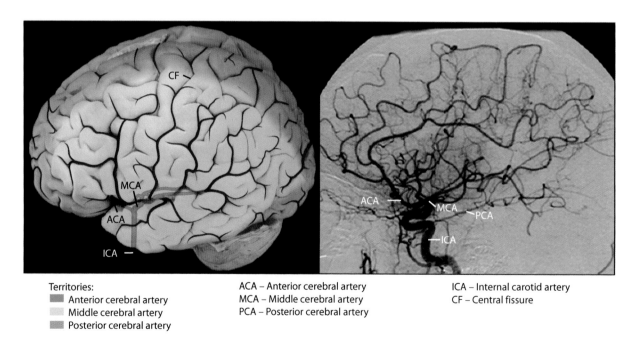

Territories:

▓ Anterior cerebral artery

░ Middle cerebral artery

▓ Posterior cerebral artery

ACA – Anterior cerebral artery
MCA – Middle cerebral artery
PCA – Posterior cerebral artery

ICA – Internal carotid artery
CF – Central fissure

FIGURE 8.4: Blood supply to the brain – lateral view. Blood supply to the brain: a lateral view of the brain with the territories of the ACA, MCA and PCA shaded and a corresponding left carotid angiogram.

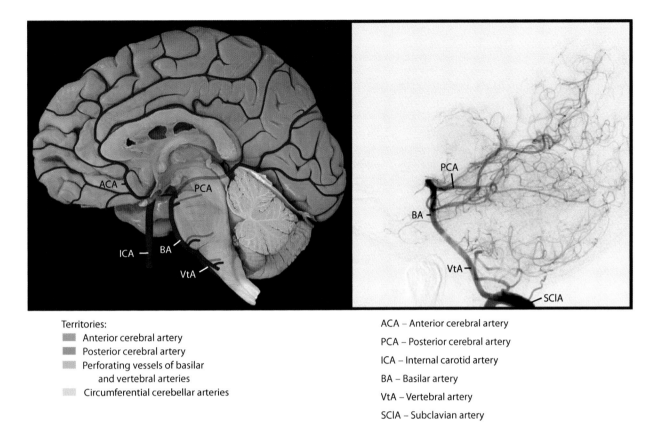

Territories:
- ▬ Anterior cerebral artery
- ▬ Posterior cerebral artery
- ▦ Perforating vessels of basilar
      and vertebral arteries
- ▦ Circumferential cerebellar arteries

ACA – Anterior cerebral artery

PCA – Posterior cerebral artery

ICA – Internal carotid artery

BA – Basilar artery

VtA – Vertebral artery

SCIA – Subclavian artery

FIGURE 8.5: Blood supply to the brain – mid-sagittal view. Blood supply to the brain: a mid-sagittal view of the brain and brainstem showing the territories of the vertebral, basilar and PCA, which are shaded, and a left vertebral angiogram from the same perspective.

autoregulation serves to protect the brain from the deleterious effects of systemic blood pressure fluctuations. In general, blood vessels in areas of acute brain damage lose their ability to autoregulate and the areas served by these blood vessels are dependent on the direct arterial pressure gradient for tissue perfusion.

8.4.2 CEREBRAL ARTERY OCCLUSION

The mechanisms of how the arteries become blocked will now be considered. Any process that occludes the lumen of one of the cerebral arteries to a critical level (usually over 90%) will lead to ischemic injury to the brain tissue that the artery serves. The mechanism by which the arteries become blocked is usually thromboemboli that have been formed upstream and have floated down to occlude the narrower lumen in the more distal portion of the vessel. Atherosclerosis intrinsic to the large branches of the intracranial cerebral arteries in the brain does occur but, in general, is not the commonest cause of arterial occlusion. The commonest source of arterial occlusion is usually from thromboemboli formed at the beginning of the internal carotid artery just above the carotid bifurcation or from emboli from the heart or aortic arch (also called central source emboli).

8.4.2.1 CAROTID

The common carotid artery originates off the aortic arch on the left side and the brachiocephalic artery on the right side. At the angle of the jaw, the common carotid artery splits in to the external carotid artery, which serves the face, scalp and skull as well as the external components of the eye. At the carotid bifurcation, there is turbulent flow due to the separation of the blood flow column at the split of the common carotid into the external and internal carotid arteries. The energy dissipated by this turbulent flow causes endothelial stress and injury especially at the origin of the internal carotid artery. The combination of this shear stress and the physical and chemical factors, which are known to accelerate atherosclerosis (such as hypertension, smoking, hyperlipidemia, diabetes and genetic factors), leads to the build-up of atherosclerotic material in the subintimal region of the artery wall. The evolution of atherosclerosis in this area includes the build-up of fatty material in the affected vessel with subsequent activation of inflammatory cells such as macrophages, leading to an increase in the volume of the plaque. The progressive accumulation of this fatty material causes gradual narrowing of the lumen of the internal carotid artery, otherwise known as carotid stenosis.

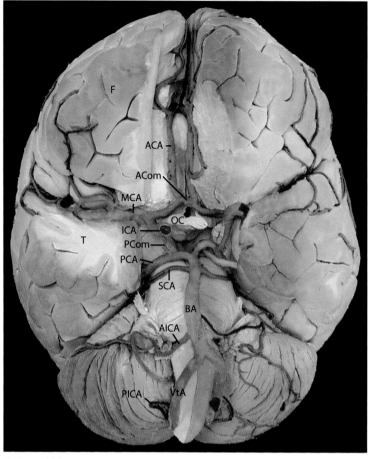

F – Frontal lobe
T – Temporal lobe (cut)
OC – Optic chiasm (cut)
ACA – Anterior cerebral artery
ACom – Anterior communicating artery
MCA – Middle cerebral artery
ICA – Internal carotid artery (cut)
PCom – Posterior communicating artery
PCA – Posterior cerebral artery
SCA – Superior cerebellar artery
BA – Basilar artery

AICA – Anterior inferior cerebellar artery
PICA – Posterior inferior cerebellar artery
VtA – Vertebral artery

Territories:
■ Anterior cerebral artery
▫ Middle cerebral artery
■ Posterior cerebral artery
■ Perforating vessels of basilar
   and vertebral arteries
▫ Circumferential cerebellar arteries

FIGURE 8.6: Blood supply to brain – basal view. Blood supply to the brain: a basal view of the cerebral hemispheres and brainstem showing the territories of the ACA, MCA, PCA, vertebral and basilar arteries. Note the contribution of the circumferential cerebellar arteries (PICA, AICA and SCA) to the blood supply of the brainstem.

The atherosclerotic material can also reach a critical point at which there is damage to the overlying endothelium, leading to a so-called plaque accident. This phenomenon occurs when there is a denuded area of the endothelial lining, which then leads to the formation and build-up of thrombus over the surface of the plaque. These thrombi then grow in size and, depending on the underlying substrate, produce platelet emboli consisting of either clot, or clot mixed with cholesterol and, occasionally, with calcium deposits. Thrombi formed at the origin of the internal carotid artery are inherently mechanically fragile and can then detach from the arterial surface to migrate downstream to become lodged in smaller arteries, producing

arterial occlusion and, thus, neurological deficits. The neurological consequence of the release of this thrombus (which has now become an embolus) depends on which arterial territory the embolus occludes. The middle cerebral artery (MCA) tends to be the preferential end point for emboli originating from the internal carotid artery, but the anterior cerebral artery (ACA) or both the MCA and ACA can be occluded if the embolus is large.

8.4.2.2 CENTRAL SOURCE EMBOLI

Central source emboli are formed in the heart or the aortic arch and then travel from the heart to the brain either through the carotid arteries, the vertebral arteries or both. Any patient presenting with an acute focal deficit with a computed tomography (CT) or magnetic resonance imaging (MRI) showing ischemic lesions old or new, in multiple territories or on both sides of the brain, should then trigger a search for a central source of emboli.

The commonest cause of central source emboli is a thrombus, which formed in the left atrium due to stasis of blood flow associated with atrial fibrillation. Atrial fibrillation may not be present on the electrocardiogram (ECG) on admission, and therefore, prolonged Holter monitoring or event triggered loop recording (monitoring the heart for arrhythmias) is now extensively used to determine if occult atrial fibrillation has been the cause of the embolic event. These investigations are essential in order for the clinician to make the decision to start the patient on long-term systemic anticoagulant medication such as warfarin to prevent further embolic events. Besides warfarin, there are now other choices of systemic anticoagulants known as non-vitamin K antagonist oral anticoagulants. These agents have a better safety and efficacy profile than warfarin without the need for routine blood work. The disadvantage is that these new agents do not have easily available agents to reverse the systemic anticoagulation.

Other sources of emboli include prosthetic valves, acute anterior wall myocardial infarction, endocarditis, atrial myxoma and atherosclerosis with visible thrombus on the aortic arch. Another important source of central source emboli is from the extremities passing through a Patent Foramen Ovale (PFO) or a ventriculoseptal defect (VSD). Transthoracic and transesophageal echocardiography (TEE) are usually able to identify and localize these sources of central source emboli as well as right to left shunts.

8.4.3 CEREBRAL ISCHEMIA

Interruption of blood supply to any area of the brain or spinal cord – known as 'ischemia' – will lead to a series of increasingly serious physiological consequences depending upon the location, intensity and duration of the interruption of blood flow. In addition, different areas of the brain have different susceptibilities to localized or generalized ischemia. For instance, in adults, the hippocampal cortical areas and the Purkinje cells in the cerebellar cortex are most susceptible to ischemia, whereas the white matter areas are relatively less susceptible to ischemic damage.

Acute ischemia is due to a sudden loss of supply of oxygen, glucose and other critical metabolic substrates by blockage of one of the cerebral arteries or its branches. The effect of this acute interruption of cellular metabolism initially leads to disturbances in the electrical function of the neurons, glial cells and transmitting axons. Following this, there is biochemical breakdown of the integrity of the neurons and axons that, if not corrected, is followed by irreversible cell death. In the face of an acute ischemic event,

there are generally three affected populations of neurons or axons in the area of ischemia to consider. The first group of cells are those that have been most severely affected and have undergone irreversible changes; these are usually at the centre of the area of ischemia and are effectively dead cells and thus will never recover function. The second group are less severely affected and are still alive but not properly functioning due to a suspension of their electrical membrane functions. This second area surrounds the region of most severely affected neurons and axons and is known as the *ischemic penumbra*. This is the group of damaged cells that can potentially be salvaged from permanent damage by restoring blood flow to the area. The third group of neurons and axons are outside of the penumbra and are usually fed by other branch arteries proximal to the occluded artery or are supplied by one of the adjacent cerebral arteries.

For the clinician, these concepts are important in that, when a patient arrives in the emergency department with acute ischemic damage, it is important to know the duration and extent of the ischemia as well as to determine the magnitude of the ischemic penumbra; this information will help to decide the choice of treatments to restore blood flow to the damaged areas. There are various neuroradiological imaging techniques available to assist the clinician to determine the extent of damage and the amount of brain tissue that might be salvageable. For CT scanning, these are known as perfusion maps; for MRI scanning, they are known as diffusion weighted images and apparent diffusion coefficient maps.

8.4.4 MCA

The most common and devastating of all strokes is an occlusion of the left MCA in a left hemisphere-dominant (usually right-handed) person. This type of stroke is most damaging because of its effect on the centres that support language and motor function on the dominant side. Both Broca's area (responsible for the fluent output of spoken content of language) and the more distributed area for speech recognition (including Wernicke's area[a]; see Chapters 2 and 3 for a revised definition of this eponym) are served by the MCA through two different branches (Figure 8.4).

The MCA originates from the internal carotid artery in the middle cranial fossa. It runs along the outer surface of the thalamus and internal capsule, giving off small branches, the lenticulostriate arteries, which supply these structures. The MCA then breaks up into branches that resemble candelabra; of these, there are two larger branches: one anterior, whose branches serve the lateral frontal lobe including Broca's area, and one posterior, whose branches serve the lateral temporal and parietal lobes, including the more distributed area that serves speech recognition.

---

[a] Located in the superior gyrus of the temporal lobe as well as other areas in the dominant parietal lobe.

Thus, the amount of damage to the brain and the associated neurological deficits will depend on the location of the occlusion in the MCA and its branches.

An occlusion of the MCA close to its origin in a dominant hemisphere will give severe deficits of motor and sensory function in the face, arm and leg on the contralateral side, as well as loss of receptive and expressive language function.

Damage to the frontal lobes also affects the frontal eye fields (see Figure 1.10c), which control the upper motor neurons mediating contralateral conjugate eye movement. If there is damage to the frontal eye fields in one hemisphere, the eyes will respond to the output of the undamaged hemisphere, resulting in conjugate deviation of the eyes towards the damaged hemisphere, the so-called phenomenon of 'looking towards the lesion'.

Damage to the optic tract, lateral geniculate body of the thalamus and optic radiations can lead to homonymous (on the same side) hemianopia. This deficit relates to damage to the white matter and relay structures that transmit information concerning the visual fields from the retina. The visual system will be discussed in detail in Chapter 9. Damage to these areas causes loss of perception of vision in the contralateral visual field in both eyes.

An occlusion of small penetrating blood vessels serving the internal capsule or lateral thalamus will result in pure motor weakness or sensory loss on the contralateral side. A small round area of damage can be seen on CT or MRI scan; this is sometimes called a *lacune* or 'small lake' (Figure 8.7). Lacunes, which occur in the lateral thalamus, cause pure sensory deficits on the contralateral side. Lacunes in the posterior limb of the internal

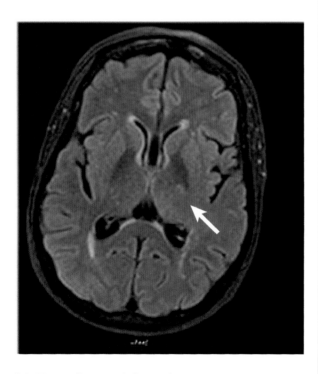

FIGURE 8.7: Lacunar infarct – internal capsule.

capsule cause pure motor deficits on the opposite side. There is a homuncular distribution to the fibers in the posterior limb, starting at the genu with the face, arm, trunk, leg in an anterior to posterior sequence.

Branch occlusions to arteries that primarily supply areas of cortex lead to wedge-shaped areas of damage visible on CT scan or MRI, with deficits specific to the location. For instance, a branch occlusion to the arterial supply to the left angular gyrus will lead to a very specific syndrome of deficits called Gerstmann syndrome: the patient is not able calculate simple arithmetic, loses the ability to perceive left from right and is unable to identify which finger is being touched by the examiner.

If the occlusion is distal to the perforating lenticulostriate branches, thus sparing the internal capsule, there will be less motor deficit in the contralateral face, arm and, to a lesser extent, the leg; nevertheless, movement of the face and arm and, if in the dominant hemisphere, language, will be affected.

More distal occlusion of either the anterior or posterior branches will give selective damage. Occlusion of the anterior division results in weakness of the face and arm with a Broca's aphasia (see also Chapter 14). Occlusion of the posterior division causes mild upper motor neuron weakness on the contralateral side, with a receptive Wernicke's aphasia and contralateral cortical sensory loss (e.g. loss of stereognosis – see Chapter 2).

Isolated occlusions of some branch vessels will produce deficits specific to the function of the location damaged.

8.4.5 ACA

The ACA originates from the internal carotid artery in the middle cranial fossa; it runs up and over the corpus callosum, serving the corpus callosum and the cerebral cortex of the medial frontal gyri, cingulate gyrus and part of the medial parietal lobe.

Occlusion of the ACA close to its origin will cause damage to a strip of cortex that it serves running longitudinally back along the parasagittal area. Damage to this area leads to motor weakness, spasticity and hyperreflexia in the leg on the opposite side (see the motor homunculus, Figure 4.2). Occlusion of this vessel is relatively uncommon, accounting for less than 10% of all strokes.

8.4.6 PCA

The two PCAs originate at the termination of the basilar artery at the top of the brainstem; the PCA on each side runs back to the parietal and occipital lobes but also serves the medial and inferior portions of the temporal lobe. Close to its origin, the PCA gives off small branches to the medial thalamus (Figures 8.5 and 8.6). There is a forked artery close to the origin of the PCAs called the artery of Percheron, which has a single origin and serves the medial surface of both thalami adjacent to the third ventricle.

Occlusion of this artery can lead to a medial thalamic syndrome characterized by apathy and anhedonia (inability to experience emotional pleasure), similar to a frontal lobe syndrome.

Occlusion of the PCA close to its origin will result in destruction of all the regions that it serves (including the thalamus, parietal, medial temporal and occipital cortices), leading to hemianopia (loss of perception of the visual fields on the opposite side), contralateral cortical sensory loss and thalamic dysfunction.

Occlusion of branches distal to the section of the artery serving the medial temporal and parietal lobes will result in pure occipital lobe damage and isolated hemianopia without other sensory changes.

Interruption of blood flow to the medial temporal lobes can lead to partial visual field defects and to memory loss. The hippocampus is situated right at the border zone of the arterial supply from the MCA and PCA. Occlusion of branches of either of these arteries can lead to hippocampal dysfunction and loss of short-term memory.

The two PCAs have a communicating branch between them and the internal carotid arteries at their origin; these are called the PCom arteries and are also part of the circle of Willis. They join the PCAs to the internal carotid artery (ICA) in combination with the AComs, allowing for redundancy of blood flow from one hemisphere to another in the event of obstruction of a major vessel below the level of Willis' circle (e.g. the internal carotid on one side). The circle of Willis has significant variability, and its ability to provide collateral circulation depends on the state of vessels involved.

8.4.7 BASILAR ARTERY

The basilar artery originates from the joining of the two vertebral arteries at the base of the skull. The basilar artery runs up the anterior surface of the brainstem from the medulla to the pons and ends at the top of the midbrain to form the two PCAs (Figure 8.6).

The basilar artery gives out arcuate branches along its path, which serve the medial and lateral aspects of the brainstem and cerebellum at each level. Occlusions of the basilar artery are usually branch occlusions, which produce one of the six brainstem syndromes described in Table 8.2.

Brainstem syndromes depend on the level and the laterality. As was noted in Chapter 1, there are basically three levels of the brainstem: midbrain, pons and medulla, with a medial and lateral syndrome for all three levels.

Table 8.2 summarizes the various stroke syndromes at each level of the brainstem.

Arcuate branch occlusions of the basilar artery lead to the medial and lateral pontine and midbrain syndromes. Occlusion of these branches by intrinsic atherosclerosis is common in patients with diabetes and hypertension.

Midbrain lesions involve distal sensory and motor deficits in the face and limbs as well as abnormal function of CN III and sometimes CN IV.

TABLE 8.2: Brain Stem Localization Signs and Symptoms

Level	Symptoms	Signs	Structures	Investigations
Midbrain				
Medial (Weber's)	Diplopia	Loss of elevation, loss of elevation, depression and adduction, contralateral hyperreflexia	CN III, corticospinal tract to opposite side	MRI, CT/CT angiogram
Lateral (Claude's)	Loss of sensation, ataxia	Decreased pin prick, vibration to opposite side	Spinothalamic tract, medial lemnicus, superior cerebellar peduncle	
Pons				
Medial (Millard–Gubler)	Diplopia, loss of sensation, weakness	Loss abduction, loss of pinprick and vibration sense	CN VI, medial lemnicus, spinothalamic tract, corticospinal tract	MRI, CT/CT angiogram, brainstem evoked potential, blink reflex, ENG
Lateral (Foville's)	Ataxia, hearing loss, facial weakness	Loss of balance, facial weakness, loss of sensation to face, ataxia	CN VII, vestibular nucleus, mid-cerebellar peduncle	
Medulla				
Medial	Numbness, weakness bilateral	Loss of vibration, proprioception sensation, weakness	Medial lemnicus bilateral, corticospinal tracts bilateral	MRI, CT/CT angiogram, Apnea test
Lateral (Wallenberg)	Ataxia, loss of sensation on opposite side	Loss of sensation to pin to ipsilateral, face, contralateral arm and leg, ipsilateral Horner's, ipsilateral ataxia	Spinothalamic tract, descending sympathetic tract, inferior cerebellar peduncle	

Weber's syndrome is the medial midbrain syndrome; it involves CN III and the cerebral peduncle leading to a unilateral CN III palsy and contralateral hemiparesis of face and arm.

Claude's syndrome is the lateral midbrain syndrome. It involves the medial and lateral lemnisci, the spinothalamic tracts and the superior cerebellar peduncle. The clinical findings include contralateral sensory loss and ataxia of the ipsilateral limbs.

Lesions of the pons affect distal sensory and motor function as well as producing dysfunction of CN V, VI, VII and VIII.

The lateral pontine syndrome (Foville's syndrome) includes loss of hearing, balance and facial sensation as well as limb ataxia. The nuclei of CN V, VII and VIII are affected as well as the middle cerebellar peduncle.

The medial pontine syndrome (Millard–Gubler syndrome) includes loss of function of CN VI, VII and contralateral hemiparesis of the arm and leg. The medial longitudinal fasciculus (MLF) may also be involved; it connects the CN VI to CN III nuclei to coordinate eye movements. An MLF lesion leads to an internuclear ophthalmoplegia (INO) in which the patient can abduct the ipsilateral eye (CN VI) but not adduct the contralateral eye (CN III).

Lesions of the medulla affect distal sensory and motor function as well as producing dysfunction of CN IX, X, XI and XII.

The two vertebral arteries give off branches that serve the medulla and inferior surface of the cerebellum; these are known as the posterior inferior cerebellar arteries (PICAs). Occlusion of a PICA will lead to the lateral medullary syndrome.

The lateral medullary syndrome (or Wallenberg syndrome) consists of damage to the lateral structures of the medulla and often the inferior surface of the cerebellum. The anatomical components damaged include the inferior cerebellar peduncle, the descending sympathetic tract, the crossed spinothalamic tract, the descending tract of CN V and the lower end of the vestibular nucleus. Depending on the size of the damage to this area, the patient experiences ipsilateral facial sensory loss, ataxia, Horner's syndrome (see Glossary, available on the Web site, and Chapter 2, Section 2.3.1.4) and loss of pain and temperature sensation on the opposite side of the body.

The medial medullary syndrome (which is rather rare) includes loss of the pyramids and both medial lemnisci, leading to the bilateral extremity weakness and loss of the sensation for vibration and proprioception.

The most serious brainstem syndrome occurs with a complete occlusion of the basilar artery, leading to extensive ischemic necrosis that is most concentrated in the ventral pons (see Figure 8.8). This leads to a catastrophic condition called *locked-in syndrome* in which the patient has lost all motor function to the rest of the body below CN III but with preserved sensory, auditory, visual and

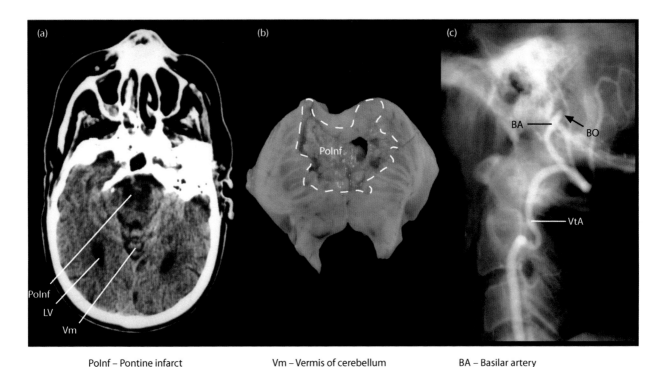

Polnf – Pontine infarct   Vm – Vermis of cerebellum   BA – Basilar artery
LV – Lateral ventricle   VtA – Vertebral artery   BO – Basilar obstruction

FIGURE 8.8: Ischemic Infarct–Pons: (a) CT scan. (b) Pathology specimen of the same area as seen in the CT scan. (c) Cerebral angiogram shows no flow in the basilar artery with injection of the left vertebral artery.

cognitive function. These individuals are usually ventilator dependent and require total care. Unable to communicate except through eye blinks, they have a poor quality of life; this situation leads to difficult ethical questions with respect to end-of-life decisions.

Central source emboli can originate from the heart or aortic arch and travel through a vertebral artery to the basilar artery. Often, these emboli will fragment as they travel up the basilar artery, sending small emboli to its branches and giving multi-focal brainstem and cerebellar deficits. If the emboli reach the top end of the basilar artery, the embolus often will fragment into several pieces going into the PCAs on both sides and resulting in infarction in the territories of the PCAs. This is the so-called 'Top of the Basilar Syndrome'. The clinical effect of this condition is to cause multiple infarctions in the brainstem and often in both occipital lobes depending on where the fragments of the embolus interrupt arterial flow.

8.5 LOCALIZATION

Following the defining event, the completed Expanded Localization Matrix would appear as shown in Table 8.3.

As can be seen from the matrix, muscle, neuromuscular junction and peripheral nerve disorders may be eliminated due to the presence of signs at or above the spinal cord level such as facial weakness, cognitive impairment, hemianopia and hyperreflexia. The major, more likely localizations target the right frontal, parietal and occipital areas. Brainstem and spinal cord localizations are also ruled out due to the cognitive impairment and hemianopia, which have to be above the level of the tentorium. The cerebellum as a primary localization is ruled out due to the presence of upper motor neuron weakness.

In terms of localization, prior to the major crisis, Etienne suffered an initial event affecting the motor system in the right frontal lobe, producing transient left facial weakness and dysarthria. The presenting symptoms at

TABLE 8.3: Expanded Localization Matrix for Etienne

	Neuro Exam		MUS	NMJ	PN	SPC	BRST	WM/TH	OCC	PAR	TEMP	FR	BG	CBL
Attention	N													
Memory	U													
Executive	U													
Language	Dysarth							O			O		
Visuospatial	Left hemi					O	O	O	O	O			
	Right	Left												
CRN I	N	N												
CRN II	N	N												
CRN III, IV, VI	Gaze to right					⊘					O		
CN V	N	↓					⊘							
CRN VII	N	Lower face					⊘							
CRN IX, X	N	↓					⊘							
CRN XI	N	↓					⊘							
CRN XII	N	Dev L					⊘					O		
Tone	N	↑				⊘	⊘	O	⊘			O	⊘	⊘
Power	N	↓	⊘	⊘	⊘	⊘	⊘	O	⊘	O		O	⊘	⊘
Reflex	N	↑	⊘	⊘	⊘	⊘	⊘	O	⊘	O		O	⊘	⊘
Involuntary Mvt	N	N												
PP	N	↓	⊘	⊘	⊘	⊘	⊘	O	⊘	O				
Vib/prop	N	↓	⊘	⊘	⊘	⊘	⊘	O	⊘	O				
Cortical sensation	N	↓						O	⊘	O				
Coord UL	N	Abs											⊘	⊘
Coord LL	N	Abs											⊘	⊘
Walk	NA	NA												
Toe/heel/invert/evert	NA	NA												
Tandem	NA	NA												
Romberg	NA	NA												

Abbreviations: Abs, absent; Dev L, deviated to the left; Dysarth, dysarthria; Left hemi, left hemianopsia; N, normal; NA, not attempted; U, untestable.

that time, localizing to the right frontal area, suggest that there was a transient disturbance of flow in a branch of the right MCA probably due to a small embolus originating in the right internal carotid artery.

The localization suggested by a sudden transient episode of left face and arm weakness, had Etienne mentioned it to anyone, should have triggered a search for the source of this event. The vascular territories supplying the right hemisphere can be retraced sequentially starting with the MCA to the internal carotid artery, to the common carotid artery, the aorta and then the heart.

The discovery of a high degree of stenosis (>70%) in the symptomatic right internal carotid artery could have then prompted a rapid response to initiate medical treatment and to perform a right carotid endartectomy or angioplasty and stenting procedure. This could have prevented the more serious event several weeks later.

8.5.1 INTRACEREBRAL HEMORRHAGE

The location and causes of intracerebral hemorrhage are more varied than that of ischemic disease. One method of localizing hemorrhages is to divide the location of the hemorrhage anatomically into cortical, subcortical, brainstem and cerebellum.

In general, cortical hemorrhages are related to arterial rupture from aneurysms, vascular abnormities such as arteriovenous malformations (AVMs) or cavernous angiomas (cavernomas), trauma or degenerative conditions such as congophilic angiopathy. Cortical hemorrhages can also be caused by cerebral venous occlusion.

Subcortical hemorrhages are usually due to rupture of small penetrating arteries that have undergone degenerative changes from hypertension.

Hemorrhages of the brainstem are usually associated with vascular abnormities such as AVMs or cavernomas.

Hemorrhages of the cerebellum are most often caused by hemorrhagic transformation of large infarction, AVMs or tumors such as hemangioblastomas. Unstable hemorrhages of the cerebellum are considered a neurosurgical emergency due to the potential expansion of the hematoma, causing life-threatening brainstem compression.

Other medical conditions can predispose to hemorrhage, such as systemic anticoagulation, low platelet count, hemorrhagic coagulopathies and septic emboli caused by endocarditis. Hemorrhage into metastatic tumors needs to be considered in the etiological diagnosis, especially in a patient with a pre-existing metastatic malignancy.

Endocarditis can cause septic emboli to infect cerebral blood vessels causing mycotic aneurysms, which are mechanically fragile and can lead to cerebral hemorrhage and abscess formation due to weakening of the arterial wall from the metastatic infection.

Modern imaging techniques such as angiography, CT and MR angiopathy allow rapid identification of bleeding sources within the brain. Specialized MRI sequences such as gradient echo and susceptibility weighted imaging can identify the presence of previous episodes of bleeding due to their sensitivity to detect hemoglobin breakdown products.

8.6 ETIOLOGY – ETIENNE'S DISEASE PROCESS

In terms of the History Worksheet and Etiology matrix, the two events can be categorized as acute in nature. Since the localization exercise clearly points to a cerebral blood vessel disorder, this section will focus on acute disease processes affecting cerebral vessels.

8.6.A PAROXYSMAL DISORDERS

Acute hemiplegia can present as a manifestation of a paroxysmal disorder. Epileptic seizures (see Chapter 11) can cause prolonged unilateral weakness in the post ictal phase a condition called Todd's paralysis. This condition usually clears after 12–24 hours. There is usually a history of a preceding seizure, but sometimes, the seizure can be unwitnessed or can occur during sleep.

Migraine headache can be associated with hemiplegia in a condition called hemiplegic migraine. These spells are often recurrent and there may be a family history. When an episode occurs for the first time, however, it is essential to perform vascular imaging so as not to miss a serious vascular problem. In Etienne's case, it is unlikely that a hemiplegic migraine attack would occur for the first time at age 57; the disorder is primarily seen in children and young adults.

8.6.B TRAUMATIC VASCULAR INJURY

Traumatic injury to cerebral blood vessels can occur with sports injuries, causing direct trauma to the carotid artery under the angle of the jaw: a typical example is a blow from a flying object such as a baseball or golf ball. Dissections of either carotid or vertebral vessels have been caused by falls, swimming, pitching a baseball or therapeutic neck manipulation.

Direct trauma to the top of the head can injure vertebral arteries in the neck, especially if there are degenerative changes in and around the course of the vertebral arteries in the transverse processes of the vertebrae.

In military environments with flying debris such as shrapnel, one may encounter direct rupture of blood vessels or vascular spasm related to blast injury.

Clearly none of these conditions would apply to Etienne.

8.6.C VASCULAR DISORDERS

Vascular disorders include infarction or hemorrhage secondary to large or small vessel occlusion or rupture. This is the disease category that best fits Etienne's story. In the preliminary incident, there was a transient small vessel occlusion due to an embolus that formed in the internal carotid artery more proximally. The presence of a single small vessel event should trigger a search for the offending large vessel. An alternative would be small vessel disease due to diabetes mellitus or vasculitis; in such patients, the disease is usually chronic and the patient has other systemic manifestations.

The commonest cause of occlusive vascular disease is atherosclerosis, in this case of the right carotid artery distal to its origin. As previously explained, the bifurcation of the common carotid into the internal and external carotid arteries leads to turbulent flow of blood; this then causes shear injury to the wall of the internal carotid artery at its origin. This regional increased risk for damage, coupled with Etienne's multiple risk factors for atherosclerosis, led to the formation of a complex plaque at the origin of his right internal carotid artery.

In Etienne's case, the first embolus probably formed over complex plaque at the origin of the right internal carotid artery and travelled to the MCA, where it transiently blocked a small branch artery serving the motor strip on the right. Because of the short duration and neurological recovery from the event, it would be classified as a TIA (see Chapter 3) involving a branch of the right MCA.

The second event that caused the left-sided weakness, eye deviation and visual and sensory loss involved damage to the frontal, temporal and parietal lobe areas served by the both the MCA and ACA. This would suggest an occlusion of the right internal carotid artery.

Therefore, if Etienne had sought medical advice after his event, given his multiple risk factors for cerebrovascular disease, he might have been worked up for a vascular cause of his symptoms. A Doppler or CT angiogram would have revealed a high-grade stenosis of the right internal carotid artery, probably of 90%, with a complex plaque. He could have then been referred for neurosurgical or neuroradiological interventional therapy before his stroke occurred 10 days later.

As was mentioned in Section 8.4, central source emboli refer to embolic material formed in the heart or great vessels either directly or indirectly. These sources include the left atrium due to stasis from atrial fibrillation, diseased or prosthetic heart valves or infected thrombi forming on heart valves (endocarditis).

Sources of emboli outside the heart that subsequently pass through it include deep vein thrombosis (e.g. in the legs or pelvis) and tumor or fat and air bubbles from open fractures; the emboli then pass through right to left shunts such as PFO, atrial septal defects and ventricular septal defects, or through intrapulmonary vascular shunts.

Vascular disorders causing hemorrhage include conditions such as arterial berry aneurysms, AVMs and cavernomas. These vascular abnormalities are generally present at birth but expand to rupture and bleed later in life; whether or not they eventually rupture is influenced by risk factors such as age, hypertension and smoking.

The initial small branches of the MCA, known as the lenticulostriate arteries, undergo degenerative changes with age and with poorly controlled blood pressure. This may lead to rupture of these arteries leading to hemorrhages involving the internal capsule, lateral thalamus and basal ganglia. These are referred to as hypertensive gangliothalamic hemorrhages.

Blood vessel degenerative diseases that tend to cause hemorrhage include congophilic angiopathy, in which there is infiltration of vessel walls with amyloid material resulting in weakening of the arterial walls. This condition leads to recurrent cortical hemorrhages with focal neurological deficits according to location, and to progressive dementia. This condition is also part of the larger process of amyloid accumulation elsewhere in the brain in Alzheimer's disease.

Arterial dissection refers to a situation where there is a weakness or injury to the intima of a vessel. If the weakness leads to structural failure of the vessel wall, then blood under arterial pressure creates a false passage between the intima and adventitia through which there is blood flow. This false channel then terminates by usually re-entering the main lumen. Complications of this situation in cerebral vessels are a restriction of flow of the affected artery and formation of thrombus at the distal end of the false lumen created by the dissection.

8.6.D TOXIC INJURY

In general, medications or recreational drugs do not cause focal neurological deficits, although drug addicts who mainline substances that are contaminated with particulate matter may develop emboli that can lodge in any of the cerebral arteries. Progressive visual loss, often starting in just one eye, may develop in the context of an individual who smokes heavily and has poor nutrition, a disorder known as tobacco amblyopia. The damage to the retina is thought to be due the combined effects of vitamin deficiencies and of toxic chemicals in cigarette smoke.

With some exceptions, drugs and medications rarely lead directly to vascular damage causing ischemic or hemorrhagic stroke. Any medication that might have a prothrombotic potential must be considered; examples include L-asparaginase, a chemotherapeutic agent, and IV immunoglobulin. More commonly, the withdrawal of medication, in particular anticoagulants for central source emboli, can result in acute infarction. Patients on anticoagulants for established central source emboli are thus at high risk of recurrence if these medications are stopped or mismanaged. Patients on systemic anticoagulation for any reason have a 2–3 times higher risk of intracerebral hemorrhage than the normal population.

Life-threatening gastrointestinal and genitourinary hemorrhages are obvious indications for stopping systemic anticoagulation. Careful consideration, however, must be given to the risks versus the benefits of stopping – as against resuming – the anticoagulants once the cause of the hemorrhage has been found and resolved.

Medications that cause a precipitous rise in blood pressure can lead to both hemorrhagic and ischemic strokes. These substances can be simple over-the-counter cold medications; in patients with severe hypertension, the use of cocaine, amphetamines and other vasopressor agents may lead to infarction or hemorrhage. Medications that cause a precipitous lowering of blood pressure (such as medications for hypertension, erectile dysfunction and angina) can lead to cerebral and ocular hypoperfusion with ischemic damage in border zone areas.

Since the advent of MRI scanning, a new syndrome has emerged involving the dysregulation of the posterior cerebral circulation resulting in ischemic changes and, on occasion, infarction in the occipital lobes. The posterior reversible encephalopathy syndrome (PRES) has been associated with the use of medications that induce or aggravate hypertension. Sometimes, the damage from this injury becomes permanent – PERMAPRES.

8.6.E INFECTION

Infections of blood vessels are usually caused by metastatic infection from material released by infected heart valves (endocarditis). These infections are usually bacterial, with the commonest organisms being staphylococcus and enterococcus. The source of the infection of the heart valve itself must be determined; possibilities include dental infections or occult malignancy in sites such as the colon or cecum. Tertiary syphilis is an example of a chronic infection of blood vessels of the brain resulting in the formation of micro-abscesses or gumma adjacent to the infected blood vessels.

8.6.F METABOLIC INJURY

Most metabolic disorders cause diffuse cerebral dysfunction and damage in a global rather than a focal fashion. However, sometimes, extremes of metabolic disturbance can present with hard focal findings indistinguishable from those of stroke. Extreme hypoglycemia associated with insulin overdose or insulinoma may present in this way.

8.6.G INFLAMMATORY/AUTOIMMUNE

Inflammatory diseases of blood vessels such as systemic lupus erythematosus, temporal arteritis and primary cerebral arteritis are all associated with multi-focal vessel wall inflammation. These disorders can come on quickly but would not resolve as quickly and completely as did Etienne's original attack. Vasculitis is often associated with systemic symptoms such as headache, arthritis, skin changes, liver, lung and renal dysfunction.

A disorder known as 'reversible cerebral vascular spasm' can present with features of acute neurological deficit associated with severe thunderclap headache and spasm of multiple cerebral arteries on angiography. The cause is not well understood but may be related to uncontrolled hypertension, certain forms of chemotherapy or to a recent immune challenge such as an infection or surgical intervention.

Vascular etiologies not previously mentioned that may also lead to complete occlusion of the carotid, anterior, middle, posterior or basilar arteries include inflammatory diseases such as Kawasaki disease, primarily seen in children. Moyamoya disease refers to an occult occlusion of the intracranial carotid arteries accompanied by the development of extensive regional collateral vessels that appear on cerebral angiograms as a 'puff of smoke' – the latter term in Japanese is 'Moyamoya'. The cause of this disorder is not known but may be related to qualitative abnormalities of collagen in the cerebral vessels.

Other non-infectious inflammatory disorders that can cause ischemic stroke include rheumatoid arthritis, sarcoidosis, Sjögren's syndrome and Reiter's syndrome. These diseases are often associated with systemic symptoms.

8.6.H NEOPLASTIC DISORDERS

Neoplasms of blood vessels are exceedingly rare. So-called glomus tumors of jugular veins or carotid arteries are probably embryonic rests of specialized neural crest cells that are located in these vessels and subsequently undergo neoplastic change.

Neoplasms of the heart such as atrial myxoma may be the source of recurrent emboli to the brain. These can cause characteristic murmurs and are best identified with TEE.

8.6.I DEGENERATIVE DISORDERS

Degenerative diseases of blood vessels other than atherosclerosis are being increasingly recognized; examples include vascular leukoencephalopathy, a disorder in which there is premature degeneration of small blood vessels in the white matter. Cerebral autosomal dominant arteriopathy with subcortical infarcts and leukoencephalopathy is an example of a progressive small vessel disease of genetic origin also affecting white matter.

8.6.J GENETIC DISORDERS

There are several genetic markers for the occurrence of intracranial aneurysms that should prompt the need for surveillance of first-degree relatives of any patient who has a cerebral aneurysm either ruptured or not. The risk of rupture is further increased by the presence of other risk factors such as hypertension, hyperlipidemia and especially cigarette smoking.

Genetic abnormalities of blood vessels include abnormalities of formation of blood vessels such as

Sturge–Weber syndrome, a neurocutaneous disorder in which there is a congenital facial vascular nevus accompanied by abnormal development of cerebral vessels on the same side and by early-onset unilateral cerebrocortical infarction resulting in hemiparesis and epileptic seizures.

Fibromuscular dysplasia refers to a disorder of blood vessels, usually involving the renal and cerebral arteries. There are fibrotic changes in the blood vessel walls that cause tortuosity and stenosis. Other complications include aneurysm formation and dissection.

Another familial disorder of blood vessels is the familial cavernoma syndrome, in which affected members have multiple cavernomas throughout the brain that can bleed and cause seizures.

Other genetic disorders that can lead to either spontaneous vascular occlusion or dissection include homocystinuria and Marfan's disease.

8.7 CASE SUMMARY

In summary, Etienne presented with acute left hemiplegia, left-sided sensory loss and visual inattention following a previous transient episode of left face and arm weakness probably caused by an embolus originating from an ulcerated plaque on the right internal carotid artery.

In terms of localization at the time of his stroke, the combination of acute profound weakness of the left side (including face, arm and leg), gaze preference to the right and visual inattention to the left is indicative of widespread cortical and white matter dysfunction in the right frontal, temporal and parietal regions due to lack of blood flow in both the anterior and middle cerebral arteries. The common source vessel for all of these areas is the internal carotid artery (or, proceeding even further upstream, the common carotid artery).

The thrombus built up over a plaque occupying approximately 90% of the cross-section of his right internal carotid artery gradually increased in size to the point that it completely occluded the small amount of lumen left. Occlusion of the right internal carotid artery then interrupted flow to the right anterior and middle cerebral arteries – serving the frontal, temporal and anterior parietal lobes – and leading to the neurological deficits. In Etienne's case, the ACom and PCom arteries in the circle of Willis were too small or diseased to allow for significant shunting of arterial blood from the left internal carotid and basilar arteries.

The acute occlusion of his right internal carotid artery by fresh thrombus provided the opportunity for his health care team to give thrombolytic therapy to re-establish blood flow and to prevent and even reverse neurological damage. In this case, he had complete large proximal vessel occlusion, which is more amenable to intra-arterial (IA) therapy than to IV therapy. The common practice, as in this case, is to use an initial dose of IV tPA and, if there is no immediate clinical improvement, to then use clot extraction methods coupled with IA tPA. tPA catalyzes

the conversion of plasminogen to plasmin, resulting in the lysis of newly formed clots. If given within 4.5 hours of the onset of symptoms, this treatment is now considered the standard of care for acute ischemic stroke.

8.7.1 CASE EVOLUTION

Etienne was seen in the emergency department within 29 minutes and had a rapid clinical assessment, blood work, CT scan and CT angiogram within one hour of the onset of his stroke.

The CT scan showed an evolving infarct involving the right MCA and ACA territories without evidence of hemorrhage; the CT angiogram showed an acute internal carotid occlusion.

The neurologist on-call notified the neuroradiologist on-call that this represented a case of a large vessel occlusion and that Etienne was a possible candidate for IA tPA and clot removal. It was agreed that IV tPA should be started (up to 2/3 of the normal full dosage) and a decision with respect to clot extraction and IA tPA then be made based on the patient's response to the initial therapy.

Etienne had no history of gastrointestinal or genitourinary bleeding, and no recent surgery or myocardial infarction. His blood work showed a normal platelet count, international normalized ratio (INR) and partial thromboplastin time (PTT). It is important to check for abnormalities in platelet quantity and in the clotting system as these could increase the effect of tPA and cause bleeding.

The risks and benefits of both IV, IA clot extraction and IA tPA treatment were presented to Etienne, who, after consulting with family, signed the consent to proceed with treatment. The consent was also cosigned by his wife. If his vascular occlusion had instead occurred in the left cerebral hemisphere and had produced a global aphasia, Etienne's wife, as his next of kin, or another individual with legal power of attorney for medical decision making would have been required to sign the consent on his behalf.

At 1 hour 45 minutes from the onset of his symptoms, IV tPA was started. After 35 minutes and 2/3 of the total dose given, he was reassessed and found not to have recovered from any of his deficits. He was then taken to the angiogram suite, where a selective right carotid angiogram showed no flow in the right internal carotid artery. The artery was opened using angioplasty and a clot retrieval device was used followed by IA tPA. After several attempts, flow in the artery was re-established. A self-eluting stent was then deployed across the stenotic segment.

His neurological examination showed gradual improvement over the next 24 hours, during which he regained most of the strength on his left side. A follow-up CT scan at 24 hours showed a medium-sized right frontal infarct.

After 24 hours, he was started on medical treatment to prevent recurrence of stroke, including Clopidogrel (a platelet paretic agent), Ramipril (for hypertension), Atorvastatin (to treat his elevated cholesterol) and nicotine substitutes to minimize the effects of withdrawing from nicotine. He was also referred for physiotherapy, occupational therapy, speech therapy assessments as well as a quit smoking program.

When seen in the neurologist's office one month after discharge from hospital, he had some weakness and clumsiness of his left arm as well as a mild dysarthria. His follow-up CT scan showed that the infarct seen at 24 hours had become somewhat smaller but more radiolucent (indicating permanent tissue loss). The follow-up CT angiogram showed that the stent was patent.

With medication, Etienne's blood pressure became controlled, and his cholesterol returned to the normal range. He was referred for a sleep study, which showed severe obstructive sleep apnea with an Apnea/Hypopnea Index of 65 per hour with desaturations down to 79%. He responded to autotitrating nasal continuous positive airway pressure (CPAP) in a range from 8 to 18 cm of water; this completely controlled the obstructive sleep apnea events and associated desaturations. His symptoms of fatigue, lethargy and excessive daytime sleepiness were significantly improved with the CPAP treatment.

He continued to follow his rehab program, lost 30 pounds, stopped smoking and returned to work with a reduced schedule three months after his stroke.

This scenario represents one of the most serious clinical presentations of stroke, in this case with a gratifying outcome.

8.8 E-CASES

8.8.1 CASE 8E-1: BJ, AGE 67

Background: BJ is a genial fellow now retired who volunteers to drive people to their medical appointments notwithstanding his medical condition; this includes an occasional irregular heart rhythm for which he is being treated.

Chief complaint: Sudden onset of dizziness and unsteadiness six hours ago

History:

- Long history of episodic rapid, irregular heartbeat; during episodes, heartbeats would vary in intensity while heart rate fluctuated from very fast to quite slow; has been on a cardiac medication which has been partially helpful
- History of chronic alcoholism – now on the wagon except for an occasional slip
- Today, while driving to the store, BJ developed a sudden-onset spinning sensation accompanied by ringing in his left ear; tried to get out of the car but was extremely off-balance, tending to veer to the left; also nauseated
- BJ managed to drive home where he tried to rest; after several hours, he was much worse: unable to stand up and severely nauseated; his wife called an ambulance
- Only medications are ASA and metoprolol; he has not taken the latter drug for several weeks because it made him feel tired

Examination:

- Blood pressure (BP) = 100/60; heart rate (HR) = 130, irregularly irregular; ECG shows atrial fibrillation
- Alert and cooperative; the slightest head movement provokes a subjective spinning sensation and nausea; speech is slurred but no evidence of dysphasia
- Normal visual acuity and visual fields
- Full range of extraocular movement but nystagmus on horizontal gaze, worse toward the left side
- Left-sided miosis and mild ptosis; decreased pinprick sensation in the left side of the face (see Figure 8.9)

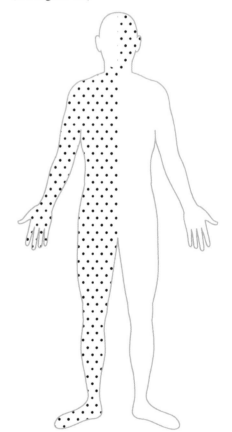

∵∴ Decreased pinprick and temperature sensation

FIGURE 8.9: Sensory map for case 8e-1.

- Normal muscle power in all four limbs; mild hypotonia and hyporeflexia on the left side
- Decreased and altered pinprick and temperature sensation in the right arm, trunk and leg; normal vibration sense (256 Hz) bilaterally (see Figure 8.9)

8.8.2 CASE 8E-2: AMY, AGE 72

Background: 'Aunt' Amy, as she is known in the neighbourhood, is a retired unmarried civil servant librarian who has consistently followed a healthy lifestyle but unfortunately still has medical issues for which she is being treated.

Chief complaint: Fell down in the street a short time ago; unable to move her right side

History:

- History of type 2 diabetes mellitus and hypertension
- Lives alone; no children; does not smoke or consume alcohol
- One hour ago, she collapsed on the street while walking home from church; found by

- Slowed rapid alternating movements of the left forearm; impaired finger tapping and heel-shin testing on the left side; exaggerated rebound on the left side
- Unable to sit in bed without support, veering to the left; attempt at walking showed severe ataxia

her neighbour; unable to speak or to move the right arm and leg
- Medications are ASA, Norvasc, Glyburide and Simvastatin

Examination:

- BP = 150/90; HR = 86/minute, regular
- Normal general physical examination; no neck bruits
- Completely unable to speak; difficulty in understanding some verbal commands
- Lower quadrantic visual field defect on the right side; normal fundi
- Moderate weakness of the right lower face, right arm and leg with hypertonia (Figure 8.10a)
- Reflexes: Figure 8.10b

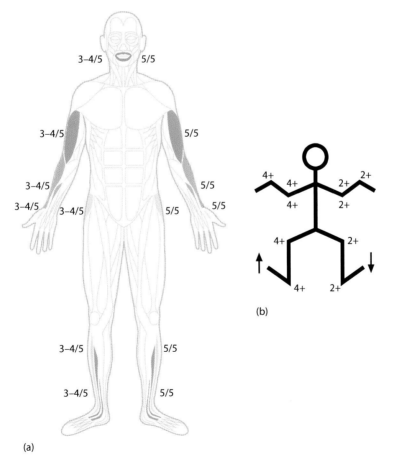

(b)

(a)

FIGURE 8.10: (a) Motor map and (b) reflex map for case 8e-2.

8.8.3 CASE 8E-3: ANTONIO, AGE 78

Background: This jovial, sociable retired restaurant owner, Antonio, thought he was in good health and only occasionally went to see his family physician – that is, until earlier today.

Chief complaint: Sudden-onset headache a short time ago, followed by inability to move his right side or to speak

History:

- History of high blood pressure; on an oral diuretic and low-dose ASA
- During breakfast today, Antonio abruptly developed a severe left-sided headache and fell off his chair; immediately unable to move his right face, arm or leg
- His wife called 911; he was brought to the hospital ER 1½ hours after the first symptom began

Examination:

- BP = 210/110; HR = 90/minute, regular; ECG shows sinus rhythm with left ventricular hypertrophy but no sins of acute ischemia
- Drowsy and uncooperative; Antonio is unable to answer questions or to obey simple commands; a few dysarthric vocalizations
- Responds to visual threat on both sides; full range of eye movement to head-turning

- Severe weakness of the lower half of the right face; absent gag reflex; no contraction of the right sternocleidomastoid or trapezius (Figure 8.11)
- Dense right hemiplegia, arm and leg (Figure 8.11)

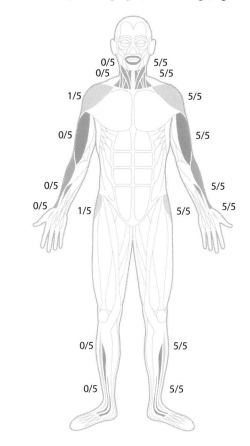

FIGURE 8.11: Motor map for case 8e-3.

8.8.4 CASE 8E-4: JOHN, AGE 67

Background: John is a hard-living, hard-working (delivering parcels) macho kind of guy and sadly a heavy drinker and heavy smoker. He is an avid (TV) football and hockey fan but never neglectful of his family responsibilities.

Chief complaint: Numbness and weakness of the left side beginning over thee hours ago

History:

- Previously a heavy smoker and drinker
- At 7 AM, having been up for two hours, John complained to his wife of numbness on the left side; decided to 'sleep it off'
- He re-awoke at 10:30 AM with profound weakness on the left side; was able to speak clearly, complaining of severe right-sided headache

- He arrived by ambulance in the ER slightly over four hours after his symptoms began
- His father and two brothers died in their 50s of either stroke or heart attack

Examination:

- BP = 180/100; HR = 90/minute, regular; ECG– sinus rhythm
- Alert and cooperative; moderate dysarthria but no dysphasia
- No response to visual threat from his left side
- Eyes are conjugately deviated to the right side but move reflexively to the left with passive head rotation (doll's eye manoeuvre)
- Left lower facial weakness; decreased sensation in the skin of the left face (Figure 8.12a and c)

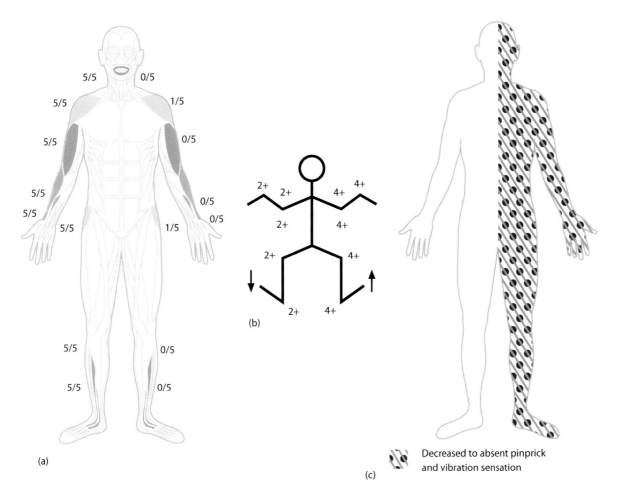

FIGURE 8.12: (a) Motor map, (b) reflex map and (c) sensory map for case 8e-4.

- Weak gag reflex; no contraction of the left sternocleidomastoid or trapezius muscles
- Profound left hemiplegia with hyper-reflexia; patient is unable to stand or walk (Figure 8.12a and b)

- Absent pinprick/vibration sense in the left arm, trunk and leg (Figure 8.12c)
- Normal rapid alternating forearm movements, finger tapping, finger-nose testing, toe-tapping and heel-shin testing on the right side

8.8.5 CASE 8E-5: VIC, AGE 8

Background: An orthopaedic surgeon has asked you to see Vic for assessment of his right leg. The surgeon was seeing Vic about a possible surgical correction of a foot deformity when he noted that there was an extensor plantar response in the involved foot. Vic is accompanied by his father and grandfather.

Chief complaint: Toe-walking with his right foot since he began to walk

History:

- Normal term pregnancy and delivery; no neonatal difficulties

- Vic started sitting at age 7 months, crawled at 9 months and walked at 16 months; he was speaking in sentences by age 2; currently in Grade 3 and doing well
- He has always walked on the toes of his right foot, being unable to put his foot flat on the floor despite a regular program of physiotherapy; with a recent growth spurt the problem has become worse
- Despite his toe-walking, Vic has learned how to play baseball and soccer; he is able to skate and plays goalie on a local pee-wee hockey team
- Vic is left handed; both his parents are right handed; there is no family history of left handedness

Examination:

- Cooperative, attentive and articulate for age 8
- Cranial nerve examination normal
- Muscle strength, tone and bulk are normal in both upper limbs; rapid alternating forearm movements and independent finger movements are impaired in the right hand, normal on the left
- Right leg is 1.5 cm shorter than the left; foot also smaller

- Increased muscle tone in right plantar flexors and hamstrings; mild weakness of dorsiflexion and eversion of the right foot (Figure 8.13a)
- Unable to wriggle the toes of his right foot or to tap the right forefoot on the floor
- In the gait cycle, he has a normal heel strike on the left and a flat-foot strike on the right (calcaneal gait)
- Reflexes: Figure 8.13b
- Touch, position and pinprick examination is normal in all four limbs

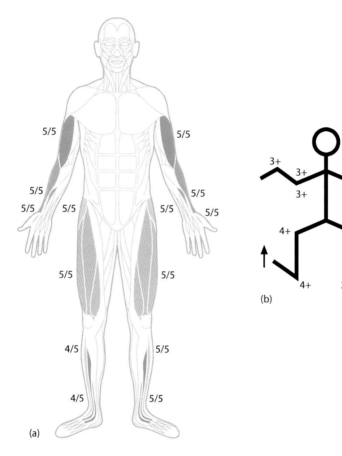

(a)

(b)

FIGURE 8.13: (a) Motor map and (b) reflex map for case 8e-5.

8.9 SUMMARY OF KEY NEUROANATOMICAL AND NEUROPHYSIOLOGICAL INFORMATION

Neuroanatomy:
- Internal carotid artery divides into anterior and middle cerebral arteries – the branches form the anterior circulation of the brain.

- Vertebral arteries unite to form the Basilar artery (at the lower border of the pons).
- Basilar artery divides (at the top of the brainstem) into two PCAs – the branches from the posterior circulation of the brain.
- The arterial circle of Willis is formed by the ACom artery (joining the anterior cerebrals) and the two PCom arteries (joining the posterior cerebral with the internal carotid arteries).
- The arterial territories overlap in the area where the dorsolateral aspect of the

hemispheres becomes the medial surface – this region is called the arterial border zone and is vulnerable to ischemia.

Neurophysiology:

- Cerebral blood flow is normally controlled by a process known as autoregulation.
- Turbulent flow conditions exist at the carotid bifurcation, making this area prone to atherosclerotic plaque formation and subsequent plaque accidents and emboli.
- Central source emboli (e.g. from the heart) can also produce strokes.
- The core area of a stroke is surrounded by potentially viable brain tissue, called the ischemic penumbra.

8.10 CHAPTER-RELATED QUESTIONS

1. Describe and draw the principle cerebral blood vessels and their branches.
2. What are the six major brain stem ischemic syndromes?
3. What are the principle medical measures of primary prevention of ischemic stroke?
4. What are the major risk factors for cortical hemorrhages?
5. What are the common genetic forms of risk factors for stroke?

SUGGESTED READING

Adams, H.P. Jr. et al. Guidelines for the early management of adults with ischemic stroke. *Stroke* 2007. 38: 1655–1711. Principle guideline statement for the use of thrombolytics in acute stroke.

Chalela, J.A. et al. Magnetic resonance imaging and computed tomography in emergency assessment of patients with suspected acute stroke: A prospective comparison. *Lancet* 2007. 369(9958): 293–298. Review of methods of DWI, ADC maps for MRI and CT perfusion mapping.

Feldmann, E. et al. The Stroke Outcomes and Neuroimaging of Intracranial Atherosclerosis (SONIA) trial. *Neurology* 2007. 68(24): 2009–2106.

Golledge, J., Greenhalgh, R.M., Davies, A.H. The symptomatic carotid plaque. *Stroke* 2000. 31: 774–781. A description of the clinical symptoms associated with carotid stenosis.

Gong, I.Y., Kim, R.B. Importance of pharmacokinetic profile and variability as determinants of dose and response to dabigatran, rivaroxaban, and apixaban. *Can J Cardiol* 2013. 29: S24–S33.

Hacke, W.M. et al. Thrombolysis with alteplase 3 to 4.5 hours after acute ischemic stroke. *N Engl J Med* 2008. 359(13): 1317–1329.

Husted, S., de Caterina, R., Andreotti, F. et al. Non-vitamin K antagonist oral anticoagulants (NOACs): No longer new or novel. *Thromb Haemost* 2014. 111: 781–782.

Lee, J.M., Zippfel, G.J., Chio, D.W. The changing landscape of ischemic brain injury mechanisms. *Nature* 1999. 399 (suppl): A7–A14. A review of the sequence of ischemic damage and ischemic cascade.

Lyden, P. Thrombolytic therapy for acute stroke – Not a moment to lose. *N Engl J Med* 2008. 359(13): 1393–1395.

Mahoney F.I., Barthel D. Functional evaluation: The Barthel Index. *Maryland State Medical Journal* 1965. 14: 56–61.

Mehdi, A., Hajj-Ali, R.A. Reversible cerebral vasoconstriction syndrome: A comprehensive update. *Curr Pain Headache Rep* 2014. 18(9): 1–10.

Modified Rankin Scale: http://www.strokecenter.org/trials/scales/modified_rankin.pdf. Another commonly used scale to score disability.

NIH Stroke Scale: http://www.ninds.nih.gov/doctors/NIH_Stroke_Scale_Booklet.pdf. Most commonly used scale to score stroke severity in acute ischemia to determine use of thrombolytics. The scale has its limitations in that deficits have unequal weighting, e.g. aphasia can score equally to facial weakness yet its contribution to disability is significantly greater.

Scaglione, F. New oral anticoagulants: Comparative pharmacology with vitamin K antagonists. *Clin Pharmacokinet* 2013. 52: 69–82.

Chapter 9

Cheryl

Objectives

- To understand the anatomy of the cranial cavity and its contents other than the brain, including the meninges, the cerebrospinal fluid (CSF) and its circulation, and venous drainage, as well as the involvement of these components in clinical disease states
- To comprehend the visual pathway and the usefulness of the clinical assessment of vision in the diagnosis of intracranial lesions

9.1 CHERYL

Cheryl is a joyful, almost boisterous person who loves to talk, smoke and eat. Living alone without a partner, she is always invited to family get-togethers in the countryside around Calgary, particularly at the holiday season.

Those who knew Cheryl as a young girl remember her as a slim, somewhat shy figure. Over the years, she has become more and more outgoing but at the same time quite overweight. There were a few opportunities for her to 'get involved' with a member of the opposite sex but somehow things never worked out; there were obligations to her dogs, or to a project she was working on, and so Cheryl remains 'unattached'. Now that she is almost 40 years old, everyone has stopped pestering her to get married or trying to 'fix her up' with some male of dubious character.

About three years ago, Cheryl began having sporadic headaches, sometimes associated with her menses. Her family physician had noted the occurrence of these and the fact that her blood pressure readings were averaging about 160/95. She instructed Cheryl to watch her diet, to avoid foods with a high salt content and to cut down on her smoking – all to no avail.

The physician felt that that the headaches were 'migrainous' in nature, as there is a positive family history of migraine, and recommended that, when the headache strikes, she take two extra-strength acetaminophens and stay away from noisy and brightly-lit rooms; no migraine-specific medication has been prescribed. About six months ago, hydrochlorthiazide was prescribed to control her

blood pressure, but Cheryl admits that she has only taken the medication from 'time to time'.

Over the past few months, the headache frequency has gradually increased and Cheryl has noted a few 'episodes' where the headaches have not gone away for over a day, despite taking acetaminophen every four to six hours. She says that the headaches are generally in the forehead area, although the most recent ones have been located mostly at the back of her neck; there has been no radiation of the neck pain into her arms. Flashes of light or other visual phenomena have not preceded the onset of the headaches, nor are they accompanied by nausea or vomiting.

Her current visit has been precipitated by a recent worrisome event: she was driving home from a family gathering (it was snowing and the country road was totally dark) and her car slipped off the road and into a ditch. Luckily, there was enough snow to cushion her car from serious damage, but Cheryl was not wearing her seat belt (it had become too uncomfortable over her heavy coat). Fortunately, she was found by a passing motorist driving a four-wheel-drive panel truck.

Initially, Cheryl did not respond to the shouting of her 'rescuers' but quickly regained consciousness. The driver of the truck and his friend easily managed to pull the car out of the ditch; afterwards, they all went back to his nearby farmhouse, where there was a friendly gathering. Cheryl was soon back on the road and arrived home safely. Her mother, hearing the story, insisted that she seek immediate medical attention and Cheryl is now in the emergency room (ER) at the local community hospital, the morning following the incident.

9.1.1 PHYSICAL EXAMINATION

On examination, Cheryl presents as a pleasant if somewhat subdued individual who is clearly worried about what could have happened to her. On recounting her story, it seems that she had failed to see a sign indicating that there was a curve in the road. She had consumed only one beer at the party (early in the evening, followed by lots of finger food) and avows that she never has more than this to drink if she knows that she is going to drive home afterwards. Her retelling of the previous night's incident is quite clear and consistent, and she does remember her car sliding towards

178

the ditch. Her next recollection is some person yelling at her and reaching over her to shut off the car engine. She has no idea how much time elapsed between the moment the car slid off the road and the time when she was found.

Cheryl is clearly very overweight (body mass index 38 kg/m²; see also Etienne, Chapter 8) and her blood pressure is 170/105 sitting and 165/95 reclining. She has a slight bruise on the top of her forehead, on the right side, just at the hairline. Her mental status examination is completely within normal limits – she is fully oriented in person, time and place. She is fully cooperative with the examiner.

9.1.2 CRANIAL NERVE EXAMINATION

CN I: Smell testing is normal.

CN II: Using the confrontation technique, no visual field defect is noted. Visual acuity with her glasses is within normal limits. The pupillary reaction to light is normal and equal in both eyes.

CN III, IV, VI: Movements of the eyes in the cardinal positions are within normal limits.

CN V: Sensory – touch and pain are normal in the three divisions of the trigeminal nerve; motor – teeth clenching is symmetrical and the jaw reflex is normal.

CN VII: No facial asymmetry is noted and facial movements are normal.

CN VIII: No defect in hearing between the two sides is noted with tuning fork testing.

CN IX, X: The gag reflex is normal and no hoarseness of voice is noted.

CN XI: Head turning and shoulder shrugging are normal and symmetrical.

CN XII: Tongue movements are normal.

9.1.3 MOTOR EXAMINATION

Muscle strength is appropriate and symmetrical in all four limbs. Reflexes are somewhat 'sluggish' but symmetrical; the plantar reflex is downgoing bilaterally. There is no ankle clonus.

Coordination and gait are within normal limits. The Romberg sign is not present. In this test, the patient is asked to stand with both feet together. (Note that the examiner must stand nearby.) If the patient begins to fall, then there may be a cerebellar problem. If the patient is able to maintain a standing posture with eyes open, she/he is then asked to close his/her eyes; if the person then begins to fall, she/he must have been relying on vision to maintain balance, indicating a possible problem in the vestibular or spinal proprioceptive system.

9.1.4 SENSORY EXAMINATION

No deficits in sensory testing are noted over the limbs and trunk; in particular, there is normal vibration sense and position sense in all four limbs.

9.1.5 FUNDOSCOPY

With the lights in the room dimmed, the disc margins are found to be definitely blurred bilaterally (compare Figure 9.1a and b).

The assessment is that Cheryl has *bilateral papilledema*, which is considered an emergency and requires an immediate consultation. An arrangement is made for her to see the senior ophthalmologist at the regional tertiary referral hospital early that afternoon. The bilateral papilledema is confirmed along with the presence of retinal venous engorgement. Cheryl is sent immediately for testing of the visual fields and for neuroimaging (computed tomography [CT] scan), and with a strong likelihood of needing a lumbar puncture (an LP).

Normal

Papilledema

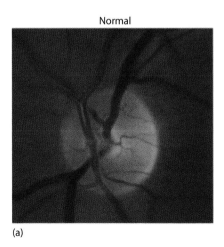 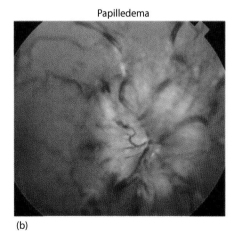

(a)   (b)

FIGURE 9.1: Optic disc: Photographic images of the retina as seen with the ophthalmoscope. (a) A normal optic disc with its whitish appearance and blood vessels, both arteries and veins. Note the clearly defined disc margin. (b) An abnormal disc in a patient with raised intracranial pressure. Note the 'swollen' appearance of the disc and the absence of any clear margins. (Courtesy of Dr. M. O'Connor.)

Visual fields can be thoroughly and accurately mapped using a technique called perimetry. With this method, the patient focuses on the centre of a rounded screen (the same shape as the eyeball) and a small light stimulus is projected at random on the screen; the patient is requested to give a response the moment a new light stimulus has been perceived. Each eye is tested separately and the visual field is mapped and charted in a standardized manner. The testing can be done by an automated machine, available at most major hospitals, and takes 10–15 minutes; testing can also be done by a technician if the automated testing is not satisfactory or in order to confirm any abnormal findings. Cheryl's visual field testing reveals a surprise! A significant enlargement of the physiological blind spot is found, eye in both eyes (compare a normal eye in Figure 9.2a with an abnormal eye in Figure 9.2b). In effect, there is a large region just lateral to the fovea where Cheryl can see nothing whatsoever.

9.2 CLINICAL DATA EXTRACTION

At this point, the student should proceed to the Web site and complete the history, physical examination, localization and etiology worksheets, utilizing the information just provided concerning Cheryl's problem.

9.3 REVIEW OF THE MAIN CLINICAL POINTS

- This is a 37-year-old obese female with a history of hypertension and a long-standing neurological complaint of headache.
- There has been a recent motor vehicle 'incident', associated with a period of unconsciousness of unknown duration.

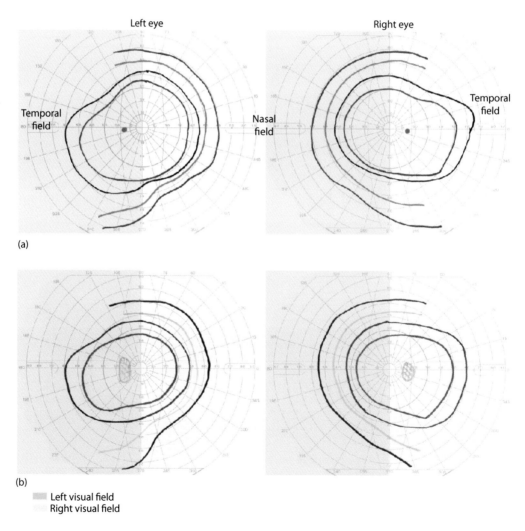

FIGURE 9.2: Visual field (Goldmann) perimetry. In presenting the visual fields, the convention is to display them in the same way that the patient is viewing (right field on the right and left field on the left). (a) Normal visual fields, with the blind spot indicated (in orange) in the temporal field. (b) Visual fields of a patient with the same disorder as Cheryl. Note the enlarged blind spot (bilaterally, in yellow). (Courtesy of Dr. M. O'Connor.)

- There is a small bruise over the right forehead.
- Significant papilledema is present while the rest of the neurological examination is within normal limits.
- Formal visual field testing has revealed enlarged blind spots bilaterally.

9.4 RELEVANT NEUROANATOMY

In order to understand Cheryl's complaint of headache, we need to know which structures of the head and in the interior of the skull are pain-sensitive. Pain can originate from any of the structures of the head, including the skin, teeth, mucous membranes (including the gums), the sinuses and from the eye and orbit. The brain itself has no pain fibers, but the brain coverings, the meninges, are pain sensitive, as are the walls of the major cerebral blood vessels. Any irritation (infection, blood) or stretching (pulling) on the meninges or cerebral vessels will give rise to pain.

In the light of this information, it will be necessary for us to review the anatomy of the cranial and spinal meninges. Secondly, as part of the meningeal compartment, the CSF and its circulation need to be considered. Thirdly, in order to appreciate the significance of bilateral papilledema and enlarged blind spots, the visual system and its pathway should be understood. Finally, we must also consider further aspects of the cerebrovascular supply to the skull and brain, both arterial and venous.

9.4.1 MENINGES AND CSF

The meninges are the connective tissue coverings of the brain and offer a certain measure of protection to the underlying brain tissue. The meninges consist of three layers – the dura, the arachnoid and the pia – with spaces or potential spaces between the layers. The CSF lies within the meninges such that the brain is in fact 'floating' in fluid.

9.4.1.1 CRANIAL MENINGES

The outermost layer of the meninges is the dura, a thick, strong sheet of connective tissue. Within the cranial cavity, the dura and the skull bones adhere somewhat to one another, leaving a potential gap, the *epidural* space (Figure 9.3a). Bleeding from the large artery that supplies the meninges, the middle meningeal artery, occurs between the skull and dura (Figure 9.3b) and is appropriately called an *epidural hemorrhage*. This is usually caused by trauma to the side of the head in the temporal region of the skull. The dura is composed of two layers, the periosteal and the meningeal.

The next layer is the arachnoid. There is a potential gap between the dura and arachnoid, the *subdural* space. The cerebral veins pass through this 'space'. These bridging veins, as they are called, course from the surface of

the brain into the venous sinuses, particularly the superior sagittal sinus (Figure 9.3b). It is here that they may be disrupted by trauma to the head, in which case blood leaks into this space, producing a *subdural hemorrhage*. This venous bleed usually occurs slowly over time (subacute or chronic) but may also present acutely; this type of bleed is more common in the elderly, either because of the increased fragility of the veins, brain shrinkage with age or both.

The innermost layer, the pia, lies on the surface of the brain and follows all its folds. The subarachnoid space, between the arachnoid and pia, contains the cerebrospinal fluid, the CSF; large arteries and veins are also found in this space. The larger cerebral arteries course through the subarachnoid space on the way to their destinations. (The arterial supply of the brain has been considered in Chapter 8.)

Pathological localized dilations or bulges of the arterial wall of arteries, called (cerebral or berry) aneurysms, may occur on the larger arteries, particularly those of the circle of Willis; these may bleed. Usually, this occurs precipitously, as a *subarachnoid hemorrhage*, resulting in an intense headache and frequently by a loss of consciousness. The CSF is tinged red, as it is filled with arterial blood. If CSF is taken several days after the hemorrhage, the CSF appears yellow or xanthochromic due the breakdown of the red cells into metabolic byproducts, which give the CSF a yellow color.

All three layers of the meninges, including the CSF, continue onto the optic nerve and the dural layer merges with the outermost layer of the eyeball. From the developmental perspective, the retina is an 'extension' of the brain and the optic nerve is in fact a CNS tract.

9.4.1.2 SPINAL MENINGES

The meninges continue around the spinal cord within the vertebral canal (Figure 9.4). The spinal cord dura is separated from the periosteum of the vertebra (and the intervertebral discs) by a space, the epidural space, which is filled with fat in the lower vertebral region (shown in Figure 5.4, an MRI of the spinal column and spinal cord) and which contains a plexus of veins. A sleeve of pia, arachnoid and dura accompanies the ventral and dorsal roots of the spinal nerves, until they come together in the intervertebral foraminal region to form the mixed spinal nerve (Figure 9.4c).

The CSF continues within the subarachnoid space around the spinal cord. The spinal cord with its pia ends at the vertebral level of L2, whereas the dura and arachnoid end at the level of S2. The large cistern in the vertebral canal below the level of the spinal cord is called the lumbar cistern (see Section 9.4.2) and is the site for sampling of CSF (discussed below).

9.4.2 THE VENTRICULAR SYSTEM AND CSF CIRCULATION

The cerebral ventricles of the brain and the central canal of the spinal cord are found within the nervous tissue and are the remnants of the original neural tube from which

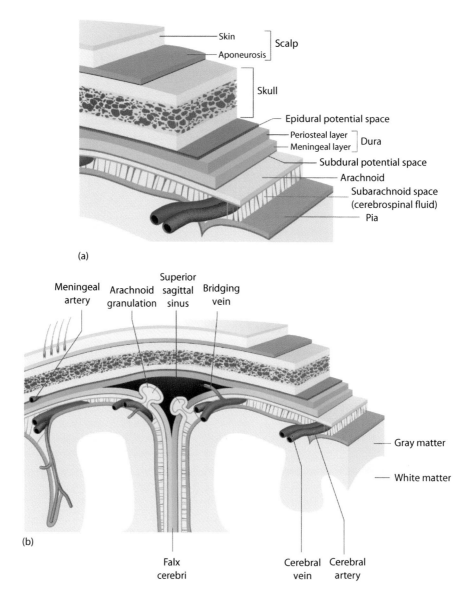

(a)

(b)

FIGURE 9.3: Cranial meninges. (a) This view includes the scalp, the skull and the layers of the meninges with the subarachnoid (CSF) space. Note the potential spaces between the meningeal layers. (b) The meninges of the cranial cavity are shown in a coronal view, including the superior sagittal venous sinus and the CSF compartment. Note the bridging vein and the arachnoid granulation, as well as the location of the meningeal artery.

the nervous system developed. There are four ventricles: one in each of the cerebral hemispheres (ventricles 1 and 2), also called the lateral ventricles (see Figure 1.11), the third ventricle in the thalamic (diencephalic) region and the fourth ventricle in the brainstem region (Figure 9.5).

CSF is formed within the ventricles from specialized tissue, the *choroid plexus*, located in each of the ventricles. The composition of CSF differs from plasma because of a blood – CSF barrier; the site of this barrier is the cells lining the choroid plexus. Notwithstanding this, the cells lining the remainder of the ventricular system do allow a free exchange between the extracellular space of the brain and the CSF fluid. As a result, the CSF will accumulate certain proteins with specific neurological diseases, such

as multiple sclerosis; therefore, sampling of CSF may aid in determining some disease processes of the CNS.

There is a circulation of CSF within the ventricles, with the flow being from lateral ventricles to the third ventricle, and then via a narrow aqueduct in the midbrain to the fourth ventricle in the pontine region (black arrows in Figure 9.5). At the lower end of this ventricle, CSF 'escapes' into the subarachnoid space, the cerebellopontine cistern, known generally as the cisterna magna. Cisterns are enlargements of the subarachnoid space, and several others are found around the brainstem; the largest of these is the cisterna magna, located behind the brainstem and below the cerebellum, in the posterior cranial fossa, just above the foramen magnum (see Figure 9.5).

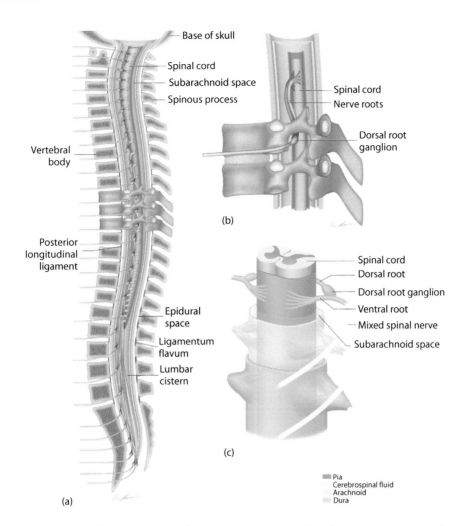

Base of skull
Spinal cord
Subarachnoid space
Spinous process
Vertebral body
Spinal cord
Nerve roots
Dorsal root ganglion
(b)
Posterior longitudinal ligament
Spinal cord
Dorsal root
Dorsal root ganglion
Ventral root
Mixed spinal nerve
Subarachnoid space
Epidural space
Ligamentum flavum
Lumbar cistern
(c)
(a)

Pia
Cerebrospinal fluid
Arachnoid
Dura

FIGURE 9.4: Spinal meninges. (a) This longitudinal view of the spinal cord within the vertebral column includes the layers of the meninges and the actual epidural space. The CSF space below the termination of the spinal cord is the lumbar cistern, with the nerve roots – the cauda equina. (b) This is a focused view of the location of the spinal nerve with the dorsal root ganglion (DRG) located in the neural foramen. (c) The meninges of the spinal cord including the subarachnoid (CSF) space are shown in an axial view. The meningeal layers continue just beyond the DRG, where the dorsal (sensory) and ventral (motor) roots unite to form the mixed spinal nerve.

CSF flows from the cisterna magna within the subarachnoid space around the brain and also downwards around the spinal cord, where it accumulates in the lumbar cistern, below the termination of the spinal cord. It is almost always the lumbar cistern that is used to sample CSF and to take the CSF pressure, a procedure called a *lumbar puncture*, or LP. CSF pressure is measured when the needle initially penetrates into the lumbar cistern and is normally 7–25 cm H_2O or 5–19 mm Hg (mercury).

The normal amount of CSF in the ventricles and craniospinal subarachnoid spaces is estimated to be around 150 ml and is replaced roughly every 6–8 hours, which indicates that there is a continuous process of production and absorption of CSF, in effect a (slow) CSF circulation. CSF is returned to the venous circulation via the arachnoid granulations (see Figure 9.3a), which protrude into the venous sinuses, particularly into the superior sagittal sinus (blue arrows in Figure 9.5). A small pressure differential is thought to account for the

transport of CSF across these villi and into the venous sinuses, thus completing the circulation of CSF.

Blockage of CSF flow, for example at the level of the midbrain where the aqueduct connecting the third with the fourth ventricle is very narrow, would cause an increase in size, and of pressure, in the cerebral ventricles. This blockage, called obstructive or non-communicating *hydrocephalus*, would be visualized with CT or magnetic resonance imaging (MRI) as an enlargement of the lateral ventricles of the hemispheres. In an adult, because the sutures of the skull are fused, this process would be accompanied by raised intracranial pressure (ICP), whereas in a young child (e.g. the first two years) with non-fused cranial sutures, the head itself will enlarge and there would be a separation of the bones of the skull; at a very early age, the anterior fontanelle would bulge.

Blockage of the CSF flow can also occur at the level of the arachnoid granulations, and in fact, this does occur

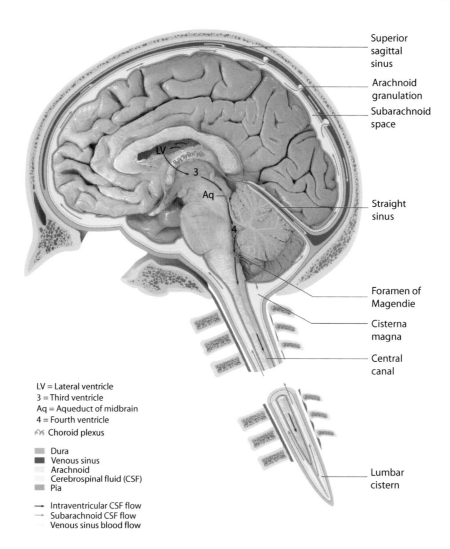

Superior
sagittal
sinus

Arachnoid
granulation

Subarachnoid
space

Straight
sinus

Foramen of
Magendie

Cisterna
magna

Central
canal

Lumbar
cistern

LV = Lateral ventricle
3 = Third ventricle
Aq = Aqueduct of midbrain
4 = Fourth ventricle
ᐭ Choroid plexus

▨ Dura
▨ Venous sinus
▨ Arachnoid
▨ Cerebrospinal fluid (CSF)
▨ Pia

→ Intraventricular CSF flow
→ Subarachnoid CSF flow
→ Venous sinus blood flow

FIGURE 9.5: CSF circulation: mid-sagittal view: CSF flows though the ventricular system, as indicated by the arrows (black). At the lower end of the IVth ventricle, CSF enters the subarachnoid space and flows around the brain and down and around the spinal cord, and into the lumbar cistern (blue arrows). Note the choroid plexus in the ventricles, and the arachnoid granulations in the superior sagittal sinus where the CSF is 'returned' to the venous circulation. The flow of blood is indicated (white arrows) in the straight and superior sagittal sinuses.

following meningitis (further discussed in the etiology section of this chapter). If the villi are non-functional or blocked, or if there is a blockage of the venous sinuses (e.g. a venous sinus thrombosis) or a stenosis of the sinuses, then CSF can no longer return in a sufficient manner to the venous circulation, resulting in an increase in CSF pressure. Hydrocephalus developing from CSF flow obstruction at a point outside the brain is called communicating hydrocephalus.

9.4.3 CEREBRAL VASCULATURE (VENOUS)

Two dural sheaths separate the parts of the brain from each other and 'hold' the cerebral hemispheres in place. The first is the *falx cerebri*, situated in the midline between the two cerebral hemispheres, above the corpus callosum (Figure 9.6a). The other is the *tentorium cerebelli*, a horizontal sheath of dura that lies between the occipital lobe of the hemispheres above and the cerebellum below

(Figure 9.6c). Both the falx and the tentorium are attached to the bone on the inner aspect of the skull.

At this point, it is relevant to consider the venous system. The venous sinuses are the venous channels within the dura formed along the attachments of the dura to the bones of the skull. The two dural layers, the periosteal and meningeal, separate to form these large venous spaces (see Figure 9.3b). The venous return from the brain courses via a superficial system over the surface of the hemispheres and a deep system within the substance of the brain tissue. Both systems drain into the venous sinuses of the skull.

A major venous sinus is the *superior sagittal sinus*, which is found along the upper (attached) border of the falx cerebri, in the midline (Figure 9.6a, also Figure 9.5 and Figure 9.6d). Most of the superficial veins of the hemispheres empty into the superior sagittal sinus. This sinus continues posteriorly, and at the back of the

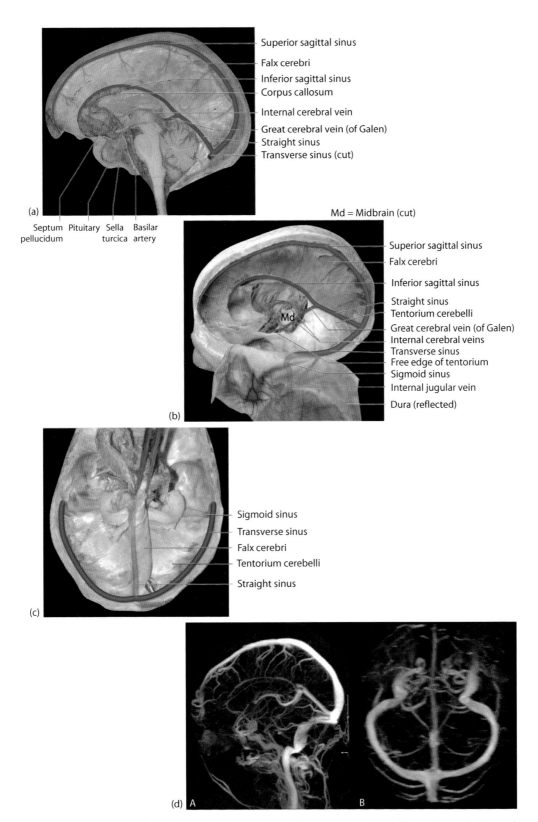

FIGURE 9.6: The venous sinuses of the brain are shown in these in situ preparations – with overlays – in three views: (a) Mid-sagittal: including the superior and inferior sagittal sinuses, at the upper and lower margins of the falx cerebri, (b) oblique view, showing the internal cerebral veins joining to form the great cerebral vein (of Galen) and the straight sinus and continuing as the transverse and sigmoid sinuses, becoming the internal jugular vein, and (c) axial (horizontal) view with the tentorium cerebelli, including the shadow of the straight sinus and the transverse sinuses with their continuation to the sigmoid sinuses. (d) The corresponding views of the venous sinuses are shown neuroradiologically, called an MRV. A: mid-sagittal view. B: axial view. (Courtesy of Dr. C. Torres.)

interior of the skull is joined by the straight sinus (see white arrows in Figure 9.5) and then divides to become the laterally positioned *transverse venous sinuses*, one on each side, attached to the skull at the lateral edges of the tentorium cerebelli. An exception to the rule of the location of the venous sinuses is the *inferior sagittal sinus*, located in the inferior margin of the falx cerebri, where there is no attachment to bone.

The system of veins that drain the deep structures of the brain emerges medially as the *internal cerebral veins*, one from each hemisphere (Figure 9.6b). These veins join in the midline in the region behind the diencephalon to form the *great cerebral vein* (of Galen). At this point, there is another exception to the rule of the formation of venous sinuses. A sinus is located in the midline, where the falx splays out laterally to form the tentorium. This is known as the *straight sinus* (Figure 9.6b and c). The great cerebral vein becomes continuous with the straight sinus, in the midline lying above the cerebellum. At this point, it is joined by the *inferior* sagittal sinus. At the back of skull, the straight sinus joins with the *superior* sagittal sinus. The venous sinuses now divide, and the blood flows into the transverse sinuses (Figure 9.6c and d). The venous blood exits the skull via the sigmoid sinus, becoming the internal jugular vein (Figure 9.6b and d), one on each side of the neck.

The venous system can be visualized radiographically using MRI by tracking the blood flow in time (time-of-flight) without injecting some of the paramagnetic dye gadolinium, or by injection of this dye to illuminate the veins; this technique is called a magnetic resonance venogram, an MRV (Figure 9.6d).

9.4.4 VISUAL SYSTEM

Vision starts in the retina with the photoreceptors, the rods and cones. The fovea, a small area of the retina in the central axis of visual input, is the visual area required for fine vision, including reading and colour vision (with the cone photoreceptors). The vast peripheral region of the retina captures peripheral vision, using the rod photoreceptors, and is used in conditions with poor illumination.

The cones and rods are specialized receptor cells located in the deepest part of the retina and are activated by light. They connect with first-order sensory neurons, also in the retina; after processing by other neurons within the retina, the messages are passed on to the ganglion cells, the second-order neurons in the visual system, which are also located in the retina. It is these neurons whose axons form the optic nerve, CN II.

The axons of the ganglion neurons exit the eye at the optic disc (see Figure 9.1a), an area of the retina with no photoreceptors and hence responsible for the physiological blind spot (see Figure 9.2a, also Figure 2.1). It marks the beginning of the optic nerve; CN II is in fact a pathway of the CNS, with the myelin of the nerve formed by oligodendrocytes. In its path through the orbit, it is sheathed by the meninges of the brain, with a typical subarachnoid space containing CSF. Any increase in ICP may be reflected via this space onto the optic nerve and cause its compression, as well as compromising the blood vessels supplying the retina (the arteries and veins) running within the nerve.

9.4.4.1 PATHWAY

After exiting the eyeball, the *optic nerves* cross the orbit and pass through the optic foramen and enter the interior of the skull. In the area above the pituitary gland, the nerves undergo a partial crossing (decussation) of fibers in a structure called the *optic chiasm*. (There is no synapse in the optic chiasm.) The fibers from the nasal retina on one side cross the midline and join with those from the temporal retina from the other eye (which do not cross) to form the *optic tract*. Thus, the image of the visual world, which started in different parts of the retina of the two eyes, is now brought back together in the optic tract (Figure 9.7a, b and c).

The visual world, usually called the visual field, is divided into quadrants, temporal and nasal, superior and inferior, for each eye. Because of the pinhole effect of the pupil, the upper visual field projects to the lower retina (and the converse for the lower visual field), while the temporal visual field projects to the nasal retina, and the nasal visual field, to the temporal retina.

Using a specific example, imagine looking through a keyhole. To see the world on the left, you have to position your head on the right of the keyhole and vice versa for the world on the right. Similarly, you have to look from the bottom of the keyhole to see the world above and vice versa for below. This results in the visual world being represented backwards and upside down on the retina in addition to occupying a much smaller area on the retina than the real object. This property allows objects much larger than our eyes to be viewed, in addition to providing wide peripheral vision. The brain eventually 'flips' our perception of the image so that it appears right side up. (Suggestion: Making a sketch diagram of this verbal description is a simple and effective way of understanding the visual pathway; also see Figure 9.7a and b.)

Within the cranial cavity, because of the optic chiasm, the information from the nasal portions of each retina – representing the temporal halves of both visual fields – crosses to the opposite side. The optic tract is now formed, bringing together information from the contralateral visual world, consisting of the temporal visual field of the contralateral eye and the nasal (medial) visual field of the ipsilateral eye.

Using a specific example (see Figure 9.7a), the right optic tract will carry information from the temporal retina of the right eye (which receives light from the right

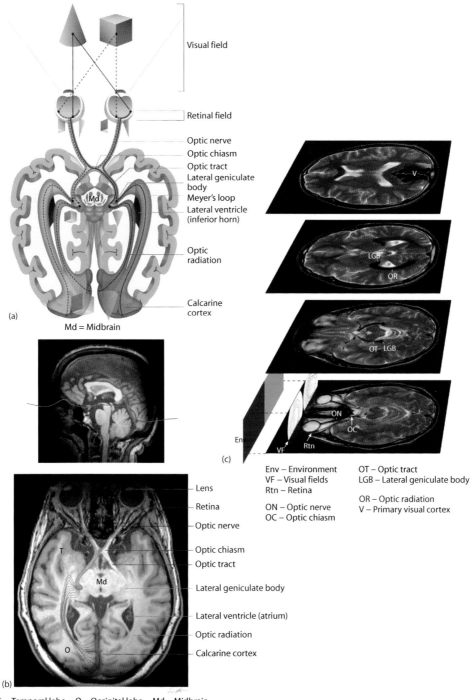

(a)

Visual field

Retinal field

Optic nerve
Optic chiasm
Optic tract
Lateral geniculate body
Meyer's loop
Lateral ventricle (inferior horn)

Optic radiation

Calcarine cortex

Md = Midbrain

V

LGB
OR

OT LGB

ON
OC

En

VF Rtn

(c)

Env – Environment	OT – Optic tract
VF – Visual fields	LGB – Lateral geniculate body
Rtn – Retina	
	OR – Optic radiation
ON – Optic nerve	V – Primary visual cortex
OC – Optic chiasm	

Lens
Retina
Optic nerve
Optic chiasm
Optic tract
Lateral geniculate body
Lateral ventricle (atrium)
Optic radiation
Calcarine cortex

T
Md
O

(b)

T = Temporal lobe   O = Occipital lobe   Md = Midbrain

FIGURE 9.7: (a) Visual pathway I: The complete visual pathway is shown from the visual field, to the retinal field, and then the optic nerve, optic chiasm and optic tract. After a synapse in the lateral geniculate nucleus (of the thalamus), the optic radiation splits and terminates in the primary visual area or the calcarine cortex of the occipital lobe. Note the partial crossing of fibers in the optic chiasm. (b) Visual pathway II: This is a radiological reconstruction of the visual pathway. The upper illustration shows the various levels needed for this reconstruction, which is shown in one plane in the lower illustration. As is customary with MR images, the right cerebral hemisphere is on the left side of the image. Note that the cortex of the right hemisphere receives lateral visual field of the left eye and medial visual field from the right eye – both from the opposite side. (c) The visual pathway III: MRI overlays of the brain. An image in the left visual field – of each eye – is conveyed to the retinal fields, the nasal portion in the left eye and the temporal portion of the right eye. Because of the crossing of the nasal retinal fibers in the optic chiasm, both images are united in the right optic tract. From here, the image relays in the LGB of the thalamus and projects as the optic radiation through the white matter of the hemispheres to the visual cortex in the occipital lobe. (Courtesy of Dr. R. Grover.) Note that this illustration is animated on the text Web site.

eye's medial or nasal visual field – representing the left side of visual space) and the crossed fibers from the nasal retina of the left eye (which receives light from the left eye's lateral or temporal visual field of the left eye – also representing the left side of visual space). This explains how the right optic tract is carrying information from the contralateral left visual fields of each eye, and includes the fibers comprising one-half of each retina.

The visual fibers in the optic tract have a specific relay nucleus (the *lateral geniculate nucleus* or body, the LGN or LGB, in the thalamus (see Figures 1.8 and 13.3); this is the site of the third-order neurons whose axons project as the optic radiations. Some of the fibers project directly posteriorly deep in the white matter of the parietal lobe to the visual cortex in the occipital lobe, the calcarine cortex; other fibers from the lower retina (from the upper visual field) project from the thalamus forwards in the temporal lobe before proceeding posteriorly to the occipital cortex (a route also known as Meyer's loop). Note that the fibers that carry the information for the pupillary light reflex leave the optic tract before the thalamic relay nucleus and head for the midbrain, where the centre for the light reflex is located (as described in Chapter 2, see Figure 2.2).

The final destination for the visual fibers is the cortex along the calcarine fissure of the occipital lobe, located on the medial surface of the brain; this is the primary visual area, V1 or calcarine cortex, also known as area 17 (see Figure 1.10d). Cortical areas adjacent to the calcarine cortex further process the visual information; additional visual regions in other areas of the brain process specific aspects of vision: more ventral areas deal with form perception (recognizing faces, objects, etc.) and more dorsal areas for visuospatial and motion processing.

Note that the visual pathway – from the cornea to the calcarine cortex – extends through the whole brain (excluding the frontal lobe), hence its importance in the assessment of nervous system integrity.

9.5 LOCALIZATION PROCESS

Cheryl's headache symptoms, the presence of papilledema and the associated enlargement of the blind spot all clearly point to something occurring within the skull, affecting the visual system but not the motor or sensory functions and not affecting cognition, Cheryl's personality or her behaviour. The structures within the skull, other than brain tissue, include the meninges, CSF (the ventricles and cisterns and CSF circulation) and the arteries and veins. All of these structures must be considered in the localization exercise.

Let us now consider the localization implications for the three main clinical features of Cheryl's case: headache, papilledema and enlarged blind spots.

9.5.1 HEADACHE

An analysis of headache characteristics by itself does not usually contribute much to the localization of a neurological problem. (Note that the opposite is true for the diagnosis of certain types of migraine headache.) On occasion, a localized head pain will reflect the presence of pathology in the immediate region of the pain, e.g. a left frontal lobe tumor in someone with left frontal head pain, the stretching of the regional meninges and major blood vessels stimulating selected CN V sensory fibers whose input can be localized by thalamic VPM and somatosensory cortex. In general, however, a bilateral frontal headache like Cheryl's could result from pathology anywhere inside the head.

9.5.2 PAPILLEDEMA

Optic disc edema may result from a variety of causes, including optic nerve vascular compromise, inflammation, infiltration, and infection. Raised ICP is also transmitted to the optic nerves, resulting in optic disc edema. Optic disc edema caused by raised ICP is known as papilledema. Because the raised ICP is usually transmitted to both optic nerves, papilledema is almost always bilateral; in contrast, optic nerve inflammation (papillitis) is often unilateral. The cause of this increased pressure could be anywhere within the skull.

In its early stages, papilledema is not typically associated with significant impairment in visual acuity. In contrast, other causes of optic disc edema such as papillitis commonly produce early, severe loss of visual acuity. In Cheryl's case, the finding of bilateral optic disc edema with normal visual acuity is consistent with a diagnosis of papilledema due to elevated ICP.

In summary, bilateral disc oedema without any obvious impairment in visual acuity suggests the presence of increased ICP but does not, by itself, help one to localize the cause of the pressure elevation.

9.5.3 ENLARGED BLIND SPOTS

Having previously reviewed the anatomy of the visual pathways, we must determine the part of that pathway that, if damaged, would present with selective enlargement of the blind spots. By a rapid process of elimination, this turns out to be quite simple. A unilateral optic nerve lesion would result in problems in one eye (e.g. complete loss of vision or a large central scotoma); bilateral optic nerve lesions would produce, if sufficiently severe, almost total blindness. Lesions of the optic chiasm affect vision in the temporal fields (the crossed fibers) while sparing the nasal fields, a visual defect known as *bitemporal hemianopia*. Lesions of the optic tract, the lateral geniculate bodies or the optic radiations, would produce a homonymous visual field defect on the side opposite to the hemisphere affected (otherwise known as a *homonymous hemianopia*). (Suggestion: Using your sketch drawing, place the lesions in the visual pathway and note the visual field deficits.)

Clearly, none of the earlier patterns of visual deficits correspond to those of our patient. With papilloedema, while there is no loss of visual acuity, the size of the blind spot is increased as the disc edema disrupts photoreceptors surrounding the optic disc. Eventually, damage to the nerve can occur resulting in progressively restricted peripheral vision; this is quite difficult to detect during a routine clinical neurological examination. Thus, Cheryl's visual field abnormalities are likely related to the presence of raised ICP and resultant papilledema.

In summary, the absence of any focal neurological deficit does not enable one to localize the problem to any specific area of the brain. The location, however, for the chronic state that led to the papilledema must be intracranial, leading to an increase in ICP. This includes any intracranial mass, often called a space-occupying lesion, leading to a mass effect. Clearly, however, if Cheryl has such a mass, it cannot be located within CN II, the optic tracts, thalamus, optic radiations or the visual cortex.

There is no indication, as yet and without any investigation, as to where the problem is localized within the cranial cavity. Raised ICP may result from a volumetric enlargement of any one of the following intracranial components (or a combination thereof):

- The brain tissue
- The meninges
- The vasculature, either arterial or venous
- The ventricles and/or the quantity or pressure of CSF (CSF circulation)

In other words, any pathophysiology or lesion that leads to an increase in the volume of the contents of the intracranial cavity will cause a mass effect and lead to an increase in ICP. Since the cranial cavity is a closed 'box' of fixed size in the adult, the general term often used is a *space-occupying lesion*, but, as we will see, the term itself may divert one's thinking away from some intracranial pathology which would not fit under the category of a space-occupying lesion (discussed in the Etiology section). An LP is necessary to measure whether there is or is not an increase in CSF pressure (discussed in detail in Section 9.7.2).

9.6 ETIOLOGY – CHERYL'S DISEASE PROCESS

In considering the possible etiology in this patient, we are now certain that the raised ICP is of longstanding duration, i.e. chronic, whereas the trauma is an acute event.

Papilledema takes time to develop, and one would not associate its presence with a lesion that occurred acutely, such as Cheryl hitting her head as a result of the car incident the previous night. In other words, as causes of papilledema, paroxysmal (A) and traumatic (B) disorders are ruled out.

Our summary exclusion of paroxysmal and traumatic disorders as potential causes of Cheryl's neurological problem is but the first of a number of similar exclusions in Chapters 10 and 11. In Chapter 3 (page 59), we noted that the Etiologic Diagnosis sections of Chapters 4–13 would use a lettering (rather than numeric) system for each disease category in order to provide a classification framework for the reader. Beginning in this chapter and continuing through Chapters 10 and 11, the nature of the disease or disorder affecting the chapter case is such that some of the 10 disease categories are so unlikely that they do not merit specific consideration. Thus, rather than having 10 subsections (A–J), this chapter's Etiology section will only have 7 (C–I), and Chapters 10 and 11 will only have 5 each, both in different combinations.

Of the various disease categories with a potentially chronic course, which ones could be associated with an increase in ICP (see Figure 4.8)? The etiologic possibilities associated with the acute event will be considered subsequently.

Disease processes related to the meninges should also be taken into account when discussing raised ICP, including any interference with CSF production, circulation or reabsorption.

9.6.C VASCULAR

In the description of the meninges, mention was made of the possibility of bleeding between the various layers. Although vascular disease processes usually produce acute-onset symptoms, bleeding from the veins draining the brain may occur over a prolonged period of time in the form of a chronic subdural hematoma; again, this is a space-occupying lesion, which might give rise to headache and raised ICP. This condition usually affects the elderly, possibly because of the fragility of the veins at this age associated with an age-related shrinkage of the brain, and may occur following even a minor head trauma. A chronic subdural hematoma is most unlikely in a 40-year-old woman without a history of a remote head injury or of alcoholism. The possibility of an acute subdural will have to be revisited because of the history of the acute head trauma.

Another vascular-related possibility is that CSF circulation can be obstructed due to a problem with CSF reabsorption; this could be impaired because of a reduction in venous return, such as a venous sinus thrombosis, leading to an increase in ICP. Spontaneous venous sinus thrombosis can occur in individuals with a high fever, a hereditary predisposition to clotting (e.g. among many disorders, protein C and protein S deficiencies; Factor V Leiden mutations) or with chronic autoimmune vasculitides. Recent studies indicate that narrowing or stenosis of a venous sinus, rather than complete occlusion, may be associated with intracranial hypertension.

9.6.D TOXIC

Under toxic conditions, it is well known that a drug used for the treatment of acne, isotretinoin (a form of vitamin A), has been associated with an increase in ICP; Cheryl is beyond the usual age for this condition.

It should be noted that other agents have also been associated with idiopathic intracranial hypertension (IIH): antibiotics such as tetracycline and chlortetracycline (both used to treat acne) or nitrofurantoin and nalidixic acid (used to treat urinary tract infections), as well as hormones (human growth hormone, anabolic steroids). Cheryl is not using any of these medications.

9.6.E INFECTIOUS

Meningitis is an inflammation of the meninges accompanied by an inflammatory reaction of neutrophils (with most bacterial infections) or lymphocytes (with viral meningitis) that are found in the CSF. Although most cases are acute, bacterial meningitis can be subacute to chronic in nature. Sometimes, these infectious processes can interfere with the CSF circulation, i.e. reabsorption, by obstructing the arachnoid granulations in the walls of the venous sinuses (discussed earlier).

A chronic intracerebral abscess from an infectious source (e.g. from an infection on the heart valves) may cause a cerebral mass anywhere within the brain; there is, however, no history of fever or of an infectious process elsewhere in the body in this case, and no heart murmur (which might be expected, for example, in the context of rheumatic fever causing a diseased mitral valve). Parasitic cysts, sometimes multiple, can also occur within the brain and produce raised ICP (e.g. cysticercosis). Such diseases are relatively rare in North America but common in Central and South America, and in Africa; in North America parasitic infections of the brain must be considered in recent immigrants from endemic areas who present with raised ICP. Parasitic infections are highly unlikely in this case as Cheryl has never travelled abroad, and these usually present with an acute onset epileptic seizure.

9.6.F METABOLIC CONDITIONS

Rarely, endocrine abnormalities (hypocortisolism/Addison's disease, hypoparathyroidism) and hypercapnia (from respiratory disease or sleep apnea) can be associated with increased ICP.

9.6.G INFLAMMATORY/AUTOIMMUNE

Inflammatory/autoimmune disease processes that would give rise to chronic increased ICP without focal neurological signs are very rare and do so through spontaneous clotting in the cerebral venous sinuses or chronic aseptic meningitis (similar mechanism to infectious meningitis); examples include systemic lupus erythematosus and Behçet's disease. Acute disseminated encephalomyelitis (ADEM) often presents with raised ICP (see Chapter 10) but, as the name implies, does so acutely (days) and with focal neurological signs or diminished consciousness.

9.6.H NEOPLASTIC

With the chronic history in this case, the possibility of an intracranial tumor must be considered, either primary (a meningioma or glioma) or secondary (a metastasis from elsewhere). A tumor may produce increased ICP either because of its size or because it obstructs the circulation of the CSF within the ventricular system, leading to progressive ventricular distension. In addition, one could explain the 'accident' as having occurred because of a brief seizure caused by a tumor (see Section 9.6.2.A). Neuroimaging studies will be needed to rule this possibility in or out.

9.6.I DEGENERATIVE CONSIDERATIONS

Degenerative brain conditions would not likely be associated with an increase in ICP. In the elderly, where there is significant cerebral atrophy, an obstruction to CSF flow may produce progressive hydrocephalus without symptoms of raised ICP or papilledema; the atrophic brain simply collapses or distorts to accommodate the enlarged CSF compartment. This disorder is known as *normal pressure hydrocephalus* and typically presents with urinary incontinence, gait apraxia and dementia (see Chapter 12).

9.6.1 FURTHER CONSIDERATIONS

9.6.1.1 SPACE-OCCUPYING LESION

Both a brain abscess and a tumor, as a space-occupying lesion, would produce a mass effect within the cranial cavity and lead to increased ICP and bilateral papilledema. It would be expected, however, that either would produce focal signs, some indication of parenchymal (brain) pathology, such as a sensory, motor or language deficit; none of these have been found clinically. Association regions of the brain, such as the frontal lobes, could harbour a lesion, without producing any focal or language deficit. A lesion in this area might be linked with a change in personality or behaviour and there has not been any hint of this, insofar as one can tell from interviewing Cheryl.

An intraventricular colloid cyst could be considered a space-occupying lesion if it obstructed the outflow of

CSF from the lateral ventricles into the third ventricle or from the third ventricle into the aqueduct thereby causing increased ICP.

9.6.1.2 CSF

The CSF and its circulation need to be considered in arriving at an etiologic diagnosis since pathology here could occur quite possibly without any focal signs. Any enlargement of the cerebral ventricles due to an obstruction (e.g. a tumor) to the CSF flow within the ventricular system would lead to an increase in ICP and to papilledema; this would be readily seen with neuroimaging as enlarged ventricles.

An alteration of CSF pressure, caused by an increased production or some change in the flow dynamics of the CSF, could lead to an increase in ICP, and this would not necessarily be seen with routine CT or MRI. There is a disease entity that fits with this category, known as *idiopathic intracranial hypertension* ('idiopathic' meaning without known cause); this disease was previously called pseudotumor cerebri, as it mimics a tumor in producing an increase in ICP.

9.6.2 THE ACUTE INCIDENT

With respect to the acute event the evening before, there are a number of additional possibilities that need to be considered. Under acute (see Figure 4.8), one should consider paroxysmal and vascular etiologies, in addition to the obvious traumatic event.

9.6.2.A Paroxysmal

Could there have been an epileptic seizure that caused Cheryl to lose control of her automobile? Isolated seizures do occur, often as a prelude to or indicative of other diseases. For example, tumors may cause irritation and/or compression of adjacent normal tissue and do cause seizures. Bleeding into a tumor may lead to its rapid enlargement and cause a seizure. Under such circumstances, however, Cheryl would not have returned to normal behaviour by the time she was helped out of her car. In addition, the fact that Cheryl remembers the car sliding off the road into the ditch argues against there having been a seizure.

9.6.2.B Traumatic Injury

The presence of a scalp bruise suggests that an additional intracerebral lesion could have occurred due to the motor vehicle incident, since it seems that there was a period of unconsciousness, of unknown duration. One must consider the possibility of some degree of brain bruising (contusion), or a concussion, however minor. Any brain contusion or concussion can lead to headaches and difficulty with

concentration, and occasionally mood and behaviour problems, lasting from days to months. In Cheryl's case, however, the headaches developed long *before* the head injury occurred.

Not to be forgotten is the possibility of an acute subdural associated with the head trauma, but this would be expected to lead to a rapid deterioration in her level of consciousness over several hours. Neuroimaging would assist in sorting this out.

9.6.2.C Vascular Disorders

Could there have been a vascular event causing a brief loss of consciousness? It would be important to ask Cheryl whether she is taking any medication including birth control pills, over-the-counter medication or 'herbal' remedies. There is a known increased risk of cerebral vascular occlusion leading to a transient ischemic attack (TIA) or a stroke in women on birth control pills, especially if that person is also a smoker. Nothing in the history or physical suggests a TIA or stroke; Cheryl did not have any extremity weakness or speech disturbance when initially pulled from her automobile.

Could there have been an acute bleed either from an aneurysm (a subarachnoid hemorrhage) or from an arterio-venous malformation, with a brief loss of consciousness? Both of these would result in an acute intense headache and a patient who is obviously ill. Neither of these possibilities is at all likely in view of the patient's normal mental status and relatively benign neurological examination the following morning. Besides, none of these vascular events would account for the bilateral papilledema.

A more urgent consideration is an intracranial bleed from a tearing of the middle meningeal artery, which usually occurs with a head injury on the side of the head (in the temporal region), resulting in an epidural hemorrhage. The typical case has a brief period of unconsciousness followed by a lucid interval before a rapid deterioration in consciousness, due to the acute arterial bleeding inside the skull and concomitant cerebral compression. With an epidural hemorrhage, however, Cheryl would not have remained conscious for more than one to two hours.

9.6.2.E Infectious (acute)

Some acute infectious diseases may produce raised ICP and papilledema if left undiagnosed and untreated for several days, the classic example being acute bacterial meningitis. Again, our patient would be seriously ill long before she entered her automobile.

9.6.2.F Inflammatory/Autoimmune

There is an acute disorder in this disease category that may also present with raised ICP and papilledema:

acute hemorrhagic leukoencephalitis. This disease is an extremely aggressive form of ADEM (see case 6e-5). Again, a patient with this disease would already demonstrate a diminished level of consciousness to the extent that driving would be impossible.

9.7 CASE SUMMARY AND SUPPLEMENTARY INFORMATION

In summary, the major finding in this case is papilledema, which needs time to develop. Its cause is an increase in ICP due to either a space-occupying lesion or to a process interfering with the normal flow of CSF (including its reabsorption). Out of the various etiological possibilities, the following emerge as the most likely:

- Intracranial hypertension caused by a space-occupying lesion
- Venous sinus thrombosis associated with intracranial hypertension, or
- Idiopathic intracranial hypertension (IIH), meaning the cause is unknown

9.7.1 EVOLUTION OF THE CASE

Since none of the previous etiological possibilities can be entirely eliminated simply on the basis of logic or reasoning, further investigations are required to specify the disease causing the papilledema so that one can hopefully find a way to manage the problem before it worsens.

The cause for the vehicular incident and the associated brief loss of consciousness remain speculative at this time, although both Cheryl's visual field deficit and the weather seem to be the major contributing factors.

Could Cheryl's visual deficit, with her enlarged blind spots, have caused her to miss seeing the road sign? This is an interesting possibility and might be considered should there have been any significant damage to her vehicle resulting in an insurance claim. On a more legal note, if Cheryl had been involved in an accident causing damage to another car and/or injury to a person, the neurologist and ophthalmologist might have been asked to give evidence as to the significance of the visual impairment in the causation of the accident. In fact, the enlarged blind spots, while seen with monocular viewing, would not impact the binocular visual field and should not affect her vision with both eyes open. The binocular visual field is the most important for driving, and despite her enlarged blind spots, there is no visual impediment to her ability to drive. Likely, the vehicular incident is in fact a true accident caused by the poor driving conditions that night and by impairment of night vision.

License to drive a car: In any event, Cheryl has been ordered off the road, i.e. not permitted to drive a car; the local Ministry of Transportation was notified concerning her visual field deficit. Although this had a serious impact on Cheryl's life in the short term, she was assured by friends that they would take her to places most times and they did; at other times she had to use a taxi or public transportation.

9.7.2 INVESTIGATIONS

A CT scan was ordered immediately and the following was reported: no intracranial mass was seen, and the ventricles (lateral and IIIrd) appeared reduced in size with a normal configuration. This eliminates several of the possibilities, including a tumor or obstructive hydrocephalus. Following this, an MRI (Figure 9.8) revealed posterior globe flattening, protrusion of the optic nerve heads into the vitreous, expansion of the CSF space surrounding the optic nerve and an 'empty' sella turcica (the pituitary fossa), all consistent with the diagnosis of IIH.

Based upon the normal CT, it was decided that an LP investigation must be done next. Note that in this clinical context, the CT study must precede the LP: should there be an intracranial mass, an LP, by reducing the pressure below the level of the mass, may cause the brain to descend and herniate into the foramen magnum of the skull. Such an event would impair brainstem regulation of respiration and cardiac control, thereby endangering the life of the patient.

An LP was performed with some difficulty (due to Cheryl's obesity) but successfully. The opening pressure was measured at 42 cm H_2O (31 mm Hg), exceeding the normal range for CSF pressure (see Section 9.4.2); the CSF was clear and colourless. Laboratory testing of the CSF was entirely within normal limits, with a normal glucose level, no white blood cells, and a normal protein concentration.

Additional neuroimaging was carried out a few days later, including a magnetic resonance arteriogram (MRA; which visualizes the major arteries of the circle of Willis) and a contrast enhanced magnetic resonance venography (MRV) (an MR of the cerebral venous sinuses with gadolinium dye).

- MRA: This is an MRI of the major blood vessels of the circle of Willis, often using contrast enhancement. No abnormalities were seen in the vessels of the circle of Willis.
- MRV: The view of the major venous sinuses is obtained later during the same examination. No occlusion or abnormality of the venous sinuses was seen.

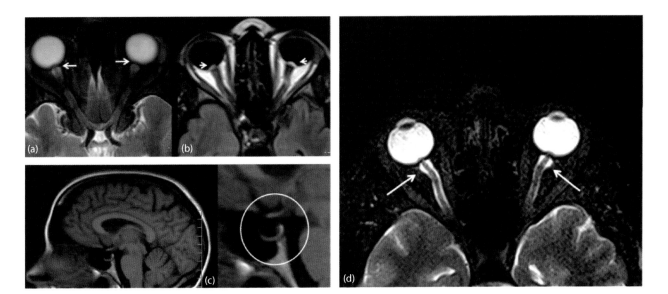

FIGURE 9.8: Radiographs of a patient with IIH. (a) Axial T2 weighted sequence shows posterior globe flattening (arrows). (b) Axial FLAIR sequence post contrast demonstrates protrusion of the optic nerve heads into the vitreous with associated post contrast enhancement (short arrows). (c) Sagittal T1 weighted sequence demonstrating a CSF filled sella turcica with a flattened pituitary gland – lower and higher magnifications. (d) Axial T2 weighted sequence showing expansion of the CSF space surrounding the optic nerve (arrows) and protrusion of the optic nerve heads into the vitreous. (Courtesy of Dr. C. Torres.)

9.7.3 DIAGNOSIS, MANAGEMENT AND OUTCOME

At this stage, the most likely diagnosis, by a process of exclusion, appears to be *IIH*, a disease of still unknown pathophysiology but found to occur predominantly in obese women between the ages of 20–40.

As this disease is being investigated more thoroughly using current imaging techniques, more and more cases previously diagnosed as idiopathic are now being found to be due to a venous sinus thrombosis or some structural abnormality (such as narrowing) of the venous sinuses. Nevertheless, the association of this disease entity with obesity and the female gender remains unclear.

Having eliminated all possibilities, we are left with a disease which has no 'known' cause. The major danger of progression of the disease process, i.e. papilledema, is irreversible blindness in up to 10% of these cases; treatment must be actively considered in order to reduce the pressure on the optic nerve and/or to relieve the intracranial hypertension, as well as the associated headache.

Pharmacologic reduction of the production of CSF may be helpful, such as the use of the carbonic anhydrase inhibitor acetazolamide (Diamox). In emergency situations, one could consider doing repeated therapeutic LPs to relieve the CSF pressure, followed eventually by the shunting of CSF, either from the lateral ventricles or from the lumbar cistern to the peritoneal cavity. A newer procedure called a third ventriculostomy is also an option. This procedure (done through a small burr hole in the skull)

allows the surgeon to enter the third ventricle and make a small fenestration in the ventricular floor, allowing CSF to drain directly from the third ventricle into the subarachnoid space (see Figure 9.5).

In order to reduce the pressure on the optic nerve, a surgical procedure has been devised that makes small holes in the dura and arachnoid of the meningeal sleeve extending along the optic nerve. This operation is called optic nerve sheath fenestration. This procedure would not be done in a scenario where there were only enlarged blind spots (medical treatment would be tried first). It would be reserved for cases with significant optic neuropathy (marked field loss or acuity loss).

Venograms sometimes reveal an unequal flow of blood in the transverse sinuses that is caused by a narrowing or stenosis of one of them. If this is severe and thought to predispose to increased venous pressure (and subsequent increased ICP), procedures are now in place (done by interventional neuroradiologists) to 'stent' the transverse sinus involved.

In addition, Cheryl agrees to undergo a change in her lifestyle. She starts counselling sessions with a nutritionist so that she can lose weight and eat properly to sustain the reduced weight status. Along with this, she is advised by a physiotherapist on an exercise program in order to maintain the weight loss. On the advice of friends, she hires a personal trainer for a six-month period until she develops the lifestyle modifications necessary to change her way of living and to cope better with stress. These measures have been found helpful not only to alleviate the intracranial hypertension but also to reduce the vascular hypertension.

She also enrolls in a smoking cessation program and has successfully reduced her smoking habit, hoping to quit entirely by the end of the one-year contract. All of this has been accomplished with tremendous support from her friends and family.

Six months after the accident, Cheryl visits her family doctor for follow-up. She now has lost over 50 kg, her blood pressure is 125/85, and she is a much changed person, both in her appearance and in her attitude. The headaches are all but gone and the physician discontinues the anti-hypertensive medication (tapering over a period of four weeks).

She is also seen at about the same time by the ophthalmologist, who notes that the disc margins are still not sharp but that there is less venous engorgement. The Goldmann perimetry is redone *and there has been no further progression in the size of the blind spot.*

Her driving license is restored but with the restriction that she is not to drive at night (starting 30 minutes before sunset and extending to 30 minutes after sunrise); Cheryl readily complies with this limitation.

Cheryl will be followed on a regular basis by both physicians. One wonders what life will have in store for her in the future.

9.8 E-CASES

9.8.1 CASE 9E-1: OLGA, AGE 44

Background: A healthy and physically fit female executive with a clothing manufacture company is brought to hospital by ambulance, having collapsed during her lunchtime workout with her personal trainer.

Chief complaint: The 'worst headache I have ever had in my life', beginning a few minutes ago

History:

- *Sudden-onset headache while lifting heavy weights with her quadriceps during a work-out at a local fitness centre*
- The headache involves her entire head and the back of her neck and is 'exploding' in character; no associated vomiting, vertigo or visual disturbance
- Previously in good health; two to three times a year, during her menstrual periods, Olga develops a bilateral pounding frontal headache accompanied by nausea, vomiting, vertigo, light intolerance and blurred vision – altogether lasting four to five hours and relieved by sleep
- Olga's two daughters develop headaches and vomiting during long automobile trips; her mother has migraine; her mother's brother died at age 50 from a 'brain hemorrhage'

Examination:

- Pulse = 65/minute; respirations = 22/minute; blood pressure = 150/90; temperature = 36.8°C

- Olga is lying on her back with her neck extended and her eyes closed; markedly intolerant of bright light
- Oriented to place but not time-of-day or date
- She drifts off topic when answering questions; some difficulty finding words
- *Marked resistance to passive neck flexion*; knees are semi-flexed – attempts to straighten them cause severe back pain (*positive Kernig sign*)
- Cranial nerve examination normal; no papilledema
- Normal grip strength bilaterally but *pronounced drift of the right arm* when she is instructed to keep both arms elevated in front of her with eyes closed; foot dorsiflexion strength is normal bilaterally
- Reflexes: see Figure 9.9

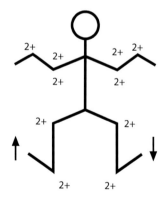

FIGURE 9.9: Illustration of reflex findings in case 9e-1 (Olga).

9.8.2 CASE 9E-2: JIŘI, AGE 52

Background: At the end of your shift in the ER, you receive a call that a middle-aged male is being rushed in from a rural community, with a somewhat confusing history of 'headache'; thankfully, his wife arrives with the ambulance.

Chief complaint: Progressive confusion and speech difficulties following a fall one month ago

History:

- While staying in a two-storey apartment hotel suite during a business trip, Jiři fell down a flight of stairs while sleep-walking

- Awoke with a terrible headache, lacerations and bruises and a badly swollen left ankle
- Taken to a local hospital for treatment; *cranial CT study reportedly normal*; discharged after 24 hours' observation
- Over the next month, he had increasingly incapacitating left-sided headache and extreme distractibility; unable to work
- The headache is continuous, sledgehammer in quality, with associated nausea and dizziness but no vomiting; no history of headaches prior to the fall
- *Increasing lethargy and apathy; for the past two days has been speaking in a slow, halting fashion*

Examination:

- Oriented in person and place but not time
- Slow, halting speech; sometimes misinterprets instructions; no difficulty repeating short sentences; *unable to repeat more than three numbers forwards; unable to reverse number sequences*
- No neck stiffness or Kernig sign

9.8.3 CASE 9E-3: SRINIVAS (SRINI), AGE 2 DAYS

Background: A newborn boy is transferred to your neonatal intensive care unit (NICU) accompanied by extremely distraught parents, recently moved from another city; this is their first child.

Chief complaint: A massively enlarged head

History:

- Srini was born by C-section at 33 weeks gestational age (GA) after his mother went into premature labour; he was transferred to NICU for management of prematurity; there is consideration today as to whether he should be placed on a *palliative care program*
- An abnormally large head was identified on routine abdominal ultrasound examination at 20 weeks GA; parents were living in another city at the time
- Subsequent foetal MRI showed *markedly enlarged lateral and third ventricles with an extremely thin cerebral mantle*; the parents were offered the option of terminating the pregnancy but refused

Examination:

- Body weight = 3.22 kg; head circumference = 42 cm (normal for a *term* infant is 32–35 cm); no respiratory difficulty and normal vital signs
- Disproportionately huge, heavy head; anterior and posterior fontanelles are abnormally large and *bulging*

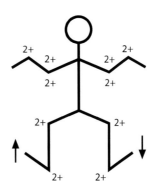

FIGURE 9.10: Illustration of reflex findings in case 9e-2 (Jiři).

- *Bilateral papilledema*; normal visual acuity, eye movement, facial movement
- Normal muscle strength, bulk and tone in all four limbs
- Normal finger-nose testing bilaterally but unable to tandem walk
- Tendon reflexes: see Figure 9.10

- Srini is alert and sucks well
- *Both eyes tend to deviate downwards in their resting state, like partially setting suns*; normal reflex horizontal and vertical eye movements (doll's eye movements)
- He moves all four limbs actively and has normal muscle tone for GA (much lower than would have been seen at term)
- *No tendency to fist his hands when in a calm state* (persistently fisted hands in a neonate is a sign of upper motor neuron pathology)
- Tendon reflexes very brisk; plantar responses are extensor (*normal in neonates*)
- Post-natal MRI study confirms the presence of severe hydrocephalus (see Figure 9.11; *absent*

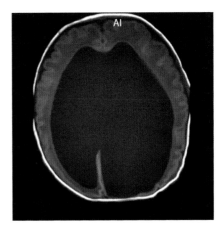

FIGURE 9.11: Post-natal cranial MRI study for case 9e-3 (Srinivas [Srini]; also on the Web site). See text for description.

septum pellucidum; no visible aqueduct; normal fourth ventricle

Question: In your opinion, is Srini's long-term prognosis as poor as the parents had been told at the time of

the foetal MRI study, and should he be placed on palliative care?

Answer will be on the Web site as the student works through the problem.

9.8.4 CASE 9E-4: EDUARDA, AGE 17

Background: You are a third year medical student and are sent to see a female teenager who has been admitted to hospital from emergency; there is a recent history of leg stiffness and an acute loss of vision.

Chief complaint: Increasing difficulty walking over the last month; rapidly deteriorating vision for the past four days

History:

- Three- to four-week febrile illness two months ago accompanied by abdominal pain, nausea and vomiting
- One-month history of *numbness and tingling in the hands and feet progressing to involve the whole body below the nipple line*

- Intermittent *urinary incontinence*
- Increasing *stiffness and awkwardness of gait*
- Progressive *difficulty in reading street signs and notes on the school blackboard* over the past four days

Examination:

- Alert, cooperative, oriented
- *Distance vision 20/100 bilaterally; near vision 20/70 in the left eye, 20/50 in the right eye*
- Normal fundi, visual fields and pupillary reactions to light
- *Mild asymmetric muscle weakness*: see Figure 9.12a
- *Increased muscle tone in plantar flexor and hamstring muscles bilaterally*

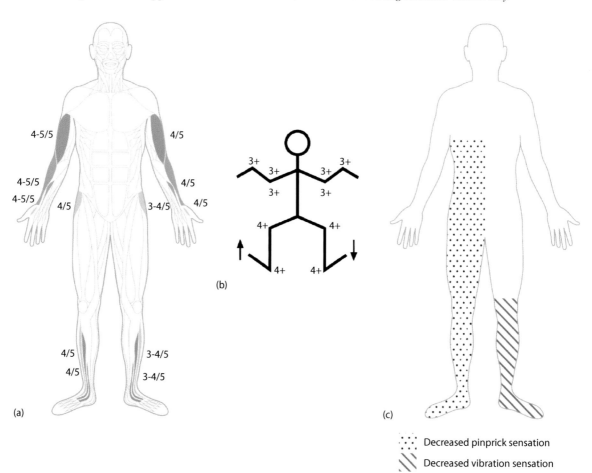

FIGURE 9.12: (a) Motor map, (b) reflex map and (c) sensory map case 9e–4 (Eduarda).

- Normal rapid independent finger movements but slow toe-tapping speed bilaterally, worse on the left
- *Reflexes*: see Figure 9.12b
- Normal finger-nose testing
- *Mild spastic gait, tendency to walk on her toes*

- *Decreased pinprick sensation on the right side below T8 level; impaired vibration sense in the left leg* (see Figure 9.12c)
- Normal rectal examination

9.8.5 CASE 9E-5: MINH, AGE 14

Background: This young teen-aged boy is brought to the emergency late one afternoon complaining of excruciating headaches; he is the top student in his high school class as well as an all-around athlete and is also musically gifted.

Chief complaint: Headaches of increasing severity for the past two months

History:

- In good health until about two months ago, when he began to experience severe, pounding bi-frontal headaches accompanied by vomiting and intolerance of bright light
- At first, headaches occurred every two to three days but now occur daily; they recur on and off over the course of the day and are *so severe that Minh may hold the sides of his head and scream; headaches also wake him out of sleep at night*
- Occasional complaints of blurred vision
- *Unable to attend school for the past week*
- No family history of headaches

Examination:

- Alert, cooperative and fully oriented; *mild hypertension (140/90)*
- Pupils are 6 mm in diameter and *do not react to bright light; they constrict normally with accommodation* (see Figure 9.13)
- *Bilateral papilledema*
- Normal horizontal gaze with no nystagmus but *Minh is completely unable to look upwards –* tries hard and wrinkles his forehead with the effort but his eyes do not move; when he does so, his eyes converge and slightly retract in a rhythmic fashion (*convergence/retraction nystagmus*) (Figure 9.13)
- Neurological examination is otherwise normal except for a *slightly unsteady gait and difficulty with tandem walking*

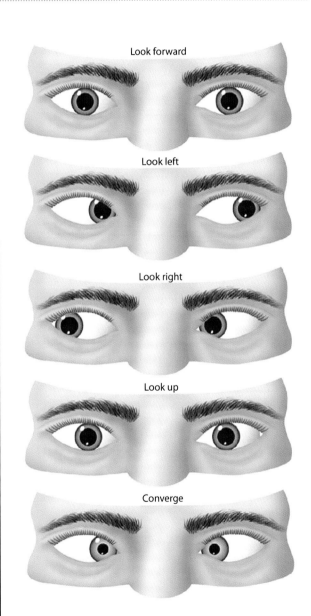

FIGURE 9.13: Illustration of the eye findings in case 9e-5 (Minh) (also on the Web site). Note that the patient is unable to look upwards. His pupils are unreactive to light but constrict normally to accommodation.

9.9 SUMMARY OF KEY NEUROANATOMICAL INFORMATION

Meninges and CSF:

- There are three meningeal layers surrounding the brain: the *dura*, *arachnoid* and *pia*. Proceeding inwards from the inner bony table of the skull, we encounter, in order, the *epidural space* (containing the meningeal arteries), the *outer* periosteal layer and *inner* meningeal *dural layers* (between which are the venous sinuses), the *subdural space* (through which pass the draining cerebral veins), the *arachnoid*, the *subarachnoid space* (containing CSF and cerebral blood vessel) and the *pia*. The latter is apposed directly to layer I of the cerebral cortex. The epidural and subdural 'spaces' are only potential spaces unless they become filled with blood from, respectively, ruptured meningeal arteries and damaged (torn) cerebral veins (Figure 9.3).

- The dural coverings of the spinal cord are the same as those of the brain except that there is an *actual epidural space* between the dura and the vertebral periosteum and intervertebral discs; this space is narrow in the cervical region and very large in the lumbar region, containing blood vessels and fatty tissue (Figure 9.4).

- While the brain is *insensitive to pain*, the surrounding meninges and the walls of major cerebral blood vessels are pain-sensitive.

- The brain contains within its structure four CSF-filled chambers, the ventricles. Each cerebral hemisphere contains a *C-shaped lateral ventricle* (ventricles 1 and 2); these connect directly with a diencephalic midline *third (3rd) ventricle* via the *foramina of Monro*. In turn, the *third* ventricle connects with a midline *fourth (4th) ventricle* (located between the pons and the cerebellum) via a narrow canal in the dorsal midbrain, the *aqueduct* (Figure 9.5).

- CSF is actively manufactured within all four ventricles in the *choroid plexi*. From the *4th ventricle*, CSF passes out of the ventricular system via the midline *foramen of Magendie* (and the two lateral foramina of Lushka) into the subarachnoid space of the posterior fossa, the *cistern magna*. CSF then percolates through the spinal and cerebral subarachnoid spaces and is primarily (but not exclusively) absorbed through the *subarachnoid granulations* located adjacent to the midline into the superior sagittal *sinus* (Figure 9.5).

Venous system:

- Venous blood from the cerebral cortex and subcortical white matter over the cerebral hemispheric convexities drains into the *superior and inferior sagittal sinuses*. From there, blood passes posteriorly into the *transverse venous sinuses*, then the *sigmoid sinuses* and finally into the *internal jugular veins* (Figure 9.6).

- Venous blood from the central regions of the cerebral hemispheres exits medially into the *internal cerebral veins*; these travel posteriorly and join to create the midline *great vein of Galen*. In turn, the great vein terminates in the midline *straight sinus*, which combines with the superior sagittal sinus in a confluence (termed the *torcula*), from which the transverse sinuses originate (Figure 9.6).

Visual system:

- Visual images from the external world are initially processed by the *rod and cone layer* in the retina. From there, visual data are passed via an intermediate cell layer to the more superficial *ganglion cell layer* adjacent to the vitreous. Ganglion cell axons pass into the optic nerves and proceed posteriorly to a junction between the two nerves, the *optic chiasm*. At the chiasm, axons from the *nasal retina* cross the chiasm to join with *temporal retina* axons from the opposite eye (which *do not* cross the chiasm). This arrangement serves to focus information from one visual field (e.g. for the right visual field, the left temporal and right nasal retina) to the *opposite cerebral hemisphere* (Figure 9.7).

- From the optic chiasm, the visual fibers pass posteriorly through the *right and left optic tracts* to synapse in the corresponding *lateral geniculate nucleus* (or body) of the thalamus. From these nuclei, axons from third-order neurons pass posteriorly in the *optic radiations* to reach the *visual (calcarine) cortex* in the medial occipital lobes.

- Information from the lower visual fields passes through the *parietal lobe* optic radiation, while data from the upper visual fields loop through the *temporal lobe* optic radiation (*Meyer's loop*).

- Ganglion cell layer axons sub-serving the *pupillary light reflex pathway* pass from the optic tracts directly into the *midbrain tectum*, where they synapse with neurons that project to the *Edinger–Westphal nuclei* (Figure 2.2).

9.10 CHAPTER-RELATED QUESTIONS

1. You have an epileptic patient who requires removal of a benign tumor (a ganglioglioma) from the left frontal lobe convexity cerebro-cortical and subcortical region. Once your patient's head has been shaved, what anatomical structures and spaces (real or potential) will the neurosurgeon have to cut through, in sequence, in order to gain access to the cerebral cortex in the region of the tumor?

2. Imagine that you are a small quantum of CSF just secreted by the choroid plexus in the right lateral ventricle. Describe the shortest course you would have to take in order to reach the superior sagittal venous sinus, listing the anatomical structures you would have to pass through, one-by-one, in order to get there.

3. a. As we did in question 4 of Chapter 4, let us imagine that you are now an axonal growth cone of a ganglion cell layer neuron in the nasal retina of the developing right eye in a six-week gestation human embryo. As you progressively increase in length, state the route that you would have to take through the CNS in order to eventually make synaptic contact with the second-order visual pathway neuron in the thalamus.

    b. In what way would your elongation pathway be different if, instead, you were the growth cone of a retinal ganglion cell layer neuron that will be contributing to the pupillary light reflex pathway?

4. A 25-year-old man comes to the emergency department with a three-week history of daily, increasingly severe headaches and repeated vomiting. Physical examination reveals a somewhat drowsy patient with bilateral papilledema; otherwise, the neurological examination is normal. A CT scan of his head reveals the presence of significant enlargement of the lateral ventricles (including the temporal horns) and, to a lesser extent, of the third ventricle. The fourth ventricle appears normal in size. Where is the most probable site of the pathology that is producing his symptoms?

5. Similar story – different disease process…

    This time our patient is a 13-year-old boy. He also has had a three-week history of increasingly severe daily pounding headache along with occasional vomiting. Physical examination reveals a wide-awake boy who also has bilateral papilloedema and no other neurological findings. An additional and striking abnormality, however, is that his face is very flushed and

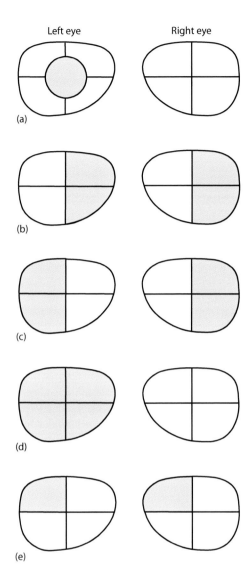

Left eye   Right eye

(a)

(b)

(c)

(d)

(e)

FIGURE 9.14: Visual field findings for question 6 (a–e).

features many, dilated subcutaneous veins in the region of his forehead, around his eyes, down his cheeks and extending into his neck. He also gets a CT study of his head, which turns out to be entirely normal. How can you best explain the history and examination findings?

6. What is the probable location of the pathology responsible for the 5 visual field defects shown in Figure 9.14?

SUGGESTED READING

Ball, A.K., Clarke, C.E. Idiopathic intracranial hypertension. *Lancet Neurol* 2006. 5: 433–442.

Friedman, D.I., Liu, G.T., Digre, K.B. Revised diagnostic criteria for the pseudotumor cerebri syndrome in adults and children. *Neurology* 2013. 81: 1159–1165.

See also Annotated Bibliography.

Chapter 10

Patty

Objectives

- Learn the principal anatomical components of the central nervous system involved in the maintenance of consciousness
- Learn definitions for coma, stupor and confusion
- Become familiar with the main causes of a comatose state
- Learn the pathophysiology of hepatic coma and the basic steps required to treat this condition

10.1 PATTY

During one of your shifts in the emergency department, a 37-year-old woman is brought in by ambulance, accompanied by her male partner. According to her companion, Patricia (but known to everyone as Patty) has been sleeping most of the time during the past couple of days, and when he tried to get her out of bed this afternoon, he could not rouse her.

Appearing dishevelled, apparently unshaven for days, and reeking of whiskey, the friend leaves the hospital after about 20 minutes saying that he has an 'urgent appointment'. You are left with an unconscious patient and a promise from Medical Records that the sixth and last volume of her chart has been located and will arrive shortly.

During the brief period before his disappearance, the friend had provided scattered fragments of information about Patty's recent history. Apparently, she has not been feeling well for several weeks, eating poorly and complaining of abdominal pain. For the past week, she has been acting in an increasingly 'strange' fashion, not remembering the content of conversations held a few minutes previously, often asking the same questions repeatedly. Patty's speech has been slurred and the content disjointed; she frequently changes the subject, often in the middle of a sentence, as if she could not remember what she had wanted to communicate. Her mood had fluctuated rapidly from morose silence to agitated combativeness. Shortly before her descent into a state of almost continuous sleep, she had mentioned in passing that her bowel movements were black.

Immediately prior to his abrupt departure, you had started pressing the friend about Patty's drinking habits. After considerable hesitation, he suggested that she had the occasional glass of whiskey as well as some wine with meals but nothing, he added, 'during the past couple of days'. To direct questioning, he denied that she had ever consumed methanol.

In the absence of both her friend and the medical chart, you decide to examine Patty, hoping the written record will help to fill in the numerous blanks.

Patty's vital signs are as follows: temperature 36°C, pulse 110/minute and regular, blood pressure (BP) 108/50, respirations 28/minute. Her breathing pattern is deep and regular, with an inspiratory noise in the upper airway. Her breath has a peculiar fruity odour.

On your general physical examination, you find a number of abnormalities. She appears underweight for height with reduced subcutaneous fat – evidence of chronic malnutrition. There is mild scleral icterus. Patty has scattered spider-like collections of dilated cutaneous vessels over the nose and cheeks; her palms are bright red. The chest and cardiac examinations are unremarkable. The abdomen reveals many prominent, dilated veins around the umbilicus. Her liver is mildly enlarged, with a hard irregular border; the spleen tip is palpable. As well, the abdomen appears distended and bowel sounds are hyperactive. There is prominent pitting edema of the feet and ankles. There are no needle marks on the extremities. A rectal examination reveals tar-coloured stools that, upon testing, react strongly for the presence of blood.

The neurological examination shows that Patty does not respond to being shaken gently or having her name called; her eyes remain closed at all times. Throughout the examination, she makes no sounds, even in response to pain. Painful stimulation of her limbs causes local withdrawal but no change in facial expression. There is no resistance to passive neck flexion.

The pupils are 2 mm in diameter; they react normally and consensually to light. When you hold Patty's eyes open, they have a tendency to look downwards. With passive head movements, however, the eyes move conjugately both horizontally and vertically. The fundi are normal, with no suggestion of papilledema. The corneal reflexes

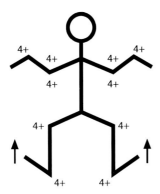

FIGURE 10.1: Reflex map for Patty, the chapter case.

are sluggish and strictly unilateral, never consensual. Compression of the cheekbones elicits a slight, symmetric facial grimace. There is no apparent gag reflex.

Examination of the limbs reveals a generalized increase in muscle tone, particularly striking in the elbow flexors, knee extensors and ankle flexors. The tendon reflexes are extremely brisk and there is prominent, bilateral ankle clonus. The plantar responses are extensor; in fact, just touching the foot causes the great toe to extend (see Figure 10.1).

Having completed your examination, you quickly calculate her Glasgow Coma Scale score (see Table 2.4); it is 6/15.

10.2 CLINICAL DATA EXTRACTION

Patty's story differs from those you have encountered in previous chapters in that, although there is a systematic description of the clinical findings on physical examination, the historical information concerning the presenting illness is fragmentary at best. This departure from the earlier chapters is important, however, as the scenario we are presenting typifies the common experience of urgent care physicians who try to make the best of a sketchy history (if any at all) pertaining to the sudden presentation of a very sick patient.

More clinical history will be provided later in the chapter, but for now, you must try to emulate your emergency department colleagues and attempt to piece together the available historical data in order to work out what part or parts of the nervous system have been sufficiently compromised to render Patty completely unconscious (or comatose) and what disease processes must be considered that, based on what little history you have, can best explain Patty's clinical state. In going through this exercise, as in previous chapters, please complete (as far as you are able) the localization and etiology worksheets.

10.3 MAIN CLINICAL POINTS

Put simply, Patty's neurological disorder comprises a week-long history of increasing confusion and combativeness followed by a descent into an unresponsive or comatose state.

10.3.1 HISTORY

- Recent (several weeks) of poor appetite, abdominal pain and black stools
- One-week history of slurred, disjointed speech, memory impairment, inability to carry on a coherent conversation and wide mood swings
- Two-day history of increasing sleepiness culminating in a state of complete loss of arousal
- Based on possibly understated information provided by Patty's partner, and by his appearance and behaviour, a history of alcohol abuse

10.3.2 PHYSICAL EXAMINATION

- Marked (almost total) unresponsiveness to any form of stimulation
- Deep, rapid breathing; breath has a fruity odour
- Mild scleral icterus and dilated facial capillaries (so-called 'spider angiomata'); palmar erythema; evidence of chronic malnutrition
- Enlarged, hard, irregular-bordered liver; splenomegaly
- Distended abdomen with dilated periumbilical veins; bilateral ankle edema
- Tar-coloured stools that test positive for blood
- Tendency for eyes to deviate downwards but full range of passive, conjugate extraocular movement
- Sluggish, ipsilateral corneal blink reflexes; absent gag reflex
- Increased muscle tone in response to passive stretch but normal limb movement patterns (flexion withdrawal) in response to local painful stimulation
- Hyperactive tendon reflexes with bilateral ankle clonus; bilateral extensor plantar responses

10.4 RELEVANT NEUROANATOMY

Given that Patty's main presenting problem is unconsciousness and almost complete lack of arousal to any kind of stimulation, we must preface any further discussion by considering the neuroanatomical basis for the creation and maintenance of the conscious state.

Our understanding of how the brain maintains a state of vigilant awareness has evolved from a consideration of what parts of the brain, when damaged, are accompanied by loss of consciousness, or *coma*.

Coma may be defined as a persistent sleep-like state from which a person cannot be aroused. In fact, a state of coma is the far end of a gradation of impaired consciousness from full awareness to complete absence of awareness. From the functional standpoint, in a fully conscious patient, we encounter someone who is awake, alert, capable of answering simple questions and able to obey simple commands. Descending from this level we encounter:

1. *Confusion*, in which an individual appears awake but is slow thinking and incoherent and has impaired or absent memory for events occurring during the confused state
2. *Stupor*, in which an individual appears asleep, can be partially roused, but is unable to respond appropriately, or to follow instructions
3. *Coma*, as defined above

Clearly, the historical tidbits provided by Patty's partner suggest that, in the past week, Patty has passed through all three of these states of disturbed consciousness, beginning with confusion and proceeding through a stuporous state to her present comatose condition.

From a clinicopathological analysis of comatose patients, it has been recognized that coma may result from damage or loss of function in the upper brainstem (pons, midbrain), both thalami, or both cerebral hemispheres (cerebral cortex and subcortical white matter, diffusely). Damage to or dysfunction of one cerebral hemisphere or thalamus is *not* normally accompanied by loss of consciousness; the only exception, on occasion, is an extensive injury to most of the dominant hemisphere.

What specific components of the brainstem, thalami and cerebral hemispheres comprise the system, or network, that maintains the conscious state? Briefly put, the components are as follows:

a. Pons and midbrain – the so-called reticular activating system, or reticular formation
b. The 'non-specific', or intralaminar and lateral reticular thalamic nuclei
c. The diffuse thalamocortical projection system, connecting the non-specific thalamic nuclei and the cerebral cortex
d. A corticothalamic feedback system connecting the cerebral cortex back to the non-specific thalamic nuclei

Collectively, these components are often referred to as the *ascending reticular activating system* (ARAS). Let us consider some of the components of ARAS in more detail.

a. Brainstem reticular formation

Located in the central portions of the pontine and midbrain tegmentum, the reticular formation, as the name implies, is (in histological preparations) a lacy network of neurons of varying sizes, for the most part without obvious neuronal aggregations ('nuclei'). The reticular formation is highly conserved through vertebrate evolution and is therefore the most 'primitive' part of the brainstem (see also Chapter 1, Figure 1.6c).

The location of the reticular formation is shown in Figure 10.2. Although only the pontine and mesencephalic portions are shown, the reticular formation also extends into the central medulla. At this level, however, the reticular formation is not concerned with the mechanism of arousal, but with projections to cerebellum and spinal cord involved with motor, sensory and autonomic functions (cardiorespiratory, gastrointestinal [GI] and many others).

As Figure 10.2 illustrates, there are some aggregates of larger neurons within the reticular formation that are recognizable at the microscopic level and that have specific functions. Examples include the nucleus magnocellularis, the median raphe nucleus and the nucleus of the locus ceruleus. The latter two nuclei are involved with the regulation of consciousness insofar as they play a role in sleep–wake regulation, among other functions. They form important components of the so-called monoaminergic projection systems and will be reviewed in some detail in Chapter 13 when we consider disorders of 'higher function'.

The reticular formation neurons participating in the consciousness system are usually small and have a characteristic morphology. They have an extensive local dendritic network, allowing them to receive simultaneous input from a variety of sources. It is important to note, for example, that second-order neurons in the somatic and visceral pain pathways send collaterals to the reticular formation. These have an obvious function, as signals of noxious stimuli, in increasing levels of arousal so as to put in motion behaviours designed to avoid or remove the source of pain.

Reticular formation axons may be short, with local connections, or long, projecting to the thalamus, or even higher. In either case there is an elaborate network of terminal branches. This allows a single reticular formation neuron to potentially modify the function of many neurons at a higher level, whether within the consciousness system (e.g. thalamic intralaminar nuclei) or without (e.g. specific thalamic nuclei, basal ganglia, hippocampi, cerebral cortex). Considering this arrangement, it is easy to see how a small, localized painful stimulus, such as a bee-sting on one foot, could result in a massive, diffuse effect on cerebral function.

b. Non-specific thalamic nuclei

We have already considered some of the components of the thalamus, the largest gray matter complex in the

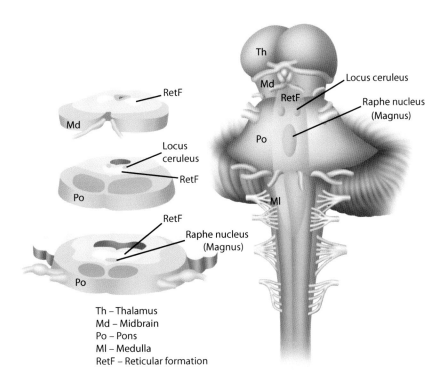

Th – Thalamus
Md – Midbrain
Po – Pons
Ml – Medulla
RetF – Reticular formation

FIGURE 10.2: Illustration of the location of the raphe nucleus (magnus) and the locus ceruleus within the brainstem. The reticular formation is shown in pale yellow, the two named reticular nuclei in darker yellow.

diencephalon, in previous chapters. The ventral posterolateral (VPL) and ventral posteromedial nuclei of the thalamus relay sensory information from the body and face, respectively, to the somatosensory areas of the cerebral cortex. Chapter 7 included a discussion of the ventral anterior and ventral lateral nuclei, both involved in the initiation and coordination of motor function. All of these thalamic nuclei, as well as others to be considered later, have 'specific' functions in that they are connected to discrete regions of the cerebral cortex.

'Non-specific' thalamic nuclei, on the other hand, project in a diffuse fashion to the entire cerebral cortex, as might be predicted for structures involved in the regulation of consciousness. As is demonstrated in Figure 10.3, the non-specific thalamic nuclei consist of three main components:

1. A midline thalamic area adjacent to the third ventricle medially and the dorsomedial thalamic nucleus laterally
2. The intra-laminar nuclei, located in the Y-shaped intermediate band of tissue separating the medial and lateral thalamic nuclear clusters, and the anterior nucleus (in the 'fork' of the Y) from all the others
3. A lateral 'reticular' nucleus encapsulating the lateral aspect of the thalamus

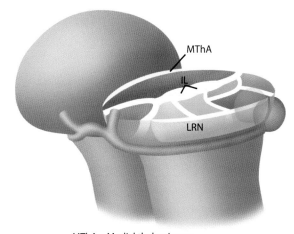

MThA – Medial thalamic area
IL – Intralaminar nuclei
LRN – Lateral reticular nucleus

FIGURE 10.3: Axial section of the thalamus showing the location of the non-specific 'reticular' nuclei.

The non-specific thalamic nuclei receive input from both the pontine and mesencephalic reticular formation and from the cerebral cortex, among other sources. Neurons in the various reticular nuclei of the thalamus resemble those of the brainstem reticular formation in having elaborate dendritic trees and widespread axonal projections to higher levels.

c and d. Thalamic and cortical reciprocal connections

Figure 10.4 shows the interconnection between the non-specific thalamic nuclei and the cerebral cortex. Essentially, there is a feedback loop involving the two brain regions, a diffuse reciprocal interaction that is crucial in the regulation of states of vigilance varying from sleep to focused, task-oriented awareness.

Axons derived from non-specific thalamic nuclei terminate in the most superficial layer of the cerebral cortex, establishing contact with dendrites from cortical neurons whose cell bodies are located in deeper layers of the cortex. The main cerebral cortical cell layers are illustrated in Figure 10.5, the version shown being that seen in 'typical' or relatively non-specific, multipurpose cortex such as one might see in the prefrontal lobes. Some cortical areas (such as motor cortex and visual cortex) are highly specialized and have atypical layering arrangements.

In brief, most cerebrocortical areas have six fairly distinct layers defined by a preponderance of certain cell types. Proceeding from outside-in we have the following:

Layer 1 – relatively free of cell bodies, primarily consisting of non-specific thalamic axon terminations and cortical neuronal dendrites
Layer 2 – small neurons primarily involved in local cortical circuitry

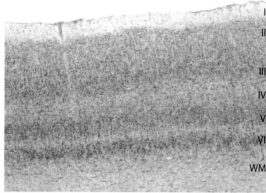

WM – White matter

FIGURE 10.5: Photomicrograph of human cerebral cortex, cresyl violet stain, showing the six architectural cell layers in roman numerals. (Courtesy of Dr. J. Michaud.)

Layer 3 – medium to large neurons whose axons project to adjacent cortical areas, homologous areas of the opposite cerebral hemisphere (via the inter-hemispheric callosal connections) and thalamic nuclei
Layer 4 – small neurons involved in the receipt of input from *specific* thalamic nuclei, e.g. VPL nucleus for somatosensory information from the

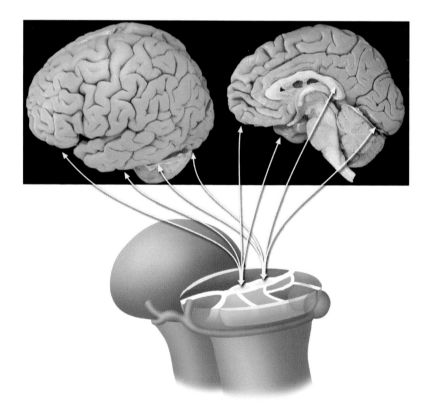

FIGURE 10.4: Reciprocal connections of the non-specific thalamic nuclei with the cerebral cortex.

neck down; lateral geniculate body for visual
information

Layer 5 – large neurons whose axons project primarily to brainstem and spinal cord, e.g. upper
motor neurons

Layer 6 – variably sized neurons also involved in
intracortical circuitry

The prominence or thickness of each of the six cortical layers varies from one cortical region to another,
depending upon the primary function of that region. For
example, in the primary motor cortex of the precentral
gyrus, layer 5 – containing the greatest concentration of
motor neurons – is unusually thick. In contrast, the primary visual cortex adjacent to the calcarine fissure has a
prominent layer 4, the termination point of the optic radiations; indeed, this layer of the visual cortex contains so
many myelinated nerve fibers that it appears as a white
band within the cortex of the post-mortem brain. This so-
called line of Gennari is the reason the primary visual
cortex is sometimes referred to as *striate cortex*.

It is also important to note that, in addition to being
distributed in discrete layers, cortical neurons are also
organized in columns extending the full width of the six-
layered cortex. Within each column, there are extensive
connections among the varying neuronal types constituting the column; as discrete interconnected neuronal
aggregations, these columns represent the basic functional
units of the cortex.

For our present purposes, it is the relationship between
the axon terminals from non-specific thalamic neurons
and dendrites from layer 3 (thalamic projection) neurons
that requires our attention. Electrical activity in the thalamic axon terminals induces, through synaptic connections with cortical layer 3 dendrites, fluctuating degrees of
electrical charge in the dendrites, referred to as *dendritic
potentials*. Dendritic potentials are unlike axon potentials
in that they are not all-or-nothing phenomena: they fluctuate locally without necessarily committing their cell
body (and axon) to a complete membrane depolarization,
or action potential. The phenomenon of the resting membrane potential (or 'negative charge'), and the way neurons communicate with one another, has been considered
briefly in Chapter 1 and will be revisited in more detail in
the next chapter.

Located as they are in the most superficial layer of the
cerebral cortex and, therefore, in many areas, quite close
to the scalp skin surface, dendritic potentials in the cortex can be recorded and amplified from electrodes placed
on the scalp. Fluctuations in cortical dendritic potentials
form the basis of background rhythms in electroencephalography (EEG).

In a relatively low vigilance state (such as lying awake
with eyes closed), there is a relatively large rhythmic
oscillation in cortical dendritic potentials. This reflects
a rhythmic fluctuation in the thalamo–cortico–thalamic
feedback loop we have just described (see Figure 10.6).
With the eyes open, and a state of enhanced, focused
attention, the rate of oscillation in dendritic potentials
increases while the amplitude decreases; i.e. the scalp
electrode recording becomes *desynchronized*. In contrast,
in a sleeping state, the rhythmic oscillation in dendritic

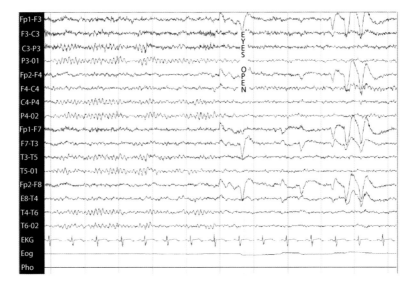

FIGURE 10.6: Normal EEG tracing in the awake state, showing the pattern change in the posterior head regions accompanying
eye opening. The letters and numbers on the left of the image refer to electrode positions on the scalp. F = frontal, C = central, T =
temporal, P = parietal, O = occipital; odd numbers apply to the left hemisphere, even numbers to the right. (Courtesy of S. Bulusu
and Dr. S. Whiting.)

potentials slows while the amplitude increases: a scalp electrode pattern of *hypersynchronization.*

These varying degrees of hypo- and hypersynchronization of cortical dendritic activity reflect changes in activity of the thalamo–cortico–thalamic feedback loop and are the pedestrian equivalents of changes in the level of consciousness.

Thus, if the proverbial bee-sting to the foot occurs while one is asleep, the sudden, massive volley of pain data surging up the somatosensory pain pathway will spread to the pontomesencephalic reticular formation and the non-specific thalamic nuclei. This enhanced level of reticular formation activity will disrupt the slow, rhythmic oscillation in the thalamo–cortico–thalamic consciousness system feedback loop, leading to a sudden desynchronization of cortical dendritic activity accompanied by an abrupt state of focused, panic-stricken awareness!

10.5 LOCALIZATION

Although the information we have provided thus far suggests that Patty has not been well for some time, and that alcohol may be a relevant factor in determining her present state, her neurological symptoms are only a week old and have progressed rapidly, particularly in the past 48 hours. The main symptoms have included impaired recent memory, slurred speech, increasingly severe confusion and periodic agitation, proceeding eventually to somnolence and finally coma. There are a number of findings on neurological examination that may have potentially localizing value: sustained hyperventilation; a tendency for the eyes to deviate downwards; diminished, strictly ipsilateral corneal reflexes; an absent gag reflex; increased muscle tone; hyperreflexia; bilateral extensor plantar responses.

As usual, in considering the localization of the disease process, it is best to start with an analysis of the initial symptoms. Memory, speech and attention difficulties all point to the cerebral hemispheres but do not implicate one hemisphere more than the other. The history does not suggest that Patty was dysphasic: her speech difficulty consisted of poor pronunciation (dysarthria); neither fluency nor comprehension appears to have been selectively affected. Her inability to speak coherently reflected a severe disturbance in attention and memory, not an inability to formulate phonemes. Overall, Patty's story does not suggest a specific left hemisphere lesion but a disease process compromising both cerebral hemispheres, particularly the frontal lobes (attention) and the temporal lobes (memory).

The next element to consider is our patient's increasing somnolence and eventual coma. As we noted earlier, coma may result from a disease process involving the midbrain reticular formation, the thalami or both cerebral hemispheres. Based on the results of our deliberations so far, it would be logical to assume that the bilateral cerebral hemispheric disease process became worse, leading to loss of consciousness. This may well be the case, but we obviously cannot eliminate the possibility that the disease affects the cerebral hemispheres, the diencephalic structures and the midbrain, or all three structures in combination.

Do the results of your neurological examination help resolve this issue? Unfortunately, in this instance, they do not. To support this conclusion, we must look at each abnormal finding in turn.

A tendency for the eyes to deviate downwards in a comatose patient is not uncommon and seldom has much localizing value. A complete absence of up-gaze (as detected by passively flexing the neck of the unconscious patient while holding the eyelids up) usually results from a lesion in the centre for vertical eye movement control. This centre is located in the most rostral portion of the midbrain tectum, above the superior colliculi. Isolated absence of up-gaze in a conscious patient is seen most often in association with pineal region tumors and in non-communicating hydrocephalus (see Srini and Minh in Chapter 9 e-cases). In these instances, the superior tectal region is compressed by, respectively, the downward-expanding tumor and the dilated posterior third ventricle.

A complete absence of the corneal reflexes would have definitely suggested a pontine lesion. The corneal reflex pathway involves an afferent loop via the first and second divisions of the trigeminal nerve, the descending nuclei and tracts of trigeminal nerve in the medulla, and an efferent component via the facial nerves. A lesion of the trigeminal nerve, the descending nucleus of cranial nerve V, the facial nucleus or the facial nerve will therefore abolish the corneal reflex.

In our patient, however, the corneal reflexes are depressed, not absent, and are strictly ipsilateral rather than consensual, as would normally be the case. The consensual nature of the normal corneal reflex – stimulation of one cornea causes both eyes to blink – appears to result from a secondary, long-loop pathway that ascends from the trigeminal nucleus through the midbrain to an as-yet unidentified locus in the diencephalon, then back down to *both* facial nerve nuclei. Selective abolition of the consensual aspect of the corneal reflex suggests the presence of a lesion somewhere in the midbrain or diencephalon (or both).

With the absent gag reflex, we must direct our attention to the medulla. For this reflex, the afferent limb is the glossopharyngeal nerve (cranial nerve IX, the sensory innervation of the pharyngeal wall); the efferent limb is the vagus nerve (cranial nerve X, specifically the voluntary motor fibers originating in the nucleus ambiguus) (see Figure 2.5), and terminating in the pharyngeal musculature.

Next, we come to the findings on the motor system exam. Although there is no obvious evidence of muscle weakness – Patty withdraws her limbs and grimaces the facial muscles in response to local pain – there is increased muscle tone, striking hyperreflexia and the presence of extensor plantar responses. These are classic upper motor neuron findings and, given their symmetry, suggest the presence of bilateral pathology in the descending motor pathways. Based on the limited information we have, the motor pathway pathology could be located in the subcortical white matter, internal capsules, subthalamic area, cerebral peduncles, pons or the medulla. Since our localization exercise thus far has previously indicated dysfunction in the cerebral hemispheres, in the thalami or midbrain and in the medulla, the motor pathway abnormality could be situated at any one of these levels, or all combined.

There is one final piece of information that, despite its apparently non-specific nature, may have important localizing value: the patient's rapid, deep respirations. Such an observation could result from pulmonary or cardiac disease, or from a severe systemic metabolic acidosis. The former possibility is not supported by the results of your physical examination; the latter possibility would be addressed by an immediate blood gas determination. If we accept, as was the case, that a metabolic acidosis was not present (on the contrary, there was a respiratory alkalosis, presumably secondary to over-breathing), we must consider the possibility that Patty's hyperventilation was of neurogenic origin.

The anatomical basis of the neural control of respiration is extremely complex and beyond the scope of this book. Suffice it to say that clinical experience has taught us that central neurogenic hyperventilation in a comatose patient usually results from a lesion in the *midbrain*. The abnormal respiratory patterns that often accompany coma-producing pathology residing at different levels of the nervous system will be discussed in more detail later in this chapter (Section 10.9).

In summary, our localization exercise leaves us with one of two alternatives. Either the disease process is multifocal (with discrete lesions in the two cerebral hemispheres, the diencephalon, the midbrain and the medulla) or it is of a diffuse nature, shutting down functions to a greater or lesser extent throughout the cerebral hemispheres and the brainstem.

10.6 ETIOLOGY

Since Patty's comatose condition clearly reflects the presence of a neurological emergency, your first decision – immediately following your general and neurological examinations – is to summon the staff of the intensive care unit (ICU) to come to the emergency and commence the process of urgent intervention prior to admission. This gives you an opportunity to review the most recent volume of Patty's hospital chart which arrived in the emergency department while you were doing your initial clinical assessment.

10.6.1 ADDITIONAL HISTORY

Well known to the hospital's mental health unit, Patty has had several admissions during her late teens and her twenties for post-traumatic stress disorder, severe depression and three suicide attempts. From ages 6 to 10, she had been repeatedly raped and tortured by a male relative.

Alcohol and drug abuse began in the mid-teens. Previously an excellent student, Patty dropped out of high school; several attempts to complete high school through adult programs and correspondence course ended in failure. She now lives in poverty, supported by a meagre medical disability pension.

Since the age of 20, when the presence of substance abuse could no longer be concealed, there have been repeated attempts – by Patty and her parents – to address the alcohol and drug addictions through out-patient and residential treatment programs. Although these programs were invariably successful for a few weeks, there was an inevitable return to baseline, usually commencing with a protracted drinking binge. Sometimes, these binges would be triggered by a reversal in an attempt to find work, or by an imagined slight from a friend or relative; more often, however, Patty would start drinking again as this was the only effective way she had found to deal with the demons that haunted her every day.

About 10 years ago, Patty abandoned, for all intents and purposes, her attempts to control her alcohol addiction. Drug abuse became a rare event, largely because alcohol was less expensive. Her alcohol consumption, although virtually continuous, fluctuated in amount according to a monthly cycle. During the first half of each month, following receipt of her disability payment, she drank heavily, mostly whisky and white wine. As her money ran out, her alcohol intake declined but did not stop, primarily because she became adept at 'borrowing' money from a succession of boyfriends and from neighbours.

In recent years, she had presented to the emergency department on a number of occasions in an inebriated, dehydrated state following a period of intractable vomiting due to alcoholic gastritis. On several occasions, she had vomited small amounts of blood. Her enlarged liver had been recognized several years ago and dire warnings were given as to what might happen if the liver disease continued to progress.

10.6.1.1 ETIOLOGICAL CONSIDERATIONS

Clearly, this etiology exercise will have to consider, at some point, our patient's declining health in the years prior to the onset of her neurological symptoms. For the time being, however, we will confine ourselves to working out the neurological diagnosis.

With a mere seven days of progressive neurological symptoms, the most likely disease categories, as suggested by Figure 4.8, are toxic (D), infectious (E), metabolic (F) and inflammatory and autoimmune (G). In addition, for reasons to be explained in the following, we will consider the possibility of an atypical, slow-onset disorder that belongs in the vascular category (C).

10.6.C VASCULAR

As just suggested, there is a diagnostic possibility not suggested by a review of Figure 4.8, perhaps reinforcing the old adage about there being exceptions to every rule! The exception here belongs in the vascular disease category, in which disease evolution usually occurs over minutes to hours, not days. We are here referring to the possibility of a *chronic subdural hematoma* (see Chapter 9 e-cases). Chronic alcoholics are particularly prone to this complication, typically triggered by a fall and a blow to the head while in an intoxicated state. With the closed head injury, one or more cerebral veins traversing the parasagittal subarachnoid and subdural spaces en route to the superior sagittal sinus are disrupted. This results in low-pressure bleeding into the subdural space and the insidious development of an extracerebral mass over several days to even weeks. The gradually enlarging hematoma may lead to headache, personality change, confusion, hemiparesis and even coma secondary to raised intracranial pressure. If hemiparesis and papilledema are present, the diagnosis is more obvious; if not, a high index of suspicion is required. Although a much less common event, because of its proximity to the motor cortex regions devoted to the lower extremities (see Figure 4.2), a bilateral parasagittal subdural hematoma could produce an impaired level of consciousness accompanied by hyperreflexia in the legs and extensor plantar Babinski responses. This latter explanation for Patty's condition would not explain the down-gaze and corneal reflex abnormalities. Nevertheless, if the patient's history or physical examination findings suggest the possibility of a chronic subdural hematoma, a cranial imaging study would be necessary.

10.6.D DRUGS/TOXINS

A *toxic encephalopathy* is a definite possibility. Our patient had a long history of alcohol and drug abuse. It is conceivable that her confused state could have resulted from protracted, heavy drinking and the 48 hours of extreme somnolence from the rapid ingestion of a large volume of whiskey. Alternatively, the coma could have resulted from the supplementary consumption of an overdose of

barbiturates or acetaminophen, from intravenous narcotics or from methanol consumption (unfortunately a common 'substitute' for ethanol in chronic alcoholics, often with devastating consequences). These possibilities would have to be addressed with an urgent drug screen. Coma might also result from a deliberate overdose of a large amount of acetaminophen; in this case, the comatose state would result from the metabolic consequences of a severe hepatic injury.

If the partner's history is to be believed, however, Patty has been almost continuously asleep for the past two days; this suggests that Patty was too stuporous to have succeeded in either consuming large volumes of alcoholic beverages or to self-administer other drugs. A narcotic overdose was unlikely any way – on clinical grounds – because this class of drugs typically produces miosis and our patient's pupils were of normal size. Finally, while a drug overdose might suppress the corneal and gag reflexes, it would likely not produce signs of a symmetric upper motor neuron lesion – unless, of course, there had been a respiratory arrest or hypotensive episode leading to an anoxic or ischemic brain injury.

10.6.E INFECTIOUS

As we have seen in a previous e-case (4e-5), an infectious disease can produce a confusional state leading to coma. For example, bacterial meningitis caused by pneumococcus is a well-described complication in individuals with chronic alcohol abuse. It should be noted, however, that acute bacterial meningitis would have evolved over a period of one or two days, as a rule, and would have been accompanied by fever and neck stiffness, neither of which was manifested by our patient. The one-week time course would certainly be consistent with viral encephalitis, but the absence of fever would not. A brain abscess could evolve over several weeks, sometimes in the absence of an obvious fever, but there would probably have been prominent unilateral symptoms and signs such as epileptic seizures and hemiparesis; papilledema would also have been likely had the patient proceeded to a comatose state. Thus, a brain abscess is unlikely.

10.6.F METABOLIC

Within this disease category, we can immediately conclude that, at age 37, our patient probably does not have a hereditary metabolic disorder leading to coma. Such scenarios are common in infants and young children; they include, among many others, urea cycle disorders (with hyperammonemia), organic acidemias and disorders of fatty acid oxidation. In an adult, an acquired metabolic disorder is more likely. Thus, a stupor proceeding to coma may occur in the context of renal failure, hepatic failure, severe hypothyroidism and an acute adrenal crisis.

Of these possibilities, *hepatic encephalopathy* is the obvious choice, given our patient's history of alcohol abuse and well-documented, long-standing hepatomegaly. There are many findings on clinical examination consistent with post-necrotic cirrhosis, portal hypertension and hepatocellular dysfunction: a hard, knobby liver edge; splenomegaly; dilated periumbilical veins; GI bleeding due to esophageal varices; jaundice, ankle edema, facial spider angiomas and palmar erythema.

For various reasons, the other three acquired metabolic disorders are less likely to account for Patty's current state. While chronic renal failure could account for her ankle edema, edema in end-stage renal failure would produce severe generalized edema and pallor due to anemia. As well, seriously considering renal failure as an option would be to fly in the face of the obvious features of hepatic dysfunction, in particular scleral icterus. Finally, uremic encephalopathy is often accompanied by epileptic seizures. While isolated renal failure is therefore an unlikely cause of Patty's coma, renal function will need to be assessed, as renal failure sometimes accompanies hepatic failure (hepatorenal failure).

Severe hypothyroidism may produce a stuporous state but seldom profound coma; as well, this disorder would be associated with sluggish, 'hung-up' tendon reflexes, not with hyperreflexia and extensor plantar responses. An acute adrenal crisis sufficient to cause coma would present with hypotension and severe circulatory collapse, neither of which are present at the time of Patty's assessment.

Finally, we must consider the possible complications of chronic *hyponatremia*, a condition that is often encountered in chronic alcoholics with severe malnutrition. Na+ levels may gradually fall to as low as 95–100 mmol/L, in which case relatively higher intracellular (neuronal and glial) Na+ levels will draw water into these cell populations and cause diffuse *cerebral edema*, accompanied by progressive stupor leading to coma, seizures and a cerebral herniation syndrome (see case 10e-1). Although there is no suggestion that she has had a seizure, hyponatremic coma must be considered in the differential diagnosis of Patty's presenting state.

Another potential sodium-related cause of coma and upper motor neuron deficits in a patient with chronic alcohol abuse is *osmotic demyelination syndrome* (formerly known as *central pontine myelinolysis*). In this disorder, there is edema and demyelination in white matter regions containing complex, crisscrossing white matter tracts – in particular the central pontine region, the internal capsules and the periventricular cerebral white matter (see Figure 10.7). The bilateral central pontine pathology is the most common and leads to rapid onset stupor, coma and spastic quadriparesis. CPM is typically triggered by

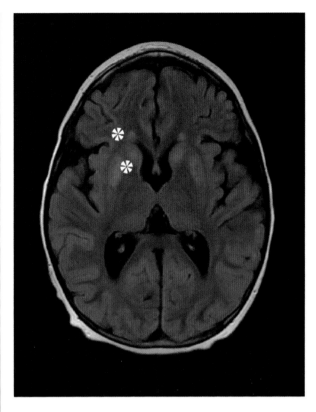

FIGURE 10.7: T2-FLAIR axial MR image of the cerebral hemispheres in a patient with cerebral osmotic demyelination and abrupt-onset coma following rapid correction of severe hyponatremia secondary to diabetes insipidus. The stars indicate the (bilateral) presence of increased signal (edema) in the putamen, caudate head and periventricular white matter.

excessively rapid correction of hyponatremia with the use of intravenous isotonic or hypertonic saline. As such, it is a potential cause of rapid-onset coma in an alcoholic *following* admission for signs and symptoms of hyponatremia.

10.6.G INFLAMMATORY/AUTOIMMUNE

The most frequently encountered disorder in this category, multiple sclerosis, is common in young women and, because of the presence of multiple, demyelinating lesions in the cerebral white matter, brainstem and spinal cord, is typically associated with clinical signs suggesting multifocal pathology. Multiple sclerosis, however, is a chronic disease with acute relapses, not a subacute disease, and would not produce a comatose state this early in the course.

A more likely candidate in the inflammatory disease category is *acute disseminated encephalomyelitis*, often abbreviated as ADEM. This is a relatively rapid-onset disease, evolving over a period of several days, with large multifocal, patchy demyelinating lesions in cerebral white matter, brainstem and, often, the thalami. Usually triggered by a recent viral illness, ADEM may produce a comatose state associated with prominent upper motor neuron signs and, sometimes, with brainstem dysfunction. In the context of a history of chronic hepatic disease, ADEM would be unlikely as it would challenge the law of parsimony. Nevertheless, it remains a distinct possibility that could only be addressed with a cranial imaging study.

In this discussion, we have considered only the causes of coma that would most likely explain Patty's clinical state, given her history and current findings. There are many other causes of coma with which you will need to become familiar. The most important causes of coma in adults and children are listed in Table 10.1. Some causes of coma in infants and small children are unique to this age group and are dealt with separately in Table 10.2.

TABLE 10.1: Main Causes of Stupor and Coma in Adults and Children

Disease Category	Disorder
Paroxysmal	Non-convulsive status epilepticus (i.e. coma without signs of seizures)
	Complex partial status epilepticus
	Migrainous confusional state
Traumatic	Accidental and non-accidental head injury (falls, blunt trauma, motor vehicle accidents, gunshot wounds etc.)
	Concussion
Vascular	Arterial ischemic stroke (midbrain, pons, bilateral thalamus, large left hemisphere)
	Intracerebral hemorrhage (hypertensive, aneurysm, vascular malformation etc.)
	Subarachnoid hemorrhage
	Hypertensive encephalopathy
	Acute epidural hematoma
	Acute or chronic subdural hematoma
	Superior sagittal sinus thrombosis
	Severe systemic hypotension of any cause
Toxic	Ethanol
	Methanol
	Barbiturates
	Opiates
	Tricyclic antidepressants
	Acetaminophen
	Gamma-hydroxybutyric acid (GHB) and other "date-rape" drugs
Infectious	Bacterial meningitis
	Viral meningitis (herpes simplex etc.)
	Brain abscess
	Subdural empyema
	Cerebral malaria
Metabolic	Anoxia (near-drowning, post-cardiac arrest etc.)
	Carbon monoxide poisoning
	Hypoglycemia
	Diabetic ketoacidosis
	Hyper/hyponatremia
	Hypothyroidism
	Acute adrenal crisis/hypotension (Addison's disease)
	Hepatic encephalopathy
	Uremic encephalopathy
	Severe hypothermia
Inflammatory/autoimmune	Acute disseminated encephalomyelitis (ADEM)
	Autoimmune encephalitis (anti-NMDA receptor; voltage-gated channel; Hashimoto's encephalitis etc.)
	Isolated central nervous system vasculitis
Neoplastic	Malignant and benign brain tumors or metastatic tumors causing brainstem compression through tentorial or foramen magnum herniation
	Intrinsic brainstem tumors
	Large malignant cerebral tumors with bilateral extension (butterfly tumors)
	Bilateral thalamic tumors

TABLE 10.2: Causes of Stupor and Coma Unique to Infants and Small Children

Disease Category	Disorder
Traumatic	Non-accidental head injury ('shaken baby/impact syndrome')
Vascular	Intraventricular hemorrhage with acute hydrocephalus (very low birth weight infants)
	Perinatal hypoxic-ischemic brain injury/acute neonatal encephalopathy (probably partly infectious/endotoxic as well as hypoxia-ischemia)
Toxic	Accidental overdoses (alcohol following parental cocktail parties; sedatives; antidepressants; antiepileptic drugs)
Metabolic	Maple syrup urine disease (branched chain ketoaciduria)
	Isovaleric academia (sweaty feet syndrome)
	Propionic academia
	Methylmalonic academia
	Glutaric aciduria
	Nonketotic hyperglycinemia
	Urea cycle defects: carbamyl phosphate synthesis deficiency; ornithine transcarbamylase deficiency; citrullinemia
	Medium chain acylCoA dehydrogenase deficiency

10.7 CASE SUMMARY AND SUPPLEMENTARY INFORMATION

Patty is brought to the emergency department with a one-week history of confusion progressing to stupor followed by deep coma. Her physical examination has revealed abnormalities consistent with dysfunction in both cerebral hemispheres, in the descending motor pathways, in the midbrain and in the medulla, without any signs suggesting a preponderance of pathology at any one of these levels. Given the history of alcohol addiction and the obvious clinical features of significant liver disease, the most probable diagnosis remains that of hepatic encephalopathy.

Wait a moment, you say, if our patient indeed has a metabolic encephalopathy of hepatic origin – a disorder that would be predicted to cause diffuse neurological dysfunction, how do we account for the focal signs just mentioned: the upper motor neuron signs, the central neurogenic hyperventilation and the depressed corneal and gag reflexes? As it turns out, this particular pattern of seemingly focal neurological signs is frequently encountered in patients with hepatic encephalopathy, for reasons that are poorly understood.

We have already stated that corneal and gag reflexes – and, for that matter, even respiratory drive – are often suppressed in patients with encephalopathies due to barbiturate and narcotic overdoses. In a sense, hepatic encephalopathy is an endogenous toxic encephalopathy (see the next section). The spasticity and hyperventilation are more specific for hepatic encephalopathy and presumably result from selective vulnerability of specific cerebral and brainstem structures to the particular cocktail of accumulating neurotoxic compounds associated with severe hepatic dysfunction.

10.7.1 HEPATIC ENCEPHALOPATHY

Patty's story was presented with two separate ends in mind: to illustrate for students the typical picture of a metabolic encephalopathy and to make clear in vivid terms the context surrounding and problems associated with addiction.

Hepatic encephalopathy is a complex disorder whose pathogenesis is only partly understood. It may develop rapidly over a few weeks in a previously normal individual, usually in the context of an acute severe viral hepatitis or exposure to a hepatotoxic compound. This scenario is usually termed *fulminant hepatic failure* and has a very poor prognosis. Alternatively, hepatic encephalopathy may develop in the context of chronic liver disease and is then sometimes referred to as *portal-systemic encephalopathy* (*PSE*). Alcoholic cirrhosis, as in our patient, is a common example of PSE. In general, hepatic coma in the context of PSE has a somewhat better prognosis – especially if the hepatotoxic agent can be eliminated; even so, the prognosis of deep coma in PSE is still much worse than for milder cases.

A discussion of the variegated etiologies of acute and chronic liver disease is outside the scope of this book. We will confine ourselves to a brief review of what appears to be the most important etiologic factors in the production of neurological symptoms.

One of the key etiologic factors, and the most studied, is hyperammonemia. Blood ammonia levels increase in patients with acute and chronic liver disease due to impairment in the ability of liver cells to convert circulating ammonia to urea (in consequence, blood urea levels in individuals with hepatic encephalopathy are typically abnormally low). A second mechanism of hyperammonemia involves the obstruction of the portal circulation in chronic liver disease and the secondary development of portal-systemic collateral vessels – dilated periumbilical

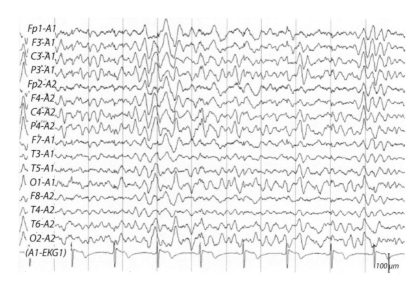

FIGURE 10.8: EEG showing triphasic slow waves, a characteristic finding in patients with hepatic coma.

veins and esophageal variceal veins are good examples. One of the normal functions of the portal vein is to deliver ammonia and other potentially neurotoxic compounds generated by enteric bacteria, then absorbed, to the liver for detoxification. Ammonia (and many other compounds) is shunted through the portal-systemic collaterals and are then available for immediate delivery to the brain.

That elevated arterial ammonia levels contribute to the generation of hepatic encephalopathy is suggested by the fact that ammonia levels correlate directly with the depth of coma. In addition, administration of ammonium salts to experimental animals produces a reversible encephalopathy. Finally, ammonia-forming substances in the GI tract (e.g. blood from bleeding varices, as in our patient; a high protein meal) are well known to precipitate coma in patients with chronic liver disease.

Other potentially encephalopathic agents in liver disease include false neurotransmitters generated from absorbed aromatic amino acids (e.g. octopamine), short chain fatty acids, gamma-aminobutyric acid, glutamine (generated by the amination of glutamate), endogenous benzodiazepines and manganese deposition in the brain.

The diagnosis of hepatic encephalopathy is made on clinical grounds, with the support of simple laboratory tests. Our patient's story is characteristic: a long history of chronic alcoholism and documented liver disease; a period of increasing confusion and intermittent agitation followed by coma, evidence of portal-systemic shunting (periumbilical venous tortuosity, splenomegaly, GI bleeding, spider nevi, palmar erythema) and evidence of hepatocellular dysfunction (jaundice, peripheral edema). As will be seen in the next section, she also demonstrated a typical collection of biochemical abnormalities including hyperammonemia, low blood urea concentration, hyperbilirubinemia and hypoalbuminemia. Many

patients also have hypoprothrombinemia, another hepatocellular synthetic defect, and may have pathological bruising.

Notwithstanding the presence of supporting laboratory evidence of hepatic encephalopathy, it would also be necessary to perform other laboratory tests to eliminate the possibility of other potentially treatable conditions that might ensue in a patient with chronic ethanol abuse, e.g. methanol poisoning and acetaminophen or acetylsalicylic acid overdoses.

Cranial imaging studies are typically normal in cases of PSE but may be required to exclude alternative diagnoses such as chronic subdural hematoma and ADEM. In acute fulminant hepatic failure secondary to viral hepatitis, cerebral edema may be present and confirmed on imaging. An EEG performed in a patient with deep coma typically shows high-voltage diffuse slowing with bilateral frontal triphasic waves (see Figure 10.8), as was present in our patient. It must be noted that triphasic waves, while characteristic, are not specific for hepatic encephalopathy.

10.7.2 LABORATORY INVESTIGATIONS

Blood was immediately drawn in the emergency department, with the following findings: moderate microcytic anemia; Na+ 127 mmol/L (low); K+ 4.8 mmol/L (borderline high), Cl– 94 mmol/L (low); glucose 4 mmol/L (normal); urea 1.8 mmol/L (low), creatinine 121 mmol/L (normal); aspartate aminotransferase 450 u/L and gamma-glutamyl transpeptidase 1050 u/L (both very high); albumin 25 g/L (low); total bilirubin 53 mmol/L (high); venous blood gases showed a mild respiratory alkalosis (rather than a metabolic acidosis as would be seen in methanol poisoning); ammonia 270 mmol/L (very high); blood alcohol level 0. These findings are all consistent with the diagnosis of chronic and acute liver disease and of *hepatic encephalopathy.*

A computed tomography (CT) scan of Patty's head was performed with normal results.

Later on, in the ICU, an EEG was performed, revealing the presence of diffuse, high-voltage slowing with prominent, frontal triphasic waves.

The drug screen drawn on admission eventually revealed the presence of citalopram (an antidepressant agent) but was negative for barbiturates, benzodiazepines, narcotics, acetaminophen, acetylsalicylic acid and commonly used street drugs.

10.7.3 TREATMENT/OUTCOME

Notwithstanding her severely compromised neurological state at the time of admission, Patty responded well to urgent interventions and improved markedly over the next few days – to the point where she was wide awake and able to carry on a normal reciprocal conversation.

Immediately following admission, she required major supportive measures including intubation, mechanical ventilation and intravenous isotonic fluids – the latter to treat dehydration and correct the hyponatremia. Given her malnourished state, she was loaded with thiamine and folate. In addition, while waiting for the results of the drug screen, she was given intragastric charcoal with a view to the sequestration of any possible drugs ingested. The most important interventions were aimed at the treatment of hyperammonemia, especially by the reduction of intestinal ammonia absorption. Ammonia-producing bacteria in the gut can be reduced by the use of cathartic agents such as lactulose, and by the administration of enemas. Lactulose has the additional advantage of acidifying the colonic contents, thereby inhibiting urease-producing bacteria (and ammonia production) as well as drawing ammonia from the systemic circulation into the gut.

Another method of reducing ammonia absorption is the suppression of ammonia-producing intestinal flora with the use of antibiotics. At the moment, the most effective antibiotic appears to be Rifaximin. Arterial ammonia levels can also be reduced by the administration of sodium benzoate.

With respect to the observed consequences of impaired liver synthetic functions, hypoproteinemia would be treated with albumin infusions (if necessary) while hypoprothrombinemia would be managed with injections of vitamin K.

Finally, recent evidence suggests that treatment of patients with hepatic encephalopathy with probiotics may improve clinical recovery, partly by reducing ammonia absorption and also by decreasing hepatic inflammation and reducing oxidative stress in the liver.

But what of Patty's mid-range and long-term prognoses?

Some degree of improvement in hepatocellular function will occur over time, provided Patty abstains indefinitely from ethanol consumption; the jaundice and peripheral edema would gradually clear. On the other hand, having already developed portal-systemic shunting and probable esophageal varices, Patty will remain at risk for future GI tract bleeding episodes, events that, by themselves, could trigger another attack of hepatic encephalopathy.

Finally, given the frequent presence of malnutrition in chronic alcoholics, and once the hyperammonemia has resolved, Patty will need a high-protein diet as well as dietary supplementation with vitamin B complex and, possibly, folate. For further details concerning the treatment of hepatic failure, you should consult the references provided.

Of necessity, given the original cause of Patty's mental health problems, this section must conclude on a note of ambiguity. Patty will certainly leave the hospital in vastly better condition than when she arrived. Hopefully, arrangements can be made for her to work with a psychologist who has experience in dealing with drug and alcohol addiction in persons with post-traumatic stress disorder. Unless a way can be found to satisfactorily address the reasons why Patty became addicted to alcohol, however, it is only a question of time before she either returns to the emergency department in a similar state or develops terminal liver failure.

10.7.4 BASIC APPROACH TO THE LOCALIZATION OF BRAIN LESIONS CAUSING COMA

The clinical assessment of unresponsive patients for localization purposes is obviously a challenge; many aspects of the neurological examination outlined in Chapter 2 are simply not feasible. Nevertheless, there are a number of clinical clues that are often helpful in pointing to specific locations within the neuraxis where the coma-producing lesion is likely to be found.

Sometimes, the location of the lesion causing the patient's unresponsiveness is obvious upon brief scrutiny. This is particularly true for comatose states secondary to trauma, whether witnessed or unwitnessed. In addition to scalp lacerations, bruises and scalp depressions (a sign of a depressed skull fracture), there are two other less obvious findings that clearly point to a hidden skull fracture. First, the presence of a purple discoloration in the skin over one mastoid should prompt a careful assessment of the ipsilateral tympanic membrane. Often, one will find evidence of bleeding behind the eardrum, an indication of a fracture in the petrous temporal bone, often accompanied by contusion of the overlying temporal lobe. The presence of a subcutaneous hematoma in the mastoid region – an extension of hemorrhage from the fractured temporal bone – is commonly referred to as a 'battle sign'. Second, the presence of clear fluid draining from one nostril in a comatose patient suggests the presence of a cerebrospinal fluid (CSF) leak through a fracture in the cribriform plate of the skull base below the frontal lobes and a tear in the underlying meninges, again also often associated with contusion of the overlying brain tissue. A CSF leak can also occur with petrous temporal bone fractures accompanied by tearing of the adjacent meninges and rupture of the tympanic membrane. In the case of a nasal CSF leak, the fluid may be distinguished from clear nasal secretions by the fact that CSF tests positive for glucose while nasal secretions typically do not.

A very helpful introduction to the clinical approach to the diagnosis of lesions causing coma and stupor may be found in the iconic monograph by Plum and Posner (see reference list). Although this slim volume was originally published several decades ago, it remains a compelling outline of the clinical approach to the comatose patient and is well worth the few hours it will take for you to read.

The most important elements of the neurological examination that are helpful in localizing lesions causing coma are as follows: pupillary light reactions, reflex extraocular movements, respiratory pattern, limb movement patterns and reflex postures.

10.7.4.1 PUPILLARY LIGHT REACTIONS

The presence of a unilateral dilated, unresponsive pupil in a comatose individual clearly points to a lesion of the oculomotor nerve parasympathetic fibers in the ipsilateral midbrain region. Such a lesion could be secondary to transtentorial temporal lobe herniation (see case 10e-1) or to an intrinsic lesion of the midbrain, as may occur in the context of a head injury or mesencephalic hemorrhage.

Bilateral dilated unreactive pupils strongly suggest that the lesion causing the comatose state is intrinsic to the midbrain region or upper pons.

In contrast, the presence of bilateral pin-point (miotic) pupils implies an absence of sympathetic autonomic input to the pupillary dilator muscles. In a comatose patient, bilateral miotic pupils are typically secondary to disruption of the first-order sympathetic neurons at the level of the mid to lower pons. Thus, a pontine infarction due to an occluded basilar artery or a pontine hemorrhage may be suggested, in the appropriate clinical context, by the presence of bilateral pin-point pupils. It must be remembered, however, that coma and bilateral miosis can also result from a pharmacological mechanism such as an opiate overdose.

10.7.4.2 REFLEX EXTRAOCULAR MOVEMENTS

The clinical assessment of reflex extraocular movements has been described in Chapter 6 (Section 6.4.2.2.1)

and comprises both the oculocephalic reflexes (doll's eye manoeuvre) and caloric testing. Demonstration of these reflexes depends, respectively, on the integrity of the semicircular canals, the vestibular component of the eight cranial nerve, the pontine centres for coordinated horizontal eye movement and the oculomotor and abducens nerves.

In a patient whose comatose state derives from bilateral pathology in the cerebral hemispheres or thalami, reflex extraocular movements to passive head turning or ear canal ice-water irrigation are fully preserved. If the comatose state is due to pathology (infarction, hemorrhage, tumor) in the midbrain or pons, oculocephalic reflexes are often absent altogether. Please remember that the use of oculocephalic reflex testing in awake patients is largely pointless as an awake individual will fixate on adjacent objects, thus blocking the reflex eye movements altogether. In some patients who have no reflex eye movement with the doll's eye manoeuvre, a partial response may still be obtained following the more potent stimulation of ice-water irrigation.

If cold-water caloric testing is performed on a comatose patient whose reflex pathway is intact, the opposite-direction beating nystagmus will be absent (see Section 6.4.2.2.2) and the eyes will tonically deviate towards the irrigated ear. In situations where the pontine component of the reflex arc is preserved but the mesencephalic component is damaged, the ipsilateral eye will abduct while the contralateral eye will not adduct (see Figure 10.9). Absence of adduction may be due to damage to the median longitudinal fasciculus in the upper pons or midbrain or to damage to the contralateral oculomotor nerve or its nucleus (see case 6e-1 and Figure 6.3).

It is important to note at this point that, just as Patty's hepatic encephalopathy included suppressed corneal and gag reflexes, metabolic encephalopathies can also be accompanied by suppressed extraocular movements.

10.7.4.3 RESPIRATORY PATTERN

We have already noted the complexity of the anatomical pathways for the involuntary and voluntary control; if you are interested in learning more about this topic, please consult the reference cited at the end of the chapter. For purposes of this discussion, suffice it to say that involuntary control of respiration resides in two groups of reticular formation neurons within the brainstem: the pre-Bötzinger complex in the medulla and the pneumotaxic centre in the pons. These centres will maintain automatic control of breathing in a comatose individual unless one or both centres are damaged or suppressed.

There are several distinct abnormal respiratory patterns in a comatose patient that, if present, will provide some localizing information as to the possible source

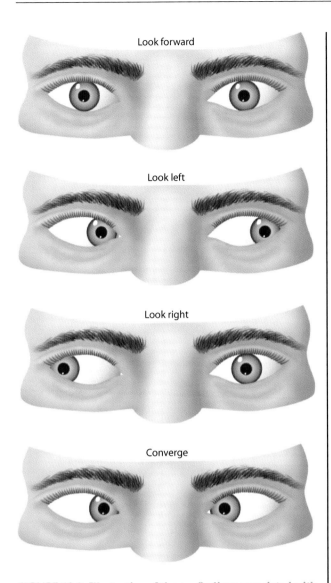

Look forward

Look left

Look right

Converge

FIGURE 10.9: Illustration of the eye findings associated with a lesion involving the median longitudinal fasciculus in the midbrain. This finding is called internuclear ophthalmoplegia.

of the comatose state. The most important of these telltale patterns are outlined here and illustrated in Figure 10.10:

1. *Cheyne–Stokes respiration* – This pattern consists of a fairly rhythmic oscillation between regular, deep breathing to a maximum amplitude peak followed by an equally gradual decrease in depth to the point where breathing is almost imperceptible. The cycle then repeats itself. This breathing pattern in a comatose individual, named after the physicians who originally described it, usually points to the presence of bilateral cerebral hemispheric pathology, whether structural or metabolic in type. A Cheyne–Stokes breathing pattern can also be seen in someone with a large middle

cerebral artery (MCA) territory stroke or in the presence of severe congestive heart failure.

2. *Central neurogenic hyperventilation* – This respiratory pattern consists of regular, deep, rapid breathing without regards to circulating partial pressures of oxygen or carbon dioxide. In consequence, a comatose individual with this respiratory pattern will gradually develop hypocarbia and a metabolic alkalosis. Patients with this pattern of breathing most often have lesions in the midbrain (e.g. trauma or hemorrhage). As we saw in Patty's case, however, a systemic metabolic disturbance causing coma can also be associated with this type of breathing disturbance. In the specific instance of diabetic coma, hyperventilation is typical but, in this instance, is secondary to a severe metabolic acidosis.

3. *Apneustic breathing* – Typically caused by lesions in the lower pons, this pattern of breathing consists of a deep inspiration followed by a two- to three-second pause during which full inspiration is held, after which expiration finally occurs.

4. *Cluster breathing* – A pattern of breathing typified by short clusters of breaths interspersed with periods of apnea; it is sometimes observed in patients with lesions in the lower pons.

5. *Ataxic breathing* – This type of respiratory pattern is marked by the *absence* of a pattern: each breath varies in depth while the period between breaths varies in length. Formerly referred to as Biot's breathing, this type of respiration would be better named 'uncoordinated' breathing and stems from damage to or dysfunction of the medullary reticular formation. It is often seen in moribund patients shortly before a complete cessation of respiratory effort. Recall that a comatose patient with ataxic breathing would also have pathology above the medullary level as a lesion confined to the medullary reticular formation would not produce coma (see Section 10.4).

Clearly, the prognostic significance of the various respiratory patterns just described grows increasingly grave as the location of the pathology descends within the neuraxis.

10.7.4.4 LIMB MOVEMENT PATTERNS AND REFLEX POSTURES

A brief assessment of reflex limb movements in a comatose patient may provide important clues concerning the location of the lesion causing the coma. For example, a patient in a stuporous or comatose state secondary to a massive stroke involving the entire territory of the left middle cerebral artery will typically be found to have a

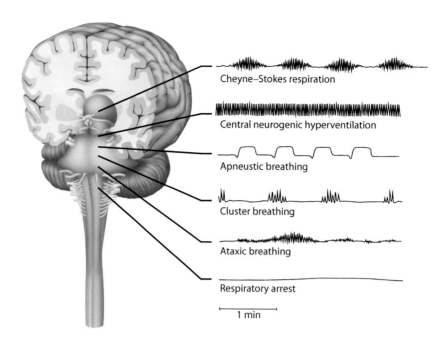

Cheyne–Stokes respiration

Central neurogenic hyperventilation

Apneustic breathing

Cluster breathing

Ataxic breathing

Respiratory arrest

1 min

FIGURE 10.10: Anatomical correlations of specific respiratory pattern abnormalities in comatose individuals.

flaccid right arm and leg, while tone on the left side along with reflexive withdrawal to painful stimulation of the left arm and leg will be preserved. In such a patient, at least in the acute phase during the first hours following the cerebrovascular occlusive event, tendon reflexes on the right side of the body will be diminished or even absent while the plantar response is extensor. On the left side of the body, however, tendon reflexes and plantar responses will be normal. Later on in the course of such a patient, the infarcted left hemisphere will become edematous, expand dramatically in volume and begin to push (herniate) the edematous temporal lobe through the tentorial notch and compress the adjacent midbrain. In turn, this compression will lead to disruption of the descending corticospinal and corticobulbar tracts from the *right* cerebral hemisphere, resulting in a left hemiparesis as well as the pre-existing right hemiparesis (i.e. the patient now has a spastic quadriparesis). As well, the midbrain compression will result in deepening coma and central neurogenic hyperventilation (for a more detailed review of transtentorial herniation, see case 10e-1).

Bilateral severe cerebral hemispheric damage (as may be seen with an anoxic injury in the context of a cardiac arrest or near-drowning) may lead to a characteristic posture of the limbs that becomes more exaggerated when a painful stimulus is delivered – such as forcible rubbing of the patient's sternum with the examiner's knuckles. This posture consists of bilateral fisting of the hands, flexion of the elbows, extension of the knees and plantar flexion and inversion of the feet. Termed *decorticate posturing*, this phenomenon results from virtual absence of cerebral hemispheric inhibition of descending motor pathways originating in the midbrain and pons: the rubrospinal tracts (responsible for the flexed upper limbs and fisted

hands) and both the vestibulospinal and reticulospinal tracts (responsible for the fixed lower limb extension posture) (see Figure 10.11a). Such a patient would typically demonstrate normal pupillary light reactions, intact oculocephalic reflexes and Cheyne–Stokes respiration.

In contrast, a hemorrhagic or traumatic lesion in the midbrain or upper pons is typically associated with a different characteristic limb posture: *extension* of the elbows, internal rotation of the shoulders, fisting of the hands and a lower limb posture that is similar to that seen in decorticate states. Termed *decerebrate posturing*, this phenomenon is also made worse by a painful sternal rub (see Figure 10.11b). Decerebrate posturing results from bilateral interruption of the descending corticospinal and corticobulbar pathways – as well as the rubrospinal tracts – in the midbrain or upper pons. The four-limb extension posturing is the consequence of the unopposed action of the vestibulospinal and pontine reticulospinal tracts.

10.7.5 GENERAL COMMENTS CONCERNING ASSESSMENT OF COMA LEVELS

It is important to recognize that true coma (unconscious sleep-like state in an unarousable individual whose eyes are continually shut) is a relatively short-lasting condition that either changes after a few hours to a couple of weeks to an increasingly awake state or evolves into one of two related disorders in which the patient appears to be awake but is profoundly impaired:

1. *Chronic vegetative state* – In which the patient's eyes are open but there is no evidence of a consistent behavioural response to any

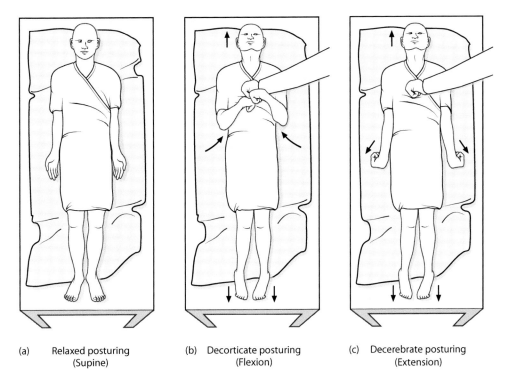

(a)   Relaxed posturing
       (Supine)

(b)   Decorticate posturing
       (Flexion)

(c)   Decerebrate posturing
       (Extension)

FIGURE 10.11: (a) Assessment of limb posture changes in response to painful stimulation in a comatose patient; normal limb posture in a supine position while asleep. (b) Typical limb postures in comatose patients with 'decorticate' rigidity; the main focus of pathology in patients with this type of limb posturing is in the cerebral hemispheres. (c) Typical limb postures in comatose patients with 'decerebrate' rigidity; the main focus of pathology in patients with this type of limb posturing is in the midbrain.

type of stimulation, whether visual, auditory or peripheral sensory (e.g. sternal rubbing); brainstem reflexes are completely preserved.

2. *Minimally conscious state* – In which the patient appears to be awake and, for the most part, unresponsive but who demonstrates evidence of at least one sign of some degree of awareness, e.g. emotional response to hearing sounds of another patient crying; clearly purposeful limb movement.

A patient in a *chronic* vegetative state (i.e. remains in this state for more than one month) may remain thus indefinitely or may proceed to a minimally conscious state. Such a patient may then make no further progress towards recovery or may begin to show signs of increasing cognitive ability. In general, this kind of more dramatic improvement following a protracted period of unconsciousness is more likely to be seen in a patient whose coma follows a traumatic brain injury rather than a cardiac arrest.

Finally, you should be aware that, while the Glasgow Coma Scale you used to assess Patty's level of consciousness is widely used and extensively validated as a clinical tool, it has some limitations. First, the 'best motor response' (see Table 2.4), with six items, is more heavily weighted than 'best verbal response' (five items) or 'eye

opening' (four items). Second, the verbal response item is effectively unusable in an intubated patient, an aphasic individual or in a child aged less than 3. In the light of these concerns, several other clinical coma scales have been developed, each of which leaves out an assessment of verbal responses. Of these, the most promising appears to be the *F*ull *O*utline of *U*n*R*esponsiveness (FOUR) scale developed in the late 1990s at the Mayo Clinic. This scale uses four assessment categories (eye response, motor response, brainstem reflexes and respiration), each with five possible answers scored from 0 to 4. This scale has yet to be extensively evaluated in different centres, however, and the Glasgow Coma Scale remains the gold standard for coma scales, at least for the present time.

10.8 E-CASES

The following clinical cases are intended to broaden the scope of the material that we have introduced in this chapter. Only the case histories and physical examination findings are given here. To work through these cases, please follow the same routine as you have just done for Patty's story. The materials required, discussions of localization and etiology for each case, relevant illustrations and summative comments are all available on the Web site.

10.8.1 CASE 10E-1: HECTOR, AGE 55

Background: A worried wife accompanies her somewhat bemused husband to your office as an urgent appointment late on a Friday afternoon. She tearfully explains that she fears he is 'losing his marbles'. He is a highly successful aeronautical engineer.

Chief complaint: Personality change evolving over the past three months

History:

- Previously a highly successful, energetic, thorough professional
- Three-month history of increasing *errors and omissions at work, of irritability and apathy*; constant complaints of fatigue
- One-week history of marked apathy – not getting out of bed, not shaving, showering or brushing teeth
- Frequently complains of continuous, non-specific *headache*; no associated vomiting or visual symptoms
- *Incontinent of urine* on the day of assessment

Examination:

- Pulse (P) = 62/min; respirations (R) = 20/min; BP = 160/110; temperature = 37°C
- Speaks softly and in a monotone
- Oriented to person, place, month and year but doesn't know the exact date or day of the week
- Normal language comprehension and mental arithmetic skills but has an impaired ability to repeat number sequences forward and backwards
- *Recalls only one-third of words he was previously asked to remember for a period of 10 minutes*
- Normal visual fields and visual acuity; *bilateral papilledema* (see Chapter 9 and Figure 9.1)
- Mild *weakness of hand grip, wrist extension and foot dorsiflexion on the left* (see Figure 10.12a)
- Impaired rapid finger movements and toe-tapping on the left
- *Reflexes*: see Figure 10.12b
- Normal touch, pinprick and position sense on the left, but marked impairment of *graphesthesia* in the left hand; with simultaneous soft touch stimulation of both hands, he is unable to detect the stimulation on the left (*sensory extinction*)
- *Shortly after the end of the examination, Hector's eyes begin to jerk toward the left; he stops responding, falls over to his left, gives a guttural cry and extends both arms and legs; this is followed by rhythmic twitching of arms, legs and face for three to four minutes.*
- After the episode, he remains unresponsive and has a *dilated, poorly reactive right pupil.*

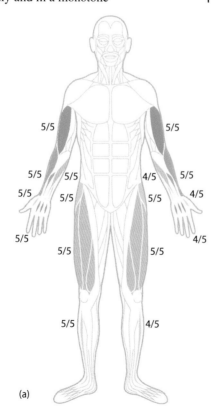

(a)

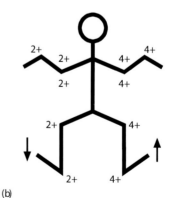

(b)

FIGURE 10.12: (a) Motor examination map for case 10e-1 (Hector). (b) Reflex map for case 10e-1 (Hector).

10.8.2 CASE 10E-2: HANS, AGE 19

Background: You are just starting the early-morning shift at the community hospital emergency room when you are confronted by the nurse and asked to see a young farm worker. His parents inform you that they have not been able to rouse him this morning and that he hit his head the day before.

Chief complaint: Unarousable from sleep on the day of presentation

History:

- At 5 pm the previous day, Hans accidentally struck his head on an overhanging post while driving a tractor; was knocked off the seat to the ground and was unresponsive for one minute
- Post head injury he was initially confused, complaining of severe head pain; no recollection of his accident
- Was able to eat dinner but complained of headache and sleepiness; went to bed after dinner

- *The next morning at 5 am* (time to milk the cows!), *Hans could not be awoken*

Examination:

- Pulse = 96/min; respirations = 22/min (stertorous); BP = 125/85; temperature (T) = 39°C
- Unarousable to deep pain; partial limb withdrawal to pain with an inarticulate groan and no eye opening
- Scalp hematoma behind his left ear; *fresh blood present behind the left tympanic membrane*
- Marked neck stiffness with attempts at passive neck flexion; *attempts to flex his neck result in involuntary knee flexion*
- Within the limits imposed by his comatose state, the cranial nerve examination, including the fundi, is normal.
- Limb withdrawal movements are symmetric; normal muscle tone, tendon reflexes and plantar responses

10.8.3 CASE 10E-3: MAGGIE, AGE 16

Background: Maggie is a student who is brought in by ambulance from the nearby high school. According to a classmate who accompanied her (with the guidance counsellor), Maggie seems to be somewhat of a 'loner' and becomes very easily upset by anything that goes awry.

Chief complaint: Confusion and vomiting following a minor head injury two hours previously

History:

- Two hours ago, during a school volleyball game, the ball was spiked off her head; no loss of consciousness
- Able to continue playing at first but, five minutes later, Maggie left the game because of dizziness.
- After the game, her team-mates found her in the dressing room looking very pale and complaining of a headache
- Maggie went to her next class but, part way through, she got up from her desk and staggered out of the room
- Found in the bathroom having vomited, sitting on the floor; she was ashen-faced and *spoke in disjointed, incoherent phrases, apparently having trouble finding words*

Examination:

- Vital signs normal except for mild tachycardia; afebrile
- Normal general physical examination; no neck stiffness
- Incapable of carrying on a conversation because of severe word-finding difficulties; followed simple instructions
- Impossible to formally test her mental status but *she does not appear to recognize that she is in a hospital*
- *Unable to see a large hand puppet placed in her right visual field* (with either eye) (see Chapter 9)
- Otherwise, the cranial nerve and motor system examination is entirely normal
- Subsequent history from Maggie's father reveals that she has a previous history of headaches at the time of her menstrual periods as well as a life-long history of motion sickness

10.8.4 CASE 10E-4: MONA, AGE 3½

Background: During the overnight shift at the pediatric hospital, the ambulance arrives with a little girl. The parents are frantic because they have not been able to rouse their child for the last two hours.

Chief complaint: Lethargy and anorexia for two days; repeated vomiting for one day; now in apparent coma

History:

- Previously well, although always a picky eater; *poor intake of dairy foods and meat*
- Three days ago, Mona fell in the driveway, hitting her head; no loss of consciousness
- The next day, she appeared tired and unwell, complaining of abdominal pain; *poor food intake; decreased urinary output*
- *The day prior to admission was similar but she vomited several times*
- That night, the worried parents had her sleep in their bed; Mona appeared to be asleep but was *continuously moaning and grinding her teeth; parents were unable to arouse her at 3:30 am and called 911*
- No sick contacts; no known drug ingestion but mother has been taking sertraline, an antidepressant, for several years
- No family history of sudden coma in childhood; six-year-old brother is normal except for strabismus

Examination:

- *No response to verbal stimuli; groaning and a slight movement of the limbs in response to pressure over her chest wall; no spontaneous eye opening*
- P = 140/min; R = 22/min, regular; BP = 107/75; T = 37.1°C
- Pupils 2 mm, react equally to light; fundi are normal; normal range of eye movements to passive head movement; no neck stiffness
- Normal corneal reflexes; symmetric facial grimace to painful stimulation; unable to test gag reflex as she bites the tongue depressor
- Normal muscle tone in all four limbs; withdraws symmetrically to local painful stimuli; *no fisting or extensor posturing*
- Reflexes: see Figure 10.13

Question: What is Mona's Glasgow Coma Score?

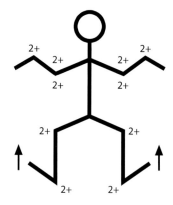

FIGURE 10.13: Reflex map for case 10e-4 (Mona).

10.9 SUMMARY OF KEY ANATOMICAL AND NEUROPHYSIOLOGICAL INFORMATION

- The central neurological apparatus for the maintenance of a conscious, alert state include the *pontine and mesencephalic reticular formation* (in the tegmentum of both), the *non-specific thalamic nuclei* and a *diffuse thalamo-cortical projection system* that reaches the entire cerebral cortex, as well as other deeper structures within the cerebral hemispheres (Figure 10.2). The whole system is known as the *ARAS*.
- Brainstem reticular neurons, for the most part, are not aggregated in discrete nuclei; they have *elaborate local dendritic networks* allowing for widespread excitatory and inhibitory input, and extensive *terminal axonal arborizations* that permit a single reticular formation neuron to influence the function of many thalamic or cerebral cortical neurons.
- There are three main thalamic regions that comprise the non-specific thalamic neuronal aggregates responsible for the maintenance of consciousness: the *medial thalamic area*, the *intralaminar nuclei* and the *lateral reticular nucleus* (Figure 10.3).
- Axonal projections from the non-specific thalamic regions terminate throughout the cerebral cortex but are particularly concentrated in the most *superficial cortical layer* (molecular layer or layer I) where they establish synaptic contacts with *apical dendrites* from neurons in deeper layers of the cortex, especially layer III (Figure 10.4).

- The vast majority of the cerebral cortex is organized into *six architectural layers* that vary in individual thickness according to the specific functions of the cortical area in question. Six-layer cortex is also referred to as *neocortex* (Figure 10.5).
- There is a feedback loop from the cerebral cortex to the non-specific thalamic nuclei; *hypersynchronization* of activity within this loop is associated with drowsiness and sleep while *desynchronization* is seen in the awake, alert state. *Dendritic potentials* within cortical layer I fluctuate in amplitude according to the level of excitation received from the projection fibers of the non-specific thalamic nuclei, i.e. activity that is not all-or-none as would be the case for axonal potentials. Layer I dendritic potentials can be recorded from electrodes placed on the overlying scalp; this constitutes the basis for *EEG*.

10.10 CHAPTER-RELATED QUESTIONS

1. Would a comatose state be anticipated in a patient with a lesion in the following brain regions (answer yes or no)?
   a. Bilateral thalamus?
   b. Right middle cerebral artery territory?
   c. Upper midbrain tegmentum?
   d. Left cerebral peduncle?
   e. Superior medulla?
   f. Bilateral basis pontis?
   g. Bilateral pontine tegmentum?

2. By what mechanism might an embolic thrombus lodged at the bifurcation of the basilar artery into the posterior cerebral arteries produce a comatose state?

3. A 25-year-old male is brought to the hospital in a comatose state immediately following a high-speed head-on collision that occurred on a country road while he was driving to work. On examination he has dilated, unresponsive pupils and absent 'doll's eye' movements but a strong gag reflex. His arms and legs are maintained in a rigid, extended posture; painful stimulation of the sternal region causes the extended posture to become worse, with hyper-pronation of both upper limbs. The plantar responses are extensor.

What is the most likely location of the lesion causing the clinical problem? Explain the probable mechanism of injury in this case.

4. A 75-year-old woman was found in bed one morning apparently awake but unable to speak and having difficulty understanding questions. She was unable to move her right arm and leg. After assessment in the emergency department, she had a CT scan of her head that revealed the presence of a thrombus in the proximal left middle cerebral artery; a subsequent diffusion-weighted magnetic resonance (MR) image showed high signal in the entire anatomical distribution of the left MCA. After 48 hours, she gradually became totally unresponsive and developed marked dilation of the left pupil.

How would you explain this woman's comatose state?

5. An 81-year-old man with a history of type 2 diabetes mellitus and hypertension is brought to the emergency department in a comatose state. On examination, his BP is 200/110, his pulse is 80/minute, and his respirations are 12/minute with long inspiratory pauses. His pupils are 1 mm in diameter and poorly responsive to light. Oculocephalic reflexes are absent in both directions. His limbs are stiff and extended, with hyper-reflexia.

In what region of the nervous system is the lesion causing his comatose state probably located? Explain the reasons for your choice. What is the most likely mechanism?

6. At the time of the usual end of his afternoon nap, a 2-year-old boy is found to be unarousable. On examination, his vital signs are normal. His eyes are closed but open briefly in response to squeezing a finger. He does not respond when his name is called in a loud voice. With pinprick stimulation of the soles of his feet, he withdraws his legs, makes a groaning sound but says nothing. Within the limits imposed by his level of consciousness, the cranial nerve and motor system examinations are normal.

The boy has been entirely well in the past; his parents are also in good health. He has a 5-year-old brother with hemiparetic cerebral palsy and focal epileptic seizures, controlled with carbamazepine (see Chapter 11 for further information concerning epilepsy).
   a. What is his Glasgow Coma Scale score?
   b. What investigations would you order?

REFERENCE

Posner, J.B., Saper, C.B., Schiff, N., Plum, F. *Plum and Posner's Diagnosis of Stupor and Coma*, 4th ed. New York: Oxford University Press, 2007.

SUGGESTED READING

Heidelbaugh, J.J., Sherbondy, M.A. Cirrhosis and chronic liver failure: Part II. Complications and treatment. *Am Fam Physician* 2006. 74: 767–776.

Kornbluth, J., Bhardwaj, A. Evaluation of coma: A critical appraisal of popular scoring systems. *Neurocrit Care* 2011. 14: 134–143.

Poh, Z., Chang, P.E. A current review of the diagnostic and treatment strategies of hepatic encephalopathy. *Int J Hepatol* 2012. Article ID 480309.

Riordan, S.M., Williams, R. Treatment of hepatic encephalopathy. *N Engl J Med* 1997. 337: 473–479.

Chapter 11

Didi

Objectives

- Review the neurophysiology of neuronal membranes
- Review the mechanisms of synaptic transmission and the types and functions of neurotransmitter molecules
- Develop the ability to distinguish the various causes of *transient* loss of consciousness
- Learn the main mechanisms involved in the production of epileptic seizures
- Review the classification of epileptic seizure types
- Learn the basis of electroencephalographic (EEG) 'brain-wave' activity and the main EEG abnormalities associated with epileptic disorders
- Review the principal mechanisms of action of the various anti-epileptic drugs

11.1 DIDI

Dimitri's presenting problem was first noted three months ago by his teacher, at the beginning of first grade. Known to family members as Didi, Dimitri (aged 6) had performed well during his senior kindergarten year, being above the class average. This year, he has been having difficulty paying attention in class and is having trouble keeping up with the work assigned. In addition, the teacher has noticed that, several times a day, Didi will abruptly stop what he is doing and stare straight ahead for a few seconds. During that time, he does not respond to his name being called or appear to be aware of activities going on around him. Just as abruptly, he then returns to his previous activity as if nothing had happened. Sometimes, however, he appears to be transiently perplexed following an episode, as if he does not know where he is or what he is doing. After weeks of observing this behaviour, and watching Didi's work evaluations continue to decline, the teacher called Didi's mother to share her concerns.

As soon as Didi's mother heard the teacher's comments, she realized with a start that she had witnessed similar episodes in the evening over the same time period.

Perhaps because, at some level, she did not want to recognize what was going on, Didi's mother ascribed the episodes she had seen as 'day-dreaming' or due to fatigue. She now realized that Didi's 'spells' were similar in nature to episodes his younger sister Tony had first developed at age 4. When Tony's staring episodes began to interfere with normal conversation, her mother realized there was a significant problem and took her daughter to the family physician. There followed a referral to a local neurologist who witnessed an episode and arranged for an EEG. The EEG revealed the presence of an intermittent abnormality (to be described later on); this finding led to Tony being started on a medication, following which the staring spells gradually disappeared.

Concerned that Didi has developed the same disorder as Tony, their mother took Didi to see the family physician, who turns out to be, *guess who*?

Upon reviewing your file on Dimitri, you remind yourself that he was a term baby following an uncomplicated pregnancy, weighing 3.1 kg at birth. He was walking by age 12 months, speaking in short sentences by age 2 and toilet-trained at age 3 years 3 months. He learned to ride a two-wheel bicycle 18 months ago and to tie his own shoelaces last summer. He now knows all the letters of the alphabet, reads four- to five-word sentences, writes his first and last names, can draw stick figures and can name all the players on the local National Hockey League team. In the past, Didi has been in good health except for several ear infections.

Your age-appropriate general physical and neurological examinations are perfectly normal. Having learned from your experience with Tony, you have Didi carry out one more test manoeuvre. You ask him to pretend that his right index finger is one of those trick birthday candles that resists being blown out. Didi is to try to blow out the candle as hard as he can and to keep puffing on it until you tell him to stop. After about 45 seconds of vigorous huffing and puffing, Didi suddenly stops blowing and stares vacantly into space. When you call his name, there is no response. You ask him to look over at you but, again, he does nothing. You notice a very fine tremor of his head. After about 10 seconds of this behaviour, he suddenly changes his sitting position, blinks once and then resumes blowing out the imaginary candle.

11.2 CLINICAL DATA EXTRACTION

Didi's neurological disorder is clearly quite different from those you have encountered previously. His symptoms are intermittent rather than relentlessly progressive; he appears to be completely normal when his symptoms are not present. Nevertheless, it is important that you begin the problem-solving process by reviewing the clinical data provided and working through the history, examination, localization and etiology worksheets available on the Web site. Once you have completed this process, you are ready to proceed with the remainder of the chapter.

11.3 MAIN CLINICAL POINTS

In comparison with many of the previous problems, Didi's history is relatively short, and summarizing the clinical features is relatively straightforward:

- Three-month history of recurrent brief episodes of unresponsiveness (seconds only)
- Episodes begin and end abruptly, without any incoordination or loss of balance
- Difficulty with paying attention in the classroom
- Unexpected problems in keeping up with class work
- Normal pregnancy and birth history
- Normal developmental milestones
- History of similar episodes in the younger sister
- Normal neurological examination
- Witnessed episode triggered by voluntary hyperventilation: blank stare, unresponsiveness to questions or commands; slight rhythmic head tremor; no confusion afterwards

In other words, Didi appears to be having intermittent disruptions in conscious awareness and, except for academic difficulties, seems to be neurologically intact.

11.4 RELEVANT NEUROANATOMY

Refer to discussion in Section 11.6.A.

11.5 LOCALIZATION

In the previous chapter, our patient had a sustained disruption of consciousness and significant abnormalities on the neurological examination performed during the unconscious state. While Didi almost certainly has a very different disease process from what we saw in Patty, the fact that Didi's disorder comprises periodic, rather than sustained, loss of consciousness should suggest to you that the localization of his disease process also involves those brain structures responsible for the maintenance of an awake, alert state, i.e. the brainstem reticular formation, the intralaminar thalamic nuclei and the cerebral cortex. In fact, given that Didi's transient episodes of unresponsiveness are not accompanied by any loss of head support or trunk posture or by any motor manifestations other than a slight head tremor, it seems likely that Didi's disease process or disorder is almost entirely confined to the so-called 'consciousness system', with no overlap to some adjacent structures as was the case with Patty.

Rather than belabour the process of localizing Didi's neurological disorder, we will now have to consider in some detail the neurophysiological basis upon which neuronal aggregates can work together to maintain a conscious state, how their function can be abruptly altered to result in transient loss of awareness and how they are equally abruptly able to restore the status quo ante.

11.6 ETIOLOGY – DIDI'S DISEASE PROCESS

Returning to Figure 4.8, we quickly recognize that Didi's history presents an ambiguity. On the one hand, the duration of each period of unresponsiveness lasts only seconds; on the other hand, he has been having these episodes for at least three months. Based on the former criterion, Didi has an acute process, based on the latter, a chronic process. While both conclusions, in effect, are correct, it is the former conclusion – that Didi has an acute, or even hyper-acute, process – that is most helpful in working out a differential diagnosis. As we discussed in Chapter 3, long-standing disorders that produce brief symptomatic bursts, separated by long asymptomatic intervals, are collectively termed *paroxysmal disorders*.

Since Didi's disorder is clearly paroxysmal in nature (A), we will abbreviate the systematic review of all 10 etiologic categories that has formed a component of most of the previous problem-based chapters in this text. As will be seen shortly, the toxic (D), metabolic (F) and genetic (J) categories may contribute to paroxysmal disorders and will be considered in varying amounts of detail.

The remaining etiologic categories may, on occasion, disrupt the consciousness system but not in a paroxysmal fashion. Trauma to the head (B) routinely interrupts consciousness, most often due to contusion of the midbrain. Vascular pathologies (C), whether ischemic or hemorrhagic, also produce coma if they extensively damage the brainstem or both thalami. Coma in viral encephalitis or bacterial meningitis (E) is secondary to diffuse cerebral dysfunction. Post-infectious autoimmune disorders (G) such as acute disseminated encephalomyelitis often present with a rapid-onset comatose state that may last several days. Other inflammatory/autoimmune disorders such as multiple sclerosis and central nervous system (CNS) lupus disturb consciousness only when they are far advanced, in the terminal phase of the illness. The same is true for neoplasia (H) and for degenerative diseases (I).

Returning to the paroxysmal category, there are a variety of such disorders, each having quite distinct mechanisms. Our task will be to consider the most likely items on the list. In working through this list, we must bear in mind that, whatever Didi's problem is, it appears to be anatomically confined to the consciousness system.

In general, CNS paroxysmal disorders may be defined according to the mechanism of transient disruption of neurologic function:

1. Disturbances in neuronal membrane function, either *hyperexcitability with inappropriate depolarization of neurons* or *hypoexcitability with inappropriate hyperpolarization*

2. Pharmacological suppression of neuronal function, e.g. drugs and toxins

3. Lack of substrates required for maintenance of neuronal function. For the most part, there are only two such substrates – *glucose* and *oxygen*

Let us consider these possibilities in turn.

11.6.A ELECTRICAL

11.6.A.1 DISTURBANCES IN NEURONAL MEMBRANE FUNCTION

In this instance, we are considering a group of disorders in which there is an intrinsic disturbance in neuronal function in which cell membranes are either unable to consistently stabilize resting membrane potentials when normal functioning neuronal networks require this, and thus produce action potentials inappropriately, or are excessively hyperpolarized and unable to generate appropriate action potentials to allow participation in normal neuronal network function.

An episodic, inappropriate, excessive electrical discharge of cortical neurons is an accepted definition for an *epileptic seizure*. Epileptic seizures, depending upon the area or areas of the brain participating, are accompanied by a variety of transient disturbances in neurological function, including twitching movements, sensory phenomena (e.g. tingling sensations, visual or auditory hallucinations), stereotyped semi-purposeful movements (automatisms) and *loss of conscious awareness*. In the latter case, the epileptic seizures either originate in or

spread to involve the consciousness system as we have defined it.

Epileptic seizures typically last seconds to minutes, occasionally longer. Thus, as a neuropathological phenomenon, they are in keeping with the transient disturbances described in Didi's history.

On the other hand, transient suppression of normal neurological function usually implies a distinctly different form of neurological disorder. While it is true that some forms of epileptic seizure, in some areas of the brain, may produce a hyperpolarization (rather than depolarization) of cells in a specific neuronal network leading, for example, to transient paralysis of an arm or leg, this is distinctly rare. A far more common scenario involving transient neuronal hyperpolarization is the phenomenon of cerebrocortical spreading depression, originally described by Leão and bearing his name.

Transient spreading depression of cortical neurons, by which is meant the slow sequential suppression of function across the surface of the cortex, is the hallmark of *classical migraine*. Migraine is a common paroxysmal disorder of genetic origin characterized by recurrent attacks of pounding headache, photophobia, nausea and vomiting. In prodromal or 'classical' migraine, prior to the onset of headache, there may be a premonitory period lasting several minutes during which there is a transient sensory disturbance (hallucination of a flickering, saw-toothed line; numbness of the face and arm on one side) known as an aura; it is this phenomenon that results from spreading cortical depression. Migraine without aura is experienced much more frequently and is often called 'common' migraine.

Some forms of migraine headache do involve disturbances in consciousness, either with confusion and loss of memory for events during the attack (confusional migraine) or with transient unconsciousness. This disorder is called *basilar artery migraine* because the symptoms suggest dysfunction in midbrain reticular formation, thalami and temporal lobes, all supplied by the basilar artery and its superior branches (see Chapter 8). Symptoms of neuronal depression in migraine, however, last for minutes or hours, not seconds.

In conclusion, our review of possible etiologies of transient, recurrent neurological dysfunction (or paroxysmal disorders) leaves us in the 'electrical' column of the etiology matrix, and the most likely diagnosis of some form of *recurrent epileptic seizures*.[a]

Before considering what form of epileptic seizure might fit best with Didi's history, however, we will need to

---

[a]   It is important to make a distinction between the terms *epileptic seizure* and *epilepsy*. According to the presently accepted definition, patients may be said to have epilepsy either if they have had two or more unprovoked epileptic seizures or if they have had one unprovoked seizure and have had an interictal EEG demonstrating potentially epileptiform abnormalities.

back up a little and consider how and why neurons develop electrical charges that 'run amok' when seizures occur. In this section, we will be elaborating basic concepts introduced in Chapter 1.

11.6.A.2 NEUROPHYSIOLOGY OF NEURONAL MEMBRANES

What enables a neuron to develop and maintain an electrical charge is its semi-permeable and highly specialized cell membrane. Neuronal membranes are completely impermeable to large organic molecules but selectively permeable, under certain circumstances, to small inorganic molecules such as sodium, potassium, calcium and chloride. In living systems, all of these molecules, large and small, are present in their ionic forms: sodium, potassium and calcium have positive charges, chloride and large organic molecules negative charges.

In fact, neuronal membranes, consisting of lipid bilayers enclosing a central protein component (Figure 11.1), are, as such, completely impermeable to Na^+, K^+, Ca^{++} and Cl^-. That they are capable of intermittent permeability to these ions is due to the presence of specialized ionic pores, normally closed but capable of briefly opening to permit passage of the ion across the membrane (Figure 11.1).

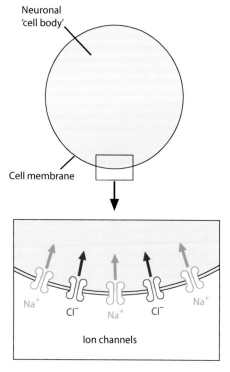

Na⁺ – Sodium entry excites cell

Cl⁻ – Chloride entry inhibits cell

FIGURE 11.1: Cartoon of a simplified neuron shown as a circle with blue 'cytoplasm', without axon or dendrites. The lower panel shows a magnified view of the 'cell' membrane revealing sodium and chloride ion channels.

There are specific pores, known as *ion channels*, for each of the four small charged molecules, i.e. specific sodium, potassium, calcium and chloride channels. Each type of ion channel is a highly individualized structure, designed specifically to permit passage, for example, of sodium ions, but not the other three ions. All four types of ion channel are constructed of a number of protein components in a stereochemical relationship that determines pore size and ion specificity. A typical sodium channel is constructed of α ($n = 2$), β ($n = 2$) and γ ($n = 1$) subunits, the three subunit types being encoded by separate genes in the nuclear genome.

In a resting (not actively discharging) neuron, several ionic gradients are maintained by ion pumps located in the cell membrane adjacent to the ion channels (see Figure 11.2):

Na^+ – high concentration outside the cell, low inside
K^+ – high concentration *inside* the cell, low outside
Ca^{++} – high concentration outside the cell, low inside
Cl^- – high concentration outside the cell, low inside
Protein⁻ – high concentration *inside* the cell, low outside

The net effect of these various ionic gradients is to create a negative charge inside the neuron with respect to the outside, typically about −75 mV. It is this negative charge that is referred to as the *resting membrane potential*.

When the neuron receives a message inciting it to discharge its accumulated voltage, a threshold is eventually crossed and two almost simultaneous events occur. The sodium channels open briefly, permitting Na^+ ions to enter the cell along the established concentration gradient. This abrupt ionic shift results in the loss of the negative intracellular charge, replaced by a weakly positive charge

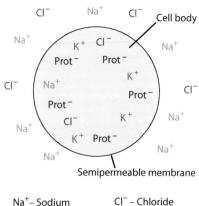

Na⁺– Sodium   Cl⁻ – Chloride
K⁺– Potassium   Prot⁻ – Protein

FIGURE 11.2: 'Neuron' cartoon showing the relative concentrations of different ions within and outside the cell; these ionic gradients generate the resting membrane potential.

(about +15 mV). The sudden depolarization of the cell creates an *action potential* that then propagates over the adjacent cell body membrane and down the axonal membrane centrifugally, with Na⁺ channels opening in succession in a rapidly spreading wave (Figure 11.3). In this manner, our sample neuron sends an electrical message to other neurons in its network.

Microseconds after the local action potential is produced, the potassium channels open, resulting in K⁺ ions quitting the cell for the extracellular fluid space and re-establishing the resting membrane potential. The calcium and chloride channels usually remain closed throughout this process.

The ionic shifts and associated axon potential have no sooner occurred than the neuron sets about re-establishing the status quo ante. Sodium ions are rapidly pumped outside the cell and potassium ions inward, to recreate the baseline state and ready the cell to repeat the process. Until this occurs, the cell is incapable of generating another action potential, i.e. it is in a *refractory* period.

The active re-establishment of the original concentration gradients requires a great deal of energy, in the form of ATP, acting on a kind of ion-exchange pump, the *Na⁺/K⁺ ATPase system*. The ionic shifts and exchanges occur incredibly quickly, permitting a typical neuron to produce action potentials, or 'fire', repeatedly in the space of one second.

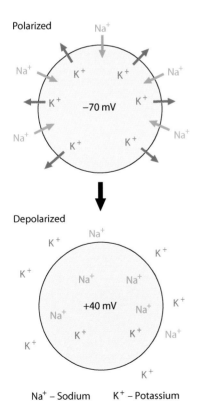

Na⁺ – Sodium   K⁺ – Potassium

FIGURE 11.3: 'Neuron' cartoon illustrating the transmembranous ionic flows generating an action potential.

You will notice that, despite their existence having been mentioned, the chloride and calcium channels did not play a role in the production of action potentials. Their participation in this story will come later, after we have considered how one neuron is able to affect the excitation status of another. This 'communication' mechanism is *synaptic transmission*.

11.6.A.3 SYNAPTIC TRANSMISSION

Action potentials propagated along our neuron's axon may influence the activity of a large number of other neurons. Towards the end of its course, the axon typically splits into a number of branches each of which may abut against dendrites or cell bodies of downstream neurons. At the point of contact, or *synapse*, the axon branch terminates in a highly specialized bulb, the *synaptic bouton* (Figure 11.4 and Figure 1.2c).

Synaptic boutons contain many mitochondria and a collection of vesicles in which are stored one or more chemical compounds designed to influence the excitability status of the adjacent neuron. These chemical compounds, or *neurotransmitters*, are synthesized in our neuron's cell body and transported down the axon to the synaptic boutons where they are packaged in the synaptic vesicles for subsequent use.

The outer membrane of the synaptic bouton is studded with calcium channels. When an action potential reaches the bouton, the calcium channels open, thus permitting calcium to enter the cell. This, in turn, triggers a complex sequence of events that terminates in the synaptic vesicles moving to the portion of the cell membrane adjacent to the contact point with the next cell, the *synaptic cleft* (Figure 11.4). The vesicles then fuse with the membrane, releasing their neurotransmitter packages into the synaptic cleft. This process is referred to as *calcium-modulated exocytosis*.

In the membrane of the dendrite or neuronal cell body opposite our synaptic bouton are complex structures known as *neurotransmitter receptors*. Typically, there are several different types of receptors in any given post-synaptic site. Receptor molecules are membrane proteins frequently linked with adjacent ion channels. The *nature* of the ion channel will determine what effect the transmitter molecule has on the downstream neuron.

Thus, receptor molecules linked with sodium channels typically, when activated by the neurotransmitter, cause the channels to open, admitting Na⁺ to the inside of the cell. This sodium influx is strictly a local process, leading to a small drop in the resting membrane potential, but not below the threshold point required to trigger an action potential. Such localized transient potential changes are called *excitatory post-synaptic potentials* (EPSP).

On the other hand, a receptor molecule linked to a chloride channel will have the opposite effect. If, in

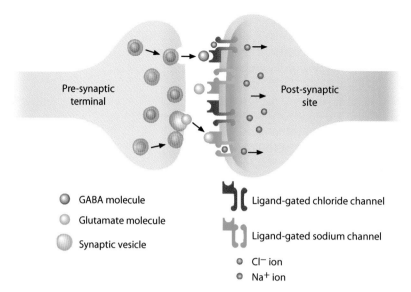

FIGURE 11.4: Cartoon of a stylized 'multipurpose' CNS synapse illustrating neurotransmitter molecule release and the generation of excitatory and inhibitory post-synaptic potentials.

response to the presence of a transmitter molecule in the synaptic cleft, chloride channels selectively open, Cl⁻ ions will be admitted to the post-synaptic cell, resulting in localized *hyperpolarization* of the membrane, and an *inhibitory post-synaptic potential* (IPSP).

Receptor molecules can also be linked to potassium channels, in which case the local effect of receptor activation will be inhibitory. Some receptor molecules are not linked to ion channels but to a different class of membrane molecule known as *G-proteins*. Activation of G-proteins leads, in turn, to modifications in cell metabolism through adenylate cyclase (cyclic AMP) that will make the cell less sensitive to excitation, thus effectively inhibiting the cell. Receptors linked to ion channels are often termed *voltage-gated receptors*, to distinguish them from G-protein-linked receptors.

Since it is the neurotransmitter receptor and its intramembranous linkages that determine whether the response to the transmitter will be excitatory or inhibitory, it follows that a given transmitter molecule may induce *either* an EPSP or an IPSP, depending upon which type of receptor is present at a given synapse. In general, axo-dendritic synapses tend to generate EPSPs (through Na⁺ channel activation), while axosomatic (cell body) synapses tend to generate IPSPs (through Cl⁻ channel activation). We have already seen a similar duality in the mechanism of action of dopamine on the accelerator and brake circuits in the motor system (Chapter 7).

If we return briefly to our representative neuron, we are now in a position to demonstrate how our neuron is persuaded to generate an action potential or, as the case may be, is dissuaded from doing the same. Our neuron, with its elaborate dendritic tree, has thousands of synaptic boutons from other neurons that impinge variously upon the dendrites and cell body, even its axon. Sufficient excitatory input from other neurons may generate enough EPSPs in the dendritic tree to cause our neuron's resting membrane potential to cross the threshold and trigger an action potential. This sequence could still be vetoed, as it were, if there were sufficient contravening IPSPs generated through chloride channel-linked receptors on the cell body. Such a veto would come about through the influence of inhibitory neurons (through inhibitory transmitters) acting to counterbalance the influence of excitatory neurons within the network to which our neuron belongs. Fluctuating countervailing synaptic influences are the basic machinery that permits normal operation of the nervous system.

11.6.A.4 NEUROTRANSMITTERS

Although transmitter molecules, depending upon the nature of the receptor on which they act, may function in an excitatory or inhibitory capacity, most transmitters tend to be primarily either excitatory or inhibitory. There are four main categories of transmitter molecules: acetylcholine (Ach), amino acids, amines and peptides. The most important transmitters, their modes of action and their principal locations in the nervous system are given in Table 11.1.

As can be seen in Table 11.1, *Ach* is primarily excitatory and functions chiefly in the peripheral motor system and in the autonomic nervous system. There is also a very important role for Ach in the CNS with respect to learning and memory, a role that will be introduced in Chapter 12 and will be further explored in Chapter 13.

The amino acid transmitters, particularly *glutamic acid (glutamate)* and *gamma-aminobutyric acid (GABA)*,

TABLE 11.1: Neurotransmitters

Transmitter Type	Location of Cell Body and Projection
a. Acetylcholine	Basal forebrain (e.g. nucleus of Meynert) → cerebral cortex
	Striatum interneurons (muscarinic)
	Brainstem and spinal cord lower motor neurons (nicotinic)
	Preganglionic autonomic (muscarinic)
	Postganglionic parasympathetic (muscarinic)
b. Amino acids	
Glutamic acid	Throughout CNS, excitatory
Glycine	Spinal cord and medulla (inhibitory)
Gamma-aminobutyric acid	Throughout CNS, inhibitory
c. Amines	
Dopamine	Midbrain (SN) → corpus striatum, limbic cortex, frontal cortex (neuromodulatory)
Norepinephrine	Pons (locus ceruleus, lateral tegmental area) → cerebral cortex, central gray, limbic system, cerebellum, cord (neuromodulatory)
Serotonin	Pons, medulla, cord, hippocampus → cortex, central gray, limbic system, cerebellum (neuromodulatory)
Histamine	Posterior hypothalamus, midbrain → forebrain cortex, thalamus (excitatory)
d. Peptides	Throughout CNS, neurenteric NS (neuromodulatory) (includes substance P, enkephalins, somatostatins, neuropeptide Y, vasoactive intestinal polypeptide, dynorphin, vasopressin, cholecystokinin)

are highly relevant to Didi's case and will be considered shortly.

The biogenic amines *dopamine, norepinephrine* and *5-hydroxytryptamine (serotonin)* play important roles in many CNS functions including basal ganglia motor circuitry (dopamine), sleep–wake cycle regulation (norepinephrine, serotonin) and regulation of attention, emotion and other higher functions (all three compounds). Their roles will also be considered primarily in Chapter 13.

Peptide transmitters are numerous, tend to influence neuronal function through G-proteins and are present throughout the nervous system, whether peripheral somatosensory, autonomic/visceral or central. They play key roles, for example, in intestinal motility, pain transmission and modulation, basal ganglia circuitry, hypothalamic–pituitary axis function and in cognition.

Returning, then, to the amino acid neurotransmitters, glutamate is the most ubiquitous *excitatory* transmitter in the CNS. We have already considered its role in the motor system, in upper motor neuron function and in basal ganglia circuitry (Chapter 7). Glutamate is also the main transmitter employed by non-motor cortical neurons projecting to adjacent or distant cortical gyri within a hemisphere or to homologous areas in the contralateral hemisphere (via the corpus callosum). As such, it plays an important excitatory role in association cortices and is implicated in learning and behaviour.

As noted in Table 11.1, glutamate has varying effects on neuronal networks depending upon the type of glutamate receptor present: *N*-methyl-d-aspartate (NMDA), kainate, α-amino-3hydroxy-5-methyl-4-isoxazolepropionic acid (AMPA) or metabotropic receptors. The first three receptors are linked to voltage-gated sodium channels, while the fourth (metabotropic) is linked to a G-protein and thus functions as a neuromodulator rather than an excitatory agent.

Glycine's activity is primarily inhibitory at the spinal cord level (in modulating motor function) and excitatory at the cerebral level. Its roles do not directly impinge on Didi's problem.

GABA, on the other hand, is the principal *inhibitory* transmitter in the CNS. Neurons producing GABA as their synaptic messenger are typically small and have relatively short axons, while glutamatergic neurons are large and have long axons (see Figure 11.5). GABAergic neurons are widely distributed in the nervous system, as befits their crucial role, and are found in cerebral cortex, basal ganglia, brainstem, cerebellum and spinal cord. GABA receptors, as might be predicted from our previous discussion, are linked to chloride channels; their activation tends to elevate resting membrane potentials, thus inhibiting the production of action potentials.

There is a close and interesting chemical relationship between glutamate and GABA, the paramount excitatory and inhibitory CNS transmitters. As demonstrated in Figure 11.6, GABA is synthesized directly from glutamate by the removal of a CO_2 moiety. This biochemical

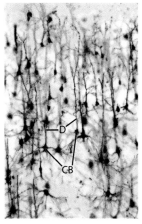
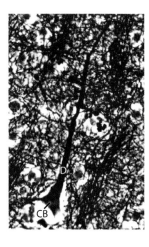

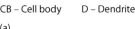
CB – Cell body   D – Dendrite

(a)   (b)

FIGURE 11.5: Photomicrographs showing typical glutamatergic (pyramidal) cortical neurons under low (a) and high magnification (b). (Courtesy of Dr. J. Michaud.)

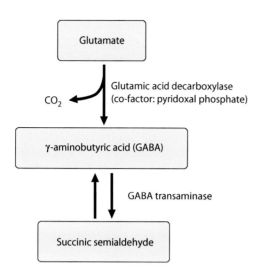

FIGURE 11.6: GABA synthesis equation. GABA synthesis biochemical pathway.

conversion is catalyzed by the enzyme glutamate decarboxylase, using pyridoxine (vitamin B6) as a cofactor. Thus, through the action of a single enzyme, a delicate balance is maintained between excitation and inhibition. On the chessboard of synaptic transmission, glutamate and GABA are like opposing queens on adjacent squares, each preventing the one side from overwhelming the other.

11.6.A.5 Mechanisms of Epileptic Seizures

Now that we have considered how neurons develop electrical charges and how, through action potentials, they are able to excite or inhibit one another, we must return to our definition of epilepsy. Given that epileptic seizures consist of abnormal, excessive electrical discharges by cortical neurons, we must ask the next obvious question: why do cortical neurons discharge excessively?

The answer to this question cannot be given in its entirety as some aspects remain unclear. For simplicity's sake, it is helpful to review the main circumstances in which humans develop epileptic seizures. These essentially break down into two main categories:

1. Focal cerebral cortical pathology in which a cluster of cortical neurons become unstable and produce synchronous, massed action potentials that interrupt normal *regional* brain functions.
2. A lowered threshold in both the cerebral cortex and the thalamic regions with which it has reciprocal connections, such that the cortex as a whole has a tendency to develop synchronous neuronal discharges that *diffusely* interrupt brain functions, including consciousness.

Both of these epileptogenic factors may produce epileptic seizures in isolation, as well as by acting in collaboration. In the case of category 1, almost any type of focal cortical pathology (whether ischemic injury, hemorrhage, trauma, infection or perturbed development) may produce focal epileptic discharge. In the case of category 2, any systemic metabolic dysfunction (e.g. anoxia, hypoglycemia, neuroexcitatory pharmacologic agent) may trigger generalized epileptic discharge, as may the existence of a genetically-determined low threshold for such discharges.

11.6.A.5.1 Focal Pathology
Focal epileptic discharges, regardless of the type of pathology, originate in large, cortical glutamatergic neurons. These neurons have a unique capacity to fire in bursts, once goaded to discharge. Micropipette recordings from single neurons of this type reveal an initial decline in resting membrane potential followed by a burst of action potentials, then a return to a normal resting membrane

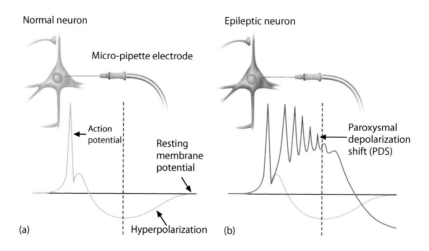

FIGURE 11.7: Normal and epileptogenic pyramidal cell neurons. The neuron on the left (a) shows a normal action potential, while the 'epileptogenic' neuron on the right (b) shows a paroxysmal depolarization shift.

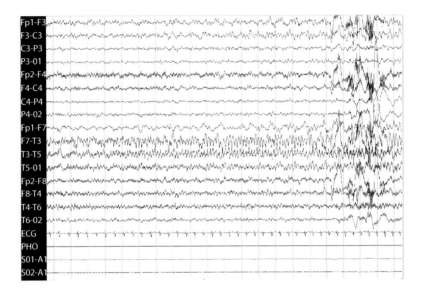

FIGURE 11.8: EEG tracing illustrating a focal-onset epileptic seizure. The beginning of the left-sided, rapid, gradually augmenting focal discharge was accompanied, in the patient, by a sudden loss of responsiveness, a blank stare and by intense facial flushing. (Courtesy of S. Bulusu and Dr. S. Whiting.)

potential. This phenomenon is known as a *paroxysmal depolarization shift* (see Figure 11.7). The mass production of synchronous paroxysmal depolarization in a large cohort of epileptogenic cortical neurons produces a summative discharge that appears via regional scalp electrodes as a focal *spike discharge*.

Isolated sharp or spike discharges may be recorded from patients with focal-origin epilepsy in the asymptomatic state. During clinical seizures, however, long trains of repeated spike discharges occur, reflecting the uncontrolled sustained synchronous discharge of thousands of large cortical neurons, the trains lasting seconds to minutes (Figure 11.8).

11.6.A.5.2 Low Epileptic Threshold

Generalized, synchronous epileptic discharges involving all cortical regions appear to require the collaboration of the non-specific thalamic nuclei and cortical neurons, acting through a feedback loop: cortico–reticulo–cortical. In essence, this means that generalized epileptic discharges are based, at least to some extent, in the upper components of the consciousness system, as outlined in our description of the ascending reticular activating system (Chapter 10, Section 10.4).

The participation of the non-specific cortico–thalamo–cortical feedback loop in the genesis of generalized epileptic seizures was elucidated through an elegant series of experiments in cats. Penicillin, a potent epileptogenic compound if administered in large doses, was administered intramuscularly to awake cats; this resulted in the development of recurrent episodes of unresponsiveness, lasting seconds, during which there was flickering of the eyelids or twitching of the whiskers. During these unresponsive episodes, surface electrodes on the

cerebral cortex recorded synchronous discharges from all cortical areas consisting of repetitive spike bursts, each individual spike followed by a high voltage slow wave. Rhythmic *spike-and-wave* discharges are the hallmark of generalized epilepsy in the penicillin cat model and in EEG recordings of generalized seizures in humans (Figure 11.9).

Sequential recordings following penicillin administration demonstrated an interesting evolution. Initially, there was a generalized high voltage rhythmic slow-wave discharge of steadily increasing amplitude, reminiscent in its appearance of what one sees in humans (and cats!) as they become drowsy and fall asleep. This increasingly hypersynchronous discharge gradually transformed into the spike-and-wave discharge, suggesting the reticulo–cortical mechanism of generalized epileptic discharge was functionally related to the normal mechanism of deep sleep.

If you have remained alert, and not narcotized by the above dissertation, you will have recognized that the staring-blinking episodes seen in the penicillin model of generalized feline epilepsy are remarkably reminiscent of the episodes that have so perturbed Didi's teacher! In fact, Didi's episodes are secondary to an abnormally low, presumably genetic, threshold in the same cortico–reticulo–cortical epileptogenic mechanism uncovered by penicillin administration in the cat.

Having finally returned from a basic consideration of neuronal membrane function and mechanisms of epileptogenesis to the intermittent behavioural abnormalities in Didi ascribed as epileptic seizures, it is now necessary to briefly look at the sundry patterns of clinical epileptic seizures that the two fundamental epileptogenic mechanisms may evoke.

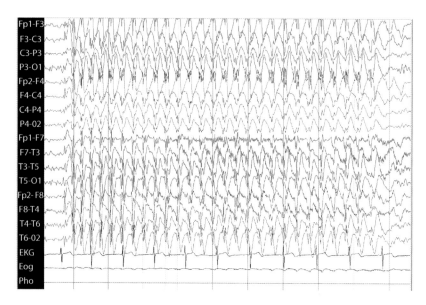

FIGURE 11.9: EEG tracing of a typical absence seizure in childhood absence epilepsy syndrome. The background activity is abruptly replaced by a generalized 3/second spike-and-wave discharge that terminates equally abruptly. (Courtesy of S. Bulusu and Dr. S. Whiting.)

11.6.A.6 Types of Epileptic Seizure

Since epileptic seizures originate in populations of excessively discharging neurons, it would be predictable that the symptoms produced during the seizure would derive from loss of normal function of the neurons involved. In general, this is what occurs. One has to distinguish, however, between *positive* symptoms, in which behaviours are seen that are abnormal or inappropriate, and *negative* symptoms, in which there is a temporary loss of normal function.

In the case of positive symptoms, there may be uncontrollable movements, sensory hallucinations or sudden, unexpected emotional experiences such as groundless fear. Negative symptoms may include temporary paresis of an arm or leg or transient loss of consciousness, as occurs with Didi.

When epileptic seizures occur in a region of focal cortical pathology, the symptoms produced will depend entirely on the normal function of the area involved. Thus, a focal seizure beginning in a cortical area devoted to hand movement may consist of uncontrolled rhythmic twitching of the thumb and fingers. In contrast, a seizure involving the visual cortex may produce a hallucination of lights, geometric forms or complex images such as a face, while seizures involving the amygdala may produce a transient, abrupt state of panic.

On the other hand, seizures originating in a hypersynchronous cortico–reticulo–cortical loop produce generalized, rather than focal, symptoms: tonic stiffening or rhythmic twitching of all four limbs; sudden loss of control of limb and trunk muscles with an abrupt fall to the ground; transient loss of awareness without loss of postural control.

At the present time, seizure patterns are classified according to a generally accepted system developed by the International League Against Epilepsy (see Table 11.2). In essence, focal-onset seizures are classified according to whether or not they also interfere with consciousness – *simple partial seizures* if consciousness is maintained, *complex partial seizures* if it is not. At the present time, the term *focal dyscognitive seizure* is often used in place of 'complex partial seizure', the reason being that, with some seizures, consciousness is at least partially retained while cognition is temporarily impaired. Simple partial seizures are also categorized according to the principal type of symptom present, whether motor, sensory, autonomic or psychic.

Generalized seizures are classified according to the type of behaviour seen by observers. Thus, a typical generalized seizure may consist of an abrupt loss of consciousness, a simultaneous uncontrolled contraction of all

TABLE 11.2: Main Seizure Patterns

Focal onset	Simple partial – motor, sensory, vegetative, psychic
	Complex partial (focal dyscognitive) – impaired consciousness or cognition
	Simple or complex partial with secondary generalization
Generalized	Tonic–clonic
	Tonic
	Clonic
	Absence
	Atonic
	Myoclonic

body musculature manifested as generalized tonic stiffening of the limbs, then a period of rapid violent twitching of all four limbs in synchrony. This sequence is usually summarized as a *generalized tonic–clonic seizure* and often referred to by the lay public as 'grand mal seizures', an archaic terminology. A generalized seizure consisting of a sudden loss of muscle control and a fall is called an *atonic seizure*, while a sudden, violent, shock-like jerk of the limbs (sometimes without loss of consciousness) is termed a *myoclonic seizure*. Finally, a transient loss of consciousness without loss of trunk support is called an *absence seizure* – 'absence' being the French term for absent awareness. The archaic, but commonly used, term for the latter is 'petit mal seizure', to contrast it with the flagrant, dramatic manifestations of the grand mal seizure. Absence type seizures are the type of epilepsy that best fit the description given for Didi's attacks.

We have already alluded to the fact that some individuals with epilepsy may have a combination of focal cortical pathology and a genetically-determined low threshold for generalized, or *corticoreticular* seizures. In such individuals, a seizure may begin in a cortical area, with focal motor or sensory symptoms, and then spread preferentially to the non-specific thalamic nuclei; this generates a second phase in which there is a generalized tonic–clonic seizure. This phenomenon is known as *focal onset with secondary generalization*. If the initial, simple partial seizure produces a transient state of fear, followed by a generalized seizure, the simple partial component, or premonitory feeling is termed an *aura*.

Alternatively, a focal-onset seizure may simply spread to adjacent cortical regions without projecting to the thalamus. If the initial symptom of the seizure consists of rhythmic twitching of the thumb, sequential localized spread will lead to twitching of the hand, arm, shoulder and ipsilateral face, a phenomenon called a *Jacksonian march*, named after Hughlings Jackson, who originally described it. Likewise, a seizure beginning focally in the amygdala (with a sensation of fear) may then spread to the hippocampus (with associated confusion and loss of memory registration) and to the insular cortex (with retching and drooling behaviour). In this case, the seizure commences as simple partial, then transforms to focal dyscognitive (complex partial).

11.6.D DRUGS AND TOXINS

Any drug capable of inducing neuronal suppression (e.g. short-acting barbiturates, benzodiazepines), as well as any chemical compound inducing hypoglycemia or anoxia (e.g. oral hypoglycemic agents, carbon monoxide, cyanide) may produce transient unconsciousness. In practice, however, such drug- or metabolically induced comatose states last for hours to days rather than seconds.

11.6.F METABOLIC – LACK OF SUBSTRATE

Glucose: All neurons require a continuous supply of glucose in order to maintain a resting membrane potential. If the supply of glucose is inadequate, neuronal synaptic function (about which more later in this chapter) ceases within seconds, followed by loss of resting membrane potentials and persistent neuronal depolarization. The severely hypoglycemic patient becomes rapidly unconscious and unable to move. If glucose is then replenished within a satisfactory period of time (minutes), normal neuronal function will resume and the patient will resume normal activities unscathed. Severe, prolonged hypoglycemia (more than 8–10 minutes duration) may result in the death of neurons and permanent deficits.

Hypoglycemia, therefore, as may occur in patients with diabetes mellitus whose insulin dose is too large for the amount of glucose consumed, may produce recurrent episodes of unconsciousness. Typically, however, such episodes are preceded by warning symptoms of impending neuronal failure: brief confusion, pallor, sweating and anxiety. In addition, the period of unconsciousness with hypoglycemia lasts for minutes, not seconds, and the patient, if sitting or standing up, will collapse, unable to maintain postural stability.

Since Didi's episodes of unresponsiveness only last about 10 seconds, during which time he maintains his sitting or standing posture, a hypoglycemic etiology seems unlikely.

Oxygen: Neuronal function is disturbed just as rapidly when the brain is deprived of oxygen as when it is deprived of glucose. Lack of oxygen may occur as a consequence of reduced atmospheric oxygen (as in sudden airplane decompression at high altitude), impaired pulmonary function (as in drowning or severe pneumonia), impaired red blood cell oxygen carrying capacity (as in severe anemia or carbon monoxide poisoning) or impaired perfusion of blood to the brain (as in infection-induced shock or in cardiac arrest).

None of these putative etiologies, however, is likely to produce brief, recurrent episodes of diffuse CNS hypoxia. On the other hand, localized blood vessel disease involving arteries supplying the brainstem reticular formation or the thalami might result in transient periods of impaired perfusion to these structures, so-called transient ischemic attacks (TIAs). TIAs associated with episodic unconsciousness are due to cerebrovascular disease in the vertebrobasilar system (see Chapter 8).

Episodic unconsciousness may also occur in the context of transient hypotension, such as may occur with intermittent cardiac arrhythmias (cardiac conduction block or Stokes-Adams attack) or with reflex hypotension in response to a strong emotionally laden stimulus in predisposed individuals: the common 'fainting spell', otherwise known as *vasovagal* or *neurovascular syncope*.

With the exception of the last-mentioned cause of transient cerebral anoxia or ischemia, however, none of these possible etiologies is very likely in a previously well 6-year-old boy! In addition, all of the objections applied to episodic unconsciousness secondary to hypoglycemia also apply to anoxia.

11.6.J GENETIC DISORDERS

As can be surmised from Didi's family history, epileptic disorders are often the result of a genetic predisposition. This is true both for seizures of focal and of generalized origin. Even if one inherits a genetic predisposition for epileptic seizures, this does not mean that seizures will invariably appear. Whether or not seizures develop is more often determined by environmental or acquired factors such as high fever, hormonal fluctuations, head injury or the development of a brain tumor. The genetic basis of epilepsy is beyond the scope of this book. Nevertheless, we will be referring to some common genetic epilepsy syndromes later in this chapter (next section).

11.7 CASE SUMMARY AND SUPPLEMENTARY INFORMATION

Didi has been experiencing frequent brief (10 seconds) episodes of staring, inattention and unresponsiveness for at least three months; they appear to be interfering with his ability to function appropriately in the classroom. His developmental status and neurological examination are entirely normal. Given that the episodes can be triggered by deliberate hyperventilation and that Didi's sister has similar episodes, the probable diagnosis is *childhood absence epilepsy*.

11.7.1 ADDITIONAL INFORMATION

Childhood absence epilepsy is a specific epilepsy syndrome whose diagnostic criteria are outlined in the International League Against Epilepsy classification system of epileptic disorders (see Table 11.3). Childhood absence epilepsy is now known to be a genetic disorder, as are the other non-localization-related 'idiopathic' epilepsy syndromes listed in Table 11.3. In the case of childhood and juvenile-onset absence epilepsy, a growing list of gene mutations has been described. To date, the majority of gene mutations implicated in absence epilepsy have either been in subunits of GABA receptors or components of calcium channels (those interested are directed to the review article by Yalçin in the chapter reading list).

TABLE 11.3: Selected Epilepsy Syndromes

	Syndrome
Localization-related	
Idiopathic	Benign epilepsy of childhood with centro-temporal (rolandic) spikes
	Benign epilepsy of childhood with autonomic features (Panayiotopoulos syndrome)
Symptomatic	Temporo-limbic epilepsy with mesial temporal sclerosis
Non-localization-related	
Idiopathic	Benign neonatal familial convulsions
	Benign sporadic neonatal convulsions
	Benign myoclonic epilepsy of infancy
	Severe myoclonic epilepsy of infancy (Dravet syndrome)
	Childhood absence epilepsy
	Astatic/myoclonic absence epilepsy (absence seizures with falls – Doose syndrome)
	Juvenile myoclonic epilepsy (Janz syndrome)
	Juvenile-onset absence epilepsy
Symptomatic	Neonatal tonic spasms (Ohtahara syndrome)
	Infantile spasms (West syndrome)
	Lennox–Gastaut syndrome

Note: Adapted from the epilepsy syndrome classification system of the International League Against Epilepsy.

Faced with a child suspected of having absence epilepsy, the physician's first task, before undertaking a treatment regime, is to confirm the diagnosis by documenting an absence episode during an EEG recording. This step is essential because, after all, why couldn't Didi simply be having day-dreaming episodes? Anyone with any experience of six-year-olds will recognize that children this age can be so self-absorbed, all environmental 'filters' fully activated, that they fail to respond to having their names called repeatedly. In addition, children with chronic airway obstruction, due either to allergic rhinitis or tonsillar hypertrophy, may have such difficulty sleeping that they are chronically sleep deprived and inattentive during the daytime. All of these alternative possibilities need to be considered.

The EEG findings in absence epilepsy are quite dramatic, even between seizures (the interictal state). While the patient is lying quietly, awake but with closed eyes, the normal background EEG rhythms are frequently interrupted by one- to two-second generalized spike-wave bursts (see Figure 11.9). Recorded synchronously over all scalp regions, and occurring at a rate of three spike-wave complexes per second, these discharges are not accompanied by any obvious change in behaviour or awareness. On occasion, as well as in response to hyperventilation, longer periods (8–10 seconds) of 3/second (or 3 Herz) spike-and-wave discharge will occur, this time accompanied by unresponsiveness to name-calling and often by rapid flickering of the eyelids.

This EEG result is typical of childhood absence epilepsy and is entirely different from what we would observe if we recorded a complex partial (focal dyscognitive) seizure. Typically originating focally in one temporal lobe, such a seizure may resemble superficially a childhood absence attack in that the child will be inattentive and stare blankly for the duration of the episode. In general, however, a focal dyscognitive seizure of the 'absence' type will last one or two minutes, not 10 seconds, and will be followed by a period of confused behaviour. Automatic behaviours such as lip-smacking and purposeless fidgeting with clothes may also occur during the seizure itself. An EEG during a focal dyscognitive seizure will reveal a sustained focal rhythmic sharp wave discharge in the temporal lobe region, with some spread to surrounding areas as the seizure proceeds (see Figure 11.8).

The mechanism of the alternating spike-and-slow-wave discharge in absence epilepsy is thought to be a rhythmic oscillation in cortical neurons, alternating between an excitatory spike potential and an inhibitory slow wave. This rapid cycling between excitation and inhibition appears to be driven by the non-specific thalamic nuclei. It is not correct to assume, however, that generalized epileptic discharge of this type originates in the thalamus; both the cortex and the thalamus, acting in tandem, appear to be necessary for this phenomenon.

If the diagnosis of childhood absence epilepsy is confirmed electroencephalographically, you may well ask whether other investigations are required, for example an imaging study of the brain. Decades of accumulated experience, both with CT and magnetic resonance (MR) imaging, however, have demonstrated that children with any form of primary generalized epilepsy nearly always have normal-appearing brains. Hence, for well-documented absence epilepsy, an imaging study is unnecessary. Imaging studies are definitely indicated, however, for most patients whose seizure patterns or EEG findings suggest a focal origin. The only exception to this dictum is a group of benign, familial localization-related epilepsies, e.g. benign epilepsy of childhood with rolandic spikes; benign epilepsy with autonomic features – see Table 11.3.

Once the diagnosis of childhood absence epilepsy has been made, the next step is to determine whether or not Didi should be started on an anti-epileptic medication. For childhood absence epilepsy, starting such medication is not an automatic step, given that patients do not fall during attacks and are at low risk of injury. In most instances, it is the frequency of the absence attacks, and their possible negative effect on attention and learning, that determine whether anti-seizure medication is to be used. As you

will recall, Didi's presenting problem is a striking decline in school performance; this and the numerous witnessed absence seizures strongly suggest that his epilepsy should be treated.

At the present time, there is a large number and variety of antiepileptic medications; the principal drugs, and their indications for use, are listed in Table 11.4. Which drug or drugs should we consider using? To some extent, our decision may be guided by a consideration of mechanisms of action of anti-epileptic medications.

There are three principal mechanisms of action of anti-epileptic drugs thus far identified:

> *Prolongation of inactivation of sodium channels in circumstances of high-frequency neuronal discharge:* In essence, this mechanism involves the stabilization of neuronal membranes and is the main mode of action of many of the older anti-epileptic medications like phenytoin and carbamazepine.
>
> *Augmentation of GABA-ergic inhibition:* Drugs acting through this mechanism attach to GABA receptors, thus augmenting chloride conduction and hyperpolarizing neuronal membranes. The benzodiazepine class of medication (e.g. clonazepam, nitrazepam, clobazam), as well as phenobarbital, falls into this category.
>
> *Inhibition of T-type calcium channels:* T-type Ca^{++} channels are particularly prominent in neurons located in the non-specific (reticular) thalamic nuclei. Inhibition of these channels appears to suppress generalized corticoreticular epileptogenic

TABLE 11.4: Anti-Epileptic Medications

Seizure Type	Main Therapeutic Options
Generalized	
Tonic–clonic	*Valproate, carbamazepine, phenytoin,* clobazam, phenobarbital, lamotrigine
Absence	*Ethosuximide, valproate,* lamotrigine
Myoclonic	*Valproate,* topiramate, clonazepam, nitrazepam, acetazolamide, levetiracetam, ketogenic diet
Atonic	*Valproate,* lamotrigine, topiramate, zonisamide
Partial	*Carbamazepine, phenytoin, clobazam,* valproate, lamotrigine, topiramate, gabapentin, vigabatrin, phenobarbital, lacosamide
Syndromic	
Benign rolandic	*Carbamazepine,* valproate, phenytoin, sulthiame
Juvenile myoclonic	*Valproate,* lamotrigine, topiramate
Lennox–Gastaut	Valproate, lamotrigine, clonazepam, topiramate
West syndrome	*Vigabatrin,* ACTH, steroids, valproate

Note: Medications shown in italics are those used most frequently.

mechanisms, particularly in the case of absence seizures, rather than generalized tonic-clonic seizures. Ethosuximide is an anti-epileptic drug that has its main effect through this mechanism.

A fourth anti-epileptic mechanism, in principle, is the inhibition of excitatory glutamate receptors; drugs designed to address this mechanism are currently being developed.

A consideration of the earlier mechanisms of action, and a perusal of Table 11.4, would lead us, for a child with absence epilepsy, to settle on ethosuximide. Alternative agents, in the event that ethosuximide were to be either ineffective or poorly tolerated, would be valproate and lamotrigine. Most patients with absence epilepsy will respond to one or more of these three medications.

11.7.2 INVESTIGATIONS

As expected, Didi's baseline EEG in the awake state showed frequent brief bursts of 3 Hz spike-and-wave discharges; no clinical absence seizures occurred.

When Didi was directed by the neurophysiology technologist to hyperventilate, however, he had a 10-second episode identical to the one we saw in your office. During that 10-second event, there was sustained 3 Hz generalized spike-wave discharge with a slowing in frequency just before the offset. The instant the spike-wave discharge ceased, the EEG resumed a normal background and Didi was immediately alert.

No other investigations were performed.

11.8.1 CASE 11E-1: OMAR, AGE 23

Background: Omar has just moved to your city to continue with his graduate studies under a very famous professor. Omar has epilepsy; he is concerned about the occasional breakthrough seizure now that he is living away from home and about the possible social stigma of being labelled an 'epileptic'.

Chief complaint: Has had focal dyscognitive seizures for most of his life; needs ongoing follow-up

History:

- Completely well and developmentally normal to age 16 months.
- During an upper respiratory tract infection with a fever of 39°C, he had a *generalized convulsion* lasting over an hour; in the last half of the seizure, twitching was confined to the right side of the body, with conjugate eye deviation to the right.
- At age 3½, he began having unprovoked (i.e. no associated illness or fever) episodes consisting of sudden onset of *staring, unresponsiveness,*

11.7.3 TREATMENT/OUTCOME

Didi was started on a small amount of ethosuximide, given in two divided doses per day. The dose was gradually increased over a period of two weeks, during which time the absence seizures declined in frequency and eventually became undetectable. At the same time, there was a dramatic improvement in Didi's school performance and he was soon functioning at his usual superlative level. Initially, he complained of feeling nauseated shortly after taking each dose of medication; this side effect disappeared after about a month.

After Didi had been seizure-free for a year, an attempt was made to taper his medication. The absence seizures almost immediately returned so the medication was restarted. A year later, the ethosuximide was again tapered, this time with no recurrence of seizures.

11.8 E-CASES

The following clinical cases are intended to broaden the scope of the material that we have introduced in this chapter. Only the case histories and physical examination findings are given here. To work through these cases, please follow the same routine as you have just done for Didi's story. The required materials, the discussions of localization and etiology for each case, relevant illustrations and summative comments are all available on the Web site.

flushing and repetitive lip smacking; duration of the episodes was one minute, followed by five minutes of confusion and drowsiness.

- After some investigations, he was started on oral carbamazepine and was almost completely seizure-free for the next six years.
- From age 10 onwards, his seizures were more difficult to control; various other anti-epileptic drugs were used, in varying combinations, including clobazam, oxcarbazepine, divalproex, topiramate, phenytoin and levetiracetam.
- At the present time, on a combination of topiramate and levetiracetam, he has seizures every 12–18 months, usually triggered by excessive fatigue – and is thus *not permitted to drive an automobile.*

Examination:

- Excellent academic record throughout high school and university; currently working on a master's degree in political science.
- *Neurological examination is entirely normal.*

11.8.2 CASE 11E-2: YUKIO, AGE 6

Background: Yukio is brought to the emergency room (ER) by ambulance early one cold winter morning accompanied by his panic-stricken single mother. After hearing strange sounds from his bedroom, she went in to find him twitching and unresponsive.

Chief complaint: First nocturnal generalized convulsion 45 minutes ago

History:

- Mother was awoken by a strange, choking sound coming from her son's bedroom.
- She found Yukio lying on his back, eyes open and rolled upwards, with the *left* side of his face contracted into a snarl and with copious drooling.
- *His breathing was harsh and stertorous; rhythmic 2/second twitching of both arms and legs for 90 seconds, after which he went back to sleep.*

- Yukio was subsequently reawakened by the emergency response team and, while somewhat bewildered by their presence in his bedroom, he was able to converse normally.
- Further history reveals that, one month ago, his mother had heard a similar noise during the night and found him sitting in bed, *awake but unable to speak*: copious saliva pouring from his mouth and rhythmic twitching of the *right* side of his mouth for about a minute, followed by transient *slurring of speech*.
- Normal pregnancy, birth and developmental history.
- Previously well except for minor head injury on a play structure six weeks ago.
- *Mother's sister had a few short generalized seizures during sleep prior to age 10; off antiepileptic medication by age 12.*

Examination:

- General physical and neurological examinations are entirely *normal*

11.8.3 CASE 11E-3: MRS. V, AGE 40

Background: Following her annual medical assessment, Mrs. V, a mother of three, has just left your office to have some blood work done in the laboratory downstairs in your medical building. A short while later, you receive an emergency call to come down to see her because of a distressing event.

Chief complaint: An apparent generalized seizure during a venipuncture

History:

- The episode occurred during the venipuncture while Mrs. V was seated; considerable difficulty in finding a vein because she is overweight.
- Technician noted that the patient had become pale and sweaty, eyes open but staring blankly; a few moments later, she *arched her back, extended her neck, stiffly extended her arms and legs followed by rhythmic jerking of all four limbs for about 10 seconds.*

- Mrs. V was quickly removed from the chair and placed on her side on the floor; she awoke within 10–15 seconds, asked what had happened and complained of a severe headache.
- Upon questioning, she recalled that the needle had been very painful, that she began to feel *nauseated and light-headed followed rapidly by darkening or loss of vision.*
- No personal or family history of epilepsy; patient's father used to faint at the sight of blood.

Examination:

- Initial pulse rate *60/minute and faint*; by the time her blood pressure could be checked, the pulse had risen to 90/minute and the blood pressure was 115/70.
- Mrs. V initially appears *extremely pale* but her colour gradually returns during the examination.
- General physical and neurological examinations are normal.

11.8.4 CASE 11E-4: FRANCESCA, AGE 22

Background: It is 7:30 am and you are just finishing your overnight ER shift when the ambulance arrives with a young woman who is having seizures. Francesca is a physiotherapy student home for the holidays and is accompanied by her terrified parents.

Chief complaint: Headache, fever and chills for the past 24 hours; three epileptic seizures during the past 30 minutes

History:

- Francesca awoke yesterday feeling unwell, with headache, chills, scattered muscle pains; her temperature was 38.8°C – she took ibuprofen every four hours while awake.
- At 6:30 am today, she awoke with severe headache and back pain; she was *irritable, argumentative and sometimes appeared not to understand what her parents were saying.*
- Parents report that her speech this morning has been *incoherent, with a tendency to use inappropriate and nonsense words.*
- First seizure occurred 15 minutes after getting out of bed: *stopped talking, stared blankly and did not respond for 20 seconds, followed by turning of her head and eyes to the right* and then by a four-minute generalized convulsion with tongue-biting and urinary incontinence.
- Post-ictally Francesca was even more confused and was *unable to use her right hand for 10 minutes*; two further similar seizures in the ambulance en route to the hospital.

Examination:

- Pulse 120/minute; respirations 25/minute; BP 130/75; temperature 39.5°C.
- General physical examination normal except for *marked neck stiffness with passive neck flexion.*
- Fluent speech but many substitutions and nonsense words; *unable to give the names of people or objects in the room.*
- Disoriented in place and time; unable to follow simple instructions or to do simple mental arithmetic tasks.
- *Unable to remember your name despite frequent reminders.*
- Cranial nerves: normal fundi and eye movements; *mild weakness of the lower part of the right face* (see Figure 11.10a).
- *Mild weakness of right hand grip and of foot dorsiflexion*; normal muscle tone bilaterally (Figure 11.10a).
- *Reflexes:* see Figure 11.10b.
- Sensory examination is not feasible.

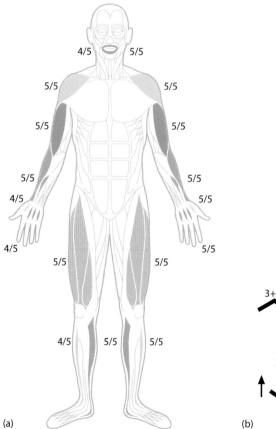

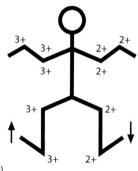

FIGURE 11.10: Motor examination (a) and reflex (b) maps for case 11e-4 (Francesca).

11.8.5 CASE 11E-5: JASON, AGE 5 MONTHS

Background: Jason is the fifth child in a working-class family living just at the poverty line. His parents have brought him for assessment of some strange behaviours at meal times that they think must be due to colic.

Chief complaint: Irritable and 'colicky' at meal times for the past three weeks

History:

- Jason has been an irritable infant since birth, but worse in the past three weeks with repeated behaviours that his parents refer to as 'colic'.
- His abnormal behaviours only occur at meal times after waking in the morning or after a nap.
- 'Colicky' episodes begin with sudden *flexion of the neck, trunk and legs, with elevation of his arms at the shoulders, all persisting for two to three seconds*; transient eye deviation to the left and facial grimace during the episode.
- Second episode identical to the first about 10 seconds after the first, followed by further episodes every 5–30 seconds for several minutes altogether; the intensity of the episodes declines over the sequence.
- *Unresponsive throughout each cluster of attacks.*
- Initially, Jason had one cluster of episodes in a day (three weeks ago); gradual increase in clusters to four to six per day.
- He has no history of constipation.
- Normal pregnancy birth and early development; over the past two weeks, *he seems less interested in his surroundings and is less sociable.*

Examination:

- General physical examination is normal except for the presence of *several oval de-pigmented macules in the skin of his trunk and legs*; his mother has two similar macules (see Figure 11.11 for example).
- Intermittent eye contact; Jason follows brightly coloured hand puppet in all directions of gaze.
- Symmetric facial movement when crying; he turns his head to soft sounds.
- Slight head lag when he is pulled from supine to sitting.
- Normal muscle tone for age (i.e. quite high in comparison with older children); withdraws all four limbs vigorously to tickle.
- Firmly grasps objects placed in either hand.
- Brisk tendon reflexes; flexor plantar responses.

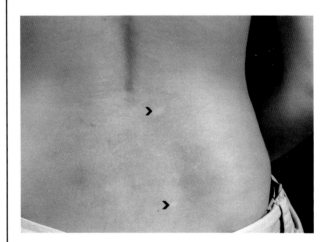

FIGURE 11.11: Depigmented skin macule in a patient with the same diagnosis as case 11e-5 (Jason).

11.8.6 CASE 11E-6: DILLON, AGE 2

Background: Dillon is the younger of two children in an affluent family. His parents have requested a second opinion regarding the management of his epileptic condition.

Chief complaint: Frequent, unprovoked generalized convulsions

History:

- Normal pregnancy and birth history; good health to age six months.
- From age 6 to 15 months, Dillon had a series (five or six) of *prolonged generalized convulsions (lasting up to one hour) triggered by febrile illnesses*; normal developmental status and neurological examination on repeated assessments during that time period; normal EEG.
- At age 18 months, onset of *unprovoked* one- to five-minute generalized convulsions (average four to six per month, followed within a few weeks by the appearance of *generalized myoclonic jerks, often in clusters.*
- Repeat EEG now very *abnormal, with generalized spike-and-wave discharges and two independent spike foci*; MR imaging study of the head was normal.
- Started on phenobarbital by family physician but no decline in seizure frequency (either type) despite increasing doses.

- *Developmental status at age 2 is now clearly abnormal*: two-word vocabulary; poor language comprehension; no knowledge of body parts; restless hyperactivity.
- No family history of epilepsy, febrile seizures or developmental delay.

- Alert, hyperactive; good eye contact; does not respond to simple instructions (look at mummy; touch your nose).
- Cranial nerve and motor system examinations are within normal limits for age.

11.9 SUMMARY OF BASIC NEUROPHYSIOLOGICAL INFORMATION

- Neurons are able to generate and maintain an electrical charge (*resting membrane potential*) because they possess highly specialized limiting membranes that contain a variety of ionic pores (*ion channels*) that are usually closed but can be transiently opened to permit passage of specific ions either into or out of the cell (Figure 11.1).
- There are four main neuronal ion channels that selectively admit *sodium, potassium, calcium and chloride* ions respectively. Each channel is constructed of several distinctive subunits that collectively determine pore size and the type of ion that is permitted to pass through. Energy-consuming intramembranous *ionic pumps* adjacent to the ion channels act in collaboration to create *ionic gradients* across the neuronal membrane: e.g. high Na^+ outside the cell, low inside; high K^+ concentration inside the cell, low outside. This phenomenon results in the establishment of the resting membrane potential, typically about -70 mV (Figure 11.2).
- Sufficient excitation of the dendrites or cell body will result in the sequential opening of *Na^+ channels*, allowing bulk entry of Na^+ along the concentration gradient into the cell and creating an *action potential* (Figure 11.3). This event is followed almost immediately by the opening of *K^+ channels* to allow K^+ to flow outside the neuron and re-establish the resting membrane potential.

At this point, the neuron is briefly refractory to further stimulation; the baseline state is rapidly re-achieved by the active pumping of Na^+ to the outside the cell and K^+ to the inside. The intramembranous ion exchange *pump* that accomplishes this feat is known as *Na^+/K^+-ATPase*.

- Neurons communicate with one another by the transformation of an electrical event – the action potential – to a *chemical* event occurring at *synaptic contacts*. At the presynaptic site, the action potential, acting via *calcium channels*, leads to the release of one or more types of *neurotransmitter molecules* that cross the synaptic cleft and attach to specific receptor molecules in the membrane of the post-synaptic neuron (Figure 11.4).
- Whether the post-synaptic neuron is excited or inhibited by the neurochemical message received from the pre-synaptic neuron is determined by the *nature of the neurotransmitter molecule released on the presynaptic site and by the nature of the receptor on the postsynaptic side*. Excitatory responses involve the opening of Na^+ channels at the post-synaptic site while inhibitory responses are accompanied by the selective opening of *chloride channels* (Figure 11.4).
- The most important neurotransmitters are *Ach, glutamic acid (glutamate), GABA, glycine, dopamine, serotonin and norepinephrine*. Ach and glutamate are primarily excitatory; GABA and glycine are primarily inhibitory. Dopamine, serotonin and norepinephrine are called *biogenic amines* and modify downstream neuronal activity through G-proteins; the effects can be excitatory or inhibitory depending on the type of aminergic receptor.

11.10 CHAPTER-RELATED QUESTIONS

1. Give the probable anatomical cerebrocortical region implicated for focal epileptic seizures that have the following initial characteristics:
   a. Twitching of the right leg and foot
   b. Twitching of the left perioral region
   c. Numbness and tingling of the left hand
   d. Hearing the first line of *The Star-Spangled Banner*
   e. Seeing coloured shapes in the left visual field
   f. Having a sudden feeling of intense, undirected fear

g. Having a squeezing or rising sensation in the throat

h. Suddenly developing the sensation of smelling burned toast

i. Sudden, uncontrollable deviation of the eyes to the left without loss of consciousness

j. 30-second-duration stereotypic episodes of trunk arching and arm thrashing while asleep

k. Sudden onset of unresponsive staring accompanied by repeatedly murmuring 'itsy-bitsy-blue'

2. Based on your review of the chapter material and the associated e-cases, determine whether the disorder in the following scenarios is most likely *(a) epileptic, (b) syncopal or (c) migrainous* in nature

a. A 35-year-old clinical psychologist, while busy seeing his fourth patient of the afternoon, begins to notice that his vision towards the left side is becoming blurred. After two or three minutes, he observes that, within the blurred region, there is a flickering, black and silver zigzag line. This line gradually elongates over the next 10 minutes, assuming an overall C-shape with the open end pointed to the right. When he closes his left eye, he notices that the visual disturbance is also present in his right eye. Although the visual phenomenon is distracting, he develops no other symptoms and is able to continue working with his patient. Over the succeeding 10 minutes, the zigzag line gradually disappears and nothing further occurs.

b. A 13-year-old girl has woken this morning with symptoms of a flu-like illness: cough, nasal congestion, sore throat and low-grade fever; she elects to spend the day in bed. At around 8 in the evening, while feeling feverish, she gets out of bed to go to the bathroom. As soon as she stands up, she feels intensely dizzy; her vision rapidly darkens and she loses consciousness, falling backwards into her dresser and slumping to the floor. Her father, having heard a loud crash, runs into her room and finds her on her side, awake and complaining of pain in her upper back.

c. A 20-year-old college student is out jogging around the campus in order to get some exercise and fresh air before his first class of the day. A bystander observes the student suddenly crumple to the ground. Running over to see if she can help, she finds him unresponsive and pale, his eyes partly open and staring blankly. After about 30 seconds, his face flushes and he awakens, although transiently disoriented and complaining of a headache. His helper insists that he stay lying down and calls 911 on her cell-phone. While they are waiting for the ambulance to arrive, he tells her that he has had one similar episode in the past – about three months ago – also while running. Subsequently, he had gone to the college medical clinic, where a physical examination had shown no abnormalities.

d. A 24-year-old bus driver has begun to have occasional episodes of abrupt-onset numbness and tingling involving the right side of her mouth and tongue. The episodes began about two months ago and occur every few days, typically later in the day. Each episode lasts 30–60 seconds and is entirely subjective: anyone present at the time sees no change in her behaviour other than the fact that, if she speaks while the symptoms are present, her speech is transiently slurred. The episodes are not followed by headache or confusion; on the other hand, during the past few months, she has begun to have pounding headaches, predominantly left sided, during her menstrual periods.

e. A 30-year-old cosmetician has a 15-year history of stereotyped sensory symptoms occurring two or three times a year, usually just prior to or during her menses. She gradually develops a tingling sensation in the perioral region, the tongue and in the fingers of both hands. The symptoms wax and wane over a period of 15–20 minutes and are then replaced by a pounding occipital-region headache accompanied by nausea and dizziness, all lasting three to four hours before subsiding. During the time that her fingers are tingling, her hands feel awkward and slightly weak such that she is temporarily unable to grasp objects firmly. By the time the headache begins, however, her grip strength has returned to normal.

f. A 16-year-old high-school student comes to your office for assessment of two episodes of unconsciousness that have occurred during the past six months. Both episodes took place while he was playing basketball. On both occasions, without any warning symptoms on his part, he suddenly fell to the ground, developed an extension posture of his limbs and trunk and then began to have rhythmic twitching of both sides of his face and all four limbs – at first very rapidly, followed by progressive slowing of the twitch rate over a two-minute period until he stopped moving, becoming limp and unresponsive. During the initial portion of the episodes, his eyes were rolled upwards and his facial colour was dusky. Within four minutes of the onset, he began to move and to respond but was drowsy and confused for 10–15 minutes. With neither episode had he either bitten his tongue or been incontinent of urine. He has been completely well in the past but has a long-standing history of learning difficulties.

REFERENCE

Yalçin, O. Genes and molecular mechanisms involved in the epileptogenesis of idiopathic absence epilepsies. *Seizure* 2012. 21: 79–86.

SUGGESTED READING

Avoli, M. A brief history on the oscillating roles of thalamus and cortex in absence seizures. *Epilepsia* 2012. 53: 779–789.

Browne, T.R., Holmes, G.L. Epilepsy. *N Engl J Med* 2001. 344: 1145–1151.

Chang, B.S., Lowenstein, D.H. Epilepsy. *N Engl J Med* 2003. 349: 1257–1266.

Gardiner, M. Genetics of idiopathic generalized epilepsies. *Epilepsia* 2005. 46(Suppl 9): 15–20.

Sillanpaa, M., Jalava, M., Kaleva, O., Shinnar, S. Long-term prognosis of seizures with onset in childhood. *N Engl J Med* 1998. 338: 1715–1722.

Wirrell, E.C. Natural history of absence epilepsy in children. *Can J Neurol Sci* 2003. 30: 184–188.

WEB SITES

National Institute for Neurological Diseases and Stroke: www.ninds.nih.gov/disorders/epilepsy/detail_epilepsy.htm

www.epilepsy.com

Chapter 12

Armand

Objectives

- Learn the basics of the information system architecture of the human central nervous system (CNS)
- Learn the principal anatomic components of the CNS related to higher cognitive function
- Learn the effects of damage on these components and the associated clinical manifestations
- Learn the major dementia types and their clinical manifestations
- Become sensitized to the medical, physical, and psychosocial implications of dementia

12.1 ARMAND

Armand had been a successful businessman in real estate over the years, building up a local firm consisting of five partners of whom he is the most senior. Two of his sons have gone into the business and are also partners.

Armand had served in the army during WWII, had married his high school sweetheart when he returned from the war, had then settled down to a comfortable life and had raised his family.

Armand is now 82; nevertheless, he still gets dressed in his suit and tie and drives himself to work every day. Two months ago, however, he had an accident with his car: when he missed a red light and hit another car. No serious damage was done but the police had charged him with dangerous driving. His excuse was that it was a rainy day and he had not seen the light change. The Department of Motor Vehicles (DMV) had sent him a notice that he needed a medical assessment by a certain date or his license would be suspended.

There have been other concerning behaviours cropping up both at home and at work. Over the past one to two years, he has become increasingly forgetful about the details of the various deals that his company has been involved in and tends to blame the other partners or his secretary if there are negative consequences to his forgetfulness.

At home, his wife has noticed increased forgetfulness for names and places. As well, there have been episodes of unexplained anger, usually directed against her; these outbursts are usually short-lived.

When he finally showed up for his medical assessment, he appeared well groomed, with a suit and tie. He kept repeating the question, 'You are not going to take away my driver's license – are you?'

A review of his medical history reveals that he has had hypertension for many years and is currently on three different antihypertensive medications. Otherwise, his neurological and systemic systems review was negative, in particular for stroke and myocardial infarction. His mother died in her eighties of 'old age'; his father died of tuberculosis in the 1920s, when Armand was still an infant.

His general physical examination showed a blood pressure of 160/90 in both arms; his heart rate was 84 and regular. Auscultation of the chest and neck revealed a slight aortic murmur and bilateral carotid bruits, left louder than right. The rest of the exam was normal for age.

Armand's neurological examination showed a Montreal Cognitive Assessment (MOCA) score of 22 (refer to Chapter 2). He lost points for delayed recall, serial sevens and naming. He became quite frustrated and angry during the mental status testing, so it was stopped and then resumed later. The neurological examination showed some decreased ability to look upwards on testing the extra-ocular movements; otherwise, the cranial nerves were normal.

The motor exam showed normal power with a slight increase in his reflexes throughout except for his ankle jerks, which were decreased. His plantar responses were flexor. He had a brisk palmomental response on both sides (a reflex contraction of the ipsilateral mentalis muscle when the palm of the hand is vigorously stroked – a frontal lobe release sign). The muscle tone was quite variable; he seems to have had trouble with understanding the instructions to relax.

There was slight decrease in vibration sensation (at 256 Hz) in his toes. His upper and lower limb coordination for finger nose and heel shin was slow.

His gait was slightly unsteady for tandem walking but otherwise normal.

12.2 CLINICAL DATA EXTRACTION

What other information would you want to gather by history?

Before proceeding with this chapter, you should first carefully review Armand's story in order to extract key information concerning the history of the illness and the physical examination. On the Web site accompanying this text, you will find a sequence of tools to assist you in this process.

With this information, the neurological examination section of the Extended Localization Matrix Worksheet can be completed.

12.3 MAIN CLINICAL POINTS

The principle clinical features of Armand's presentation are as follows:

- Degradation of intellectual and social skills
- Difficulties with memory, attention and emotional control, all gradually developing over one to two years
- Moderate systemic hypertension controlled by medication
- Bilateral carotid bruits
- Decreased score on the mental status examination
- Difficulty with upwards gaze
- Unsteady gait
- Impaired vibration sense in the toes
- Hypoactive ankle jerks

12.4 RELEVANT NEUROANATOMY, PHYSIOLOGY AND PATHOPHYSIOLOGY

Since Armand's principal deficits are in memory and attention, we must consider some anatomical and pathophysiological aspects of cognition.

12.4.1 COGNITIVE FUNCTION

Human cognitive function represents the most sophisticated information system of which we are aware amongst living creatures on earth. Other animals have interesting and diverse adaptations that are necessary for their survival and reproduction, but none rivals human cerebral function in terms of complexity, adaptability and use of language.

The basic elements of human cognitive function include the following:

- Ability to sense visual, audio, olfactory, position, acceleration and somatosensory information
- Ability to perform motor tasks of locomotion, speech and object manipulation
- Ability to comprehend and codify sensory information
- Ability to comprehend language
- Ability to express language
- Ability to remember information
- Ability to perform mathematical calculations
- Ability to multi-process
- Ability to codify basic information objects
- Ability to sense, codify and learn new situations and responses to those situations
- Ability to detect and defend against threats
- Ability to perform executive functions of command and control in order to maintain a system of priorities. This system ensures protection, growth, evolution and, above all, the adaptive capability of the organism

The performance of these tasks 24 hours a day, seven days per week, for an average 85-year life span requires an information system architecture that must have significant redundancy and survivability to be able to regenerate itself and work within the metabolic framework of nutrients and waste disposal systems provided by the circulation. These functions are accomplished by a massive parallel system of neurons and axons, which communicate amongst themselves and to muscles and nerves.

The neurons are the microprocessors of the brain. As with personal computers or large-scale systems, the neuron is essentially a central processing unit (CPU). The basic unit of information processing in any computing system is a CPU having the basic functional elements of input, memory, computation, logic processing and output. The sophistication of the processing depends upon whether the processor is designed for generalized use or for specific applications.

The processing capability of a given CPU chip depends on the instruction set that is designed into the processor. The instruction set for a given chip is called the microcode; it constitutes the smallest unit of computational work performed by the chip. For example, to do the most basic operation of adding two double precision numbers together using a high language such as C or Visual Basic, literally hundreds of operations are required at the microcode level on the chip; nevertheless, the chip does all this very quickly. In human neurons, the microcode is different from one neuron population to another, thus allowing for different functions but with much slower switching speeds constrained by the biology of membrane conduction.

Thus, the human brain consists of a vast network of interconnected microprocessors, with differences in function depending on their location and connections.

The cerebral cortex contains billions of these neuronal microprocessors; they perform complex computations on the input from all of the sensory systems and from other areas of cortex. The output of these computations is transmitted via white matter tracts either to other areas of the brain, to brainstem or to spinal cord, resulting in complex behaviours such as limb movements or language production.

The human nervous system is therefore a massive parallel processing system that uses the basic neuron as a template and allows different functionalities based on anatomical location and interconnections.

In the process of embryogenesis, the cerebral cortex forms from neurons that migrate outwards from a matrix in the periventricular region; under the control of genetic inducers, neuronal stem cells migrate and differentiate into cortical neurons. Instead of designing multiple types of neurons with different functionalities, a basic template is used, which then is chemically and electrically induced to perform different functionalities.

For example, the neurons in the occipital cortex, receiving visual information from the retina via the lateral geniculate bodies, perform a very limited and specific function of creating a matrix of the visual image, which represents the reconstitution of either the left or the right visual field. No other neurons in the CNS can perform this function, but the occipital cortical neurons cannot do anything else.

These neurons are in contrast to those in the angular gyrus; if the latter are damaged, there is impairment in the ability to calculate, know left from right and to identify by touch a given finger on the hand. These integrative functions are very different from the image processing functions of the occipital cortex, but the hardware is drawn from the same basic design template. One of the great challenges for neuroscientists is to discover what makes the functional capabilities of specific neuron populations different. How are the inner workings of each neuron – its instruction set – made different and how can we measure these differences?

There is a term used to describe this difference in function of different neurons in different areas of the brain: *neuronal cytoarchitectonics*. Under the microscope, the layers of cerebral cortical cells (described in Chapter 10) appear slightly differently arranged in each area, thus giving rise to the analogy of continental plate tectonics in the form of layers of cortex. Once laid down after embryonic life and programmed during the early years of development, most areas of cortex assume their given function, which, if destroyed, will not recover; the function in that part of the brain is lost. There are some regions of what are called hypervariable function neurons; these seem to be able to take over the function of adjacent neurons that have been damaged.

The dendrite and axons are the input and output cables of the neurons; they receive input from other neurons and direct output information either to other cortical neurons for more processing or to motor neurons of the brain stem or spinal cord for motor output. Motor output can be in the form of locomotion, use of hands and tools or vocal output via the muscles of the upper airway, chest and diaphragm.

The human cortex has been organized functionally such that certain neurons perform certain specialized functions. Table 3.9 is a summary of the localization of cortical functions that are important clinically. The dominance column assumes right-handedness with the left hemisphere dominant for language.

Cognitive function is also highly dependent on the connections between neurons; a function served by the axons and their terminal connections. Loss of axonal connections either locally or globally can lead to cognitive impairment which is often referred to as subcortical dementia.

12.4.2 IMPLICATIONS OF DEMENTIA ON FUNCTIONAL STATUS

The loss of cognitive function in one or more domains has significant impact on the functional capability of a given individual given his or her personal, social, educational, employment and general medical status. During the initial clinical evaluation of any patient, the clinician must endeavour to make a quick snapshot of the overall clinical situation and above all take measures to ensure the safety of the patient, the family and the public at large. The two most dangerous places for any patient with dementia are their car and their kitchen. If safety is an immediate concern, a responsible clinician must take immediate action to limit potential damage especially in these two areas by limiting driving and access to potential fire hazards in the kitchen.

The approach for initial assessment should focus on determining the following factors that can individually or in combination influence the patient's functional status:

- Memory
- Language
- Executive function
- Visuospatial processing
- Social interactions
- Mood
- Highest education level attained
- Interaction with neurodevelopmental disorders
- Co-existing medical problems

One of the quickest and most effective techniques to determine mental status is to ask the patient's area of past expertise or job function. Focused questions on what the person did in their job will often give a quick initial impression of their mental status.

Other tools such as the Mini-Cog can be used to provide a rapid assessment of cognitive function.

12.4.3 PATHOPHYSIOLOGY OF DEMENTIA

The definition of dementia is an irreversible loss of cognitive function due to either structural or metabolic damage to the neurons or their axonal connections. Dementia does not refer to transient states of cognitive dysfunction, delirium or confusion. However, the presence of an underlying dementia may amplify the intensity of an acute confusional state.

Loss of cognitive function therefore depends on which area of the cerebral cortex or white matter connections have degenerated. Pure cortical dementia occurs when large populations of cortical cells in one area lose function and affect the person's cognitive function in that domain. Damage to a specific area from ischemic or hemorrhagic stroke will cause abrupt focal deficits in motor or sensory function, usually on one side of the body, as well as cognitive dysfunction depending on location of damage. The presence of areas of ischemic cortical damage tends to amplify the effects of a developing cortical dementia.

The localization of areas of the brain responsible for specific function is being elucidated by means of functional magnetic resonance imaging (MRI) and other modalities such as magnetic electroencephalography (MEEG). The interactions of the various functional nodes and links within the brain can be studied and are collectively known as the 'Connectome'. Increasingly, the sum of interactions depends on widespread network connections and their interactions. The pattern of degradation of these interactions can be used to categorize various types of cortical and subcortical dementia. However, there remain basic brain anatomic localizations for various functions.

Memory can be divided into episodic, semantic, working and procedure subtypes; each subtype has different localization implications, as indicated in Table 12.1.

Language similarly can be localized based on the spontaneity, fluency, ability to repeat, prosody and ability to generate words. The localizations are listed in Table 12.2.

Executive function, which is vitally important to allow an individual to have 'situational awareness', needs to have intact functions of working memory, inhibition, the ability to change behaviour state and fluency. These functions are mainly localized in the pre-frontal cortex, with some differential functions and widespread connections via white matter to the basal ganglia and cerebellum. Any disease process that interferes with any or all of these structures can lead to deficits in executive function.

An assessment of the person's social interactions is of critical importance in order to determine the impact of the behavioural changes and the required degree of social support, especially as the dementia worsens. The recognition of 'care giver burnout' and the need for clinicians to interact with social service agencies to provide home and respite care are essential to maintain quality of life for the patient and the care givers.

TABLE 12.1: Clinically Relevant Memory Systems

Memory Subtype	Patient/Caregiver Concerns	Cognitive Testing Deficit	Relevant Neuroanatomy	Commonly Associated Neuropathology
Episodic (declarative, explicit)	Verbal: cannot remember breakfast this morning or the destination of a recent vacation Visual: cannot recall the cabinet in which the dinner plates are located or the side of the street of the local drug store	Verbal: recall of oral narrative, word list recall Visual: recall of figure copy, recall of figure location in space	Medial temporal lobes, Papez circuit Verbal left greater than right Visual: right greater than left	Alzheimer's disease, herpes simplex encephalitis, thiamine deficiency, hippocampal sclerosis, hypoxic ischemic insult, dementia with Lewy bodies
Semantic (declarative, explicit)	Cannot recall the number of weeks in a year or identify the breed of the family dog	Fund of general knowledge, picture naming, category fluency	Anterior and inferior temporal lobes	Frontotemporal lobar degeneration, Alzheimer's disease
Working (declarative, explicit)	Cannot recall a phone number immediately after hearing it, cannot perform a series of simple requests after travelling from one room to the next	Digit span, mental arithmetic	Prefrontal cortex, subcortical structures, parietal association cortex	Vascular insults, frontotemporal lobar degeneration, dementia with Lewy bodies, Parkinson's disease, dementia, traumatic brain injury
Procedural (non-declarative)	Cannot recall the technique for using a driver off a tee, cannot maintain a violin bow hold	Not routinely tested	Basal ganglia, cerebellum, supplementary motor area	Parkinson's disease, cerebellar degeneration, Huntington's disease

Source: Matthews, B., *Continuum*, 21, 2015, Table 2.1. Used with permission.

TABLE 12.2: Classic Aphasia Syndromes

Aphasia	Disorder of Language	Classical Localization	Spoken Fluency
Broca	Disruption of speech order and planning	Left posterior inferior frontal lobe involving Broca's area	Impaired: Speech is sparse and effortful, function words and bound morphemes are often missing
Transcortical motor	Disruption of speech planning and production	Left frontal cortex and white matter sparing Broca's area	Impaired: Speech is sparse and effortful, function words and bound morphemes are often missing
Wernicke	Disruption of representation of word sounds	Posterior half of superior temporal gyrus involving Wernicke's area	Normal but speech has abnormal word sound and structure (paraphasic errors)
Transcortical sensory	Disruption of representations of word sounds	Posterior temporal/parietal cortex and white matter sparing Wernicke's area	Basal ganglia, cerebellum, supplementary motor area
Global	Disruption of all language processing	Left hemisphere involving most of the perisylvian area	Impaired
Conduction	Disconnection of representation of words and the motoric speech process	Lesion of arcuate fasciculus	Mildly impaired with frequent paraphasic errors
Anomic	Disruption of the network allowing the proper sound structure of words	Does not localize well; can involve inferior parietal lobe	Intact with word finding pauses

Source: Gill, D., Damann, K., *Continuum*, 21, 2015, Table 3.1. Used with permission.

More subtle forms of dementia occur when there are multi-focal areas of damage from small vessel disease. This is usually due to hypertension, but there are also inherited forms of ischemic white matter disease such as cerebral autosomal-dominant arteriopathy with subcortical infarcts and leukoencephalopathy (CADASIL) (see Chapter 8).

There are some very rare forms of hereditary disease that may produce early-onset dementia. Examples include disorders of myelin synthesis and maintenance (known as leukodystrophies) and of neuronal mitochondrial function (leading to 'power failure') such as Leigh's disease.

The pathophysiology of primary dementias has taken on a new language and terminology. The newer terminology refers to various protein entities that tend to accumulate in these various conditions, ultimately ending in neuronal cell death and dementia. For Alzheimer's disease, the term *amyloidopathy* is now often used; for dementia with Lewy bodies (DLB), 'synucleinopathy'; for some forms of frontotemporal dementia (FTD), 'tauopathy'. Life is not that simple, however,

in that subdivisions of the more traditional disease states like Alzheimer's disease do not fit into these neat protein-type categories. Table 12.3 shows the different forms of cortical dementias with their principle neuropathological findings.

All neurons have metabolic cycles though which the various structural and functional proteins must pass and ultimately be exported or recycled by the neurons themselves or by adjacent glia. As a general rule, most proteins are exported in soluble forms for transport either into astrocytes or to the circulation for eventual reuse. If the isoform of a given protein produced by a neural cell is in an insoluble form, then the protein is not disposed of in the usual fashion and the proteins accumulate either in the neuronal cytoplasm or in the adjacent environment. This accumulation of insoluble protein often forms the various structures that are seen microscopically. There is significant controversy over the significance of these accumulations of insoluble protein, some of which are known as inclusions: do they represent a pathological abnormality or are they simply

TABLE 12.3: Neuropathological Features of Cortical Dementias

Condition	Location	Protein	Neuronal Features
Alzheimer's disease	Hippocampus	Amyloid and Tau	Amyloid plaques, tangles
Frontal variant of FTD	Frontal	Tau	Pick bodies
Semantic variant of FTD	Temporal		
Primary progressive aphasia			
Lewy body dementia	Diffuse, parieto-occipital	Synuclein	Lewy bodies
Corticobasal degeneration	Superior frontoparietal, basal ganglia	Tau	Coiled tangles
Progressive supranuclear palsy	Midbrain	Tau	Globose tangles
Creutzfeldt–Jakob	Cortex, thalamus	Prion	Spongiform changes

a defensive response? For instance, a Lewy body containing synuclein may be an attempt to protect the cell by sequestering the potentially toxic protein. On the other hand, it may just represent the result of how synuclein clumps together based on the physicochemical environment.

The end point is that these insoluble protein by-products eventually accumulate to the point where neuronal cell death occurs; if large populations of neurons die, dementia occurs.

A lymphatic drainage system, called the glymphatic system, within the brain, which is responsible for clearing approximately 8 grams of waste protein material per day has been discovered. Recent studies have shown that this system is selectively activated in sleep. Therefore, there is a basis on which chronic sleep deprivation might lead to cognitive dysfunction and possibly dementia.

Dementia frequently does not manifest as a single pure entity. It is increasingly appreciated that there is a continuum between the various primary dementias including Alzheimer's disease, vascular dementia due to small vessel disease, vascular dementia due to large vessel disease causing stroke and other less common forms of dementia such as DLB and FTD – see Table 12.3.

Table 12.3 lists the commonest forms of dementia with their localizations and neuropathological features.

12.5 LOCALIZATION

The localization matrix indicates findings suggestive of one localization in the peripheral nerves (sensory loss in the toes), but the majority of the deficits direct us to structures above the brainstem. Given that Armand's predominant deficits are in cognition, the localization possibilities include cerebral cortex, white matter, thalamus, basal ganglia and cerebellum. Within the cortex, there are specific functions based on location; these enable one to localize cognitive domains into frontal, temporal, parietal and occipital areas, with some overlap.

Based on the history and physical examination, the most likely localization in this case would be in the cerebral cortex in those areas serving short-term memory, word finding, executive function and emotional control. Anatomically, these would be located in the hippocampus of the temporal lobe for short-term memory, the superior temporal lobe and anterior parietal lobe for word finding and the frontal lobes for executive and emotional function.

Lesions that affect subcortical structures (such as the deep white matter, thalamus and basal ganglia) tend to cause motor and ideational slowing, elements of which Armand does have.

Lesions in the cerebellum can lead to difficulties with control of the rate and rhythm of motor output but it has been increasingly recognized that the cerebellum has similar modulating functions on cognitive and emotional output.

Table 12.4 indicates the most likely localization for Armand.

12.6 ETIOLOGY

12.6.1 ARMAND'S DISEASE PROCESS

In the ongoing assessment of Armand, it would be important to focus on the possibility that his decreased mental status might be due to a reversible cause.

The medication history may be extremely important, especially in the face of a new-onset rapidly progressive picture. This would include a detailed history of exactly what medications have been started and stopped during the period in which the symptoms have developed.

In terms of etiologic diagnosis, it is important to rule out other reversible causes as outlined below, utilizing simple blood tests and performing appropriate imaging studies to rule out any structural causes. Slow-growing tumors such as meningiomas can masquerade as dementia as can chronic subdural hematomas (see Section 8.5.1, Chapter 8).

Specific questioning is required to exclude the presence of systemic diseases such as hypothyroidism, B12 deficiency or alcoholism. Did Armand have exposure to toxins or heavy metals during his working life or when he was in the army? Did he ever suffer from venereal disease such as syphilis during the war?

He has some features of peripheral sensory loss. Could this be related to his cognitive decline?

12.6.A PAROXYSMAL DISORDERS

The time course of this disorder is not in keeping with a paroxysmal disorder, although a clinical picture like this may co-exist with frequent, uncontrolled seizures. A short-term encephalopathy may occur after a prolonged seizure, but this possibility obviously does not fit with Armand's clinical picture.

12.6.B TRAUMATIC BRAIN INJURY

Head trauma (either open or closed) may cause damage to gray or white matter, leading to dementia. The commonest form is severe closed head trauma causing diffuse axonal injury. This is essentially a shear injury to the large axonal bundles due to the effect of acceleration/deceleration forces on the brain. Other mechanisms of brain injury include hemorrhagic and ischemic damage. Was he exposed to significant or recurrent blast injury during the war?

Recurrent brain trauma related to sports such as hockey, football, boxing or ultimate fighting has been implicated in dementia pugilistica or chronic traumatic encephalopathy.

Recurrent falls leading to chronic subdural hematomas can result in cumulative damage to the underlying cortex and a pattern of cortical dementia. This etiology does not fit with the case under consideration.

TABLE 12.4: Expanded Localization Matrix for Armand

	Neuro Exam		MUS	NMJ	PN	SPC	BRST	WM/TH	OCC	PAR	TEMP	FR	BG	CBL
Attention	N													
Memory	Decr 3 item recall									○	○			
Executive	Decr trails											○		
Language	N													
Visuospatial	N													
	Right	Left												
CRN I	N	N												
CRN II	N	N												
CRN III, IV, VI	N	N												
CN V	N	N												
CRN VII	N	N												
CRN IX, X	N	N												
CRN XI	N	N												
CRN XII	N	N										○		
Tone	N	N				⊘	⊘	○	⊘			○	⊘	⊘
Power	N	↓		⊘	○	⊘	⊘	○	⊘	○		○	⊘	⊘
Reflex	↑	↑		⊘	⊘	⊘	⊘	○	⊘	○		○	⊘	⊘
Involuntary Mvt	N	N												
PP	↓	↓		⊘	○	⊘	⊘	○	⊘	○				
Vib/prop	↓	↓		⊘	○	⊘	⊘	○	⊘	○				
Cortical sensation	N	N												
Coord UL	N	N												
Coord LL	N	N												
Walk	N	N												
Toe/heel/invert/ evert	N	N												
Tandem	N	N												
Romberg	N	N												

Abbreviation: N, normal.

12.6.C VASCULAR DISORDERS

Damage to cortex and white matter secondary to large vessel hemorrhagic or ischemic strokes may result in dementia. As we saw in Chapter 8, strokes are usually associated with risk factors such as advanced age, blood pressure, cholesterol, smoking, diabetes and family history (many of which are present in Armand's case).

Vascular dementia is now recognized to have many different forms. The most common form is the so-called multiple infarct dementia; it results from the accumulation of multiple discrete lobar or subcortical infarctions due to either cerebrovascular disease or recurrent emboli from the heart. The effect of this cumulative damage is dementia due to cumulative loss of neurons from these multiple strokes. The effective treatment of this form of vascular dementia is prevention of the recurrence of stroke by risk factor management.

The second most common form is that of a slowly progressive form of deterioration of the small blood vessels, which supply primarily the white matter but also, to a lesser extent, the gray matter. Uncontrolled systemic hypertension is by far the commonest risk factor for this form. The genetic disorder CADASIL and others also belong in this category.

Diffuse ischemic white matter injury can also occur after cardiac arrest or prolonged hypotension in septicemic shock. Such injury may also occur in the context of cardiac surgery with prolonged periods on cardiopulmonary bypass or with perioperative hypotension.

Armand's slowness of action might suggest a subcortical vascular dementia due to microvascular degenerative changes considering his risk factor of long-standing inadequately treated hypertension. A computed tomography (CT) scan or MRI will aid in determining the extent of this process and its relationship to his dementia.

12.6.D DRUGS AND TOXINS

The commonest toxic agent causing dementia in North America continues to be alcohol or ethanol. Ethanol destroys cortical neurons and Purkinje cells in the cerebellum as well as depletes vitamins such as thiamine and folate; the latter problem can have both acute and chronic effects on cognitive function. It is indeed a difficult situation in the office when you have to tell a person in their late 60s that they have developed dementia and ataxia because of their chronic daily alcohol use. Their response is often: 'Why did somebody not tell me about this earlier?'

The role in the development of dementia of chronic exposure to other non-medical drugs such as marijuana, cocaine and crystal methamphetamine will be determined in time. See Figure 3.4.

Certain medical agents clearly lead to chronic cognitive dysfunction; these include chemotherapeutic agents such as high-dose Cyclophosphamide, Adriamycin and Cisplatin. Radiotherapy to the brain, especially when combined with chemotherapy, is known to cause cognitive decline and its deleterious effects on the pediatric population have been well documented.

Armand's history does not suggest that there are probable risk factors to support this etiology group.

12.6.E INFECTION

Infectious diseases caused by bacterial, virus, protozoan, fungal agents and prions may lead to dementia by virtue of the extent of the damage. These include syphilis, Lyme disease, Herpes simplex encephalitis, arbovirus encephalitides, human immunodeficiency virus (HIV), cytomegalovirus (CMV), Zika and West Nile virus.

There are a few specific infections of white matter that may cause dementia, usually in immunocompromised patients; these include progressive multifocal leukoencephalopathy (due to JC virus, a member of the papova group of viruses), HIV and CMV.

Prion diseases are caused by transmissible insoluble proteins and typically produce dementia; these include Creutzfeldt–Jakob disease and its variant form related to bovine spongiform encephalopathy, fatal familial insomnia (FFI), Kuru and Gerstmann–Straussler–Scheinker disease. Typically, these disorders (except for Gerstmann–Straussler–Scheinker disease) cause a rapidly progressive fatal dementia with myoclonic jerks.

A lumbar puncture would be indicated if there were a high enough suspicion of an infectious disorder. This is especially important not only to rule out a reversible cause such as syphilis but also for prognostic purposes with the presence of 14-3-3 protein marker and S100b in the cerebrospinal fluid. This marker is suggestive but is not specific for neurodegenerative disorders.

The time course of Armand's dementia might suggest a slowly progressive infection or a reactivation of a latent infection such as syphilis.

12.6.F METABOLIC INJURY

Lack of any of the major substrates for brain function such as oxygen, glucose, electrolytes and hormones may cause acute cortical and subcortical dysfunction. These disorders tend to be very acute. Chronic deficiencies of vital substrates can lead to chronic brain deterioration and dementia. Specific metabolic injuries include deficiencies of thyroid hormone and of vitamins B1 and B12 and folic acid.

The loss of sensation to vibration in the toes bilaterally might raise the possibility that there could be a metabolic injury or deficiency such as B12, which could affect both the peripheral nerves and cortical functions. A complete blood count with a smear would help to indicate if there were large hypersegmented polymorphonuclear leukocytes; blood assays for vitamin B12 and folate would show whether Armand has a vitamin deficiency.

12.6.G INFLAMMATORY/AUTOIMMUNE

Systemic inflammatory disorders such as systemic lupus erythematosus, Sjögren's disease and isolated cerebral angiitis may lead to dementia.

The commonest non-infectious inflammatory white matter disease seen in Caucasian populations is multiple sclerosis (MS). The dementia of MS occurs when the disease burden of MS plaques becomes so severe that interneuronal communication becomes severely degraded.

The time course of the history, Armand's age and the lack of focal white matter findings makes MS unlikely. For the vasculitis group, there were no specific findings of systemic involvement.

The recovery phase from any autoimmune encephalitis such as anti-N-methyl-D-aspartate (NMDA) antibody and voltage-gated K+ channel antibody encephalitis is often accompanied by permanent brain injury and severe cognitive deficits, in effect a form of rapidly acquired dementia.

12.6.H NEOPLASTIC AND PARANEOPLASTIC DISORDERS

Malignancies, whether primary CNS or metastatic, usually present with focal defects or seizures but may later produce cognitive decline and eventually dementia. Occasionally, however, dementia is the presenting symptom, by which time there is either a large tumor or multifocal primary or metastatic tumors. The most common primary tumors presenting with acute cognitive decline are malignant astrocytomas.

This disease category also includes the various paraneoplastic disorders that cause autoimmune brain injury.

12.6.I DEGENERATIVE DISORDERS

Specific criteria for the diagnosis of the various forms of dementia have been developed and are based on some common criteria, modified by specifics for the various subtypes; all of these are outlined in the *Diagnostic and Statistical Manual of Mental Disorders, Fifth Edition (DSM-V)*.

There are further refined criteria for each sub-type outlined in the National Institute of Neurological and Communicative Disorders and Stroke and the Alzheimer's Disease and Related Disorders' Association, which are listed in the glossary of resource material on the Web site.

12.6.I.1 Alzheimer's Disease

Alzheimer's disease tends to systematically destroy structures in the forebrain that use acetylcholine as the primary neurotransmitter. These structures include the nucleus basalis of Meynert and the diffuse cholinergic projections from this nucleus. The degenerative process in the neurons leads to the accumulation of several materials within the neurons, extracellular space and the blood vessels of the brain. Neurofibrillary tangles in the neuron cell bodies represent tau protein. Insoluble amyloid is exported into the extracellular space and deposited as amyloid plaques and also as congophilic material in blood vessels (Figure 12.1a). There appears to be an interaction between amyloid precursors and tau protein, which can make the pathological distinctions more difficult to separate.

Cerebral blood vessel walls infiltrated by amyloid are more friable, leading to spontaneous cortical hemorrhages. This disorder is called congophilic angiopathy.

This disorder initially tends to target the temporal and parietal lobes more than the frontal lobes. This leads to the initial clinical picture of memory loss and difficulty with the reception of speech. Eventually, as it progresses, the disease affects the cortex throughout, including the frontal lobes, with relative sparing of the occipital lobes. The individual loses the ability to perform basic activities of daily living. When Alzheimer's disease patients become bedridden and unable to look after their basic needs, they become at high risk for infections, especially pneumonia – the usual cause of death. The time from diagnosis is variable but is usually in the order of five to seven years.

The diagnosis of dementia of all types remains a clinical diagnosis based on a history of cognitive decline with impairment of function, which interferes with independence without any other cause to explain the cognitive decline. The observation of cognitive decline reported either by patients or family then needs to be correlated with documentation of a significant decline by some form of testing of cognitive function such as the Mini Mental Status Exam, Montreal Cognitive Assessment, Ascertain Dementia 8-item Informant Questionnaire (AD8), Mini-COG™ Instructions for Administration & Scoring or formal neuropsychological testing. It is important to use variations of the same testing instrument to provide consistency and avoid a training effect.

The clinical diagnostic criteria detailed in the NCDS-ADRDA guidelines and *DSM-V* combined with neuroimaging techniques such as CT scan, MRI scan and positron

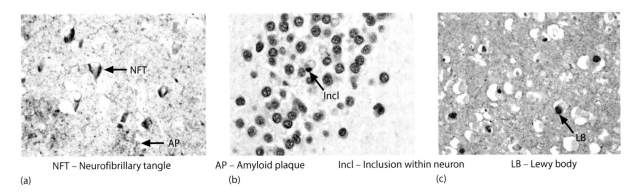

NFT – Neurofibrillary tangle   AP – Amyloid plaque   Incl – Inclusion within neuron   LB – Lewy body
(a)   (b)   (c)

FIGURE 12.1: Pathology of dementias. (a) Alzheimer's dementia. Amyloid plaques and neurofibrillary tangles: tau immunostaining (hippocampus). (b) FTD: FTLD-U: TDP-43 staining (dentate gyrus) inclusion = arrow: notice loss of normal nuclear staining in cell with inclusion. (c) DLB. Cortical Lewy bodies: Synuclein staining (cingulate gyrus). (Courtesy of Dr. J. Woulfe.)

emission tomographic (PET) scans can improve diagnostic certainty. PET scanning of the brain for dementia uses a radionuclide analogue of amyloid called the carbon-11 labelled Pittsburgh compound B to determine the location and degree of disposition of amyloid protein in a given patient's brain. PET scanning is not readily available in daily clinical practice, and the sensitivity and specificity of the test has not been found to be helpful in early cases of cognitive impairment to predict the future severity of the disease.

The introduction of acetylcholinesterase enzyme inhibitors specific to the brain and NMDA receptor antagonists have led to some success in stabilizing cognitive performance, delaying the need for institutional care. These agents may slightly alter the pathophysiological progression of the disease and have been shown to delay the need for chronic care. There has been extensive research to discover specific agents that actually slow the progression of Alzheimer's disease; no such agent has yet been found despite multiple clinical trials and some failures, which have led to fatalities.

There is increasing emphasis on non-pharmacological treatment strategies for patients and their caregivers to allow for proper understanding of the disorder and the anticipated behavioural and care issues. This allows patients, families and care providers a framework for planning future care. For example, education and instruction in the use of redirection and other behavioural therapies can be beneficial and avoid use of sedating medications for behavioural control.

There should be a review of the patient's legal affairs and end-of-life issues early in the course of the disorder while the patient still has legal capacity to make such decisions. The determination of legal capacity to carry out personal and business legal affairs varies greatly between jurisdictions and caregivers should be aware of the process in their given jurisdiction. An abnormal mental status examination does not constitute a legal examination for decision-making capacity.

Armand's clinical picture is consistent with Alzheimer's disease with a component of vascular dementia.

12.6.1.2 FTD

This form of dementia, previously called Pick's disease, is now known to have at least three specific subtypes, each having different clinical, radiological and pathological findings. These are the frontal variant of FTD, semantic dementia and the non-fluent aphasia subtype.

The frontal variant of FTD is characterized by behavioural changes that include disinhibition and/or apathy, dietary changes, social withdrawal and depression. This form has pathological changes mainly in the left premotor and right dorsolateral frontal cortex. Neuroimaging shows mainly bilateral frontal lobe atrophy. The pathology for this type is shown in Figure 12.1b.

The semantic or temporal variant manifests with compulsions, mental rigidity, emotional withdrawal,

breakdown of semantic knowledge, long-term memory loss and agnosia for faces and objects. The imaging for this type shows asymmetric atrophy in the temporal lobes.

The third type, non-fluent aphasia, presents with significant deficits in verbal fluency, grammar and word repetition, and with social withdrawal and depression; it occurs later than the other two subtypes. The imaging for this type shows asymmetric atrophy in the frontal lobes.

As yet, there are no specific medications for FTD. Often, the agents for Alzheimer's disease are tried with the hope that there might be some crossover effect.

Armand's history did indicate some emotional lability, but since most of his deficits were in the area of memory and comprehension, this makes FTD less likely.

12.6.1.3 DLB

This form of dementia is characterized by the accumulation of Lewy bodies in cortical and limbic neurons as well as in their neurites (axons and dendrites). The protein associated with this disorder is alpha-synuclein, which accumulates in the neurons mentioned previously as well as in specific populations of monoaminergic neurons in the brainstem (Figure 12.1c).

The core features that distinguish DLB from the other dementias include fluctuation of cognitive function, recurrent visual hallucinations and parkinsonism.

Other clinical features include the REM behaviour disorder (RBD). In this disorder, symptoms occur while the patients are in the rapid eye movement (REM) stage of sleep. They appear to be acting out their dreams. During REM sleep, normal individuals are effectively paralyzed except for the muscles of respiration. There is functional paralysis of the spinal cord produced by a descending pathway from the subcerulean nucleus in the pons to the spinal cord lower motor neurons. This pathway is thought to be mainly glycinergic.

In RBD, the tract originating in the subcerulean nucleus becomes dysfunctional because of alpha-synuclein deposition that destroys its functionality. This leads to a loss of the normal state of paralysis that occurs in REM sleep, leading to the behaviour of appearing to act out dreams.

Fortunately, there is a very effective medical treatment for this. A small dose of clonazepam, a benzodiazepine with some glycinergic properties, is very effective in stopping the distressing symptoms of RBD (especially for the bed partner). This medication has to be used with caution as it may cause difficulties with balance and cognition as features of Parkinson's disease or multi-system atrophy (MSA) develop.

Patients with DLB tend to be very sensitive to neuroleptics and have repeated falls, transient disturbances of consciousness and autonomic dysfunction. There is also occipital lobe hypometabolism seen on PET scan and focal posterior slowing on EEG.

There is no specific treatment for DLB except supportive care. Given the Parkinsonian features, special attention is required in order to ensure the safety of the patient from falls. The parkinsonian features are treated in a manner similar to that used in someone with idiopathic Parkinson's disease; it is important to avoid overtreatment.

Overall, Armand's clinical history and presentation are not consistent with a diagnosis of DLB.

12.6.1.4 MSA

This disorder represents an extreme form of alphasynucleinopathy in which the cortex, basal ganglia, brainstem, cerebellum and autonomic nervous system are all progressively affected. Clinically, the deficits start in one system and then spread to the others at varying rates. The treatment of the depression, hallucinations and delusional behavioural disturbances can be very difficult in that these patients are very sensitive to classical antidepressants and neuroleptics. Agents such as selective serotonin reupdate inhibitors, risperidone, clozapine and olanzapine are useful but must be started in small doses and titrated slowly while watching for side effects.

Armand does not have any of the cardinal features of MSA.

12.6.1.5 Subcortical Dementia

The term *subcortical dementia* is used to represent disorders of progressive cognitive decline secondary to primary disease of white matter. This is really a group of disorders that can be categorized using the etiology matrix.

White matter damage may be caused acutely by drugs or toxins causing white matter damage (e.g. organic solvents) and subacutely or chronically by drugs like cyclosporine and Amphotericin-B.

Careful management of blood pressure and other vascular risk factors are the principle basis of treatment to prevent progression of vascular subcortical dementia.

12.6.1.6 Sleep and Dementia

Recent studies suggest that the brain has specific pathways mediated by astrocytes and specialized vascular channels for clearing breakdown products such as beta-amyloid in the brain. These mechanisms appear to become more active during sleep. It is therefore suggested that chronic sleep deprivation, which interferes with this hypothesized mechanism of clearing toxic breakdown products, may be a risk factor for the development of subsequent dementia.

In summary, the clinical manifestations of a given dementia depend on which area of the brain is involved and its given function and the type of protein that is accumulating. Collateral factors such as the presence of large or small vessel ischemic disease and the presence of other medical or metabolic problems common in the elderly all tend to amplify the effect of the neuronal cell death on the behavioural performance of the individual.

Treatment modalities for patients should include appropriate medical and non-pharmacological modalities as well as managing vascular risk factors. Early education and involvement of family members and other caregivers are essential to provide support for behavioural problems and avoid caregiver burnout.

Management of sleep disorders such as obstructive sleep apnea and RBD are also part of the overall strategy to management dementia.

12.6.J GENETIC DISORDERS

Pure genetic forms of dementia are rare; the best known is Huntington's disease (see Chapter 15). In this autosomal dominant disorder, there is degeneration of medium-sized cells in the caudate nucleus and more diffusely in the cortex. It presents in an individual in the mid-20s with involuntary movements called chorea, characterized by small amplitude sudden jerking of distal muscles. At first, it can be very subtle and the individual tries to conceal the movement by incorporating the involuntary movement into a voluntary action. Later, there may be writhing movements of the proximal upper and lower limbs, referred to as athetosis. The dementia can be difficult to quantify at first, but usually by age 35, there is a measureable decrement in the mini-mental status or MOCA exam; by this age, significant deficits can be found on neuropsychological testing.

Huntington's disease patients tend to become progressively disabled by their dementia, more so than by the movement disorder, and are usually institutionalized by their forties or fifties. Since the advent of reliable genetic testing, it is now possible to determine if a given fetus is affected.

Other genetic forms of early-onset dementia include abnormalities of lysosomal metabolism including Tay–Sachs disease and the so-called ceroid lipofuscinoses. Mitochondrial genetic disorders cause dementia due to cumulative damage of either gray or white matter; examples include Leigh's disease, mitochondrial encephopathy with ragged red fibers or mitochondrial encephalopathy, lactic acidosis and stroke (see Table 3.13).

Genetic disorders of white matter also include the various leukodystrophies (such as adrenoleukodystrophy and metachromatic leukodystophy). These tend to declare themselves during childhood, but a few forms can have adult onset and present as a dementia.

Armand does not have a family history of premature dementia or clinical features of any of the most common genetic disorders causing dementia.

12.7 CASE SUMMARY

In summary, Armand is an 82-year-old man with a one- to two-year history of cognitive decline with features of memory loss for recent events and difficulty with word finding, executive function and emotional control. There is a long history of treated hypertension, although his blood pressure is quite high in the office on his first visit.

The clinical exam confirms impairment in the cognitive domains of memory and executive function and also suggests some slowness in his motor coordination and gait. There were no discrete focal signs suggestive of previous stroke. He has mild sensory loss in his feet.

12.7.A CASE EVOLUTION

A metabolic workup showed no obvious reversible cause of dementia such as hypothyroidism, vitamin B12 deficiency or syphilis.

Armand's MRI (Figure 12.2) demonstrated diffuse cortical atrophy, more in the temporal and parietal lobes. There were, however, many small areas of increased signal in the white matter bilaterally consistent with small vessel microangiopathic disease. There was no sign of tumor or subdural hematoma.

He also had some findings of decreased upwards gaze and mild sensory loss in his extremities. These latter two findings were probably age related and, by lab investigations and imaging, excluded as being related to the major clinical problem – his memory deficits.

The conclusion was that Armand was probably suffering from a mixed form of dementia: primarily Alzheimer's disease but with a component of vascular dementia resulting from the progressive involvement of multiple small cerebral vessels.

He was referred to a specialized Memory Disorder Clinic in his community; this provided multidisciplinary diagnostic and treatment support. He had a formal neuropsychological assessment as well as an evaluation by social services and by the local geriatric outreach program. He was started on a cholinesterase inhibitor medication, which seemed to help his memory for a couple of years, after which it seemed to have no benefit. He was told not to drive and told that the local DMV licensing authority would be contacted (as required by law) to assess his ability to continue driving. After reviewing the specialist's report, the DMV subsequently suspended his license on medical grounds.

Armand remained at home with support of family and social services for five years after his diagnosis, during which time there was progressive deterioration in his activities of daily living and the appearance of incontinence. He eventually became bedridden and was then transferred to a long-term care facility, where he died of pneumonia 18 months later.

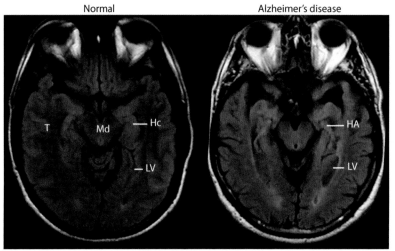

T – Temporal lobe	Hc – Hippocampus (normal) HA – Hippocampal atrophy (bilateral)
Md – Midbrain	LV – Lateral ventricle (inferior horn)

FIGURE 12.2: MRI scan of a brain with Alzheimer's disease: an axial view through the temporal lobes of a normal brain compared to a brain with Alzheimer's dementia; note the atrophy of the cerebral cortex, with enlargement of the ventricles and selective severe atrophy of the hippocampus on both sides in the Alzheimer brain.

12.8 E-CASES

12.8.1 CASE 12E-1: LOUISE, AGE 61

Background: Louise is brought to the emergency department one morning by her worried son. His mother, who looks older than her age, is now complaining of one-sided weakness and numbness. He has noted that she has been 'slowing down' recently and is aware of a family history of 'mental problems'.

Chief complaint: Progressive numbness and weakness of the left side for one day; increasingly severe memory problems over the past several years

History:

- Numbness and weakness of the left arm and leg (arm > leg) developed gradually over the past 24 hours.
- Minor forgetfulness began several years ago; she was thought to be depressed but was not treated.
- More recently, she began having difficulty in carrying out household chores, in grocery shopping and in cooking ('too much trouble').
- More socially withdrawn; she has stopped playing bridge with her friends as she is no longer able to follow the game.

- Her family has noticed that she moves around more slowly, especially while walking her dog.
- Her father developed memory problems and slowness of movement in his 60s and died at age 70; her mother is still alive in her 90s.

Examination:

- Alert and cooperative but tearful and uncertain in her answers; she often looks at her husband for him to help with historical details.
- Mini-Mental Test result at the bedside is 21/30 – thought to be unreliable because of her emotional state.
- Mild dysarthria; mild left lower facial weakness.
- Distal weakness of the left arm and leg (Figure 12.3a).
- Reflexes: see Figure 12.3b.
- Decreased pinprick and vibration sensation over the left side of the body, more obvious in the left arm (Figure 12.3c).
- Coordination testing results are slow bilaterally, worse on the left.
- Tends to fall to the left while trying to walk.

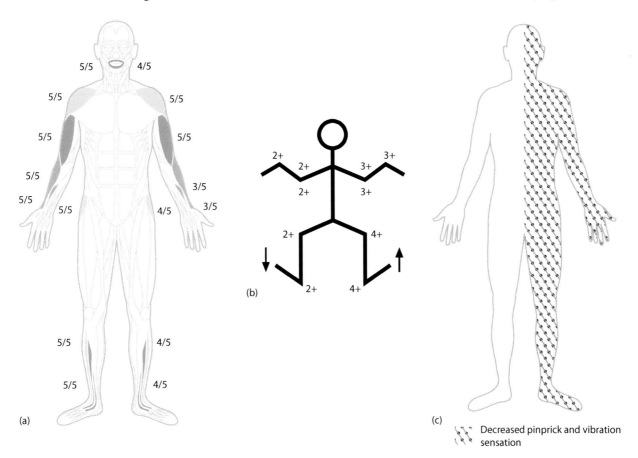

(a)

(b)

(c)

Decreased pinprick and vibration sensation

FIGURE 12.3: (a) Motor, (b) reflex and (c) sensory maps for case 12e-1 (Louise).

12.8.2 CASE 12E-2: JACK, AGE 86

Background: Jack started having difficulties about six years ago with 'spells' and some falls and recently lost his driver's license. His wife of 61 years has noted that she has heard and seen him (they now sleep in separate bedrooms) seemingly 'acting out his dreams'.

Chief complaint: Increasing problems with intermittent impairment in balance and spatial orientation since age 80

History:

- He stopped driving at age 80 when his family sold his car to prevent him from driving; he had had two recent at-fault motor vehicle accidents.
- Intermittent episodes of impaired balance and orientation in space; e.g. while taking the dog outside for a walk, he mistook the position of the stairwell and fell down the stairs, hurting himself badly; at other times, his balance has been normal.
- Frequent abrupt-onset naps during the day; at times, during the night, he is found to be punching and kicking, seemingly acting out his dreams.
- No significant memory problems – could read the newspaper or watch the television news and discuss the contents appropriately.
- No reduction in visual acuity, but he has increasing difficulty in finding the words on a page; lighted magnifying loops seem to help.
- Non-insulin-dependent diabetes mellitus for 20 years, fairly well controlled; mild coronary artery disease.
- He participated in World War 2 and in the Korean war; injured his left knee in Korea when struck by a wooden beam following an enemy artillery explosion; he had chemical and biological warfare training and had been exposed to defoliants.
- No family history of a similar disorder.

Examination:

- Friendly and attentive; very distractible, using humour to disguise this fact

- Mini-Mental test result is 24/30 – points lost, among others, for delayed recall, serial sevens, three-step commands and for intersecting pentagons
- Normal cranial nerve examination; visual acuity is 20/30 (corrected) in both eyes; normal extraocular movements
- General loss of muscle mass but good strength; normal, symmetric tendon reflexes
- Decreased vibration and pinprick sensation below both knees (Figure 12.4)
- Slowing of rapid movements bilaterally; gait is slow and stooped; difficulty making corrections to control his posture; unable to tandem walk

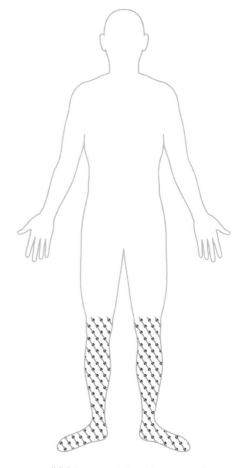

✎ Decreased pinprick and vibration sensation

FIGURE 12.4: Sensory examination map for case 12-e (Jack).

12.8.3 CASE 12E-3: SHARON, AGE 82

Background: Sharon was a well-known and well-loved veterinarian who retired at age 65. Over a period of time, she lost her ability to speak and developed other neurological problems. She is still living at home but needs additional support by family and a hired health care live-in aid.

Chief complaint: Progressive loss of expressive language over the past seven years

History:

- Well until age 75, when she began to stammer while speaking.

- By age 80, she had begun to lose the ability to articulate fluent speech, speaking only in single syllables and indicating her needs with gestures.
- At the present time, she is completely unable to speak, communicating with humming sounds and with gestures.
- Her neurological examination at age 75 had been normal (except for speech), after which her posture became increasingly stooped.
- By age 80, she also developed multifocal myoclonic jerks affecting mostly the face and upper limbs.
- She is cared for at home by her family with the help of a live-in health care aide.
- After her speech disturbance became more pronounced, she was tried on donepezil and memantine, with no appreciable improvement.

Examination:

- A thin woman who is able to walk to the examination room with some assistance.

- Accurate Mini-Mental examination is difficult as she is unable to generate any meaningful speech; she obeys simple commands.
- Cranial nerve examination is normal.
- Diffuse muscle wasting but normal power for mass; diffusely increased tendon reflexes.
- Rapid, shock-like jerks in muscles of the face and shoulder-girdle regions.
- She is able to perceive pinprick distally in all four extremities.
- Coordination testing is not feasible; she has a flexed posture when walking and tends to lean to the right.

(Note that the Web site Localization discussion focuses on the temporal lobe dementia variant with fluent aphasia while this patient had initially a non-fluent aphasia.)

12.8.4 CASE 12E-4: GEORGE, AGED 75

Background: George, a mostly retired professor of anatomy, has woken to find himself suddenly totally blind. Following the guidelines for a 'brain attack', his wife calls 911 and he arrives by ambulance.

Chief complaint: Overnight total loss of vision

History:

- Previous medical problems include hypertension, type 2 diabetes mellitus and mild coronary artery disease; plays golf and tennis.
- On awakening today, he complained to his wife that he could not see; this was confirmed by his stumbling into the bedroom furniture on getting out of bed.
- He has worked in a neuroanatomy laboratory for 30 years doing comparative anatomy of humans, sheep and cows.

Examination:

- Anxious, repeatedly inquiring as to why he is unable to see.

- Blood pressure = 155/90; pulse = 84/minute, regular.
- He answers questions appropriately and follows instructions; speech is fluent.
- Visual acuity is reduced to light perception in both eyes, with some appreciation of shapes; the remainder of the cranial nerve examination, including fundoscopy, is normal.
- Normal muscle power and tone; normal tendon reflexes; limb movements are slow and tremulous.
- After admission, his sight returns to normal over several days, after which a proper Mini-Mental assessment reveals the presence of a moderate dementia.
- Over the next six months, he has several more episodes of transient visual loss and his dementia becomes much worse with apathy, paucity of movement and the appearance of myoclonic jerks.

12.8.5 CASE 12E-5: LUC, AGE 10

Background: Luc is brought in by his father with a history of loss of hearing and some behavioural problems for about the last three months; his family has an Acadian background.

Chief complaint: Three-month history of changed behaviour, school failure and possible hearing loss

History:

- Previously well; three months ago, he began having difficulty paying attention in class, ignoring what the teacher was saying and disrupting classroom activities.
- Two months ago, his teacher phoned the parents to suggest they have his hearing checked as he

often seemed not to understand what she said, even when she was standing directly in front of him; parents had observed the same problem.

- Family physician performed some hearing screening tests – no abnormality found; audiology assessment booked but not yet done; parents find that they can get him to follow some instructions if they use simple gestures.
- Recently, his teacher suggested he be withdrawn from the class as he had learned nothing in the last three months; at home and school, he is restless and anxious, subject to outbursts of anger and screaming.
- Two older brothers and a sister, all in good health; mother has a mild, slowly progressive paraparesis, whose cause is thus far undetermined.

Examination:

- General physical examination normal except that, while parents show a typical mid-winter pallor, Luc is unexpectedly tanned.
- Restless, periodically agitated and bursting into tears; able to speak in short coherent sentences and to play a hand-held computer game.
- Understands no spoken questions or verbal instructions; cooperates for the neurological examination by imitating actions shown by

the examiner; reads aloud at age level; solves simple math problems presented in written format; writes his name and phone number upon written request.

- Cranial nerve examination normal; with his eyes closed, he hears soft sounds in both ears (coins rubbed together, paper crinkled, soft voice); when asked in writing ('what is that sound?'), he does not know.
- Mild increase in tone in hamstrings and plantar flexor muscles.
- Reflexes: Figure 12.5.
- Slightly unsteady gait; unable to stand on one leg.

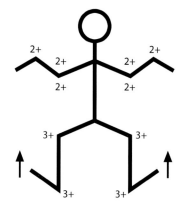

FIGURE 12.5: Reflex map for case 12e-5 (Luc).

12.9 SUMMARY OF KEY NEUROANATOMICAL INFORMATION

Higher mental functions are organized in the cerebral cortex.

- The cerebral cortex operates as an integrated functional collective with interconnecting white matter (axonal) association bundles.
- The hemispheres intercommunicate via the corpus callosum (and other commissures).
- Specialized areas include the following:
  - Lateral prefrontal cortex → executive functions
  - Orbitofrontal and medial prefrontal cortex → limbic connections
  - Frontal cortex → motor control
  - Parietal cortex → visuospatial (multisensory)
  - Occipital cortex and inferior temporal regions → visual
  - Lateral temporal cortex → auditory
- The cerebral cortex is connected to the basal ganglia, cerebellum, brainstem and spinal cord via projection fibers (ascending and descending).

- Limbic functions – including memory – are located in the medial temporal areas, including the hippocampal formation.
- The dominant hemisphere (usually always the left) is defined by language function; many areas of the cortex are involved and two are well defined:
  - Broca's area for expressive language (lower frontal cortex).
  - Wernicke's area (temporal lobe, upper aspect) for phoneme retrieval used in language production.
  - These two areas are connected by specific fiber bundles.

12.10 CHAPTER-RELATED QUESTIONS

1. List the commonest forms of dementia and their known causes.
2. What are the clinical differences between an organic and pseudo-dementia?
3. What are the immediate concerns when faced with a patient with moderately severe dementia?

4. List the commonest forms of familial dementia.

5. List the different dementias causes by prions and their known causes.

REFERENCE

Diagnostic and Statistical Manual of Mental Disorders, 5th ed. Washington, DC: American Psychiatric Publishing, 2013.

SUGGESTED READING

Bogaerts, V., Engelborghs, S., Kumar-Singh, S., Goossens, D., Pickut, B., van der Zee, J., Sleegers, K., Peeters, K., Martin, J.J., Del-Favero, J., Gasser, T., Dickson, D.W., Wszolek, Z.K., De Deyn, P.P., Theuns, J., Van Broeckhoven, C. A novel locus for dementia with Lewy bodies: A clinically and genetically heterogeneous disorder. *Brain* 2007. 130(Pt 9): 2277–2291.

Cerami, C., Della Rosa, P.A., Magnani, G., Santongelo, R., Marcone, A., Cappa, S.F., Perani D. Brain metabolic maps in mild cognitive predict heterogeneity of progression to dementia. *Neuroimage Clin* 2014. 7: 187–194.

Chabriat, H., Vahedi, K., Iba-Zizen, M.T., Joutel, A., Nibbio, A., Nagy, T.G., Krebs, M.O., Julien, J., Dubois, B., Ducrocq, X. et al. Clinical spectrum of CADASIL: A study of 7 families. Cerebral autosomal dominant arteriopathy with subcortical infarcts and leukoencephalopathy. *Lancet* 1995. 346: 934–939.

Emre, M., Aarsland, D., Brown, R., Burn, D.J., Duyckaerts, C., Mizuno, Y., Broe, G.A., Cummings, J., Dickson, D.W., Gauthier, S., Goldman, J., Goetz, C., Korczyn, A., Lees, A., Levy, R., Litvan, I., McKeith, I., Olanow, W., Poewe, W., Quinn, N., Sampaio, C., Tolosa, E., Dubois, B. Clinical diagnostic criteria for dementia associated with Parkinson's disease. *Mov Disord* 2007. 22(12): 1689–1707.

Flanagan, E.P., McKeon, A., Lennon, V.A., Boeve, B.F., Trenerry, M.R., Tan, K.M., Drubach, D.A., Josephs, K.A., Britton, J.W., Mandrekar, J.N., Lowe, V., Parisi, J.E., Pittock, S.J. Autoimmune dementia: Clinical course and predictors of immunotherapy response. *Mayo Clin Proc* 2010. 85(10): 881–897.

Hodges, J.R., Davies, R.R., Xuereb, J.H., Casey, B., Bak, T.H., Kril, J.J., Halliday, G.M. Clinicopathological correlates in frontotemporal dementia. *Ann Neurol* 2004. 56(3): 349–354.

Jessen, N.D., Munk, A.S., Lungaard, I., Nedegaard, M. The glymphatic system: A beginner's guide. *Neurochem Res* 2015. 40(12): 2583–2599.

Knopman, D.S., DeKosky, S.T., Cummings, J.L., Chui, H., Corey-Bloom, J., Relkin, N. Small, G.W., Miller, B., Stevens, J.C. Practice parameter: Diagnosis of dementia (an evidence-based review). Report of the Quality Standards Subcommittee of the American Academy of Neurology. *Neurology* 2001. 56(9): 1143–1153.

McKhann G.M., Knopman, D.S., Chertkow, H., Hyman, B.T., Jack Jr., C.R., Kawas, C.H., Klunk, W.E., Koroshetz, W.J., Manly, J.J., Mayeux, R., Mohs, R.C., Morris, J.C., Rossor, M.N., Scheltens, P., Carrillo, M.C., Thies, B., Weintraub, S., Phelps, C.H. The diagnosis of dementia due to Alzheimer's disease: Recommendations from the National Institute on Aging Alzheimer' Association workgroups on diagnostic guidelines for Alzheimer's disease. *Alzheimer's Dement* 2011. 7(3): 263–269.

Mendez, M.F., Shapira, J.S., Woods, R.J., Licht, E.A., Saul, R.E. Psychotic symptoms in frontotemporal dementia: Prevalence and review. *Dement Geriatr Cogn Disord* 2008. 25(3): 206–211.

Neary, D., Snowden, J.S., Gustafson, L., Passant, U., Stuss, D., Black, S., Freedman, M., Kertesz, A., Robert, P.H., Albert, M., Boone, K., Miller, B.L., Cummings, J., Benson, D.F. Frontotemporal lobar degeneration: A consensus on clinical diagnostic criteria. *Neurology* 1998. 51(6): 1546–1554.

O'Donnell, J., Ding, F., Nedegaard, M. Distinct functional states of astrocytes during sleep and wakefulness: Is norepinephrine the master regulator? *Curr Sleep Med Rep* 2015. 1(1): 1–8.

Parikh, S., Bernard, G., Leventer, R.J., van der Knaap, M.S., van Hove, J., Pizzino, A., McNeill, N.H., Helman, G., Simons, C., Schmidt, J.L., Rizzo, W.B., Patterson, M.C., Taft, R.J., Vanderver, A.; GLIA Consortium. A clinical approach to the diagnosis of patients with leukodystrophies and genetic leukoencephalopathies. *Mol Genet Metab* 2015. 114(4): 501–515.

Rascovsky, K., Hodges, J.R., Knopman, D., Mendez, M.F., Kramer, J.H., Neuhaus, J., van Swieten, J.C., Seelaar, H., Dopper, E.G., Onyike, C.U., Hillis, A.E., Josephs, K.A., Boeve, B.F., Kertesz, A., Seeley, W.W., Rankin, K.P., Johnson, J.K., Gorno-Tempini, M.L., Rosen, H., Prioleau-Latham, C.E., Lee, A., Kipps, C.M., Lillo, P., Piguet, O., Rohrer, J.D., Rossor, M.N., Warren, J.D., Fox, N.C., Galasko, D., Salmon, D.P., Black, S.E., Mesulam, M., Weintraub, S., Dickerson, B.C., Diehl-Schmid, J., Pasquier, F., Deramecourt, V., Lebert F., Pijnenburg, Y., Chow, T.W., Manes, F., Grafman, J., Cappa, S.F., Freedman, M., Grossman, M., Miller, B.L. Sensitivity of revised diagnostic criteria for the behavioural variant of frontotemporal dementia. *Brain* 2011. 134(9): 2456–2477.

Ropper, A.H., Brown, R.H., Eds. Cerebrovascular diseases, in *Adams and Victor's Principles of Neurology*. New York: McGraw-Hill, 2005.

Sachdev, P., Kalaria, R., O'Brien, J., Skoog, I., Alladi, S., Black, S.E., Blacker, D., Blazer, D.G., Chen, C., Chui, H., Ganguli, M., Jellinger, K., Jeste, D.V., Pasquier, F., Paulsen, J., Prins, N., Rockwood, K., Roman, G., Scheltens, P.; International Society for Vascular Behavioral and Cognitive Disorders. Diagnostic criteria for vascular cognitive disorders: A VASCOG statement. *Alzheimer Dis Assoc Disord* 2014. 28(3): 206–218.

Seeley, W.W., Carlin, D.A., Allman, J.M., Macedo, M.N., Bush, C., Miller, B.L., Dearmond, S.J. Early frontotemporal dementia targets neurons unique to apes and humans. *Ann Neurol* 2006. 60(6): 660–667.

Simon, M.J., Iliff, J.J. Regulation of cerebrospinal fluid (CSF) flow in neurodegenerative, neurovascular and neuroinflammatory disease. *Biochim Biophys Acta* 2016. 1862(3): 442–451.

Zaccai, J., McCracken, C., Brayne, C. A systematic review of prevalence and incidence studies of dementia with Lewy bodies. *Age Ageing* 2005. 34(6): 561–566.

Mickey

Objectives

- Learn the principal anatomical components of the central nervous system involved in the maintenance of attention, impulse control, learning and mood regulation
- Review the neuromodulatory transmitter systems implicated in memory, mood and behaviour
- Develop the ability to distinguish the various causes of long-standing brain dysfunction presenting with learning and behavioural disturbances
- Learn the main behavioural criteria for attention-deficit/hyperactivity disorder, Tourette's disorder, specific learning disorder and obsessive-compulsive disorder
- Review the pharmacological management of the various symptoms associated with Tourette's disorder

13.1 MICKEY

Michael ('Mickey') is 21 and still lives with his parents, who are at their wit's end over what to do about their son's behaviour. His family physician recently retired and his parents have arranged an appointment in your office. He is accompanied for this assessment by his mother.

Concerns about Mickey's behaviour have been hanging over the family like storm clouds for years. Although he was born at term without difficulty and appeared normal as a toddler, he began to show signs of hyperactive behaviour by the time he was 4. At that time, his parents found him to be fidgety and restless, always on the go. He never seemed to like to play quietly with his toys but was invariably outside in all weathers racing around on his scooter or tricycle from dawn to dusk. When he started kindergarten, his teacher found him difficult to deal with; he had trouble staying in his seat and frequently disrupted the rest of the class with vocal outbursts.

During the kindergarten years, Mickey's restless behaviour was managed, barely, with the help of teacher's aides to help keep him on task. Once he got to grade 1, however, his school progress began to suffer because he simply could not concentrate on a specific assignment for more than five minutes. Typically, he would get distracted by extraneous events such as his friend Piggy Martin scratching his backside, air bubbling in the classroom fish tank and city buses passing by the classroom window every few minutes.

Concerns about Mickey's inattentiveness and poor school performance prompted a psychological assessment at the end of second grade. The school board psychologist found that Mickey's overall intellectual abilities were in the high average range but that there was a marked discrepancy in subtest scores: verbal subtest scores were in the superior range, while nonverbal scores were in the low average to borderline range. Furthermore, the Connors Rating Scale demonstrated findings consistent with the diagnosis of attention deficit hyperactivity disorder (or, using its commonly used acronym, ADHD).

At the psychologist's suggestion, Mickey was started on the stimulant medication methylphenidate, with doses given on school days only, at breakfast and lunch time. The use of this medication was associated with a significant improvement in behaviour: Mickey became quite calm, at least during school hours; he ceased disrupting the class and was able to complete assignments successfully without being distracted. The baseline level of activity and distractibility returned in the evenings and, of course, was present on weekends. By the middle of grade 3, thanks to the methylphenidate, he had caught up to his classmates in all subjects except mathematics. While he had no difficulty with abstract maths concepts, he found written maths problems a daunting prospect.

At age 8, Mickey developed a new set of symptoms in the form of nervous tics. The problem appeared gradually over several months and took the form of repetitive blinking, nose wrinkling, mouth opening and brusque turning movements of the head to one side. At the same time, Mickey began to produce a variety of noises, including grunting, throat-clearing and repetitive sniffing. Eventually, all of these stereotyped movements and noises occurred on a daily basis, waxing and waning in frequency, in varying combinations. They occurred more frequently in the evenings, when he was tired and when he was nervous or stressed. When his exasperated parents told him to 'stop making those silly noises', he would comply, but never for more than five minutes, following which they would return. At school, some of his classmates began making fun of him, and he was the object of occasional bullying by older students.

When the tics became a major issue for the parents, the family physician wondered whether the methylphenidate might be the cause of the problem and suggested stopping the medication. Unfortunately, this did not help matters: the tics continued unabated and there was a catastrophic deterioration in school performance accompanied by marked restlessness, verbal aggression towards teachers and other students and frequent fighting in the schoolyard. Within two months, the experiment was terminated and the stimulant medication restarted. School performance returned to the previous level, but the arguments and the fighting continued, leading to visits to the principal's office and occasional day- to week-long expulsions from school.

Pressured by the parents to 'do something!' about the tics, the family physician tried Mickey on small doses of haloperidol. This produced a major reduction in tic frequency but made Mickey sleepy and lethargic, 'like a zombie'. Again, his school work suffered, so the medication was abandoned.

After the age of 12, the frequency of the movement and voice tics began to decline spontaneously. In the classroom, they were rarely seen as Mickey had learned to control them, more or less, until he got home, at which point there was a veritable explosion of twitching, blinking and noise for an hour or so. As the tics decreased, they appeared to be replaced by a number of more complex stereotypic or compulsive behaviours. These included constant twiddling of pencils and pens, thrumming his fingers on his desk, compulsive manipulation of knobs on electronic equipment and cooking appliances and shooting imaginary baskets above each door he passed through (Mickey had become very accomplished at basketball and was a valued member of the school team, a fact that helped to offset the damage to his self-esteem resulting from his classroom performance).

For a while, in his early teens, Mickey's image at school improved considerably. He had learned, for the most part, to avoid confrontations and to stay out of fights. He had a very bubbly personality, a quick wit and an ability to make friends easily (although keeping friends was more difficult). He continued to keep up academically by dint of hard work, and with the help of a mathematics tutor.

At 17, however, the behavioural problems grew worse again, particularly at home. As often happens with teenagers, Mickey grew increasingly exasperated with his parents who, he felt, were trying to control him too much. They were always taking him to task in front of his friends and were obliging him to spend long hours doing homework assignments when he wanted to be shooting baskets

at the Community Centre. Mickey began having frequent shouting matches with his parents, particularly his mother (his father was often absent on business). These arguments gradually transformed into sustained outbursts of violent behaviour during which Mickey would scream and swear, kicking holes in the wall and smashing furniture. Although the violence was never directed towards his mother, she became increasingly afraid of him. After the explosions of anger were over, Mickey was always very contrite and filled with self-loathing.

At the same time, the parents have become concerned about increasing obsessive thinking and behaviours, which reached the point of becoming disabling. Mickey is constantly counting and checking things, such as his video game collection, his coloured pencils and the ranking of his favourite NBA teams in the newspaper. At bedtime, he has an invariable ritual consisting of checking the locks on the front and back doors, then checking under his bed and in his clothes closet and finally checking to make sure the bedroom window blinds are completely closed. When doing assigned homework, he will often read the same sentence over and over, claiming not to understand what it means, and loudly proclaiming that he is 'stupid'. Attempts by the parents to interfere with his rituals are one cause of his rage attacks.

At age 18, Mickey dropped out of his final year of high school, stating that he hated it there and always had. On occasion, he has worked for a few weeks at a time (in grocery stores or gas stations) but has invariably been fired for insubordination. He has also tried taking courses in adult education programs but always drops out after attending two or three classes. Recently, he has been refusing to look for work, stating that he is 'too stupid' and 'can't do anything useful'. Most of his days are spent alone in his room playing videogames, sometimes until the wee hours of the morning. He no longer goes out to play basketball and socializes with his friends less and less.

Before closing her practice, the previous family physician had tried Mickey on two different medications in an attempt to deal with aspects of his behaviour. To reduce the violent rage outbursts, she tried him on clonidine; this produced a modest reduction in rage attacks but made him so sleepy that he eventually refused to take it. The obsessive behaviours were approached with a trial of the selective serotonin reuptake inhibitor (SSRI) paroxetine. Unfortunately, this made him very agitated, even in the presence of methylphenidate, and was withdrawn after a month. (The roles of these various medications – and others – will be reviewed later in this chapter.)

After gleaning all of this historical information from his parents, you inquire about other medical problems. It turns out that Mickey has really been in good health all his life. He has never required hospitalization for any reason and has never had any convulsions, febrile or unprovoked. His only visits to hospital emergency departments have been for scalp lacerations and for a broken finger incurred while playing basketball. He has knocked heads a few times while driving for the hoop but has never had a serious head injury.

The family history turns out to be potentially relevant. Mickey's father and older brother were both apparently pretty hyperactive as children, although not to the same extent as Mickey. In addition, Mickey's mother recalls that her older son had a few nervous tics when he was in grade school: a nose-wrinkling tic and a compulsive tendency to clear his throat when nervous. The tics disappeared by the time the brother entered high school. As well, Mickey's father has always had a tendency to blink rapidly when he is stressed. Despite having had some initial difficulties in school, both father and brother are well educated, with degrees in aeronautical engineering and business administration, respectively. Finally, Mickey's mother, an accountant by training, is something of a cleanliness nut, keeping her entire house (minus Mickey's room) spotless, vacuuming all the carpets in the house and washing the kitchen floor every day of the week. This practice contributes to the friction with Mickey, who, if permitted, would file his socks under the bed and his shoes all over the house.

During your assessment of Mickey, you notice that he is fidgety, constantly crossing and re-crossing his legs and scratching his scalp, particularly when listening to his mother describing his wall-breaking proclivities. He still has the occasional tic in the form of a rapid head flick to the right. Throughout the visit, he is very pleasant and cooperative, often quite amusing and insightful. His general physical and neurological examinations are completely normal; there are no dysmorphic features.

13.2 CLINICAL DATA EXTRACTION

At this point, please proceed to the text Web site and complete the worksheets. Once you have developed hypotheses concerning the region of the nervous system implicated, the probable disease categories and a possible diagnosis or diagnoses, you are ready to continue with the text.

13.3 MAIN CLINICAL POINTS

From the mother's perspective, the *chief complaints* would be *inattention*, *learning problems*, *obsessiveness* and *violent behaviour*, all of long duration. She might also mention nervous tics or 'habit movements' but, a decade after their decline, may well have forgotten about the problem.

From *Mickey's* standpoint, the chief complaint, assuming that you remember to ask him, is that he is 'stupid and can't do anything right!'

The key findings in Mickey's history are as follows:

- A long-standing problem with attention span and impulse control, associated with hyperactive behaviour, dating back to age 4
- Stereotypic uncontrolled facial and limb movements, as well as vocalizations, dating from age 8, peaking at around age 12, then declining but never entirely disappearing
- Complex stereotypic compulsive behaviours, beginning in the early teens and persisting
- Selective learning difficulties, particularly for mathematics, despite normal to superior intelligence, dating from the kindergarten years and persisting
- Obsessive thinking and disabling complex rituals, beginning in the mid-teens
- Attacks of uncontrollable rage triggered by arguments over trivial matters or by the parents' attempts to stop his rituals, also dating from the mid-teens
- Poor self-esteem, long-standing
- Positive family history of hyperactivity, tics and obsessive-compulsive behaviours
- Except for the occasional motor tic, a completely normal neurological examination

13.3.1 CHARACTERIZING MICKEY'S DISORDER

We can begin by carefully analyzing the main features of Mickey's history and attempting to categorize his main symptoms:

1. *Impaired selective attention* – Mickey has trouble concentrating on one task to the exclusion of others. He is thus easily distracted: witness the difficulty he had doing his classroom work when city buses went by the windows.

2. *Poor impulse control* – This is really a corollary of impaired attention in that, if you are unable to ignore extraneous stimuli while attempting a task, you are more likely to respond to the stimuli and proceed to do something irrelevant to the task at hand. Thus, Mickey might abandon the solution of a maths problem, get out of his chair and go to the classroom window, disrupting classroom activities in the process. Persistent disruptive, irrelevant activity of this type is what is meant by the rather vague but popular term *hyperactivity*.

3. *A tic disorder* – A tic is defined as a stereotypic complex movement or vocalization involving the synchronous, organized action of many

muscle groups, over which the individual has imperfect control. Tic-associated movements such as jaw-opening or head-turning would be considered normal purposeful movements under certain circumstances but, as tics, are inappropriate and irrelevant to the existing circumstances, e.g. being required to give an oral presentation in class. Tics can be voluntarily suppressed, on request, but tend to resume as soon as the person is distracted – an obvious linkage to impaired selective attention. Given that tics can be suppressed, they are clearly distinct from choreic movements (encountered in Chapter 7), which cannot. In addition, choreic movements are more disorganized – chaotic – and would not be considered as 'normal' movements under any circumstances.

4. *A learning disorder* – Defined as the state of being two or more academic years behind age level in learning specific material (e.g. reading, mathematics) despite normal intelligence and appropriate opportunities to learn. We have seen that Mickey has measured intellectual ability in the high average range but has great difficulty learning maths. Obviously, Mickey's poor attention span has contributed to his learning difficulties, but this problem alone would not explain why Mickey cannot do maths problems despite being able to read at his age level.

5. *Compulsive behaviours* – Whether shooting imaginary baskets, twiddling knobs, counting coloured pencils or repeatedly checking the bedroom closet for interlopers, these behaviours are, in a sense, uncontrolled actions akin to motor tics but far more complex. They could be construed simply as manifestations of poor impulse control but stem from an *inner, dysfunctional drive* rather than an uninhibited response to an irrelevant external stimulus.

6. *Obsessive thinking* – In essence, this term refers to unwanted, recurrent thoughts that cannot be suppressed and therefore interfere with normal thinking and reasoning. In being unwanted and poorly suppressed, they are the 'thinking' equivalent of a motor or vocal tic. Far more than tics, however, obsessions – translated into the kinds of compulsive behaviours described previously– interfere with the normal activities of daily life.

7. *Rage attacks* – As explosions of furniture smashing, these attacks are essentially a combination of unbridled anger and poor impulse control. The new element they introduce to Mickey's smorgasbord of behavioural problems is a disturbance of *mood control*. Mood is defined as a *state of mind or feelings*, the predominant

examples being happiness, sadness, anger, anxiety and tranquillity. We all live with fluctuations in these feelings, some predominant one day, others the next; overall, however, they remain in a state of balance, without one mood predominating over time. In Mickey's case, anger is never far from the surface and, once in the foreground, cannot be resolved without resort to antisocial and destructive behaviour.

If we distil this list of symptoms further, we can postulate that Mickey's core 'pathology' is an incontinence or instability of vigilance, thought, mood and behaviour. This common thread of incontinence of mental states suggests the possibility of a single underlying mechanism that could produce, in varying combinations at different stages of brain development, inattention, tics, compulsions, obsessions and rage. We will return to this hypothesis later in the chapter.

Attention, learning, thought, mood: these are all examples of the most complex functions of which the human brain is capable. This being the case, the localization of Mickey's neurological disorder would appear to be perfectly obvious: there has to be a problem involving the cerebral hemispheres – a process located in the brainstem, spinal cord or peripheral nervous system would be illogical. Furthermore, one would be tempted to echo the fictional detective Hercule Poirot in localizing the problem to the 'little gray cells' or, in neuroanatomical parlance, the cerebral cortex. As you will see, this conclusion is too simplistic but does turn out to be half right.

Our next task, therefore, will be to review the neuroanatomical basis for selective attention, impulse control, learning, thought sequencing and mood.

13.4 RELEVANT NEUROANATOMY AND NEUROCHEMISTRY

13.4.1 NEUROANATOMICAL SUBSTRATES OF BEHAVIOUR

The brain structures involved in attention, learning and behaviour are numerous and comprise a widely distributed network. The main components of this network are as follows:

1. Prefrontal cortex
2a. Basal ganglia, in particular the caudate nucleus and the ventral striatum (also known as the nucleus accumbens), and
2b. Thalamus, especially the dorsomedial and anterior nuclei

3. Cerebellar hemispheres
4. Limbic system
5. Monoaminergic and cholinergic diffuse projection systems

Obviously this, is a formidable-appearing list and the inter-relationships between the various components are exceedingly complex. Since this text is designed for novices in neuroanatomy, we will confine our discussion to the essential elements.

13.4.1.1 PREFRONTAL CORTEX

By this term, we mean those portions of the frontal lobe anterior to the various motor cortices, i.e. the precentral gyrus, premotor cortex, frontal eye fields, supplementary motor area and Broca's area (see Figures 13.1 and 1.10 and the relevant material in Chapters 7, 10 and 12). The prefrontal cortex is extensive and is distributed over the three surfaces of the frontal pole: the lateral frontal convexity, the midline parasagittal area and the orbital-frontal region above the eyes (Figure 13.1).

Based in part on the study of patients with prefrontal area lesions, it is clear that the principal roles of the prefrontal cortex are selective attention, judgement and working memory. Frontal lobe lesions are accompanied by inattention, distractibility, poor impulse control and impaired ability to retain number sequences (as in looking up a telephone number and remembering it long enough to dial the required

digits). Patients with frontal lobe damage (especially children) may be hyperactive or apathetic, sometimes both, by turns. It is customary to refer to the prefrontal areas as the seat of *executive function*; if you will, they represent the 'chief executive officer' of the brain.

We are not suggesting that the other association cortices in temporal, parietal and occipital lobes are unimportant in learning processes – as they clearly are – simply that the prefrontal cortex has an overarching, supervisory role in all forms of learning. The cellular architectural layering and modular functioning of the association cortices are reviewed in Chapters 10 and 12.

13.4.1.2 BASAL GANGLIA CIRCUIT

Just as the precentral gyrus cannot direct the performance of complex movements without the support of a variety of subcortical structures, the prefrontal area requires the collaboration of components of the same subcortical structures in deciding (for example) whether it is safe to chase an errant baseball into a busy street.

In Chapter 7, we noted that the initiation and cessation of movement appear to be centred in a looping circuit comprising the motor and sensory cortices, the putamen, the globus pallidus, the substantia nigra, the subthalamic nucleus, the lateral thalamus and finally the motor cortex (see Figures 7.4 and 7.10). The prefrontal area, in the performance of its functions, participates in an analogous circuit that involves similar, but geographically distinct,

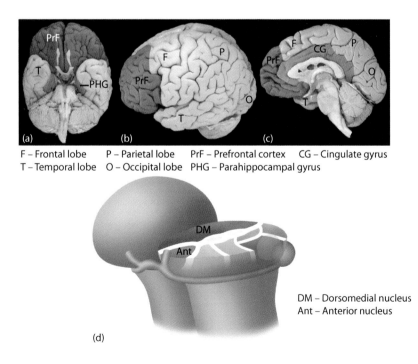

F – Frontal lobe   P – Parietal lobe   PrF – Prefrontal cortex   CG – Cingulate gyrus
T – Temporal lobe   O – Occipital lobe   PHG – Parahippocampal gyrus

DM – Dorsomedial nucleus
Ant – Anterior nucleus

FIGURE 13.1: External view of the brain (upper panel) with the prefrontal and limbic cortices highlighted. (a) Ventral view; (b) lateral view, left hemisphere; (c) medial view, right hemisphere. The lower panel (d) shows the thalamus sectioned in the horizontal (axial) plane, demonstrating the thalamic nuclei most relevant to attention, memory and emotion (dorsomedial and anterior nuclei).

portions of the striatum and thalamus (as well as the other structures just named). It is not necessary for you to know the details of this circuit. We will simply draw your attention to the two most important components of this prefrontal loop.

The first is the head of the caudate nucleus (see Figures 13.2, 1.9 and 1.11); in the prefrontal loop, it is the analogous structure to the putamen for the motor loop. The importance of the caudate head in executive functioning is revealed by the fact that its destruction (as, for example by hemorrhage or ischemic stroke) leads to deficits in attention and judgement similar to those seen in patients with lesions of the prefrontal cortex.

The second component is the dorsomedial thalamic nucleus (see Figures 13.1 and 1.8); it is analogous to the ventral-anterior and ventral-lateral thalamic nuclei for the motor loop. In the prefrontal loop, the dorsomedial thalamus projects back to the prefrontal cortex, having received input from the caudate head via distinct components of the globus pallidus and indirectly from the subthalamic nucleus and substantia nigra. Neurobehavioral disturbances similar to those seen with prefrontal cortex and caudate head lesions may also be present in patients with isolated damage to the dorsomedial thalamic nucleus.

In the course of this text, we have repeatedly made reference to the thalamus, its various nuclei and their projections to specific cortical areas. Previous thalamus–cortex illustrations have focused on the particular function(s)

relevant to the clinical problem. Figure 13.3 summarizes these anatomical relationships for the thalamus as a whole.

13.4.1.3 CEREBELLAR CIRCUIT

In Chapter 7, we saw that the coordination of movement required the participation of an integrative, feedback loop involving the motor cortex, the contralateral cerebellar hemisphere (via the pontine nuclei) and the ventral-lateral thalamic nucleus (thence back to motor cortex). Suffice it to say that there is an analogous circuit from the prefrontal area to different elements of the contralateral cerebellar hemisphere, returning via the dorsomedial thalamic nucleus.

The combined basal ganglia and cerebellar circuits for the prefrontal area are outlined in a box diagram in Figure 13.4.

13.4.1.4 LIMBIC SYSTEM

The limbic system comprises a group of cerebral hemispheric structures located at the junction of the hemisphere with the diencephalon. From the embryological standpoint, they form a ring, or fringe (*limbus*, in Latin), around the foramen of Monro (see Chapter 9) at the origin of the telencephalon (the primordial cerebral hemisphere). During the evolutionary process, the limbic elements of the cerebral hemispheres were the first to appear; indeed, in reptiles the cerebral hemisphere is almost entirely 'limbic' in function.

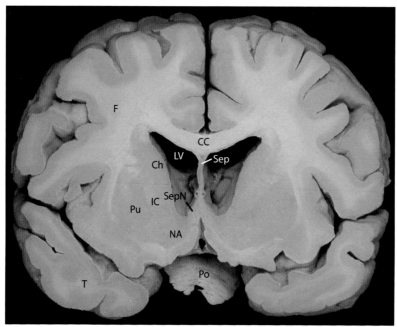

F – Frontal lobe
T – Temporal lobe
CC – Corpus callosum
LV – Lateral ventricle
Sep – Septum pellucidum
Ch – Caudate (head)
IC – Internal capsule
Pu – Putamen
SepN – Septal nuclei
NA – Nucleus accumbens
Po – Pons

FIGURE 13.2: Coronal section through the frontal lobes at the level of the head of the caudate nucleus and the nucleus accumbens.

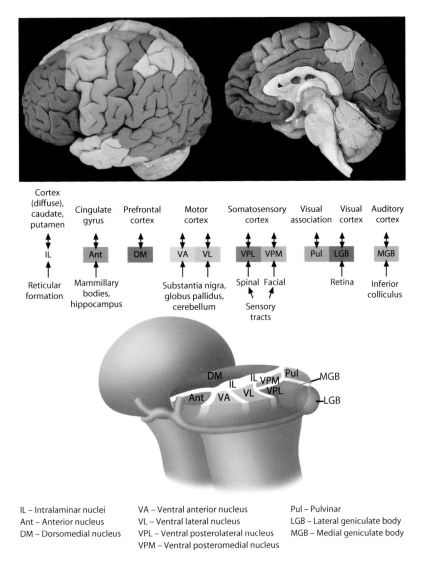

FIGURE 13.3: Reciprocal connectivity between the cortex and thalamus, including the major inputs to the various thalamic nuclei.

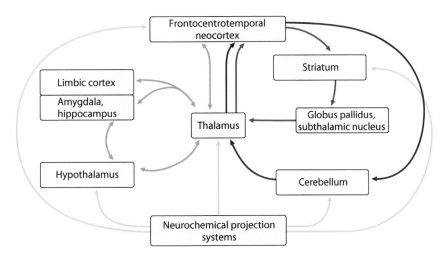

FIGURE 13.4: Schematic diagram illustrating the main CNS circuits involved in attention, learning and emotion.

There are both cortical and subcortical components to the limbic system, the most important of which are the following:

a. Cortical – cingulate gyrus, parahippocampal gyrus, hippocampal formation, septal area

b. Subcortical – fornix, mammillary body, anterior thalamic nucleus, amygdala, ventral striatum (nucleus accumbens)

Of these structures, the nucleus accumbens is illustrated in Figure 13.2, in relation to the caudate head, while the remaining structures are illustrated in Figure 13.5. Figure 13.5 shows the 'ring'-like configuration of the limbic cortex, the ring consisting of the cingulate gyrus, septal area and parahippocampal gyrus. The hippocampal formation is located medial to and is partially concealed by the parahippocampal gyrus.

Some of the subcortical elements of the limbic system also form a ring around the attachment of the cerebral hemisphere to the diencephalon. As shown in Figure 13.5, these include the fornix and the mammillary bodies.

The functions of the limbic system are only partially understood. The most important functions are memory registration (in which the hippocampal formation is a crucial element), social interaction (including sexual interaction), the response to external threats and mood control (in which the amygdala is particularly important). In animal species, the limbic system is implicated in the ability to distinguish friend from foe and in the emotional reaction to and ability to cope with an external threat – the so-called 'fight or flight' response.

Like the reticular formation and non-specific thalamic nuclei (see Chapter 10), the limbic system receives extensive input from peripheral pain receptors – witness the characteristic motor and emotional responses to being stung on the foot by a bee. The limbic system also has constant sensory input from visceral pain receptors, as demonstrated by the predominant emotions of anxiety and fear in response to the squeezing chest pain associated with myocardial ischemia.

Output from the limbic system projects diffusely to the cerebral cortex, diencephalon and brainstem – in keeping with the importance of emotional states in helping to direct motor function, decision making, language, learning and sensory perception. Output to the adjacent hypothalamus results in the characteristic autonomic nervous system reactions to emotional states; an obvious example would be the tachycardia, tachypnea, elevated blood pressure and intense perspiration that immediately follow a near-miss high-speed collision at a traffic intersection.

As might be expected, many of the components of the limbic system are functionally linked by sweeping loops of fibers with reciprocal connections; they are analogous to those already described for the motor system (Chapter 7). Although limbic circuitry is of great interest to neuroanatomists and neurophysiologists, the subject is beyond the scope of an introductory, clinically based text.

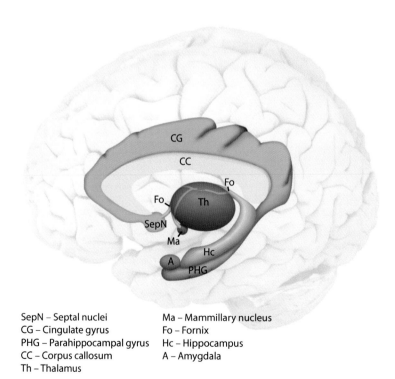

SepN – Septal nuclei
CG – Cingulate gyrus
PHG – Parahippocampal gyrus
CC – Corpus callosum
Th – Thalamus

Ma – Mammillary nucleus
Fo – Fornix
Hc – Hippocampus
A – Amygdala

FIGURE 13.5: Illustration of the location and relationship of the main components of the limbic system, both cortical and subcortical.

13.4.1.5 Monoaminergic and Cholinergic Diffuse Projection

The final components of the network are diffuse neuronal projection systems from brainstem and basal forebrain; their function is to modify the level of activity in the remaining components. Activity may be stimulated or inhibited depending upon the type of balance between the various transmitters at a given moment.

The four most important transmitter systems are serotonin (5-hydroxytryptamine), dopamine, norepinephrine and acetylcholine. Of these, the cells of origin of the first three (all monoamines) are located variously in the pons and midbrain, while those of the cholinergic system are in the basal forebrain inferior to the ventral striatum. Although the nuclei containing the monoaminergic and cholinergic cell clusters are relatively small, their axons project widely to cerebral cortex, basal ganglia, thalamus and limbic system.

Table 13.1 outlines the location of the various nuclei participating in the diffuse projection system and the main functions regulated by each neurotransmitter. Figure 13.6

TABLE 13.1: Main Neurotransmitter Projection Systems for Attention, Memory and Behaviour

Transmitter	Principal Nuclei	Functions
Serotonin (5-OH-tryptamine)	Median raphe nuclei Midbrain, pons, medulla	Sleep–wake cycle Mood Pain inhibition-via spinal cord projections
Dopamine	Substantia nigra Ventral tegmental area	Attention Mood
Norepinephrine	Locus ceruleus nuclei Upper pons	Attention Sleep–wake cycle
Acetylcholine	Basal nucleus of Meynert (basal forebrain)	Memory, learning

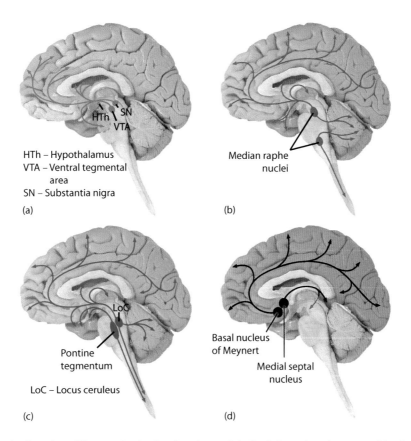

HTh – Hypothalamus
VTA – Ventral tegmental area
SN – Substantia nigra

(a)

Median raphe nuclei

(b)

Pontine tegmentum

LoC – Locus ceruleus

(c)

Basal nucleus of Meynert

Medial septal nucleus

(d)

FIGURE 13.6: Mid-sagittal section of human brain showing the nuclei of origin and regions served by the monoaminergic and central cholinergic projection systems. (a) Dopamine system, from the midbrain. (b) Serotonin system, from the raphe nuclei of the brainstem. (c) Norepinephrine system, from the locus ceruleus and other brainstem nuclei. (d) Cholinergic system, from the basal forebrain nuclei.

illustrates the main cerebral regions to which the transmitter systems project (panels a–d).

As can be seen in Figure 13.6, the serotonergic and noradrenergic systems project to the entire cerebral cortex (including limbic cortex), the basal ganglia, diencephalic structures and to the cerebellum. The dopaminergic system, whose striatal projections we reviewed in Chapter 7, largely confines its cerebral cortical projection to the frontal lobes and limbic system; these relatively restricted projections are nevertheless crucial for the regulation of attention and mood. The cholinergic system originating in the basal forebrain projects primarily to the entire cerebral cortex, the hippocampus and the amygdala.

13.4.2 NEUROCHEMICAL SUBSTRATES OF BEHAVIOUR

Having seen how the brainstem monoaminergic systems and the basal forebrain cholinergic system project variously to neocortical association areas, limbic cortex and central cerebral gray matter structures – all implicated in attention, learning, mood and behaviour – we will now briefly review the mechanisms by which these projection systems alter neuronal function.

In Chapter 11, in the context of epileptic disorders, we discussed how the amino acid neurotransmitters glutamate and gamma-aminobutyric acid (GABA) excite or inhibit neurons by opening excitatory (Na^+, Ca^{++}) or inhibitory (Cl^-) ion channels in neuronal membranes. These neurotransmitter actions are extremely rapid, the excitation or inhibition lasting only milliseconds. The excitatory effects of acetylcholine on nicotinic-type cholinergic receptors in the peripheral nervous system are equally rapid in onset and short-lasting.

In contrast, the effects on neuronal activity of the monoaminergic projection and basal forebrain cholinergic (muscarinic) systems are relatively slow in onset and long-lasting – hundreds of milliseconds or even several seconds. The term used for such prolonged modifications in neuronal activity is *neuromodulation.*

Norepinephrine, dopamine, serotonin and forebrain-derived acetylcholine do not act by directly opening ion channels (so-called ligand-gated channels), but through action on specialized neuronal membrane proteins known as *G-proteins* (see Figure 13.7). Glutamate and GABA may also act through G-proteins via metabotropic and GABA-B receptors, respectively. As demonstrated in Figure 13.7, G-proteins may then act to open adjacent ion channels or may, in turn, activate or inhibit so-called *second messenger* molecules such as cyclic AMP (adenylate cyclase). Once activated, second messengers may initiate long-lasting changes in cell chemistry such as stimulation of *protein kinase* activity or *up-regulating gene expression.* Gene transcription and associated protein synthesis are believed to be essential components in long-term memory registration and, thus, the learning of new material such as neuroanatomy.

In Chapter 7, during the discussion of the central mechanisms involved in the initiation of movement, we introduced the concept of a single transmitter chemical having either excitatory or inhibitory actions, depending upon the type of receptor present. The example given was dopamine receptors in the striatum, D1 receptors being excitatory to the downstream neuron, D2 receptors inhibitory. The same is true for norepinephrine and serotonin.

Thus, for example, there are four main norepinephrine receptor subtypes (α1, α2, β1, β2), the β subunit types predominating in the central nervous system. The list

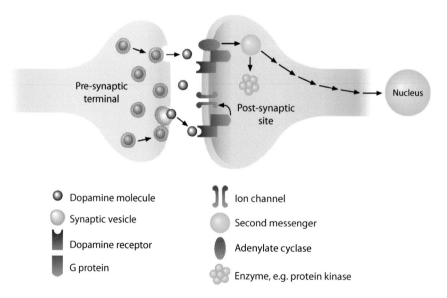

FIGURE 13.7: Schematic of a CNS synapse illustrating neuromodulation via G-proteins.

of known serotonin (5-hydroxytriptamine) subtypes is long, the most important being 5HT1A, 5HT1B, 5HT1C, 5HT1D, 5HT2A, 5HT2B and 5HT3; of these, the 5HT2 receptor subtypes appear to be the most important ones in modulating cortical and limbic function, thus modifying mood states.

The importance of normal functioning of the many neurotransmitter receptor subtypes in the regulation of behaviour is illustrated by the fact that the current pharmacological management strategies for the treatment of inattention, hyperactivity, depression, rage, obsessive-compulsive disorder and memory disintegration largely involve the manipulation of central nervous system receptor functions. Thus, hyperactivity and inattention may be treated with dopamine up-regulation, depression with 5HT2 receptor stimulation and Alzheimer-related memory loss with central muscarinic cholinergic receptor activation.

13.5 LOCALIZATION PROCESS

Fortified with an expanded fund of information concerning the array of anatomical structures – cortical, subcortical, limbic and even brainstem – you can begin to appreciate that the 'seat' of Mickey's behaviour disorders could, in principle, reside in any or all of these sites. Indeed, it is conceivable that some or all of his symptoms may originate not so much from a structural lesion at one anatomical site, but from a specific type of defect in cell-to-cell interaction involving neurons of the cortex, limbic system *and* the basal ganglia – neurons that all participate in circuits mediating attention, learning, mood, and so forth.

Earlier in this chapter, we speculated that, in a sense, nervous tics, obsessive thinking and rage outbursts could all share a common mechanism: all are stereotypic behavioural phenomena that are not wanted by the person experiencing them, cannot be easily suppressed and lead to major problems in social interaction that profoundly compromise the individual's ability to function independently. *Incontinence* was the term used; this seems appropriate given that bowel or bladder incontinence refers to a failure to develop or a loss of an essential function that allows us to participate normally in society. The development of urinary incontinence, for example, is exceedingly embarrassing to the patient, significantly curtails activities outside the home and may have devastating effects on self-esteem. The development of compulsive touching of strangers or of rage attacks (both forms of behavioural incontinence) is no different in these respects.

In other words, it is possible that stereotyped involuntary movements, compulsive thinking and uncontrollable rage outbursts may all involve a similar mechanism, possibly at the level of receptor function, residing in separate but parallel cortical – basal ganglia – thalamocortical circuits originating in supplementary motor, prefrontal and limbic cortices, respectively. With this hypothetical

conclusion, we have gone about as far as we can go in the complex task of trying to localize Mickey's disorder.

13.6 ETIOLOGY – THE NATURE OF MICKEY'S DISORDER

You may have noticed that, in this chapter, we have been using the term *disorder* in reference to Mickey's collection of behavioural disturbances rather than the term *disease*. This is a somewhat arbitrary distinction but it serves to point out that, thus far, Mickey's symptom complex is virtually life-long (thus chronic) but has shown no sign of relentless progression such as one would expect with many ongoing 'disease' processes. True, his symptoms have fluctuated over time, some waxing and others waning, but on the basis of your neurological examination two decades into the process, Mickey appears to be remarkably 'intact'. This statement by no means abrogates the fact that the effects of Mickey's behaviour disorder have had a catastrophic impact on his development as an independent adult human being and on his entire family. In the psychiatric literature, mental health dysfunctions are typically referred to as 'disorders' rather than 'diseases'; for our purposes, the two terms are interchangeable.

If, then, we refer back to the Neurological Disease Symptom Duration Grid (Figure 4.8), we see that the main disease categories capable of producing a process evolving over many years (as is the case with Mickey) are metabolic, inflammatory/autoimmune, neoplastic, degenerative and genetic.

In addition, however, there is the possibility that one of the other, more acute disease categories could have injured Mickey's brain early in life, leaving him with a static functional deficit. Such an injury could have resulted from a traumatic, vascular or infectious process; if so, however, the process would likely have occurred before birth, given that there is no history of any significant postnatal illness or injury. The term typically used to describe such a process is a *static encephalopathy of prenatal origin*, of which a classic example is cerebral palsy.

Given that, in one way or another, we will have to consider most of the 10 disease categories, we will work through the whole list, as we did in most of the previous chapters.

13.6.A PAROXYSMAL

Paroxysmal disorders of the nervous system such as epilepsy and migraine may be accompanied by significant behavioural dysfunction. Quite apart from the transient, dramatic behaviour changes that occur during complex partial seizures, the interictal personality of patients with localization-related epilepsy may be chronically disturbed: mood disorders, learning disabilities and aggressive behaviour are common. In the case of migraine attacks, there may be striking mood changes for hours prior to and following the headache symptoms. For both of these paroxysmal disorders, however, the cause of the behaviour disturbance is clear once a complete clinical history has been obtained.

13.6.B TRAUMATIC

Cerebral contusions and hemorrhages following a head injury are a frequent cause of acquired, long-standing behavioural disturbances. The prefrontal and anterior temporal cortices are particularly vulnerable to traumatic injury. It is not surprising, therefore, that head injury patients often demonstrate poor impulse control, inattention, impaired judgement, apathy and rage attacks. Such behavioural disturbances may be seen whether the brain injury occurred at birth in the context of a traumatic delivery or in the twenties following an automobile accident. In Mickey's case, there was no history of any significant head trauma.

13.6.C VASCULAR

Chronic behavioural problems are also common following ischemic or hemorrhagic strokes. In Mickey's case, there is obviously no history of such an event.

Prenatal arterial and venous strokes, however, if sufficiently small, may also lead to learning and behavioural disorders with preservation of normal intelligence. In most instances, there is an obvious motor disability, usually a hemiparesis. Frontal lobe infarcts, however, may produce no symptoms other than features of impaired executive function: distractibility, difficulty remaining on task, hyperactivity and poor impulse control. Likewise, dominant hemisphere temporal lobe infarcts may lead to difficulty with reading comprehension.

While these deficits may be found on the list of Mickey's symptoms, prenatal focal infarcts have not been typically associated with tic disorders or obsessive-compulsive behaviours. The development of epileptic seizures is also relatively common in children with prenatal stroke, particularly if the stroke is large; when present, such seizures usually appear before the age of 6.

13.6.D TOXIC

Long-standing behaviour changes may occur following a toxic injury to the brain, as in methanol poisoning or chronic alcoholism. The brain may also be damaged prenatally following exposure of the mother to neurotoxic agents; examples include ionizing radiation and, again, ethanol. In the latter instance, there is a typical combination of facial dysmorphic features and severe disturbances of learning, attention and behaviour that carries the name *fetal alcohol spectrum disorder* (FASD). Indeed, Michael shares many of the behavioural characteristics seen in

children and adults with FASD; tics and obsessions, however, are not typically seen in the latter disorder.

Toxic compounds that impede the normal development of the brain or other organs are called *teratogens*.

13.6.E INFECTIOUS

Chronic behavioural problems are one of many neurological complications of acute viral encephalitis or bacterial meningitis; these diagnoses, however, are not compatible with Mickey's history. Prenatal encephalitides may have the same behavioural consequences but are usually accompanied by varying degrees of global intellectual impairment, epilepsy and (frequently) congenital deafness. Examples of pathogens responsible for prenatal encephalitis include rubella and cytomegalic inclusion viruses, syphilis and toxoplasma, a unicellular parasite.

13.6.F METABOLIC

In adults, metabolic disorders are a common cause of behavioural and personality changes. Chronic hepatic failure may be accompanied by confusion and disinhibition (see Chapter 10), hypothyroidism by lethargy and depression, hyperthyroidism by a manic state. Consequently, an unexplained behaviour change in an older child or an adult should prompt a search for a metabolic disorder: liver function tests, urea, electrolytes, thyroid-stimulating hormone, vitamin B12 level. None of these disorders would be likely to produce a neurobehavioral disorder lasting two decades without some other features of a systemic disease.

13.6.G INFLAMMATORY/AUTOIMMUNE

Some autoimmune disorders, most notably *systemic lupus erythematosus* (SLE), may produce a chronic, fluctuating encephalopathy characterized by behavioural and mood abnormalities, even frank psychosis, without any clear-cut motor or sensory findings. It would be highly unlikely, however, that a patient with CNS lupus would have neurological symptoms persisting for two decades without the development of other classical manifestations of the disease: arthropathy, nephritis, facial rash, cardiac valvulopathy.

Another group of autoimmune encephalopathies associated with subacute or even chronic behaviour, orientation and mood disturbances without necessarily any motor or sensory deficits are the so-called autoimmune encephalitides. Only clearly identified in the past decade, these disorders stem from autoimmune attack against specific neuronal components such as the *N*-methyl-D-aspartate receptor (NMDAR). Like SLE, however, anti-NMDAR encephalitis and similar disorders would not remain clinically static for two decades (for more information, see Chapter 13 e-cases).

The only autoimmune disorder worth considering seriously in Mickey's case is *PANDAS*, an acronym for *p*ediatric *a*uto-immune *n*eurobehavioral *d*isorder *a*ssociated with *s*treptococcus. This behavioural syndrome is thought to be a cousin of *Sydenham's chorea*, a streptococcus-induced disease originally recognized in the seventeenth century by Thomas Sydenham. In Sydenham's chorea, a streptococcal infection (usually pharyngeal) is followed by the subacute appearance of chorea (whose distinction from tics has already been elucidated), sometimes lasting for months. Patients with this disorder may also develop hyperactivity, inattention, mood swings and obsessive-compulsive behaviours. In PANDAS, whose existence as a discrete disease entity is not universally accepted, the children develop, following a streptococcal infection, any or all of the behavioural problems seen in Sydenham's chorea but do not have chorea. They may, however, develop a tic disorder. While all of these symptoms have a startling resemblance to Mickey's, they typically appear in a relatively explosive fashion, following the infection, last for weeks or even months, in a fluctuating fashion, then subside. Thus, as tempting as the diagnosis of PANDAS might be, the time course of Mickey's disorder does not fit.

13.6.H/I NEOPLASTIC/DEGENERATIVE

By their very natures, neoplastic and degenerative diseases produce a slow, relentless deterioration with the gradual accumulation of new and increasingly severe neurological deficits. Thus, while they may produce major behavioural alterations along with their many other manifestations, these categories do not need to be considered any further.

13.6.J GENETIC

Genetic disorders may have behavioural consequences at all stages in life, from fetal development to old age.

As a rule, genetic disorders that compromise fetal brain development, synapse formation and synaptic plasticity (trisomy 21 – Down syndrome, fragile X syndrome, Rett syndrome, among many) also produce significant degrees of mental insufficiency – termed *mental retardation* in North America and *learning disability* in Europe, expressions that are, respectively, pejorative and unnecessarily vague.

Syndromes with X chromosome aneuploidy may be accompanied by learning difficulties but normal intelligence. These include *Turner syndrome* (phenotypic females with a single X chromosome – XO) and *Klinefelter syndrome* (phenotypic males with 2 X chromosomes – XXY). In Mickey's case, Turner syndrome is clearly not an option, while Klinefelter syndrome is typically accompanied by dysmorphic features including tall stature with excessively long arms and legs.

The genetic disorders considered thus far all seem unlikely explanations for Mickey's problem. Given the family history of similar, if less dramatic behavioural disturbances, however, we must look at the possibility of an entirely different type of genetic disorder, one whose only manifestation is behavioural.

In the Localization section (13.5), we introduced the possibility of a global neurochemical dysfunction – possibly a neurotransmitter receptor malfunction – compromising several parallel cortical – basal ganglia – thalamocortical circuits and leading to a variety of behavioural abnormalities including tics, obsessive-compulsive behaviours, rage attacks and so forth. Since an inherited mechanism is a logical explanation for such a scenario, it is time to revisit it.

13.6.J.1 GENETIC NEUROBEHAVIORAL DISORDERS

As a complex, non-progressive behaviour disorder in an adult with normal intelligence and a normal neurological examination, Mickey's disorder straddles the completely artificial separation between the clinical disciplines of neurology and psychiatry. The etiology of most psychiatric disorders is not presently known; in consequence, psychiatrists have had to define the disorders they encounter by using strict clinical criteria. As was formerly the case with many neurological disorders, it is hoped that a rigorous descriptive approach to psychiatric symptom complexes will clearly separate one disorder from another and eventually lead to the establishment of precise etiologies.

Leaders in psychiatry have collaborated in the publishing of generally accepted clinical criteria for specific behavioural syndromes. These criteria appear in the *Diagnostic and Statistical Manual of Mental Disorders*, usually abbreviated as *DSM*, with periodic updates. The most recent version is the *DSM-V* (see reference list at the end of the chapter). Since the listed criteria for each disorder are those that the experts in the field agree upon, it is not surprising that, with accumulating clinical experience and analysis, the criteria for each disorder are modified somewhat over time. This process will continue to evolve until precise neurochemical and neurogenetic mechanisms are uncovered.

With this caveat in mind, we can look again at Mickey's collection of behavioural abnormalities in the light of *DSM-V* published criteria for generally accepted psychiatric disorders. When we do this, we find that Mickey has many features of several apparently discrete disorders:

1. Attention-deficit/hyperactivity disorder:
   a. Poor attention span
   b. Easy distractibility
   c. Forgetfulness
   d. Fidgeting with hands and feet
   e. Leaving classroom seat when required to remain seated

f. Onset of symptoms before age 12

g. Inattention and restlessness present both at school and at home

h. Impairment in academic performance

i. No evidence to suggest the presence of any other neurological explanation for symptoms (for example, epilepsy)

2. Tourette's disorder:

   a. The presence of both motor and vocal tics for many years

   b. Onset of tics prior to age 18

   c. Waxing and waning frequency of tics over time

   d. No evidence that the tics are secondary to medication such as cocaine, or to a neurological illness such as encephalitis

3. Specific learning disorder:

   a. Long-standing difficulties with mathematical reasoning

   b. Academic skills in mathematics are substantially lower than expected for chronological age and limit his choice of occupation

   c. Learning difficulties with respect to mathematics were evident shortly after he began school

   d. Mickey's learning difficulties cannot be explained by an overall intellectual disability, by lack of adequate instruction, by having to learn in a language other than his mother tongue or by the presence of another neurological disorder such as epilepsy or a traumatic brain injury

4. Obsessive-compulsive disorder:

   a. Compulsive counting and checking of his possessions

   b. Rigid regime of ensuring his personal safety at bedtime

   c. Compulsive reading and re-reading of the same material

   d. Compulsions cause him distress, especially if attempts are made to stop them

   e. Compulsions are not a side effect of medication or due to an underlying medical disorder or a mental disorder such as chronic anxiety

If you review the listed criteria for these four disorders in the *DSM-V*, you will note that Mickey meets all the criteria for attention-deficit/hyperactivity disorder, Tourette's disorder (i.e. his long-standing motor and vocal tics), specific learning disorder and most of the criteria for obsessive-compulsive disorder. Faced with this plethora of simultaneous psychiatric diagnoses, you may well ask: are the four diagnoses distinct disorders that are usually present in individual patients in isolation? If so, is it possible that all four could occur in a single patient having one unitary diagnosis? The answer to both questions is 'yes!'

You will have noted that the *DSM-V* criteria for these psychiatric disorders include the important proviso that potentially causative underlying 'medical' conditions *must have been excluded*. This is a very important statement that obliges the physician to consider such conditions in any child or adolescent with any combination of inattention, hyperactivity, impulsivity, and motor or vocal tics.

Attention-deficit/hyperactivity disorder, for example, is a common behavioural complex with a strong hereditary predisposition that *appears* to affect about 10% of the pediatric population. In some suspected cases, however, it turns out that the real cause of the symptoms is a sleep disorder: poor sleep hygiene, obstructive sleep apnea due to airway obstruction from enlarged tonsils and adenoid, or periodic limb movements in sleep. In other instances, the cause is found to be excessive consumption of antihistamine medications for respiratory allergies. These are just some examples among many.

Correction of the underlying cause will alleviate the inattention and hyperactivity. When faced with a patient having these symptoms, therefore, it is important to ask the parents such questions as the following:

1. Does your child snore?

2. Does he or she get up frequently during the night?

3. Does your child get out of bed in the morning complaining of feeling tired or having a headache?

4. What medications does your child take on a regular basis?

In the case of Tourette's disorder, the diagnosis cannot be made if the patient has a history of encephalitis, traumatic brain injury or other acquired neurologic pathology. We have already considered, and excluded, the hypothetical Tourette's disorder look-alike, PANDAS.

Returning to the apparent conundrum that Mickey has symptoms that reasonably satisfy the criteria for four psychiatric diagnoses, we may ask whether three of the apparently discrete disorders may, in Mickey's case, be subsumed under the diagnostic umbrella of the fourth. Again, we can respond to this question in the affirmative, as *Tourette's disorder* patients frequently also meet the accepted criteria for attention-deficit/hyperactivity disorder, one or more learning disorders and obsessive-compulsive disorder. While some Tourette's disorder patients only have tics, others initially present with all of the features of attention-deficit/hyperactivity disorder then, several years later, begin to demonstrate motor and vocal tics – just as Mickey did. Learning difficulties are common in Tourette's disorder, but not invariable; they may be the result of an impaired concentration span but also may result from a specific learning disability that persists after the attention span has improved.

Motor and vocal tics are, in a sense, forms of compulsive motor behaviour. Many tic patients maintain that the reason they tic is an ill-defined internal pressure that requires the performance of the tic in order to release 'tension'. The same sense of pressure is present with compulsive hand-washing, counting and checking, all features of obsessive-compulsive disorder. Indeed, this psychological 'pressure' is so compelling that attempts by well-meaning relatives to prevent or interfere with compulsive behaviours may trigger an angry response or even a rage attack. Thus, rage attacks, as we have already suggested, are really a secondary behaviour disturbance derived from a combination of obsessive thinking, impulsivity and pronounced instability of mood.

Thus, we see how Tourette's disorder, as illustrated by Mickey's characteristic presentation, ties together the various symptoms and deficits he presents, all of which may be seen in isolation in other patient scenarios.

13.7 SUMMARY AND SUPPLEMENTARY INFORMATION

Mickey has a chronic, complex behavioural disorder that has troubled him, in various overlapping forms, for most of his life. The principal symptoms are poor attention span and impulse control, motor and vocal tics, complex compulsive behaviours, obsessive thinking, learning difficulties and rage attacks. Given his otherwise normal developmental status, his completely normal neurological examination and the positive family history of chronic motor tics and obsessive behaviours, the most likely diagnosis is Tourette's disorder.

13.7.1 ADDITIONAL INFORMATION

Originally described in the late nineteenth century by Gilles-de-la-Tourette, a pupil of the legendary French neurologist Charcot, Tourette's disorder (or Tourette syndrome, as it is usually termed in the neurological literature) is a common behavioural disorder, with a prevalence of about 1/200–300 in the general population. Its presumptive genetic origin is illustrated by the fact that there is often a positive family history of one or more of the Tourette behavioural spectrum: attention-deficit disorder, chronic motor tics and obsessive-compulsive disorder, each in different individuals. Tourette syndrome

and chronic motor tics are, for reasons unknown, more common in boys and men; isolated obsessive-compulsive disorder is more common in women.

Information concerning the genetic basis of Tourette's disorder is accumulating rapidly, but a detailed account of the current state of knowledge is beyond the scope of this book. Suffice it to say that numerous genes have been implicated but, in each report, only in rare cases.

The pathophysiologic basis of Tourette's disorder is unknown. While clinical evidence suggests a disturbance in the function of several neuromodulatory receptors, this may well be a second-order phenomenon. Although the fundamental defect may be a mutation or polymorphism in a gene or genes, it is clear that many other factors help determine phenotypic expression. Gender differences in rates and types of symptoms suggest an important hormonal effect. Environmental factors are also important, as revealed by the fact that, in identical twins with Tourette's disorder, the twin who was smaller at birth is usually the more symptomatic.

There are some recent data that seem to confirm the anatomical structures we have proposed as the 'seat' of the symptom complex in Tourette's disorder. Volumetric magnetic resonance imaging (MRI) studies have consistently shown a reduction is size of the caudate nuclei in comparison with controls. Neurophysiological and functional MRI studies have both implicated the premotor cortex, the supplementary motor cortex, the anterior cingulate cortex and the insular regions.

At least two neuromodulators are implicated in Tourette's disorder. Dopaminergic synapses are thought to be affected because drugs that upgrade dopaminergic function (such as methylphenidate) improve attention span and reduce impulsivity in Tourette patients. On the other hand, dopaminergic antagonists (such as haloperidol) reduce the number of motor and vocal tics. Thus, whatever the defect may be in dopaminergic transmission in Tourette's disorder, it is more complex than simple over- or under-expression.

Serotonergic synapses are also implicated in Tourette's disorder because serotonergic agonists, usually in the form of *SSRIs*, are used successfully to treat obsessive-compulsive behaviours, as well as to stabilize mood swings.

Finally, it is important to recognize that, just as Mickey's three other psychiatric 'diagnoses' have been included under the umbrella of Tourette's disorder, that diagnosis itself may overlap with other relatively distinct syndromes. The most notable example is *autism spectrum disorder*, a developmental syndrome with multiple etiologies whose main features include impaired expressive and receptive language, poor social interaction, impoverished imaginative play and obsessive fascination with moving objects or parts of objects. Many, but not all, autistic individuals are mentally handicapped. Some autistic patients with normal intellectual ability and good vocabulary may have marked obsessive traits as well as prominent motor

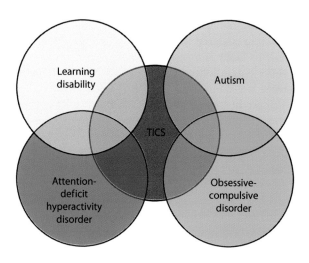

FIGURE 13.8: Illustration of the overlap between tic disorders and other developmental neurobehavioral syndromes.

and vocal tics; they meet all of the existing criteria for Tourette's disorder.

These variable overlapping neurobehavioral symptom complexes, with Tourette's disorder as the centrepiece, are illustrated in Figure 13.8 as multiple overlapping circles. This conceptualization of the relationship between the disorders will have to suffice until their exact physiological mechanisms are known. Ironically, once that achievement has been made, Tourette's disorder, as a clinical entity, may well disappear as the tic disorder will simply be seen as but one symptom of a fundamental neurochemical dysfunction.

13.7.2 INVESTIGATING AND TREATING MICKEY'S DISORDER

As already suggested, the diagnosis of Tourette's disorder depends upon the recognition of the classic clinical criteria; no supporting laboratory confirmation exists at present. MRI of the brain in Tourette's syndrome appears normal, with no consistent abnormalities. Hence, an MRI study in Tourette's disorder is not required unless the history or physical findings suggest the presence of a remote brain injury, whether traumatic, vascular or infectious.

13.8 E-CASES

13.8.1 CASE 13E-1: CHUNHUA, AGE 15

Background: This teenager is being referred to you because of recent dramatic changes in her school ability, personality and behaviour in the past 12–18 months. Formerly a well-mannered straight 'A' student, all changed following an acute infectious episode.

Similarly, there are no laboratory test abnormalities, whether biochemical or otherwise, to support the diagnosis. Given the evidence for a disturbance in monoaminergic transmitter function, it might be expected that measurement of levels of transmitters or their metabolites would reveal a characteristic pattern. In fact, however, no consistent abnormalities have been detected in large groups of patients with Tourette's disorder, whether the tissue sampled is serum, cerebrospinal fluid (CSF) or brain.

Management of the various symptoms of Tourette's disorder remains largely supportive, especially educational, psychological and pharmacologic. Behavioural interventions such as habit reversal therapy may be helpful. Some commonly used pharmacologic interventions are reviewed in Table 13.2. Clearly, the main thrust of medical management for Mickey will be directed towards his obsessive-compulsive behaviours and his rage attacks. In all probability, he would require a combination of an SSRI agent other than paroxetine and an atypical antipsychotic agent such as risperidone.

TABLE 13.2: Pharmacologic Interventions for the Symptoms of Tourette's Disorder

Attention-deficit/ hyperactivity	Methylphenidate
	Dextro-amphetamine
	Atomoxetine
	Clonidine
Tics	Clonidine
	Guanfacine
	Risperidone
	Haloperidol
	Pimozide
	Tetrabenazine
	Botulinum toxin injections
Obsessive-compulsive symptoms	Risperidone
	Clomipramine
	Fluoxetine
	Paroxetine
	Fluvoxamine
	Venlafaxine
Rage attacks	Risperidone
	Olanzapine
	Clonidine

Chief complaint: School failure and personality change

History:

- Prior to the illness, she was an excellent student with a calm, reflective personality.
- Following a mild respiratory tract infection 18 months ago, she had a two-week period of nasal congestion, bilateral frontal headaches and low-grade fever.

- Admitted to hospital in a confused state with visual hallucinations and moderately high fever.
- Computed tomography (CT) scan of the head showed bilateral frontal and ethmoid sinusitis and a large collection of pus in the subdural space outside the brain (*subdural empyema*).
- Empyema was drained surgically; postoperatively. She developed a transient right hemiparesis and expressive aphasia.
- Since that admission, she has been *uncharacteristically defiant and uncooperative; had wide mood swings; and had various risk-taking behaviours including smoking, drinking, recreational drug use and shoplifting.*
- At the same time, she has been doing poorly in school: distractible, restless, haphazard

approach, poor retention of learned material, missed assignments and *unconcerned about her poor performance.*

Examination:

- Patient is right handed.
- Formal neurological examination is entirely normal – no evidence of residual speech disturbance or hemiparesis.
- Psychological testing shows overall *intellectual capacity in the first percentile rank* (non-verbal subtest scores slightly better than verbal subtests); *severe deficiencies in working memory, speed of mental processing, verbal reasoning and reading comprehension;* normal results for visual memory and the ability to reproduce complex figures.

13.8.2 CASE 13E-2: ARIYAN, AGE 13

Background: On a Friday afternoon, which seemed to mark the beginning of a long-awaited summer, you are asked to see a teenaged girl with a history of recent school failure. Unfortunately, the history given by the parent is rather vague regarding an acute infectious event 'in her head' at the beginning of the school year.

Chief complaint: School failure

History:

- Previously a B student in all subjects.
- 10 months ago, prior to the start of the school year, she was admitted to hospital for surgical drainage of an intracranial abscess; postoperatively, she had a single epileptic seizure and was placed on carbamazepine.
- Delayed start to the school year but she eventually caught up in most subjects (except mathematics), passing with C and C+ marks.
- *Pronounced difficulty with mathematics especially geometry; incapable of grasping geometric concepts;* she also appeared to have forgotten most of the maths she had learned in the year prior to her illness.

Examination:

- Pleasant and cooperative.
- General physical examination is normal; neurological exam is also normal except that she is *unable to visually track an optokinetic drum* (see Figure 13.9) *when it rotates from left to right;* no difficulty tracking from right to left.
- Psychological testing results show overall cognitive abilities are in the average range; good

FIGURE 13.9: Photograph of an optokinetic drum, used to assess the parietal lobe function of automatic pursuit of moving objects in the visual fields.

verbal comprehension, processing speed and working memory; poor perceptual reasoning and great *difficulty in drawing shapes or in reproducing complex geometric figures.*

13.8.3 CASE 13E-3: KEVIN, AGE 12

Background: Kevin is being assessed by a variety of pediatric specialists to try to understand his behavioural outbursts (one most recently at school), as well as his verbal and mental capacities. His developmental history reveals uneven physical and verbal milestones, and he now requires a full-time teacher's aide in grade 6; he is known to excel in mathematics.

Chief complaint: Long-standing history of poor social skills; suspended from school recently for threatening another student

History:

- A fellow student was telling him an amusing story about the school gym teacher, to which the patient responded by telling the schoolmate in a threatening voice that he was 'going to KILL you for that!'; the patient's recollection is that he was just 'joking around'.
- *Patient has always been socially awkward; known to classmates as a loner and a 'weirdo'.*
- In conversation, he often changes the subject to something entirely irrelevant; he compulsively asks questions without listening to the replies; if interrupted, he becomes angry.
- Although he does very well in mathematics, he has great difficulty with most subjects because of *poor reading comprehension and ability to abstract.*
- Birth history: difficult delivery; initially limp and pale with Apgar scores of 4, 5 and 8 at 1, 5 and 10 minutes (see case 7e-4 for explanation);

required vigorous resuscitation but fed well after the first 24 hours.
- Long-standing gross motor difficulties: delayed walking (20 months) and subsequent delays in learning to run and to ride a bicycle.
- Normal early language development but delayed ability to make phrases; *initially had major difficulties with language comprehension* (but audiogram normal); improved expression and comprehension with speech therapy.
- *Has always had difficulty interpreting other people's emotional states from either tone-of-voice or from facial expression.*

Examination:

- Normal general physical examination.
- He speaks in a loud voice, as though lecturing to a large class; *eye contact is poor.*
- *Tends to repeat the last few words of the examiner's questions before trying to answer*; answers are rambling and off-topic.
- *Often interrupts conversations by asking irrelevant questions* like: 'Do you have a dog at home?' or 'Who do you think will win the World Series?'
- *Able to tell the day of the week for any date in the past 10 years as well as the next 12 months*; encyclopedic knowledge of sporting events.
- Formal motor system exam is normal except for poor rapid alternating forearm movements; hopping on one leg and tandem walking.

13.8.4 CASE 13E-4: EDDIE, AGE 14

Background: Eddie, a responsible, athletic and bright child, was hit by a van while riding his bicycle, and although he was wearing a helmet, he did suffer a severe head injury that rendered him comatose. This was followed by intensive rehabilitation, but not to his previous self. He is currently being assessed for problems with attention, behaviour and schooling, as well as physical complaints.

Chief complaint: Recurrent episodes of nausea, headache and strange behaviour for the past year

History:

- Severe head injury 21 months ago: initial Glasgow Coma Score was 4 (see case 9e-1).
- Urgent head CT scan revealed a large *hemorrhage in the right cerebral hemisphere*; surgical evacuation performed, followed by 1 month in coma.

- Upon awakening, he had a *left hemiparesis and very indistinct speech*; gradual improvement with intensive rehabilitation to the point where he could walk independently and carry on a conversation.
- He returned to school in a special class, functioning at a *grade 2 level* (mental age 7–8).
- *Marked personality change: short attention span; poor impulse control; wide mood swings* from hysterical laughter to violent anger and back within five minutes.
- 12 months ago, onset of *stereotyped episodes (sometimes several/day): abrupt complaint of nausea, facial flushing, fearful expression, repeated attempts to spit, utterance of nonsense syllables*; duration of episodes two minutes followed by complaint of intense right-sided headache.

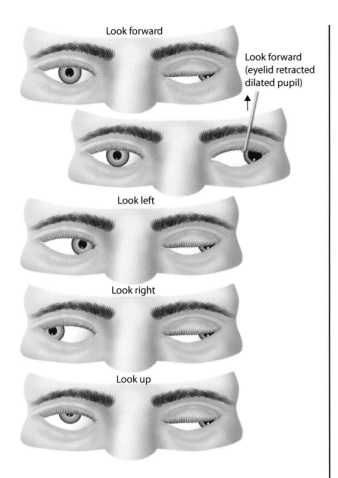

Look forward

Look forward (eyelid retracted dilated pupil)

Look left

Look right

Look up

FIGURE 13.10: Illustration of eye findings (Eddie).

Examination:

- Mental status: cooperative but with continuing wisecracks and anecdotes that disrupt conversation; *fully oriented*; *easily distracted*; difficulty with very basic mental arithmetic; *unable to remember names of three objects previously shown him after five minutes.*

13.8.5 CASE 13E-5: DAN, AGE 22

Background: You are working in a University health centre and are interviewing a senior student who is the star of his hockey team. About a month ago, he received a head injury that 'knocked him flat'. His recovery since then has been slow but continuous. Following recent guidelines, his coach has asked for an assessment for his fitness to return to active play.

Chief complaint: Headaches, dizziness and difficulty concentrating following a sports-related head injury

History:

- Senior year mathematics major at an Ivy League College; plays hockey on the varsity hockey team.

- Cranial nerves: *ptosis left eyelid; right pupil 2 mm and reactive, left pupil 6 mm and unreactive; right pupil reacts to light shone in the left eye but the reverse does not occur*; normal visual fields and fundi; *left eye deviates outwards and downwards, can neither adduct nor elevate* (see Figure 13.10); weakness of the left lower face; *rapid tongue movement impaired in all directions; pronounced jaw jerk.*
- *Left hand is constantly fisted and cannot be opened; unable to voluntarily move the left foot/toes.*
- *Marked hypertonicity of the left arm and leg.*
- *Reflexes:* see Figure 13.11.
- Normal motor system findings on the right side except for slow alternating forearm and fine finger movements.
- When both hands are stimulated simultaneously with cotton wisps while his eyes are closed, he only recognizes the touch of cotton on the right side (*sensory extinction*).

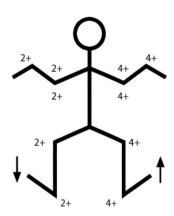

FIGURE 13.11: Reflex map (Eddie).

- *High-speed collision during a game one month ago rendered him unconscious for 15 seconds followed by headache and nausea; no memory of the collision or of the entire period of hockey during which the injury occurred.*
- Seen in the local emergency department: small left frontal hematoma and a normal neurological examination; CT scan of the head was unremarkable.
- Next day in class, he had persistent frontal headache and a sensation of turning in circles whenever he moved; *trouble solving complex maths problems.*
- Gradual improvement in symptoms over the next three weeks until he felt 'normal'; required

one to two naps a day during that period as he was constantly tired.

- After the three weeks, he tried skating but found that sudden stops made his head hurt and the dizziness briefly return; this phenomenon disappeared after one week of daily skating practice.
- When he showed up for a regular practice with the varsity squad, *his coach refused to allow him to participate until he had a letter from a physician clearing him to play.*

- Alert, articulate, very intelligent man.
- His history of the actual injury is sketchy but his recollection of events since the day of the accident is excellent.
- Formal neurological examination is entirely normal.
- *Folstein Mini-Mental Test gives him a score of 28/30*; he loses points in tests of *working memory*: repeating lists of words and of numbers.

Question: What about it, doc – does Dan get his letter?

13.8.6 CASE 13E-6: NAOMI, AGE 18

Background: This urgent transfer of a young woman from the psychiatric ward of an outside hospital occurred in the middle of the night. She had already been intubated because of her inability to maintain her own airway – caused by excessive upper airway secretions. There was a brief note regarding her mental disorder and the medications she has been receiving.

Chief complaints: Insomnia, hallucinations, paranoid ideation and bizarre behaviour for two to three weeks prior to original admission; generalized rigidity and involuntary movements for three to four days before transfer

History:

- Under constant stress for the past month due to preparation for final examinations; staying up late studying and consuming large amounts of caffeinated beverages.
- Increasing difficulty sleeping at night for two to three weeks prior to admission; on average down to one hour of sleep per night from eight hours.
- At the same time, she began to *hear pop music in her head, asked family members if they heard it too.*
- *She began to voice concerns that several of her close friends had started to hate her and wanted to hurt her.*
- One-week history of difficulty understanding what she heard in conversation; *she spoke abnormally rapidly, often repeating herself, changing the subject frequently, abruptly and illogically.*
- Admitted to local hospital with clinical diagnosis of acute psychosis; extremely agitated; treated with olanzapine, haloperidol, lorazepam and quetiapine, with some improvement.

- Several days into her admission, she began to complain of *stiffness of her limbs and had trouble walking*; treated with benztropine on the assumption that she had drug-induced dystonia, with no improvement.
- *Appearance of involuntary, stereotyped facial movements accompanied by drooling, airway congestion and by stridor secondary to vocal cord dystonia*; unable to get out of bed due to extreme rigidity of her limbs.
- Increasing agitation, fear and respiratory distress led to intubation and transfer.
- No family history of schizophrenia or of manic-depressive psychosis.

Examination:

- In a relaxed state, pulse = 93/min, respirations = 15/minute, blood pressure (BP) = 115/65, temperature = 37.9°C; *with periods of agitation, she became tachypneic, tachycardic and hypertensive (BP=160/110).*
- Normal general physical examination except for *mild liver enlargement*; no rash; no neck stiffness to passive flexion; negative Brudzinski sign.
- Despite her intubated state, she appeared alert and watched staff as they moved around the room.
- Continuous, multifocal involuntary movements: eye revulsion, shoulder shrugging, rhythmic clenching of her left hand, periodic twisting movements of her arms and legs, periods of alternating leg movements as if trying to walk while lying supine.
- Normal fundi, pupillary light reactions and eye movements.
- *Generalized rigid muscle tone to passive flexion and extension, all four limbs.*
- Normal tendon reflexes; flexor plantar responses.

13.9 SUMMARY OF KEY NEUROANATOMICAL AND NEUROCHEMICAL INFORMATION

- The principal components of the brain that contribute to networks involved in attention, learning and behaviour are the *prefrontal cortex, basal ganglia, cerebellar hemispheres, the limbic system and the monoaminergic and cholinergic diffuse projection systems.*

- The prefrontal cortex includes those extensive portions of the *dorsolateral, midline sagittal and orbital-frontal regions* of the frontal lobes that lie anterior to the main motor output regions (precentral gyrus, premotor cortex, frontal eye fields, Broca's area and the supplementary motor area) (Figure 13.1).

- Prefrontal cortex participates in a basal ganglia/thalamus circuit that is analogous to the motor loop discussed in Chapter 7. This circuit includes the head of the caudate nucleus, portions of the *globus pallidus and subthalamic nucleus*, and *the dorsomedial nucleus of the thalamus* (Figure 13.4).

- Likewise, the cerebellar hemispheres participate in the 'coordination' of attention, judgement and working memory through a loop analogous to that involved in motor coordination, i.e. *prefrontal cortex, pontine nuclei, cerebellar hemisphere cortex, dentate nucleus, dorsomedial thalamus* (Figure 13.4).

- Located at the junction of the diencephalon and the cerebral hemispheres, the limbic system is involved in memory registration, social interaction and mood regulation. The main components of the limbic system include the *cingulate gyrus, parahippocampal gyrus, hippocampus, septal area, fornix, mammillary body, anterior thalamic nucleus, amygdala* and *ventral striatum (nucleus accumbens)* (Figure 13.5). Again, some of these structures (cingulate gyrus, nucleus accumbens, anterior thalamic nucleus) participate in a go/no-go loop that is analogous to the basal ganglia motor control loop.

- Neuronal activity in the prefrontal and limbic regions, like cerebral activity elsewhere in the cerebral hemispheres, is modified by four diffuse neurotransmitter projection systems from the *brainstem* and *basal forebrain* (Figure 13.6). The three brainstem neurotransmitter projection systems, with their respective nuclei of origin, are those for *serotonin (midline raphe nuclei), dopamine (substantia nigra and ventral tegmental area)* and *norepinephrine (the locus ceruleus).* The *acetylcholine* projection system originates in the *basal nucleus of Meynert* in the basal forebrain.

- Unlike the ultra-fast effects of glutamate (excitatory via Na$^+$ channels) and GABA (inhibitory via Cl$^-$ channels), the monoaminergic and basal forebrain cholinergic projection systems exert their influence on downstream targets by the much slower mechanism of *neuromodulation.* This process involves *G-proteins* at the postsynaptic site as well as *second messenger* molecules like *cyclic-AMP* that stimulate intracellular protein synthesis by facilitation of rapid *gene transcription* mechanisms that are vital for permanent, long-term memory registration (Figure 13.7).

13.10 CHAPTER-RELATED QUESTIONS

1. In 1848, an unfortunate American railway construction worker named Phineas Gage sustained a severe, localized brain injury when a premature dynamite explosion blew an iron rod into his head. Following the injury and the prolonged recovery period, Phineas manifested a profound personality change. He became disinhibited, impulsive and extremely distractible, heedless of the appropriate norms of social behaviour that were accepted at the time. What region of his brain do you think was damaged by the iron rod?

2. The next three exercises in anatomical localization illustrate the importance of the limbic loop circuitry involved in memory registration. In trying to work out the answers to these questions, you may need to consult sources outside this text, sources that will be hinted at in the questions themselves.

   a. A 30-year-old high school teacher sustained a severe head injury while participating in a game of rugby football. After remaining in a coma for a week, he gradually awakened and regained full mobility, speech and independent living skills over a period of three months. It gradually became apparent, however, that he had a severe, permanent deficit in memory function. While his memory for previous events in his life and his fund of knowledge in his chosen field (English literature) were completely preserved, he was unable to recall any events or new people he had met since he awoke from the comatose state. In consequence, he was completely unable to return to work and could not even move from the apartment he was renting at the time of his injury. Were he to attempt to live anywhere else, he would never be able to find his way back home after going out to buy groceries! Can you explain what has happened to him?

b. For this question, we return to case 9e-4: a 13-year-old boy with a pineal region tumor, or pinealoma. Once the diagnosis had been made, he underwent a surgical removal of his tumor: a complex, difficult procedure as the pineal gland area is located in virtually the geographic centre of the head above the rostral end of the midbrain and just posterior to the roof of the third ventricle. When he awoke from the surgery, his caregivers gradually realized that he was severely amnestic. Five days after the procedure, he was unable to recall events that had occurred during the last 24 hours prior to surgery and – even more distressing – he had no recollection of anything that had happened since surgery notwithstanding the fact that he appeared alert and was able to converse normally. Over the succeeding 10 days, his ability to recall recent events (new people he had met; what he had eaten at mealtimes that day, etc.) gradually returned to normal. What do you think caused his transient disturbance in memory registration?

c. A 45-year-old woman was admitted from the emergency department for management of marked gait unsteadiness and acute agitation accompanied by visual hallucinations. A long-term alcohol abuser, she had been brought to the emergency department for repair of a large scalp laceration occasioned by a fall down a flight of stairs. Although the laceration was quickly sutured, the woman could not be discharged home because of her ataxia and her mental state. In addition, she appeared chronically malnourished, was covered in bruises and was so tremulous that she was unable to sign her consent-for-treatment form. She was treated with intravenous fluids, sedation, frequent small meals and high-dose multivitamins including a loading dose of thiamine. After her acute delirium cleared, it became obvious that she was severely amnestic for a period of several weeks prior to her admission and for much of what had taken place in hospital. Despite frequent re-introductions, she could not remember the names and faces of her various nurses, physicians and allied-health caregivers. Furthermore, possibly in an attempt to disguise her memory problems, she manufactured encounters that she believed had taken place with hospital staff. For example, when the neurologist who was assessing her memory functions asked whether she had ever seen him before (she hadn't), she said that she had run into him 'yesterday at Walmart'. Again, what do you think might be the cause of her amnestic state, and where in the nervous system is the lesion or lesions that are producing her problem?

REFERENCE

Diagnostic and Statistical Manual of Mental Disorders, 5th ed. Washington, DC: American Psychiatric Publishing, 2013.

SUGGESTED READING

Dooley, J. Tic disorders in childhood. *Semin Pediatr Neurol* 2006.13: 231–242.

Faridi, K., Suchowersky, O. Gilles de la Tourette's syndrome. *Can J Neurol Sci* 2003. 30(suppl 1): S64–S71.

Hallett, M. Tourette syndrome: Update. *Brain Dev* 2015. 37: 651–655.

Robertson, M.M. Mood disorders and Gilles de la Tourette's syndrome. An update on prevalence, etiology, comorbidity, clinical associations, and implications. *J Psychosom Res* 2006. 61: 349–358.

Serajee, F.J., Huq, A.H.M.M. Advances in Tourette syndrome: Diagnoses and treatment. *Pediatr Clin North Am* 2015. 62: 687–701.

Shavitt, R.G., Hounie, A.G., Rosario Campos, M.C., Miguel, E.C. Tourette's syndrome. *Psychiatr Clin North Am* 2006. 29: 471–486.

Singer, H.S. Tourette's syndrome: From behaviour to biology. *Lancet Neurol* 2005. 4: 149–159.

WEB SITES

American Tourette Syndrome Association: www.tsa-usa.org

National Institute of Neurological Diseases and Stroke: www.ninds.nih.gov/disorders/tourette/detail_tourette.htm

SUPPLEMENTARY CONSIDERATIONS: REHABILITATION AND ETHICS

Chapter 14

Neurorehabilitation

Objectives

- To understand the definition of rehabilitation
- To become familiar with members of a rehabilitation team and their roles in the rehabilitation process
- To be able to explain the rehabilitation approach in common neurological conditions with functional deficits
- To be aware of differences in the approach to rehabilitation in the pediatric population in comparison with that in adults
- To know useful online resources that can help in clinical practice

14.1 INTRODUCTION

Rehabilitation is an active, goal-oriented process maximizing resumption of function. It represents hope of recovery. In 2001, the World Health Organization developed the *International Classification of Functioning, Disability and Health* (Figure 14.1). This important framework reminds us of the complexity of human function and allows dissection of factors that limit maximal performance. It therefore facilitates clear goal setting and treatment.

Common neurological diagnoses that require rehabilitation services include acquired brain injury, spinal cord injury and stroke. In children, all of these diagnoses exist, but in smaller numbers. Rehabilitation principles and services are also utilized in childhood-acquired disabilities such as cerebral palsy (CP), spina bifida and neuromuscular diseases. In genetic and perinatally acquired conditions, the concept of resumption of function, or *rehabilitation*, is replaced by the process of maximizing function during growth and development through enhancement of the environment, or *habilitation*.

14.2 TEAM

Resuming function is often examined within three spheres: physical, cognitive and behavioural. It frequently requires a 'team' (Figure 14.2), with the patient and family at the centre. The size and composition of this team depend on the challenges faced by the individual participating in the rehabilitation program.

Following the onset of a newly acquired injury where rehabilitation is required, a physical medicine and rehabilitation assessment is frequently requested. This process involves the following:

1. A medical review of diagnosis, investigation and treatment
2. Definition of the neurological deficit and resultant alteration in function
3. Determination of rehabilitation readiness

Early involvement of rehabilitation specialists is important, as even prior to rehabilitation readiness, principles for early preventative care are frequently recommended. These may include muscle and joint stretching, medication or bracing to facilitate maintenance of range of motion, the use of pressure-relieving surfaces to prevent pressure sores, enhancement of nutrition, teaching early communication strategies and provision of specialized family support. The goal in early intervention is to minimize complications and maximize health and wellbeing, so the patient is in the best possible condition to undertake rehabilitation.

Centres providing rehabilitation services may have varying admission criteria. Quite frequently, these will include medical stability of the patient, patient ability to actively participate in therapy two to three hours daily, patient ability to lay down new memories and learn new information and a willingness of the individual with the disability to work with the rehabilitation team. When the clients are at this stage of recovery, they can often work with their health care providers to define rehabilitation goals and actively embrace the rehabilitation process. Examples of general goal areas, team members involved and therapeutic interventions utilized are outlined in Table 14.1.

Once general goal areas are set, each goal may be made more specific and tailored to each patient. For example, in the area of mobility, the goal may be to achieve 10 meters of independent ambulation. Specific

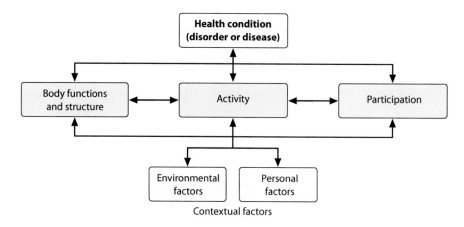

FIGURE 14.1: World Health Organization International Classification of Functioning, Disability and Health. (Courtesy of World Health Organization, Geneva, 2001.)

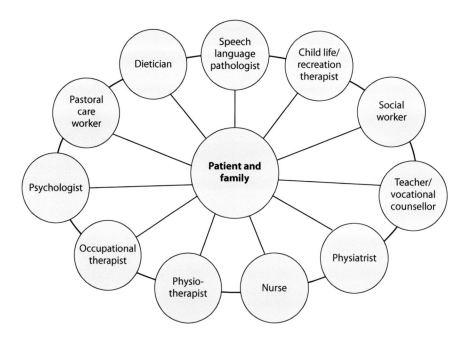

FIGURE 14.2: Rehabilitation team schematic. (Note: A physiatrist is a physical medicine and rehabilitation specialist.)

care plans can be developed, timelines are approximated, and hard work begins! Team meetings occur regularly to allow clear communication, problem solving as required and maximal team function.

Outcome measures are available to assess patients and their progress in rehabilitation. Certain tools are very general and look at global rehabilitation. The Functional Independence Measure (FIM; Hamilton et al., 1994) is one such tool that is completed with input from the whole team. Other tools are very specific and client focused, such as goal attainment scaling. An example is the Community Balance and Mobility Score (Howe et al., 2006), a tool often used by physiotherapists to look at balance and functional movement. These tools can be

extremely helpful in defining needs for caregiver support, providing feedback to clients about progress and assessing utility of specific interventions. Each tool utilized, however, must be studied carefully with respect to its strengths and limitations.

Rehabilitation does not end on discharge from the hospital. Depending on the progress, goals and the facility providing care, rehabilitation stays may range from quite short (weeks) to months and, in unusual cases, a year. On average, however, most rehabilitation stays are one to three months. As patients transition to outpatient care, they frequently continue to work their way through the hierarchy of rehabilitation-related tasks (Figure 14.3).

TABLE 14.1: Planning and Organization of Rehabilitation Services

Goal Area	Team Members (Always Includes Client and Family)	Intervention
Maximize range of motion	Nurse, occupational therapist, physiotherapist, physician	Heat, daily stretch, oral medication, bracing, botulinum toxin injections
Mobilization	Physiotherapist, rehabilitation assistant	Range of motion exercise, core strengthening, increase bed mobility and sitting, progress to walking when ready, explore alternative mobility options
Maximizing nutrition and initiation of oral feeding	Speech therapist, occupational therapist, dietician, physician	Bedside assessment, oromotor stimulation, videofluoroscopy if concerns about aspiration
Assessment of activities of daily living	Nurse, occupational therapist	Assess dressing, grooming, toileting, bathing, feeding, community mobility, prescribe equipment
Cognitive assessments with determination of beneficial learning strategies	Psychologist, occupational therapist, speech and language therapist, teacher	Formal standardized testing when able to participate, assessment in a small classroom setting
Community reintegration	Occupational therapist, nurse, child life/recreation therapist, teacher, social worker	Facilitate passes to home, school and community venues; problem solve around reintegration issues
Family support	Social worker, psychologist, pastoral care, entire team	Counselling sessions re: catastrophic injury, dealing with loss, coping with stress

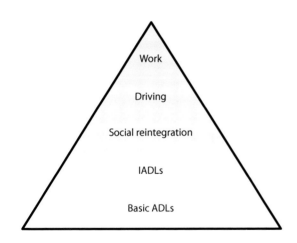

FIGURE 14.3: Hierarchy of rehabilitation goals. ADLs: activities of daily living (for example, mobility, feeding, dressing and grooming); IADLs: instrumental activities of daily living (more complex activities not always done daily but important in daily living, for example, banking and grocery shopping). (Courtesy of Dr. S. Marshall.)

14.3 SPECIFIC CONDITIONS

14.3.1 BRAIN INJURY

The learner will

- Know the common causes and significant impacts of brain injuries

- Appreciate the unpredictability of brain injury recovery
- Recognize the assessment tools utilized to evaluate the severity of the injury
- Appreciate how treatment goals are altered depending on level of consciousness
- Be aware of an excellent clinical resource for brain injury management

To put the material in this section in perspective, it would be helpful for the reader to work through problem 12e-4 (Eddie) on the text Web site.

Injuries to the brain are acquired in many ways and represent a major health concern worldwide. Common causes in North America include motor vehicle collisions, violent assaults and sporting accidents. Financial impacts are profound, with medical and rehabilitation costs, as well as loss of income.

This diagnosis presents the greatest prognostic challenge. It is extremely difficult to predict outcome. When counselling families, the term *predictably unpredictable* is useful. There is no single evaluation tool or even combination of tests that can predict ultimate outcome. Measures such as the Glasgow Coma Scale and the duration of loss of consciousness and post-traumatic amnesia all evaluate the severity of initial injury; nevertheless, it is difficult to reassure families with certainty. Neuroradiological studies and neuropsychological evaluations help to predict probable deficits but there is a lack of precision. It is simply amazing to watch individuals who have been in a coma for months suddenly start talking and eventually to watch them walk out of the hospital. It is equally frustrating to watch people stall and never regain meaningful communication with their environment.

It is notable that many individuals will quote neuropsychological evaluations at two years as being prognostic of long-term cognitive function. Although neuropsychological testing is one of the most useful guides, it is to be used carefully. There are many examples of slow-to-progress cases that make significant improvements outside this timeline. As well, it is important to remember that results of one-on-one structured cognitive tests do not always predict function within a group or social setting.

Treatment of an individual with brain injury varies depending on level of consciousness; utilization of the Ranchos Los Amigos Score can be useful in directing therapy (Table 14.2).

It must be emphasized that physical, cognitive and behavioural realms all require evaluation and treatment plans. All team members are frequently essential. Therapist involvement is usually slowly decreased over

TABLE 14.2: Rehabilitation Principles Guided by Ranchos Los Amigos (RLA Scale of Cognitive Functioning)

RLA Level	Functional Description	Approach	Goals
Level 1	No response	Stimulation oriented	Prevent sensory deprivation, elicit responses, improve attention
Level 2	Generalized response – Inconsistent and nonpurposeful	Brief sessions (15–20 minutes) with structured selective sensory input preceded by brief explanation	
Level 3	Localized response – Specifically but inconsistently to stimuli		
Level 4	Confused–agitated – Heightened state of activity with severely decreased ability to process information	– A quiet, highly structured environment – Constant supervision for safety – Calm reassurance and orientation information	– Decrease intensity, duration and frequency of irritability/agitation – Increase focus and attention to environment
Level 5	Confused–inappropriate – Appears alert and is able to respond to simple commends – Easily confused with complex information and frequent inappropriate responses	– Team goals established (physical, cognitive and behavioural) – Tasks highly structured to approach goals – Environment remains as constant as possible	– Establish focused interdisciplinary goals – Introduce home environment
Level 6	Confused–appropriate – Goal-directed behaviour but dependent on external cues/environment – Little carryover for new learning	– Home passes are short then lengthened to allow transition to home	– Initiate community reintegration
Level 7	Automatic–appropriate – Appropriate behaviour within routine structured environment – Begins to show carryover – Judgement impaired	– Slowly increase complexity of goals in the home and in the community – Transition patients from dependent in a familiar environment to independent in society	– Advance community reintegration
Level 8	Purposeful–appropriate – Alert, oriented – Evidence of new learning – Responsive to diverse environments – Cognitive and behavioural challenges may still be present	– Continue to work on specific client-centred goals	

Source: Modified from Woo, B.H., Nesathurai, S., eds., *The Rehabilitation of People with Traumatic Brain Injury*, Boston Medical Center, Boston, 2000. With permission.

two years but then may be required once again during times of transition (e.g. graduation from school, introduction into the workplace, independent living and marriage). Ensuring the appropriate intervention at the appropriate time is essential in maximizing function.

Keeping up with recommended treatment in rehabilitation is often challenging. Web-based clinical literature summaries and expert analysis can be extremely useful. Such a database exists for brain injury: Evidence-Based Review of Acquired Brain Injury (ERABI; see Web sites listing at the end of the chapter).

14.3.2 SPINAL CORD INJURY

The learner will

- Know the difference between a complete and an incomplete spinal cord injury
- Appreciate the meaning of the level of injury and tests to determine this level
- Have knowledge of key aspects of teaching and care, including management of autonomic dysreflexia (AD), neurogenic bladder, neurogenic bowel and skin care
- Gain an appreciation for timelines for recovery and reasons for hope

To help put this section in perspective, the reader is invited to revisit the story of Cletus (Chapter 5).

The first key to planning rehabilitation for an individual with a spinal cord injury is the designation of the injury as complete or incomplete. When the spinal cord reflexes have returned below the level of the lesion (signalled by return of simple sensory-motor pathways such as the anal wink or bulbocavernosus reflexes), the individual is out of spinal shock (see Chapter 5). Spinal shock can last from one day to three months post-injury, with an average of three weeks. The lesion can then be described as complete (with absence of sensory and/or motor function in the lowest sacral segment) or incomplete (having some degree of sensory and/or motor function in the lowest sacral segment). The level of the sensory and motor deficit can then be defined.

The American Spinal Injury Association (ASIA) scoring system is utilized to standardize documentation. The level of the lesion is defined as the most caudal segment that has a normal test of both sensory and motor innervation.

There are certain key muscles, the evaluation of which helps define the motor level. Each key muscle must have a grade of 3/5 or full antigravity power to be intact (Table 14.3 and Table 2.2).

The most caudal sensory segment must be graded as normal for pinprick and light touch. The face is used as the normal control point.

TABLE 14.3: Key Muscles to Test for Spinal Cord Level of Injury

Myotome	Index Muscle	Action
C5	Biceps brachii	Elbow flexors
C6	Extensor carpi radialis	Wrist extensors
C7	Triceps	Elbow extensors
C8	Flexor digitorum profundus	Finger flexors (e.g. FDP of middle finger, distal interphalangeal joint)
T1	Abductor digiti minimi	Small finger abductor
L2	Iliopsoas	Hip flexors
LS	Quadriceps	Knee extensors
L4	Tibialis anterior	Ankle dorsiflexors
L5	Extensor hallucis longus	Long toe extensors

For pinprick testing, the patient must be able to differentiate the sharp from the dull edge of a pin or neuroprobe. Each segment is scored according to the individual's response (Table 14.4).

For light touch, a cotton tip applicator is compared to the face sensation (Table 14.5).

This information can then be inserted into a comprehensive chart provided by the ASIA (Figure 14.4). This clearly defines the injury.

If the lesion is incomplete there is hope for progressive recovery. Most recovery occurs in the first six months but it can take years to determine final functional capacity.

If the lesion is complete (motor and sensory), the focus is on strengthening the innervated muscles (up to two segments below the lesion) and teaching the client about maximizing independence. This includes intensive teaching about the following:

- The medical emergency of AD in lesions above T6 (see AD definition at the beginning of this section)

TABLE 14.4: Sensory Testing by Pinprick

Scores		
0	Absent	Not able to differentiate between the sharp and dull edge
1	Impaired	The pin is not felt as sharp as on the face, but the patient is able to differentiate sharp from dull
2	Normal	Pin is felt as sharp as on the face

TABLE 14.5: Sensory Testing by Light Touch

Scores	
0	Absent
1	Impaired, less than on the face
2	Normal, same as on face

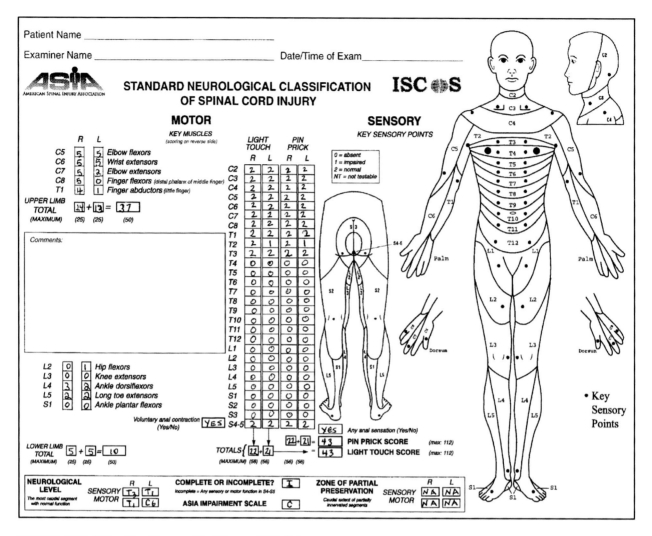

FIGURE 14.4: ASIA Classification Worksheet, 2006. (Worksheet can be freely downloaded from the Web site of the American Spinal Injury Association.)

- Spasticity management, including stretching, avoidance of provoking factors and medication options
- Care of insensate skin to avoid pressure sores, including pressure relief techniques and proper seating options
- Care of neurogenic bladder and bowel to avoid renal failure and social embarrassment

It is important to remember that very high levels of independence can be achieved in spinal cord injury, particularly in injuries affecting the cord below C6.

AD is potentially life threatening (see Table 14.6). It occurs in individuals with a spinal cord injury at level T6 or above. Noxious stimuli to intact sensory nerves below the level of injury lead to unopposed sympathetic outflow and resultant dangerous blood pressure elevation. Due to the spinal cord injury, normal counterbalancing parasympathetic outflow has been disrupted.

TABLE 14.6: Signs and Symptoms of AD

Slow pulse or fast pulse	Flushed (reddened) face
Hypertension (blood pressure greater than 20 mm Hg from baseline)	Nasal congestion
	Red blotches on the skin above level of spinal injury
Pounding headache (caused by the elevation in blood pressure)	Sweating above the level of spinal injury
Restlessness	Goose bumps
Nausea	

Parasympathetic outflow through cranial nerve X (vagus) can cause reflexive bradycardia. This does not, however, compensate for the severe sympathetic-induced vasoconstriction.

AD is a medical emergency and can lead to intracranial hemorrhage, seizures and death. When symptoms occur, the patient should be transitioned to a sitting position with the head constantly elevated. The bladder must be checked for distention, the bowel for impaction and the skin for pressure areas. Any cause of noxious stimulation should be relieved. Blood pressure should be checked every three minutes and treated medically if systolic blood pressure is greater than 150 mm Hg. Medications commonly used include nitroglycerine paste (applied above the level of the injury), nifedipine capsules (immediate release form) and intravenous

antihypertensives in a monitored setting. The blood pressure will need to be monitored for at least two hours after the resolution of an AD episode.

Bladder function is important to understand and manage (Figure 14.5). The bladder and its internal sphincter (bladder neck) have dual innervation. The stimulation of the cholinergic (parasympathetic) system allows normal voiding, causing contraction of the bladder musculature

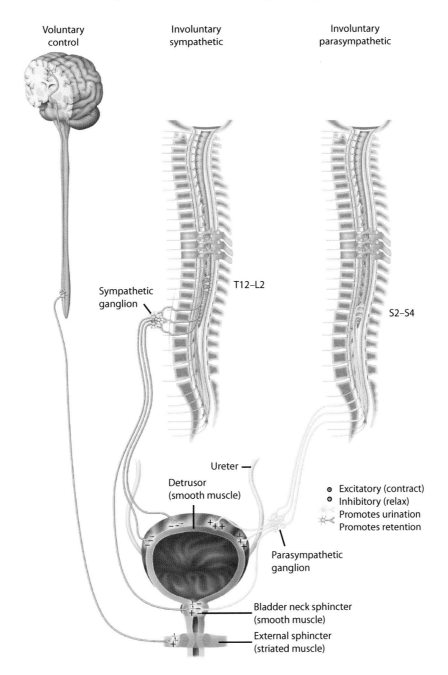

FIGURE 14.5: Bladder innervation: The effect of each system on the bladder (detrusor) muscle and the internal (smooth muscle) and external sphincters (striated muscle) is indicated by + = excitatory and − = inhibitory. The parasympathetic system (S2, 3, 4) promotes emptying of the bladder (indicated in green). The sympathetic system (T12, L1, L2) promotes urinary storage (indicated in red). The cortical voluntary pathway goes to lower motor neurons in the sacral cord (the conus medullaris) and these neurons control the external voluntary sphincter; destruction of the descending influence may lead to urinary incontinence.

and relaxation of this sphincter. The sympathetic system allows storage with relaxation of the bladder and contraction of the sphincter. It is the balance between these two systems that determines if the bladder is storing or emptying. Injury to the spinal cord can disturb this balance and result in a bladder that stores excessively and overflows or fails to store due to excessive bladder contraction or sphincter failure.

From a practical perspective, when neurological injury has interfered with bladder function, it is important to remove a continuous catheter as soon as clinically possible. This will limit the possibility of infection. Often, there is a combination of poorly coordinated contractions and incomplete emptying. Management of the bladder during recovery frequently includes the following:

1. Asking patients to try to void.
2. Checking to see if there is urine remaining in the bladder (with ultrasound or catheterization).
3. If the remaining urine is greater than 75 to 100 cc, catheterizing every four hours to maintain bladder volumes less than 350 cc.
4. If this does not achieve continence, anticholinergic medication or sympathomimetics can increase storage and prevent irregular contractions. Urodynamic studies, which define bladder filling, bladder pressures, and urine flow, will help define the need for these medications.

If recovery is not expected, catheterizing every three to four hours to maintain volumes less than 350 cc and normal pressures is recommended to maintain renal health.

In spinal cord injury, there is often evidence of upper motor neuron bowel dysfunction. In general, there is loss of spontaneous control of defecation. The key to treatment of the neurogenic bowel is training individuals to stimulate reflex evacuation and maintaining normal amounts of soft stool. This can be done using the following clinical treatments:

1. Counselling about adequate water and fiber intake
2. Regular timed toileting with attention to positioning (hip-thigh angle at 90 to 110 degrees and feet well supported)
3. Utilizing the gastrocolic reflex by toileting 15 to 20 minutes after a meal with digital stimulation of the anal sphincter if required

When these have been tried and there are still difficulties, stool softeners can be used if the stool is too firm. Oils (mineral oil or lansoyl) may be employed to help transit of the stool and bulk-forming agents may help as further distention of the bowel aides in emptying. In addition, suppositories may be used to stimulate emptying. As a last resort, bowel stimulants can be used. Usage of the latter may result in dependence over the long term. Enemas may be required initially, particularly if constipation has been prolonged.

Alteration in sexual function is also commonly experienced in patients following spinal cord injury. Individuals may be hesitant to speak of these issues and medical professionals may need to ask directly about these concerns.

In male patients, difficulty with initiating or sustaining firm erections is frequently reported. In general terms, there are two regions of the spinal cord implicated in these difficulties. Outflow from T12–L2 (sympathetic nervous system) is primarily responsible for the psychogenic component of the erection. Spinal cord injuries above T12 therefore often result in difficulties in erections stimulated by arousing thoughts originating from the brain. Reflex erections stimulated by tactile input are mediated through S2–S4 (parasympathetic nervous system via the cauda equina). Lesions below S1 therefore limit input into this reflex and diminish response. A lesion between L1 and S3 decreases neuronal communication between both centres. This communication is important for increasing the strength and sustainability of the erectile response.

Patients need to know that these problems exist and that there are a number of possible treatments for erectile dysfunction related to spinal cord injury, including physical techniques, medications and surgical procedures.

In female patients, concerns include diminished responses to tactile stimulation, decreased lubrication and increased spasticity during sexual activity. Each of these issues can also be addressed to help individuals improve their quality of life.

Fertility may also be a concern. Commonly, female fertility is unchanged when regular menses are reestablished. This usually occurs approximately one year post-injury. There are, however, some increased risks reported with pregnancy and delivery in this population. These include increased risk of AD, urinary infections, pressure sores and premature deliveries. Specialized obstetrical attention is recommended.

In males, decreased fertility is common. Although erection can frequently be attained, ejaculation occurs in only a small percentage of individuals with spinal cord injury and sperm quality is often diminished. There is hope for fertility but treatment may be required through a fertility clinic and a centre that has experience in treating individuals with spinal cord injury.

Whether an injury to the cord is complete or incomplete, there is certainly reason for optimism. There are many promising areas of research, including neuroprotection, nerve regrowth and stem cell work. Excellent care of an injured patient is essential, for both present and future opportunities.

14.3.3 STROKE

The learner will

- Appreciate the different etiologies for stroke, the variability in presentation of neurologic deficits post-stroke and, thus, the variation in needs from a rehabilitation perspective
- Gain knowledge about indicators that assist in the choice of appropriate candidates for intensive rehabilitation programs
- Be aware of the urgency of timely swallowing assessment post-stroke
- Appreciate the complexity of communication-based difficulties after neurologic injury
- Be aware of possible timelines for recovery
- Be introduced to an excellent evidence-based database for stroke rehabilitation

To help put this section into perspective, the reader is invited to revisit the story of Etienne (Chapter 8).

A stroke is the neurologic injury that occurs as a result of one of a number of vascular disease processes. Stroke is classified into two major types:

- Brain ischemia due to thrombosis, embolism or systemic hypoperfusion (approximately 80%)
- Brain hemorrhage due to direct intracerebral bleeding or to an adjacent subarachnoid hemorrhage (approximately 20%)

The challenges a stroke survivor experiences are highly dependent on the location and extent of the vascular insult. This, in turn, determines the patient's needs from a rehabilitation perspective.

The first issue that often faces a rehabilitation clinician is whether a patient is a good candidate for rehabilitation interventions. Initial cognitive assessments are very important. Individuals must learn throughout the rehabilitation process. Evidence of severe memory problems or an inability to follow instruction significantly limits one's ability to participate in rehabilitation. Other poor prognostic indicators include previous stroke, advanced age, prolonged urinary and bowel incontinence and severe visuospatial deficits.

Swallowing problems need careful assessment early in the course of this condition. Speech and language pathologists and occupational therapists will complete a bedside assessment. If concerns are noted, a videofluoroscopic assessment of the patient's swallowing in the radiology suite, with therapist involvement, may be required. This will assess coordination of swallowing and check for aspiration of solids or liquids into the lungs. Aspiration may produce surprisingly few signs and symptoms; subsequent pneumonias can be life threatening. Careful attention to swallowing issues is therefore imperative to good care.

Speech problems associated with stroke are also common and can be complex. They may be receptive, expressive or both. Frequently, the Boston classification system is utilized. The problems with language are separated into those that are fluent (with smooth flow of word production) or non-fluent (without such flow) (Table 12.2). An experienced speech and language pathologist can assist in the diagnosis of this stroke-related problem and develop a treatment plan for communication. The rehabilitation team that is working with the client will then follow these recommendations to maintain consistency and promote learning in this area. If there is evidence of receptive abilities but significant language output problems, communication tools including newly designed computerized output devices may be quite useful.

From a motor perspective, the majority of recovery occurs in the first three to six months following the stroke. Multiple hands-on techniques are used by therapists to facilitate recovery. As a clinician, it is important to support weak upper extremities with the goal of preventing subluxation of the shoulder and pain syndromes. When high tone interferes with function, stretching, avoiding nociceptive input, bracing, botulinum toxin injections and oral tone-mediating agents can be extremely useful. This will allow therapists to focus on strengthening antagonist muscles and maximizing function.

There are summaries of critically appraised data for the clinician now available on the Internet and in written form. The Evidence-Based Review of Stroke Rehabilitation (see Web sites references at the end of this chapter) is an ongoing review and synthesis of data that examines practices in stroke rehabilitation. This comprehensive information can guide best practices. The new treatments and aspects of prevention described are constantly evolving; this gives hope to survivors of stroke.

14.3.4 PEDIATRICS

The learner will

- Appreciate the differences in rehabilitation of children (versus adults) that result from ongoing growth and development
- Understand the concept of 'growing gap'
- Gain knowledge about unique features of common childhood-acquired disabilities, such as cerebral palsy (CP) and spina bifida
- Be aware of advances in pediatric rehabilitation, including classification of CP and spasticity management

To help put the material in this section in perspective, it would be helpful for the reader to work through problem 7e-4 (Deepa) on the text Web site.

Rehabilitation in the pediatric population has many unique features. A child is rapidly changing in size, cognitive skill and behaviour. Physically, children with motor deficits are at great risk for rapidly developing soft tissue contracture or orthopedic complications, such as scoliosis or hip subluxation. Appropriate stretching, bracing and spasticity management are essential to keep muscles growing along with rapidly growing bones. Frequent equipment changes are also required as needs change with development and growth.

From a developmental and behavioural perspective, there is also rapid change! Learning is a very active process; children are adaptable and studies regarding neuroplasticity hold great hope. From another perspective, children who have physical, cognitive or behavioural challenges secondary to neurological insult are at risk of developing a gap between their skills and the skills of normally developing peers. As rapid development occurs, this gap may progressively enlarge. This is referred to as the concept of *growing gap*. The role of many clinicians in the field of pediatric rehabilitation is to promote development and limit the size of this gap. Frequent reassessments and therapeutic interventions are required, particularly during times of change and transition, for example, school entry, entry to elementary school, transfer to high school, planning for after-school training or moving out of a parent's home. Each of these transitions may require environmental alterations, new training and changes to resources or equipment. Rehabilitation in children must match the rate of change.

Diagnostically, the patient population requiring rehabilitative intervention in the pediatric group is quite different from that in adults. There is a high percentage of individuals with congenital, neonatal or perinatally acquired causes of disability, such as CP, spina bifida and neuromuscular disorders. In these cases, the lack of a premorbid baseline of normal function may present a challenge. Early intervention is the standard of care, with the provision of an enriched environment and age-appropriate habilitative treatment plans. The children are not trying to regain function that was already established, but are trying to maximize the potential they have.

CP is the most common cause of complex physical disability in childhood, with an incidence of 1.5 to 2.5 per 1000 live births. CP is defined as 'a group of disorders of *development of movement and posture* causing activity limitation that are attributed to *nonprogressive* disturbances in the fetal or infant *developing brain*' (see Bax et al. in the References). The underlying motor deficit is frequently accompanied by abnormalities in sensation, cognition, communication, perception and behaviour, as well as seizures. The clinical picture is therefore complex.

Over the last few decades, researchers have established a motor classification for CP that allows prediction of whether a child will eventually walk independently or will require the support of a walker or a wheelchair for independent mobility. This tool, the Gross Motor Function Classification System ER (expanded and revised; see Palisano et al., 2007), can allow motor prediction soon after diagnosis and the classification appears to be fairly stable into the second decade of life. Any clinician working with individuals with this diagnosis should be aware of this scale, as this knowledge will help in the setting of reasonable motor goals. Classification systems are also evolving for upper extremity function (Manual Ability Classification System; Eliasson et al., 2006) and communication (Communication Function Classification System; Cooley Hidecker et al., 2011). These systems of classification promote focused assessment, efficient and clear communication among clinicians and allow grouping of patients for treatment/monitoring protocols and multi-centre research.

Management of high tone states (spasticity, dystonia and rigidity) has always been a challenge in the area of CP. Over the past few decades, management of spasticity, as well as involuntary movements (dyskinesias), has changed. Newer options such as botulinum toxin injections in hypertonic muscles, intrathecal placement of baclofen and surgical cutting of dorsal nerve root fibers have all been investigated and present alternatives for those individuals who are good candidates for such interventions. As well, oral antispasticity medications such as baclofen, dantrolene and tizanidine can be considered after careful consideration of the benefit versus side-effect ratio. The clinical key is careful assessment of patients to match the proper candidate with the appropriate intervention. Specialty clinics have been developed to facilitate this decision making and properly assess outcomes.

There have also been improvements in the management of infection, nutrition and musculo-skeletal deformities. These improvements, coupled with evidence of functional changes in the brain on functional magnetic resonance imaging (MRI) studies following therapeutic intervention, growth in the field of technology to improve quality of life and the rapidly progressive field of neurotherapeutics (for example stem cell therapy; see Ruff et al., 2013) provide reasons for hope. There is great need for clinicians to work hard to promote ongoing maximal health and well-being in these patients throughout their life spans in order to ensure an optimal level of health as breakthroughs occur.

Spina bifida, though decreasing in incidence, continues to be a significant cause of childhood-acquired disability. The complexity of this condition should not be underestimated. This population is most challenging from a rehabilitation perspective. The combination of an abnormality in spinal cord development (with motor deficits, sensory deficits, neurogenic bladder and bowel) and concomitant cerebral malformations (with cognitive limitations) results in the need for a flexible individualized habilitation program that frequently combines the concepts of spinal cord and brain injury

In this field as well, there have been significant advancements. *In utero* surgical intervention is now possible to limit the impact of hydrocephalus in the developing fetus (see Adzick, 2013). One can envision how advancements in treatments for brain injury, CP and spinal cord injury will also impact treatment for this condition. Certainly there is reason for great optimism.

14.4 CONCLUSION

In general, ongoing advancements in the field of rehabilitation lead to hope for improvement in body function and structure, activity and participation. Technology advancements, including the use of functional MRI in training, evolving virtual reality systems that allow remote programming and enhanced engagement, the use of new bracing techniques to treat contractures and robotics to support gait training, all allow highly skilled team members to develop more and more challenging treatment protocols. Internet databases with expert evaluation of literature are now becoming available to assist the clinicians in dealing with the ever-growing information explosion. Clinicians, clients and families can therefore be actively involved in the rehabilitation process as they strive for maximal outcome.

REFERENCES

Adzick, N.S. Fetal surgery for spina bifida: Past, present, future. *Semin Pediatr Surg* 2013. 22(1): 10–17.

Bax, M., Goldstein, M., Rosenbaum, P., Leviton, A., Paneth, N., Dan, B., Jacobsson, B., Damian, O.D.; Executive Committee for the Definition of Cerebral Palsy. Proposed definition and classification of cerebral palsy. *Dev Med Child Neurol* 2005. 47: 571–576.

Eliasson, A.C., Krumlinde-Sundholm, L., Rösblad, B., Beckung, E., Arner, M., Ohrvall, A.M., Rosenbaum, P. The Manual Ability Classification System (MACS) for children with cerebral palsy: Scale development and evidence of validity and reliability. *Dev Med Child Neurol* 2006. 48: 549–554.

Hamilton, B.B., Laughlin, J.A., Fiedler, R.C., Granger, C.V. Interrater reliability of the 7-level functional independence measure (FIM). *Scand J Rehab Med* 1994. 26(3): 115–119.

Hidecker, M.J., Paneth, N., Rosenbaum, P.L., Kent, R.D., Lillie, J., Eulenberg, J.B., Chester Jr., K., Johnson, B., Michalsen, L., Evatt, M., Taylor, K. Developing and validating the Communication Function Classification System for individuals with cerebral palsy. *Dev Med Child Neurol* 2011. 53(8): 704–710. doi:10.1111/j.1469-8749.2011.03996.

Palisano, R., Rosenbaum, R., Bartlett, D., Livingston, M. *GMFCS – E & R*. Ontario, Canada: CanChild Centre for Childhood Disability Research, McMaster University, 2007.

Ruff, C.A., Faulkner, S.D., Fehlings, M.G. The potential for stem cell therapies to have an impact on cerebral palsy: Opportunities and limitations. *Dev Med Child Neurol* 2013. 55(8): 689–697.

Woo, B.H., Nesathurai, S., eds. *The Rehabilitation of People with Traumatic Brain Injury.* Malden, MA: Blackwell Science, 2000.

SUGGESTED READING

Alexander, M.A., Matthews, D.J. *Pediatric Rehabilitation: Principles & Practices*, 4th ed. New York: Demos Medical Publishing, 2009.

DeCorwin, S. *Life after Spinal Cord Injury.* Montreal, Canada: Quebec Paraplegic Association (QPA), 1997.

Frontera, W.R. *DeLisa's Physical Medicine and Rehabilitation: Principles and Practice*, 5th ed. Philadelphia, PA: Lippincott Williams & Wilkins, 2012.

Howe, J.A., Inness, E.L., Venturini, A., Williams, J.I., Verrier, M.C. The Community Balance and Mobility Scale – A balance measure for individuals with traumatic brain injury. *Clinical Rehabilitation* 2006. 20(10): 885–895.

Palisano, R., Rosenbaum, P., Walter, S., Russell, D., Wood, E., Galuppi, B. Development and reliability of a system to classify gross motor function in children with cerebral palsy. *Dev Med Child Neurol* 1997. 39: 214–223.

WEB SITES

Evidence-Based Review of Moderate to Severe Acquired Brain Injury (ERABI): www.abiebr.com

The Evidence-Based Review of Stroke Rehabilitation: www.ebrsr.com

Canchild: GMFCS-ER: www.canchild.ca

Chapter 15

Ethics and Neurology

15.1 INTRODUCTION

In the previous chapters, various clinical neurological situations have been presented that challenge physicians, be they specialists or students, in their problem-solving abilities and mirror those found in the setting of bedside, office or consulting room. In real life, there are almost always issues that complicate the management of medical problems because patients are people in an environment where the illness impacts on their personal lives, on their family and on their loved ones and, not infrequently, on society. In addition, medical decisions must take into account issues of ethics. It should be emphasized that ethics is an integral part of medical care and that good ethical practice makes for good medicine, and good medicine is ethical medicine. In this chapter, we will present another neurological case, but one where the emphasis will be on some of these other features of medical practice.

While ethics is involved in all of medicine, there is a particularly close association between ethics and neurology. The relationship may be considered as a reciprocal one.

Ethical decisions are frequently required in the management of patients suffering from neurological diseases. There are many neurological diseases that invoke life-and-death situations where intervention calls ethics into play. These involve people at both the beginning of life and at the end of life, but also in the most active and productive stages of life. Examples include the treatment of anencephalic newborns and patients with fatal conditions such as amyotrophic lateral sclerosis, muscular dystrophy, multiple sclerosis and persistent vegetative states.

The understanding of the source of ethics may have its underpinnings in neurology, and neurological studies may contribute to our understanding of the basis of ethics. Humans seem to be ethical creatures and ethical behaviour is dependent on functioning of the brain. Indeed, some recent work suggests that there are areas in the brain whose function is necessary for moral and ethical decision making on the part of the individual.

The format to be followed is the presentation of a theoretical medical narrative in which there are many ethical issues. The story is presented as a vignette. The reader is encouraged to think about the narrative as it unfolds and to try to see what ethical issues may be inherent in the situation. Questions will be raised at the end of each section. Try to answer them yourself before proceeding to the discussion of these questions. Suggestions for further reading are given at the end of the chapter.

15.2 CASE STUDY, PART A

You are a consultant neurologist but have been practicing less than a decade.

A man comes to see you with regard to his mother, whom you have been treating for many years. Mr. C. Ewan Court is a 45-year-old lawyer who has developed a successful practice and has become partner in an important, well-established law firm. He comes to see you for information on the health of his mother, Mary Court, who has been suffering from peculiar neurological symptoms for more than a decade.

Mrs. Court, now in her early sixties, has been living at home alone since the death of her husband. When she was in her late forties, she developed some problems with balance and coordination, and she exhibited peculiar involuntary movements of her hands. She developed mannerisms where she would bring her hand to her face and then sweep it along the side of her head as if patting back her hair. Occasionally, she would shrug her shoulders, first the right and then the left. These peculiar movements were considered as nervous habits or tics as she was under a great deal of stress because of her husband's poor health. He was being treated for bipolar disorder and at the time was in a depressed mood. More recently, she suffered from many falls and developed a lurching gait. Initially, her condition was quite puzzling to you and you have had to do a considerable amount of reading to make a diagnosis. You remain uncertain although you have some ideas.

She is now showing not only abnormal erratic involuntary movements but seems to be getting very 'absent minded'. It is clear that in addition to the movement disorder, she has a dementing illness. She has been leaving the stove on and the doors unlocked and even forgetting to close them at night. Recently, the neighbours called the fire department because they saw smoke coming out of an open kitchen window. She had apparently left the stove on and, when a frying pan caught fire, she opened the window to let the smoke out but did not turn off the burner or remove the frying pan from the stove. The family physician had advised her son that his mother should be hospitalized for her own protection and that he should consider making plans to apply to long-term care facilities on his mother's behalf. However, she is most insistent that she remain in the house that has been her home for the past 30 years and flatly refuses to consider a move.

While Mr. Court came in to ask you about his mother, it is apparent that he is also concerned about his own health and takes advantage of the visit to consult you about his concerns.

15.2.1 QUESTIONS TO BE CONSIDERED

1. Does Mr. Court have a right to know about his mother's medical condition, or is this a private matter? What would you do? Is there any ethical way in which you feel free to discuss her health with her son? Are there situations in which confidentiality can be set aside?

2. What are the most likely causes for these symptoms? Do you know what the diagnosis is in this case?

3. How do you handle the situation where you don't know what you are dealing with or are uncertain about the diagnosis? What should you tell the patient?

4. When should you consult colleagues? Is it necessary to get the patient's permission before discussing it with colleagues?

5. At what point do you consider it necessary to recommend that a patient be hospitalized or placed in a long-term care facility? What rights does a patient have to refuse such advice?

6. Should you take on Mr. Court as a patient simply because you are his mother's physician? In general, should a specialist-consultant demand that all patients be referred by a family physician before they are seen?

15.2.2 DISCUSSION

1. At this point, the patient is Mrs. Court and not her son. The physician owes the patient privacy and confidentiality. *Privacy and confidentiality* are important ethical issues in medical practice. Every patient expects that the physician will treat any information given as being confidential and not to be divulged to any third party. This includes family members. Ethical considerations depend on certain basic principles (Beauchamp and Childress, 2012). One of the most important of basic principles in ethics is *respect for the autonomy of the individual*. This is the basis underlying the respect for privacy and confidentiality.

The observance of the rule regarding confidentiality should include family members. There are some exceptions to this rule. A child before the age of majority will usually be accompanied by a parent or guardian who is responsible for the care of the child. The physician will discuss the child's medical problem with the parent who is charged with the care of the child. The exact age where this applies varies from one jurisdiction to another and from province to province or state to state. The greatest difficulty arises in cases of sexual activity or birth control, where a young person may seek medical advice and does not wish to have parental involvement. This is a very delicate situation and requires great sensitivity on the part of the physician. As in many areas of medical ethics, there may not be an absolute right answer but a good rule is to consider the best interests of the patient, while still realizing that the family should be treated with respect.

Another exception occurs when the patient is not mentally or otherwise capable or competent and a family member is involved and responsible for the person's welfare. This may be formalized where a person is appointed as having power of attorney (POA) for health care. More often, there is an informal understanding that a close family member or members are looking after a sick and incapable family member or loved one.

In this age of multiculturalism, these matters of privacy and confidentiality, as observed in much of the western world, may have to be modified somewhat. In certain societies, it is accepted that the family plays a much more intrusive role in the affairs of all its members: sometimes, an elder or the dominant female or male assumes a decision-making responsibility for the other members. A husband may make decisions for his wife's medical care. While this may be offensive to North American practitioners, it is important that the doctor be aware of the traditions of their patients and handles them with diplomacy and sensitivity (Singer, 1999).

In the case under discussion, the son has assumed the responsibility for the care of his mother and it may be obvious to you, who have been consulting in this woman's care for many years, that the son has a legitimate interest in the health of his mother and is a responsible family member. This may not always be the case and it may be appropriate to request that the person who is your patient give a formal permission before you even speak to family members.

Occasionally, a person may come into your office and ask casually, 'I saw my neighbour, Bill Jones, in your waiting room. Doc, how's my old friend, Bill, doing?' It may be tempting to respond in an equally casual manner, but you should reply: 'I'm sorry but I don't talk about other patients. I know that you wouldn't want me to discuss your health with Mr. Jones. Why don't you ask Bill, himself? I am sure he would be glad to know that you are interested in his health'.

2. Mrs. Court has symptoms that are clearly of a neurological nature. In neurology, the traditional approach is to try and localize the possible causes for symptoms and signs, then to decide what could be the mechanism or cause and finally to try and fit them into the pattern seen in a particular disease or disorder.

Loss of balance and incoordination commonly occur together, and when they do, they may point to certain functional areas of the nervous system. Balance or equilibrium involves appreciation of position sensory information from the lower extremities, mechanisms in the inner ear and a disturbance of monitoring and modification of movement by cerebellar pathways.

Involuntary movements may arise when there is pathology in the neuro-circuitry involved in the basal ganglia or arise from pathology in the motor cortex. Abnormal movements that seem involuntary, such as tremor or tremulousness, most commonly are related to anxiety but are also seen in hypermetabolic states such as hyperthyroidism and excessive use of stimulants such as caffeine and nicotine.

Changes of behaviour are often early evidence of a dementia and suggest more diffuse involvement of the brain. Such changes, however, are not always due to organic disease; anxiety and depression may cause an 'absent-mindedness' or distractibility that has been described as 'pseudo-dementia'. However, people with depression may also show reduced activation of certain prefrontal areas of the brain on single photon emission computed tomography (SPECT) scan independent of their dementia.

What conditions present with the constellation of symptoms of movement disorder and cognitive decline? Your observations of the patient are most important. What kind of abnormal movements do you see? Indeed, are they abnormal or are they merely restlessness associated with anxiety? Would you describe it as tremor? Are the movements sudden, tic-like, or do they fit the description of chorea (dancing) or athetotic (writhing) movements. If tremor, essential tremor or the resting tremor of Parkinson's disease come first to mind, but metabolic conditions with accelerated metabolism such as hyperthyroidism may also present with tremor. People with emotional or psychiatric conditions may often present with a fine tremor. The tremor associated with anxiety will be transient and is seldom complained of by the patient and may be aggravated by the tension associated with seeing a physician. Tremor is commonly seen in people suffering from alcoholism and in those consuming excessive caffeine or taking certain medications such as lithium and valproic acid as well as drugs used as anticancer treatment.

Cognitive decline may be an accompaniment of Parkinson's disease but is usually not apparent until late in the course of the disease. On the other hand, cognitive decline may antedate the movement disorder in dementia with Lewy bodies, so named because of the accumulation of abnormal protein material (Lewy bodies) in neurons in regions of the brain that control functions like movement, memory and cognitive ability.

Also, there are a number of obscure less frequent conditions to be considered. The two most likely, although rare, are Wilson's disease (hepatolenticular degeneration) and Huntington's disease (HD) (Huntington's chorea). Other causes of chorea include the once common, but now rare, acute Sydenham's chorea and Choreo-acanthocytosis, which should not be difficult to diagnose early in life with simple blood examination. These are unlikely to be mistaken for HD. Even rarer, dentato-rubro-pallido-luysian atrophy and one of the many variants of spinocerebellar atrophy are to be considered but are unlikely. Suspected Wilson's disease can be confirmed by the finding of corneal pigmentation (Kayser–Fleischer rings) and blood tests in which abnormally low levels of the copper-binding protein, ceruloplasmin, are found and there is evidence of abnormal copper metabolism. In both HD and Wilson's disease, neuro-imaging will show abnormalities in the parts of the brain most affected. In Wilson's disease, there may be atrophy of the putamen evident on computed tomography (CT) scan and magnetic resonance imaging (MRI) will show the 'Double Panda sign' in the brainstem. In Huntington's, caudate atrophy may be apparent but is non-specific. In both Wilson's disease and HD, there is usually a positive family history. The latter can be confirmed by genetic testing; however, the diagnosis may be established on a clinical basis alone. The picture presented by Mrs. Court fits best with that of HD.

3. You may well be puzzled by this case if you are not familiar with some of the uncommon disorders that may present in this way. It may well be that having been in practice for a relatively short time, you may not have encountered some of the rarer disorders. It is important to be open and honest with the patient.

One of the most important ethical issues in medical practice is that of *fidelity or faithfulness*, which may sound like old-fashioned terms and perhaps are better described as *responsibility or trust*, of which there are many aspects. Among them is the implication that the physician has certain duties to the patient, who will have certain expectations of his or her physician. One expectation is that the physician is competent and knows what she or he is doing. The doctor is expected to be knowledgeable about diseases from which the patient may be suffering. The study of medicine is a never-ending learning process, and physicians may have to spend time boning up on conditions with which they may not be familiar or not have thought about since medical school. Wise physicians know the limits of their competence. Physicians should be honest with patients and not deceive them by implying that they know more than they do. Therefore, it is important to tell the patient of your uncertainties but assure them that you will make every effort to establish the diagnosis and to treat them appropriately.

4. Where the physician feels incapable about handling a specific condition or situation, she or he has a duty to refer the patient to a more knowledgeable or more experienced colleague or specialist. Frequently, one may simply discuss the case with a colleague, but taking great pains to hide the identity of the patient. These 'corridor consults' may save time, but often, the consultant is put on the spot and may not be able to give appropriate advice. Furthermore, some of the particular skills that a specialist has acquired include specific history taking and experience in a specific aspect of the physical examination. These provide information that cannot be easily transmitted from one physician to another. It is therefore preferable to ask for a formal consultation. One should never let personal pride get in the way of providing the best service to your patient. Patients will have more respect for you if you are honest with them.

5. There is reasonable concern that Mrs. Court may harm herself by setting her house on fire or through other inadvertent events. She plays down these dangers and resists all coaxing to move to a safer environment. While one must respect autonomy, there is another basic principle in medical ethics that may seem to be in conflict with autonomy, *beneficence*. This means that the physician should always consider what is best for the patient. She or he should always treat the patient with beneficence. In cases where the patient is not capable of acting independently, a choice may have to be made that would seem to override the autonomy of the patient in order to serve the patient's best interests. In such situations, we say that 'beneficence trumps autonomy'. In this case, although Mrs. Court wants to remain in her home, it may be necessary to override her autonomy by admitting her to a residence where she would not be in danger of harming herself. In the past, physicians assumed the controlling role and were presumed to always know what was best for a patient. This has been referred to as *paternalism* (*parentalism* would be a better gender-neutral term). Now, patients are granted greater autonomy in part because medical knowledge is more widespread. Consequently, medical decision making should be a shared activity between caregivers and those cared-for, with involvement of other interested parties such as family members and significant others as agreed upon by the patient.

Someone, usually a family member, will have the *POA for Health Care* to ensure that the patient, if considered incompetent, will be given the best care, which may include commitment against the person's will. Before one can force a patient to go into hospital or chronic care facility, one must be assured that he or she is incapable of appreciating the facts of the situation. Competency adjudicators are established in many jurisdictions. The family may be well advised to seek legal counsel in questionable situations.

Commitment to a chronic care facility may be temporarily deferred by making some compromise arrangements for Mrs. Court at home. For instance, measures to ensure safety in the kitchen such as disconnecting the stove and allowing her to prepare meals with a microwave oven and a kettle with an automatic cut-off may be a reasonable solution in the early stages of her dementia.

6. While Mr. Court is supposedly coming in to ask about his mother, it is apparent that he has some concerns about his own health and future. He asks questions that suggest that he wants you to be his physician, although he does not specifically say so. While most consultants request that a formal request be made from the family physician, it sometimes happens that a new patient will come in this way without formal referral. This is especially true where there is a familial or hereditary condition. While there may be concerns about 'cue jumping', it is often advantageous to treat the family members as a group if there is a familial or related condition present. However, you should inform Mr. Court that he should ask his family physician to make a referral and then set up an appointment when you will have the time to consider his particular case fully rather than make assumptions. It is important that the physician keep patients separate, using separate sets of records and utilizing treatments according to each individual's needs. It may be difficult for the doctor to avoid giving confidential information about one family member to another member. It is often preferable to refer one of the patients to a colleague.

In the end, it is often more a matter of local practice tradition or 'what is done' than a matter of ethics, but one must keep ethical principles of confidentiality and respect for autonomy in mind.

15.3 CASE STUDY, PART B

Mrs. Court finally agrees to go into hospital, but only for some tests. You obtain a CT scan of the brain and the scan shows diffuse cortical atrophy and considerable bilaterally symmetrical atrophy of the caudate nuclei, as may be seen in HD, but this is a non-specific finding. A battery of blood tests gives normal results.

The following week, you receive a phone call from one of the senior partners in Mr. Court's law firm Didley, Didley and Skwatt. He says that he and some of the other partners have noted some personality changes and he is aware that Mr. Court has consulted you. He says that Mr. Court had been a good partner and had made major contributions to the firm and they were even considering making him a full senior partner. He was very hardworking and often stayed late at the office. Recently, however, he has on occasion gone to a bar after work with some of the secretaries. Furthermore, he has been seen at a disco bar dancing with women who appear to be decades younger than he. Some of his partners are concerned about the appropriateness of some of this socializing with younger office staff. Although he didn't say so, you sense that he was more concerned that such behaviour would reflect badly on the firm; nevertheless, you also sense that there is some genuine concern over his colleague's well-being. You are rather surprised to receive this call and acknowledge that you have seen Mr. Court as a patient. However, you ask whether Mr. Court has authorized the caller to make these enquiries. The lawyer is somewhat taken aback and says that, he is well aware that doctors are not allowed to divulge private information to third parties and he is concerned that Mr. Court may be mentally ill and his colleagues wanted to help. Furthermore, he adds, Mr. Court is the key person in a most important deal with their major real estate client. You try to be courteous and say that you appreciate the situation but you are not allowed to divulge any information without your patient's express permission.

15.3.1 QUESTIONS TO BE CONSIDERED

7. How would you respond to this situation? What are your obligations to Mr. Court? What are your obligations to significant others on whom your actions may impact? What obligations does the physician have to society? When may your obligations to your patient be secondary to your obligations to other parties?

15.3.2 DISCUSSION

7. This brings up again the important medical ethical issue of privacy and confidentiality as has been discussed already. One can appreciate that the partners of the law firm have legitimate concerns, other than a concern over a partner's well-being, and are anxious to protect their interests. This situation presents a dilemma for the physician. He must be ever mindful of the confidentiality that he owes his patient but he also has some responsibility to society. The situation sometimes arises in which withholding privileged information may allow harm to come to a third party or even to a large number of people. One can think of situations where a physician might become aware of planned terrorist activities. The physician must make a decision as to which is the greater good and act accordingly. There may not be one right answer to the dilemma. A famous landmark case is that of Tarasoff v. Regents of University of California (1976), where a university psychologist became aware of his patient's professed intent to murder a college student but was hesitant to report it; after the murder took place, he was charged with criminal negligence. A somewhat similar case occurred more recently in Canada where an emotionally disturbed man shot his estranged wife with a cross-bow and had previously disclosed his intentions to his psychiatrist. Unfortunately, the physician did not feel that the intentions were sufficiently certain to report his patient to authorities. In these cases, the *duty to warn* is perceived as overriding a duty to confidentiality.

While our case is hardly one with such potentially fatal results, the same principle applies. One way of dealing with this situation would be to discuss the work situation with Mr. Court and enquire whether he feels confident that he is able to carry on the heavy responsibility that he has to the firm. He may well be unaware of the partner's concerns. Since he has expressed his own self-doubts previously to you, this may be seen as an opening for further discussion.

15.4 CASE STUDY, PART C

What about Mr. Ewan Court, himself?

As already mentioned, Mr. Court had expressed concerns about his own health. While he dismisses that he has a serious problem, he agrees to get a formal referral from his family doctor and you agree to see him on a later occasion.

At that visit, he admits that he has difficulty concentrating and finds that he is very impatient. He confides that an important client assigned to him involves a complicated legal battle regarding land claims. He feels unsure of his ability to meet the demands of the contract; this makes him anxious and has led to disturbed sleep. You notice that he does seem quite anxious. While he is talking to you, he is fidgeting with his pen and with the buttons on his jacket. During the functional enquiry, he confides that he has had some emotional difficulties in the past and at college had a bout of depression during which he had contemplated harming himself. He was seen by the campus physician and the diagnosis of bipolar disorder was considered at that time but he was not prescribed any medication. You carry out a thorough neurological examination but do not find any obvious abnormalities other than his nervous fidgeting. However, you start to wonder if he may be developing HD himself.

You ask yourself whether you should suggest that he have genetic testing. You also think about the implications of genetic testing and who, other than Mr. Court, might be affected by the results: who might benefit and who might be harmed.

Before considering genetic testing, you decide that you should get more information about the family. In fact, in situations such as Mr. Court's, a detailed family history is essential.

You find out the following:

Ewan Court's father died under rather mysterious circumstances. He had suffered from a bipolar disorder all his life but had been a successful businessman, in part because of his periods of untiring energy during the manic phases of his illness. Just before his death, he had suffered a significant financial loss related to an unwise business decision and had become very depressed. At the time of this decision, he was in an ebullient mood and spoke about how he expected to 'make a killing'. Despite advice from his partners against this decision, he went ahead and invested a large sum of money in what turned out to be a scam. He was on the verge of personal bankruptcy. He told the family that he wanted to be alone to consider what he should do and went to the family cottage. His body was found there several days later. The cause of death was undetermined but there was no evidence of foul play. It was suspected that he had taken an overdose of medication that had been prescribed for depression. A probable further contributor to the senior Mr. Court's disturbed emotional state was his concern over his wife's deteriorating health and the question of whether she had inherited HD. This all took place long before you had been asked to see Mrs. Court.

Ewan has a younger sister who has been living on the west coast for several decades; they have had little communication over the years. He presumes his sister is in good health. She is aware of her mother's illness. Along with her brother, Ewan, she has joint POA for their mother.

Ewan Court has been married for more than 22 years but he admits that the stress of looking after his mother has caused considerable tension in their marital relationship. He has two children: a daughter aged 21 and a son aged 16; he says that he would like to spend more time with them but that his work leaves little free time. He also admits that his wife has done most of the parenting.

He adds that the son is in high school and lives at home. His daughter is at university but is engaged and is planning to get married next summer.

As you have been the consulting physician for the senior Mrs. Court, Ewan's mother, you already are quite aware of her family history. She was born in western Canada and her family came from Ireland. Her mother was said to be rather queer and may have taken her own life. She had spent much time in mental institutions before she died. Her father was a farmer and was killed in a farm accident. Mrs. Court was the only member of her family in whom a definite diagnosis of HD had been considered. However, it may well be that her mother was also affected.

15.4.1 QUESTIONS TO BE CONSIDERED

8. What responsibilities do you have to Mr. Court's children? What are the potential harms and benefits to be considered in ordering genetic testing? Why do other family members have to be considered as well as the patient? Should Mr. Court's children be informed of the results of his genetic testing if he should agree to have it done? What would be the implications? What is your response if his daughter asks for genetic testing, if and when she learns that her grandmother has HD?

9. What is meant by POA? What is implied by the terms substitute, proxy or surrogate decision maker?

15.4.2 DISCUSSION

8. Genetic testing is readily available for people at risk for developing HD. It is important that the physician be well informed about the accuracy and implications of positive or negative test results for a disease. It is also most important to be aware of the implications of these tests for family members, not only the person being tested. Ewan's children are potentially at risk if their father tests positive but are very unlikely to develop the disease if he tests negative. Even if their father is positive, they have a chance that they will not develop the disease.

Testing minors raises a whole new set of ethical questions. While tests can be obtained by simply sending a sample to a lab, it is usually advisable, because of these very complex ethical issues, to refer the patient to a genetics specialist who will be prepared to make counselling available to the patient and his or her family. The risk of suicide in people with HD is four times that of the general population and preclinical testing of a person at risk must be done only where advanced counselling has been carried out (Farrer, 1988).

Various organizations have issued policy statements on genetic testing. The following statement beautifully summarizes the ethical dilemmas posed by HD and is taken from Johns Hopkins extensive review of the genetic basis for HD (Online Mendelian Inheritance in Man – see Web site references):

> 'Huntington disease represents a classic ethical dilemma created by the human genome project, i.e. that of the widened gap between what we know how to diagnose and what we know how to do anything about. Wexler (1992) referred to the dilemma as the Tiresias complex. The blind seer, Tiresias, confronted Oedipus with the dilemma: 'It is but sorrow to be wise when wisdom profits not' (from Oedipus the King by Sophocles). Wexler stated the questions as follows: 'Do you want to know how and when you are going to die, especially if you have no power to change the outcome? Should such knowledge be made freely available? How does a person choose to learn this momentous information? How does one cope with the answer?'

Several points are particularly important and are common to all genetic testing: Adult patients have a right to know whether they carry the gene. Testing must be done only on persons who are fully informed and aware of the implications of the testing (informed consent). There must be no coercion by family members or other parties for the person at risk to be tested (autonomy). Testing should not be done on minors. The results must be kept in strict confidence and not be revealed to third parties (privacy and confidentiality). However, the physician is obligated to make full disclosure of results of tests and should avoid using euphemistic terms or creating false impressions that may confuse or deceive the patient. *Truth-telling* is one of the most important cornerstones of a good physician–patient relationship. At the same time, the skilled physician will be sensitive to the patient's feelings and anxieties and be ready to provide reassurance and support. Preferably, counselling services should be provided by people who have special expertise and experience in genetic counselling prior to the testing being done and the results revealed by the same counsellors.

On occasion, complicated situations may arise in which the need of one family member to know his or her genetic status may contravene the autonomy rights of another family member. An example would be the granddaughter of an affected person wishing to know her own status, whereas her parent may refuse to have testing, even though that person is at risk. A positive predictive test in the grandchild would obviously implicate the at-risk parent. In such a situation, it is difficult to honour the rights of both the parent and the adult child.

Another reason why pre-testing as well as posttesting counselling is advisable is that patients may be overwhelmed by the result if positive. Some patients have attempted suicide on learning that they may well end up like their affected parent. However, it has been shown that most patients benefit from having the test results, whether positive or negative, and the benefits outweigh the potential harm. Each case must be carefully assessed. In the case of our patient, Mr. Court, this risk is increased by the history of past propensity towards depression and thoughts of harming himself. The diagnosis of bipolar illness in his father adds another risk factor. There is a 10%–15% chance that a child of a parent with bipolar disorder will develop symptoms at some time in life.

The results of his testing will have very important implications for his children. This is particularly important for his daughter, who is planning marriage and potentially having a family. The adult daughter has the right to be tested. If she is positive, the implications for her father are obvious. If he does not wish to be tested himself, he will not want to know the results of his daughter's tests, in which case it is imperative that she not reveal her results to her father. This may be a very difficult situation.

Genetic discrimination occurs when people are treated differently because they have or are perceived to have a gene mutation that causes or increases the risk of an inherited disorder. Currently, legislation is being enacted in many jurisdictions to prevent employers and insurance agencies from discriminating against people who may have a gene mutation.

The sister on the west coast may also wish to have testing although we know nothing about her clinical status.

9. Power of Attorney (POA) is a legal term whereby a person may appoint someone else to act on his or her behalf when he or she is unable to do so. POA may include delegation of authority to handle financial affairs, but in the context of this case, the POA is specific for health care decisions. POA, substitute decision maker, surrogate decision maker and proxy decision maker have various definitions and are not necessarily interchangeable terms. POA allows individuals to appoint another person or organization to handle their affairs should they become unavailable or unable to do so. Durable POA for Medical Care allows individuals to appoint another person to make decisions for medical care when they are deemed unable to decide for themselves. The legal definitions of these terms and what is entailed in their application vary from one jurisdiction to another and physicians should make themselves aware of the legislation governing their particular area. It is presumed that persons who assume POA will act in accordance with the presumed wishes and best interests of the patients they are representing and will be careful to not act in their own personal interests.

Advanced directives, commonly called 'living wills', allow people to lay out specific directions for their subsequent medical care should they become incompetent or unable to communicate their wishes. Most physicians take as a matter of duty the adherence to these expressed wishes, but there are certain caveats. The wishes may not be followed when they require the doctor to carry out treatment that is inconsistent with the physician's conscience, is against the policy of the health care institution or is deemed to be ineffective in treating the condition (see 15. Futility). In general, advanced directives are a useful guide for the health care team and for the surrogate decision makers.

15.5 CASE STUDY, PART D

After considerable debate, it is decided that Mrs. Court must be transferred to a nursing home, even against her own wishes. She is declared incompetent. She seems quite depressed and you are concerned that not only might she be a harm to herself and others through negligence but also that she might entertain the idea of harming herself and taking her own life. You go to visit her at the nursing home and find that she is extremely unhappy with her surroundings: most of the other patients have severe dementia and many are in a terminal state. She tells you that she knows that she soon will be one of them. She says she remembers how her mother was before she died and does not want to 'go that way'. She pleads with you to help her to end her life while she is still capable. You feel very sympathetic to her plight. She has almost continuous abnormal movements, cannot stand by herself and cannot feed herself. She has lost a lot of weight. The situation makes you feel extremely uncomfortable and your immediate reaction is to brush aside her request and change the subject. You mutter something like, 'Now, Mrs. Court, you really don't mean that!' although you know she does. You look at your watch and say you have to leave. You feel even more uncomfortable as you do.

15.5.1 QUESTIONS TO BE CONSIDERED

10. Under what circumstances can a patient be forced to enter hospital? Should a patient be hospitalized against his or her will when you suspect that he or she is a suicide risk?

11. How would you respond to a patient who asks for assistance to end his or her life? What are the legal, ethical and moral implications?

15.5.2 DISCUSSION

10. Hospitalization against a patient's will.

Where a physician has a good reason to suspect that the patient is at risk, one is legally obliged to commit a patient to hospital for his or her own protection. Usually, a family member who has POA will request that the physician invoke the law. The issue of committing a suicidal patient to hospital against that person's will is not as simple as it might seem. Members of the medical and legal professions are far from unanimous as to the ethics of the situation. This is particularly evident when it comes to a patient who has an incurable disease such as HD and who is still competent. One recent survey of medical and law students showed an almost even split on the subject (Elger and Harding, 2004).

11. Medically assisted suicide or euthanasia.

Many patients with progressive degenerative neurological disease consider suicide, and the physician must be mindful of the risk. A family history of suicide seems to be an added risk factor. Mr. Court has had risk factors on both sides of the family. However, Mrs. Court does not have that background and perhaps is at less risk. However, as she sees herself deteriorating and is haunted by memories of her own mother's demise, she does contemplate ending her life while she is still capable of it. She pleads with you to assist her. There are few more difficult ethical issues than those surrounding the question of assisted suicide. From an ethics standpoint, it involves truth-telling, disclosure, fidelity, respect for autonomy, non-maleficence, beneficence, patient's best interest, justice, public policy and public danger.

Volumes have been written on the subject, and the present discussion can only highlight a few of the issues. The laws vary from jurisdiction to jurisdiction and may also change over time. Furthermore, enforcement of the law may also be variable according to the personal beliefs of the authorities. Physicians must be aware of the current legal status in their area of practice. In general, the Canadian Federal law prohibited assisting suicide, but in 2015, the Supreme Court struck down this prohibition. Parliament and the Senate subsequently have enacted legislation that permits physician-assisted suicide under certain conditions.

In the United States, the law varies from state to state and there is some tension between federal and state legislation. The state of Oregon has passed legislation that would allow assisted death under very particular circumstances. In actual fact, there have been only a relatively small number of cases where assisted suicide has actually been carried out under the law. Australia also has different laws in each state. European countries have considerable variation in their approach, with the Netherlands, Belgium and Switzerland being the most lenient. In general, it is emphasized that more emphasis should be placed on palliative care for patients who are in a terminal state rather than legitimizing assisted suicide.

The ethics stance is also variable. Those who support assisted suicide cite *autonomy* of the individual as the overriding principle to be followed. Those who, on compassionate grounds, might consider assisting a person to end their life may refer to *beneficence*. Many fear that allowing assisted suicide may lead to a general devaluation of life and of individual autonomy and a slippery slope with horrendous consequences.

Religious authorities, especially Jewish, Roman Catholic and Moslem, who oppose *assisted* suicide, also oppose suicide in any form as they believe that it is thwarting God's will. They also emphasize 'sacredness of life'. Groups who represent the disabled fear that allowing assisted suicide may start society on a slippery slope towards eventually accepting assisted suicide for those who do not wish it or who cannot state their preferences but may be perceived as a drag on society.

Also under consideration is whether one, while still competent, might, as an advanced directive, express a wish for assisted death at some later time should their condition deteriorate and they might then be incompetent. This might suggest a 'Ulysses contract' where they would be unable to change their mind at a later date if they were not competent.

Various medical organizations have studied these issues, and many have developed policy statements to guide their membership. The World Medical Association (WMA), in 1983, adopted a Declaration on Euthanasia, which was revised in 2005 (Willams, 2015): 'Euthanasia, that is the act of deliberately ending the life of a patient, even at the patient's own request or at the request of close relatives, is unethical. This does not prevent the physician from respecting the desire of a patient to allow the natural process of death to follow its course in the terminal phase of sickness'.

In 2013, the WMA added the resolution: 'The World Medical Association strongly encourages National Medical Associations and physicians to refrain from participating in euthanasia, even if national law allows or decriminalizes it under certain conditions'.

Medical people are often in the position of wanting to relieve suffering, but the methods of doing so may also cause death. Thomas Aquinas is credited with introducing the *principle of double effect* in his discussion of the permissibility of self-defence in the *Summa Theologica* (II-II, Qu. 64, Art. 7). McIntyre gives an excellent explanation of the principle. Killing one's assailant is justified, he (Aquinas) argues, provided one does not intend to kill him. Aquinas observes that 'Nothing hinders one act from having two effects, only one of which is intended, while the other is beside the intention .… Accordingly, the act of self-defense may have two effects: one, the saving of one's life; the other, the slaying of the aggressor'.

This has been later applied to many medical ethical dilemmas, such as the irradiation or removal of a cancerous, but pregnant, uterus with the intention of saving the mother's life while at the same time resulting, unintentionally although still foreseeable, in the death of the fetus. This would also allow a physician to administer large, potentially fatal, doses of narcotics to alleviate pain, as long as the intent was to assuage suffering and not to cause death. The judicial systems may respond variously to these situations.

In the situation in this anecdote, Mrs. Court is not in severe or uncontrollable pain, although she may well be in severe mental and emotional anguish. She is not in danger of imminent death. Under traditional practice, the law might deal severely with a physician prescribing fatal doses of medication in such a case. Mrs. Court may well recall with horror the situation she experienced with her own mother and may also have suspicions about the events surrounding her demise. The physician's inclination may well be to simply dismiss the request with a firm 'no' and try to change the subject, pretending that the question was never asked. However, it would be preferable to discuss the situation with the patient and allow her to express her anxieties. The discussion may be continued on further occasions. It is important to reassure the patient that you will do everything to help her otherwise and will not abandon her. Fear of abandonment and the feeling of loss of control are some of the major factors leading to a person committing suicide or asking for assistance to end their life.

There is a need for *palliative care* in such situations and the patient must never be left with a feeling of abandonment. Indeed, the assurance of good palliative care may reduce the desire for assisted death.

15.6 CASE STUDY, PART E

At the nursing home.

During the next several years, Mrs. Court's condition deteriorates progressively until she seems unable to recognize or respond to those about her. She lies in bed, now immobile as the choreic movements no longer occur. Sometimes, her eyes are open and she stares blankly at the ceiling, but occasionally, she seems to look towards a stimulus such as a nurse entering the room. At other times, she seems to be in a deep sleep

from which she cannot be aroused. She has had a feeding tube inserted into her duodenum. Her state is described as 'vegetative' and a diagnosis of 'persistent vegetative state' is written in the doctor's notes in the chart. She develops pneumonia but responds to antibiotic therapy. In the past, a physician might act in a paternalistic fashion and decide whether to prescribe antibiotics or not, but in this case, the family were consulted and it was their wish to give antibiotics. The physician might argue against it but eventually give in to the family's wishes.

However, after a few more months, it is apparent that she is not showing any improvement but is still not in a state where death seems imminent. The family asks whether the tube feedings should be discontinued and active treatment withdrawn. They also ask whether aggressive treatment should be withheld, such as antibiotics, should she develop another bout of pneumonia. The social worker at the nursing home calls a family conference that is attended by the staff nurses and the attending physician. Early in the conference, someone mutters that the nursing home is very short of beds and that society cannot afford to maintain people in a vegetative state forever. This individual suggests that we should be more careful with scarce resources and that decisions have to be made on the basis of distributive justice. Others quickly dismiss this argument.

As the conference proceeds, several issues are raised. What treatments, if any, are appropriate? If Mrs. Court were to develop pneumonia, should antibiotics be given? When is treatment considered to be a matter of futility? The problems of withdrawal or withholding treatment are discussed. The family, comprised of her son and her daughter, who flew in from the west coast, has joint POA for health care. They agree that they would be honouring their mother's wishes and serving her best interests by not prolonging her life under such circumstances. She then developed pneumonia, which, untreated, led to her death. An autopsy confirmed the diagnosis of HD.

15.6.1 QUESTIONS TO BE CONSIDERED

12. What is meant by 'persistent vegetative state'? What is the neuro-pathological basis of this clinical condition?
13. What is meant by paternalism?
14. What is distributive justice?
15. What is meant by futility? Is it a useful term?
16. What is meant by withdrawal or withholding of treatment?

15.6.2 DISCUSSION

12. Persistent vegetative state is a neurological condition in which the patient has diffuse cortical damage and is believed to have lost all conscious awareness of his or her environment. 'Vegetative' functions, residing in the brain stem, remain relatively intact and cardiovascular function and respiration are maintained. As some brainstem

function is preserved, chewing and swallowing may occur. Random eye movements may be seen, and at times, there may seem to be a following of objects moving in the field of vision. With hydration, feeding and attention to intervening medical conditions, the patient may be kept in this state for months or years. Needless to say, this is a most distressing situation for both family and caregivers. Recent studies, using functional MRI, may indicate that the rare patient may have more awareness than might be suspected by physical examination alone. There is thus also an intermediate condition described as 'minimally conscious states' wherein the person may have some awareness of the environment.

13. *Paternalism.* Paternalism, or better, parentalism, is a term that was applied to the physician's tendency to act as if they had all the answers and treated the patient as they would a little child. The physician was the parent, the patient the child. It was the accepted attitude that prevailed until the past few decades but is less common now since patients are better educated and have access to medical information that allows them to share decision making concerning their own health and medical care. Physicians, however, are not obligated to carry out a treatment that they feel may be detrimental to patients' well-being or to institute a course of action felt to be futile (see 15. Futility).

14. *Justice (Distributive Justice).* Justice is one of the four principles in the oft-quoted 'Georgetown Mantra', promulgated by the ethics group at Georgetown University in Washington, United States. The others are autonomy, beneficence and nonmaleficence. In medical care, justice demands the greatest good for the greatest number. In the real-life situation where there are never unlimited resources, the individual physician is often in the difficult position of having to decide between the needs of his or her patient and those of the myriad of other patients who are equally deserving of provision of health care. Most physicians feel the strongest loyalty to their individual patient and will advocate first for that patient. Justice principles may impact on public policy, if not on formalized legislation.

15. *Futility.* The term *futility* is seldom used anymore but refers to a treatment that is not considered to be of benefit or of serving the aims of treatment. The problem is that, like beauty being in the eye of the beholder, the definition of futility varies with the point of view of the user. A physician may not consider a treatment as futile when it prolongs life for a few more months. However, the patient may feel it is futile, and therefore undesirable, if the months of extended life are spent in the hospital and require tolerating adverse effects of the medication. Of course, the reverse may be the case and some patients and families want prolongation of life at any cost. The issue of futility arises many times in the situation of patients in a persistent vegetative state.

Physicians, however, are not obligated to carry out a treatment that they feel may be detrimental to patients' well-being or to institute a course of action felt to be futile.

16. *Withdrawing and withholding treatment.* In some patients in whom the quality of life is very low or they are in a state of chronic suffering, it may seem that prolonging suffering is not beneficent and indeed may be considered maleficent or harmful. This is especially so in patients who are in a terminal state. Family members may request that aggressive medical treatment be stopped. Most empathetic caregivers would comply, especially if the patient has previously expressed (such as in a 'living will' or advanced directives) the desire not to have his or her life prolonged in such a state. While there may be a consensus that aggressive medical treatments or surgical interventions should be avoided, there is ongoing debate as to what constitutes a medical act. Is hydration and feeding a medical act or is it merely caregiving? Is there a difference between feeding or hydrating through a feeding tube and spoon feeding a patient who is able to swallow?

Indeed, whether to initiate and when to discontinue feeding and hydration is one of the more controversial issues in medical ethics. The issue often is centred on what is natural and what is extraordinary. Some authorities, including religious authorities, say that tube feeding is unnatural and a medical act and therefore may well be refused or terminated. Others maintain that nourishment and hydration are simple supportive acts and the tube is only another way of carrying these out. It is often most difficult for the nursing staff and those caring directly for a patient to see the person fade away when they feel they could be doing something seemingly helpful.

Physician's opinions are widely divided on these issues, and a survey of physicians only a decade ago revealed that there are even those who would go beyond routine medical care and provide more extensive treatments, such as flu shots or even breast or prostate examinations for patients who are in a persistent vegetative state.

15.7 SUMMARY

This case is presented to illustrate the various ethical issues that may complicate a medical condition and must be taken into account in the management. The disease (HD) has a wider impact than just on the individual affected, and management must take into account all those who are involved. This is highlighted by a quote from a family member: 'Illness and disability are a family affair. The diagnosis that made our family member need care happened to us as well. It is our diagnosis just as much as it's theirs. I have a psychosocial form of the disease, just as my husband has a clinical one'.

As in most of medical ethics, there is no absolute right answer to the questions discussed in this chapter. However, the underlying principles of doing the best for the patient must remain uppermost.

REFERENCES

Beauchamp, T.L., Childress, J.F., Eds. *Principles of Bio-medical Ethics*, 7th ed. Oxford: Oxford University Press, 2012.

Elger, B.S., Harding, T.W. Should a suicidal patient with Huntington's disease be hospitalized against her will? Attitudes among future physicians and lawyers and discussion of ethical issues. *Gen Hosp Psychiatry* 2004. 26: 136–144.

Farrer, L.A. Suicide and attempting suicide in Huntington disease; implications for preclinical testing of persons at risk. *Am J Med Genet* 1988. 24: 305–311.

McIntyre, A. Doing away with double effect. *Ethics* 2001. 111(2): 219–255.

Singer, P.A., Ed. *Bioethics at the Bedside. A Clinician's Guide*. Ottawa: Canadian Medical Association/ Association Medicale Canadienne, 1999.

Tarasoff v. Regents of the University of California 17 Cal. 3d 425 (1976).

Willams, J.R., Ed. *World Medical Association Medical Ethics Manual*, 3rd ed. Voltair Cedex: France, 2015. Available online at http://www.wma.net/ethicsunit /pdf/manual/ethics_manual.pdf

SUGGESTED READING

Bernat, J.L. *Ethical Issues in Neurology*, 3rd ed. Philadelphia, PA: Lippincott Williams and Wilkins, 2008.

Hebert, P. *Doing It Right; A Practical Guide to Ethics for Medical Trainees and Physicians*, 3rd ed. Don Mills, Ontario: Oxford University Press, 2014.

Illes, J., Ed. *Neuroethics: Defining the Issues in Theory, Practice and Policy*. Oxford: Oxford University Press, 2009.

Kay, C., Collins, J.A., Miedzybrodzka, Z., Madore, S.J., Gordon, E.S., Gerry, N., Davidson, M., Slama, R.A., Hayden, M.R. Huntington disease reduced penetrance alleles occur at high frequency in the general population. *Neurology* 2016. 87(3): 282–288.

Practice Parameter: Genetic testing alert: Statement of Practice Committee Genetic Testing Task Force of the American Academy of Neurology. *Neurology* 1996. 47: 1343–1344.

Sibbald, R., Downar, J., Hawryluck, L. Perceptions of 'futile care' among caregivers in intensive care units. *CMAJ* 2007. 177: 1201–1208.

WEB SITES

http//ww.neuroethics.upenn.edu

http://www.omim.org (The Online Mendelian Inheritance in Man)

GLOSSARY

Note to reader: The glossary contains neuroanatomical terms and terms commonly used clinically to describe neurological symptoms, as well as common physical findings detected on a neurological examination; some selected laboratory investigation techniques and clinical syndromes are also included.

Afferent Sensory – conduction toward the central nervous system.

Agnosia Loss of ability to recognize the significance of sensory stimuli (tactile, auditory, visual), even though the primary sensory systems are intact.

Agonist A muscle that performs a certain movement of the joint; the opposing muscle is called the antagonist.

Agraphia Inability to write owing to a lesion of higher brain centres, even though muscle strength and coordination are preserved.

Akinesia Absence or loss of motor function; lack of spontaneous movement; difficulty in initiating movement (as in Parkinson's disease).

Alexia Loss of ability to grasp the meaning of written words; inability to read due to a central lesion; word blindness.

Alpha motor neuron Another name for the anterior (ventral) horn cell, also called the lower motor neuron.

Amygdala Amygdaloid nucleus or body in the temporal lobe of the cerebral hemisphere; a nucleus of the limbic system.

Angiogram Display of blood vessels for diagnostic purposes, using cineradiography, MRI or CT, usually by using contrast medium injected into the vascular system.

Anopia/Anopsia A defect in the visual field (e.g. hemianopia – loss of half the visual field).

Antagonist A muscle that opposes or resists the action of another muscle, which is called the agonist.

Antidromic Relating to the propagation of an impulse along an axon in a direction that is the reverse of the normal or usual direction.

Apnea Total interruption of breathing for 10 seconds or more. Apneas can be central or obstructive in origin.

Apnea test This is a procedure conducted as part of the protocol to determine brain death. It is a test of the ability of the medulla to stimulate respiration in response to rising levels of pCO_2. With no other factors present (such as medication effects or collateral medical conditions) and a baseline pCO_2 of 40 mm Hg, failure of a patient to generate respiratory effort in response to pCO_2 levels of 55 to 60 mm Hg indicates lack of function of the medulla.

Aphasia An acquired disruption or disorder of language, specifically a deficit of expression using speech or of comprehending spoken or written language; global aphasia is a severe form affecting all language areas.

Apoptosis Programmed cell death, either genetically determined or following an insult or injury to the cell.

Apraxia Loss of ability to carry out purposeful or skilled movements (such as combing the hair or brushing the teeth) despite the preservation of power, sensation and coordination.

Arachnoid The middle meningeal layer, forming the outer boundary of the subarachnoid space.

Areflexia Loss of reflex as tested using the myotatic stretch/deep tendon reflex (DTR).

Ascending tract Central sensory pathway, for example from the spinal cord to the brainstem, cerebellum or thalamus.

ASIA Scoring System American Spinal Injury Association Scoring Scale used to document and evaluate spinal cord injury (see Chapter 13).

Association fibers Fibers connecting parts of the cerebral hemisphere, on the same side.

Astereognosis Loss of ability to recognize the nature of objects or to appreciate their shape by touching or feeling them.

Asterixis A slow, flapping tremor of the outstretched hands; sometimes seen in patients with hepatic encephalopathy.

Astrocyte A type of neuroglial cell with metabolic and structural functions; reacts to injury of the CNS by forming a gliotic 'scar'.

Asynergy Disturbance of the proper sequencing in the contraction of muscles, at the proper moment and of the proper degree, so that an action is not executed smoothly or accurately.

Ataxia A loss of coordination of voluntary movements, often associated with cerebellar dysfunction; can also be caused by loss of proprioception or by motor weakness.

Athetosis Slow writhing movements of the limbs, especially of the hands, not under voluntary control, caused by impaired function in the striatum.

Autonomic Autonomic system; usually taken to mean the efferent or motor innervation of viscera (smooth muscle and glands).

Autonomic nervous system (ANS) Visceral innervation; sympathetic and parasympathetic divisions.

Axon Efferent process of a neuron, conducting impulses to other neurons or to muscle fibers (striated and smooth) and gland cells.

Babinski response Babinski reflex is not correct; stroking the outer border of the sole of the foot in an adult normally results in a flexion (downward movement) of the toes. The Babinski response consists of an upward movement (extension or dorsiflexion) of the first toe and a fanning of the other toes, indicating a lesion of the corticospinal (pyramidal) tract.

Ballismus Involuntary, usually unilateral flinging motion of the arm or leg due to damage to the ipsilateral subthalmic nucleus.

Basal ganglia (nuclei) Gray matter structures in the deep cerebral hemispheres adjacent to the diencephalon: the caudate, putamen and globus pallidus; including functionally the subthalamus and the substantia nigra. Among many other functions, they are crucial components of the central apparatus of motor control.

Basilar artery The major artery supplying the brainstem and cerebellum, formed by fusion of the two vertebral arteries.

Blink reflex An electrical equivalent of the corneal reflex: it measures the latency of the input of an electrical stimulus to cranial nerve (CN) V and the resulting eye closure or blink output of CN VII on both sides. If the reflex response is slow or absent, this can help localize the lesion to CN V, the pons or CN VII on either side.

Boston Classification Patient classification system designed for medical rehabilitation that predicts resource use and outcomes for clinically similar groups of individuals.

Bradykinesia Abnormally slow initiation of voluntary movements (usually seen in Parkinson's disease).

Brainstem Includes the medulla, pons and midbrain.

Brainstem evoked potentials (BSEP) A neurophysiological technique involving testing the auditory portion of CN VIII using an auditory stimulus (clicking sound) and measuring the responses from CN VIII and the various brainstem structures that relay auditory information in the brainstem toward the thalamus and the cortex. This test might show a delay or slowing of the signal evoked by the sound on one or both sides, allowing for precise localization of a brainstem lesion.

Brodmann areas Numerical subdivisions of the cerebral cortex on the basis of histological differences between different functional areas (e.g. area 4 = motor cortex; area 17 = primary visual area).

Bruit French word meaning 'noise'; refers to a swishing noise heard when listening over an artery, a vein or over the head; results from turbulent blood flow in the area auscultated and may reflect critical narrowing of an arterial lumen heralding stroke or of an abnormal fistulous connection between an artery and a vein.

Bulb Referred at one time to the medulla but in the context of 'corticobulbar tract' refers to the whole brainstem in which the motor nuclei of CNs and other nuclei are located.

CAT or CT scan Computerized axial tomography; a diagnostic imaging technique that uses x-rays and computer-aided reconstruction of the brain to provide two-dimensional images of the brain and spine.

Calcarine cortex Primary visual cortex, situated just above and below the calcarine fissure in the medial occipital lobe.

Carotid artery Large artery in the anterior neck with an internal branch supplying the retina; basal ganglia, anterolateral and anteromedial cerebral hemispheres; and an external branch supplying the face and scalp on the same side.

Carotid siphon Hairpin bend of the internal carotid artery within the skull.

Cauda equina ('Horse's tail') The lower lumbar, sacral and coccygeal spinal nerve roots within the subarachnoid space of the lumbar (CSF) cistern.

Caudal Toward the tail or hindmost part of the neuraxis.

Caudate nucleus Part of the neostriatum, of the basal ganglia; consists of a head, body and tail (which extends into the temporal lobe).

Central nervous system (CNS) Cerebral hemispheres, including diencephalon, cerebellum, brainstem and spinal cord.

Cerebellar peduncles Inferior, middle and superior; fiber bundles linking the cerebellum and brainstem.

Cerebellum The little brain; an older part of the brain with important motor functions (as well as others), dorsal to the brainstem, situated in the posterior cranial fossa.

Cerebral aqueduct (of Sylvius) Aqueduct of the midbrain; passageway carrying CSF through the midbrain, as part of the ventricular system.

Cerebral peduncle Descending cortical fibers in the 'basal' (ventral) portion of the midbrain; sometimes includes the substantia nigra (located immediately behind).

Cerebrospinal fluid (CSF) Fluid in the ventricles and in the subarachnoid space and cisterns.

Cerebrum Includes the cerebral hemispheres and diencephalon, but not the brainstem and cerebellum.

Cervical Referring to the neck region; the part of the spinal cord that supplies the structures of the neck; C1 to C7 vertebral; C1 to C8 spinal segments.

Chorea A motor disorder characterized by abnormal, irregular, spasmodic, jerky, uncontrollable movements of the limbs or facial muscles, thought to be caused by dysfunction or degeneration in the basal ganglia.

Choroid A delicate membrane; choroid plexuses are found in the ventricles of the brain.

Choroid plexus Vascular structure consisting of pia with blood vessels, with a surface layer of ependymal cells; responsible for the production of CSF.

Circle of Willis Anastomosis between internal carotid and basilar arteries, located at the base of the brain, surrounding the pituitary gland.

Cistern(a) Expanded portion of subarachnoid space containing CSF, for example cisterna magna (cerebello–medullary cistern), lumbar cistern.

Clonus Abnormal sustained series of contractions and relaxations following stretch of the muscle; usually elicited in the ankle joint. Present following lesions of the descending motor pathways and associated with spasticity. When elicited as a DTR, implies a Grade of 4+ (see Table 2.3).

CNS Abbreviation for central nervous system.

Cog-wheeling This is a clinical sign of altered tone usually at the wrist and elbows. When the wrist or forearm is moved in a circular motion, the resulting tone feels like a 'ratcheting' or 'on-off' movement.

Cold caloric response Similar to the doll's eye response, this is a test of the slow component of eye movement to vestibular input in the absence of visual fixation. It requires normal function of CN VIII, the pons and the connections to CN III, IV and VI. The stimulus for this test consists of 150 cc of cold water injected slowly into the ear canal on one side (after verifying that the tympanic membrane is intact). The normal response would be slow conjugate movement of the eyes toward the side being irrigated. If performed on an awake patient with visual fixation, the fast component would follow the slow component with resulting nystagmus with the fast component away from the cold ear. The subject will complain of dizziness and often will have nausea and vomiting.

Colliculus A small elevation; superior and inferior colliculi compose the tectum of the posterior midbrain.

Commissure A group of nerve fibers in the CNS connecting structures on one side to the other across the midline (e.g. corpus callosum of the cerebral hemispheres; anterior commissure).

Concussion Brain injury due to closed head trauma associated with brief loss of neurological function, usually consciousness, with no damage found by neuroradiological imaging.

Conjugate eye movement Coordinated movement of both eyes together, so that the image falls on corresponding points of both retinas.

Connors Rating Scale Revised (CRS-R) Designed to assess attention-deficit/hyperactivity disorder (ADHD) and related problems in children.

Consciousness The state of awareness and vigilance in which the individual has full function of sensory, motor and modulating systems to interact with the environment. This implies normal waking function of the brainstem, diencephalon and cerebral hemispheres.

Consensual reflex Light reflex; refers to the bilateral simultaneous constriction of the pupils after shining a light in one eye.

Contralateral On the opposite side (e.g. contralateral to a lesion).

Coordination The integration of sensory, motor and modulating systems to provide smooth control of limb and trunk movements. Lack of coordination is ataxia.

Corona radiata Fibers radiating from the internal capsule to various parts of the cerebral cortex – a term often used by neuroradiologists.

Corpus callosum The main (largest) neocortical commissure of the cerebral hemispheres.

Corpus striatum Caudate and putamen, nuclei inside the cerebral hemisphere, including tissue bridges that connect them (across the anterior limb of the internal capsule); part of the basal ganglia; the neostriatum.

Cortex Layers of gray matter (neurons and neuropil) on the surface of the cerebral hemispheres (mostly six layers) and cerebellum (three layers).

Corticobulbar Descending fibers connecting motor cortex with motor CN nuclei and other nuclei of the brainstem (including the reticular formation).

Corticofugal fibers Axons carrying impulses away from the cerebral cortex.

Corticopetal fibers Axons carrying impulses toward the cerebral cortex.

Corticospinal tract Descending tract, from motor cortex to anterior (ventral) horn cells of the spinal cord (sometimes direct); also called pyramidal tract.

Cranial nerve nuclei Collections of cells in the brainstem giving rise to or receiving fibers from CNs (CN III to XII); may include sensory, motor or autonomic.

Cranial nerves Twelve pairs of nerves arising from the cerebrum and brainstem and innervating structures of the head and neck (CN I and II are actually CNS tracts).

CSF Cerebrospinal fluid, in ventricles and subarachnoid space (and cisterns).

CT angiogram CT scan of the brain, neck and great vessels combined with the use of intravenous contrast material to outline the arterial and venous vascular anatomy. Often used in acute stroke to determine the location and degree of narrowing of a cerebral artery.

Cuneatus (cuneate) Sensory tract (fasciculus cuneatus) of the posterior column of the spinal cord, from the upper limbs and body; cuneate nucleus of medulla.

Decerebrate posturing Characterized by reflexive extension of the upper and lower limbs in a comatose individual; lesion at the brainstem level between the vestibular nuclei (pons) and the red nucleus (midbrain).

Decorticate posturing Characterized by reflexive extension of the lower limbs and flexion of the upper limbs in a comatose individual; lesion is located above the level of the red nucleus (of the midbrain).

Decussation The point of crossing of CNS tracts, for example decussations of the pyramidal (cortico-spinal) tract, medial lemnisci and superior cerebellar peduncles.

Dementia Progressive brain disorder that gradually destroys a person's memory, starting with short-term memory, and loss of intellectual ability, such as the ability to learn, reason, make judgments and communicate; in addition, the inability to carry out normal activities of daily living; usually affects people with advancing age.

Demyelination Degeneration and loss of the myelin sheath in a region or regions of the central or peripheral nervous system.

Dendrite Receptive process of a CNS neuron; usually several processes emerge from the cell body, each of which branches in a characteristic pattern.

Dendritic spine Cytoplasmic excrescence of a dendrite and the site of an excitatory synapse.

Dentate Dentate (toothed or notched) nucleus of the cerebellum (intracerebellar nucleus); dentate gyrus of the hippocampal formation.

Dermatome A patch of skin innervated by a single spinal cord segment (e.g. T1 supplies the skin of the inner aspect of the upper arm; T10 supplies the umbilical region).

Descending tract Central motor pathway (e.g. from cortex to brainstem or spinal cord).

Diencephalon Consisting of the thalamus, epithalamus (pineal), subthalamus and hypothalamus.

Diffuse axonal injury (DAI) Injury to white matter tracts caused by shearing forces produced by trauma; this injury leads to impaired white matter transmission and is often in combination with cortical injury.

Diplopia Double vision; a single object is seen as two objects, can be horizontal, vertical or oblique.

Doll's eye response This is a test of the slow component of eye movement to vestibular input in the absence of visual fixation. It requires normal function of CN VIII, the pons and the connections to CN III, IV and VI. This test involves moving the head, usually in the horizontal direction, and observing the conjugate movement of the eyes in the opposite direction. See *Cold caloric response.*

Dominant hemisphere The hemisphere responsible for language; this is the left hemisphere in about 85% to 90% of people (including many left-handed individuals).

Dorsal column Alternate term for the posterior column of the spinal cord.

Dorsal root Afferent sensory component of a spinal nerve, between dorsal root ganglion and spinal cord.

Dorsal root ganglion (DRG) A group of peripheral neurons whose axons carry afferent information from the periphery; their central processes enter the spinal cord.

Dorsiflexion Active elevation of the foot at the ankle joint; the opposite of plantar flexion.

Dura Dura mater, the thick external layer of the meninges (brain and spinal cord).

Dural venous sinuses Large venous channels for draining blood from the brain; located within the dura of the cranial meninges.

Dysarthria Difficulty with the articulation of words.

Dysdiadochokinesia Impairment of or inability to perform rapid alternating distal limb movements, such as alternating pronation/supination of the forearm or toe tapping.

Dyskinesia Purposeless movements of the limbs or trunk, usually due to a lesion of the basal ganglia; also difficulty in performing voluntary movements.

Dysmetria Disturbance of the ability to control the range of movement in muscular action, causing under- or over-shooting of the target (usually seen with cerebellar lesions).

Dysphagia Difficulty with swallowing.

Dyspraxia Impaired ability to perform a voluntary act previously well performed, in the presence of intact motor power, coordination and sensation.

Dystonia Impaired control of limb and trunk posture, often asymmetric, with twisted postures of one or more limbs, and of the neck and face; may be focal (e.g. one foot), segmental (one limb, several joints) or generalized.

EEG Electroencephalography, a technique for recording cerebral cortical electrical activity via electrodes applied to the scalp in a standardized configuration. The EEG is a useful tool to detect the presence of seizures as well as assess the background rhythms generated by the interplay of the cortex and thalamus. Localized slowing on the EEG can reflect focal cortical damage, whereas diffuse slowing can represent widespread injury (such as in dementia) or metabolic/toxic injury, which affects the whole brain.

Efferent Away from the central nervous system; usually means motor to muscles.

Electromyogram (EMG) A technique for investigating peripheral neuromuscular disease by recording electrical activity from muscle, either by the use of needle or surface recording electrodes.

Encephalitis Inflammation of parts or all of the cerebral hemispheres; may be of infectious or autoimmune origin.

Encephalopathy Generic term for any disease affecting the cerebral hemispheres.

ENG This is a test of vestibular function performed either by movement in a rotating or by the infusion of cold water in the ear canal. The normal output of the stimulus is the production of a slow vestibular component of eye movement followed by a first corrective component.

Electronystagmogram (ENG) A diagnostic test performed to determine inner ear function of the semicircular canals; uses cold or warm water (caloric) stimulation or circular movement using a rotating chair. The velocity in degrees per second of the slow and fast eye movement components from each ear can be precisely measured. This test is used to distinguish central from peripheral causes of vertigo and ataxia.

Entorhinal Associated with olfaction (smell); the entorhinal area is the anterior part of the parahippocampal gyrus of the temporal lobe, adjacent to the uncus.

Ependyma Epithelium lining of the ventricles of the brain and the central canal of the spinal cord; specialized tight junctions at the site of the choroid plexus.

Equinovarus Abnormal posture of the foot in which it is both flexed and inverted at the ankle; commonly seen in severe upper motor neuron cerebral lesions such as stroke.

Extrapyramidal system An older clinically used term, usually intended to include the basal ganglia portion of the motor systems and not the pyramidal (corticospinal) motor system.

Falx Dural partitions in the midline of the cranial cavity; the large falx cerebri between the cerebral hemispheres, and the small falx cerebelli.

Fascicle A small bundle of nerve fibers.

Fasciculation Spontaneous (uncontrolled) discharge of a motor unit, visible as transient, irregular focal muscle twitches.

Fasciculus A large tract or bundle of nerve fibers.

Fasciculus cuneatus Part of the posterior column of the spinal cord; ascending tract for discriminative touch, conscious proprioception and vibration from upper body and upper limb.

Fasciculus gracilis Part of the posterior column of the spinal cord; ascending tract for discriminative touch, conscious proprioception and vibration from lower body and lower limb.

Fiber Synonymous with an axon (either peripheral or central).

Flaccid paralysis Muscle paralysis with hypotonia due to a lower motor neuron lesion; may also occur in the acute phase of an upper motor neuron lesion, e.g. spinal shock.

fMRI Functional magnetic resonance imaging. A technique for assessing which areas of the brain are involved in specific functions, for example language comprehension; based on the principle that, while such regions are in use, metabolic activity and regional blood flow are relatively increased.

Folium (plural = folia) A flat leaf-like fold of the cerebellar cortex.

Foramen An opening, aperture, between spaces containing CSF (e.g. Monro, between lateral ventricles and IIIrd ventricle; Magendie, between IVth ventricle and cisterna magna).

Forebrain Anterior division of the embryonic brain; cerebrum and diencephalon.

Fornix The efferent (noncortical) tract of the hippocampal formation, arching over the thalamus and terminating in the mammillary nucleus of the hypothalamus and in the septal region.

Fourth (IVth) ventricle Cavity between the brainstem and cerebellum, containing CSF.

Functional Independence Measure (FIM) A scale used in the care of patients undergoing rehabilitation. It is an 18-item ordinal scale, used with all diagnoses within a rehabilitation population. It is viewed as most useful for assessment of progress during inpatient rehabilitation.

Funiculus A large aggregation of white matter in the spinal cord; may contain several tracts.

Ganglion (plural = ganglia) A collection of nerve cell bodies in the peripheral nervous system – the dorsal root ganglion (DRG) and autonomic ganglion. Also inappropriately used for certain regions of gray matter in the brain (i.e. basal ganglia).

Geniculate bodies Specific relay nuclei of thalamus – medial (auditory) and lateral (visual).

Genu Knee or bend; middle portion of the internal capsule.

Glial cell Also called neuroglial cell; supporting cells in the central nervous system – astrocyte, oligodendrocyte and ependymal; also microglia.

Glioma Central nervous system tumor arising from glial cells (whether astrocytes, oligodendroglia or ependymal cells).

Globus pallidus Efferent part of the basal ganglia; part of the lentiform nucleus with the putamen; located medially.

Gracilis nucleus Gracilis nucleus of the medulla. Nucleus at the termination of the sensory tract (fasciculus gracilis) of the posterior column of the spinal cord; from lower limbs and pelvic area.

Gray matter Nervous tissue, mainly nerve cell bodies and adjacent neuropil; looks 'grayish' after fixation in formalin (e.g. cerebral cortex, basal ganglia, thalamus).

Gyrus (plural = gyri) A convolution or fold of the cerebral hemisphere; includes cortex and white matter (e.g. precentral gyrus).

Hemiballismus Violent jerking or flinging movements of one limb, not under voluntary control, due to a lesion of the ipsilateral subthalamic nucleus.

Hemiparesis Muscular weakness affecting one side of the body.

Hemiplegia Paralysis of one side of the body.

Hemorrhage As applies to the nervous system, leakage of blood from an artery or vein into either the parenchyma of the muscle, nerve, spinal cord, brain or anatomic spaces that overlay these structures. Hemorrhage therefore causes direct damage to the tissue affected and indirect damage to adjacent structures due to mass effect depending on the size and location of the hemorrhage.

Herniation Bulging or expansion of the tissue beyond its normal boundary (e.g. uncal).

Heteronymous hemianopia Loss of (different) halves of the visual fields of both eyes; bitemporal for the temporal halves and binasal for the nasal halves.

Hindbrain Posterior division of the embryonic brain; includes pons, medulla and cerebellum (located in the posterior cranial fossa).

Hippocampus (or hippocampus 'proper') Involved in the acquisition of new memories for names and events. Part of the limbic system; a cortical area 'buried' within the medial temporal lobe, consisting of phylogenetically old (three-layered) cortex; protrudes into the floor of the inferior horn of the lateral ventricle.

Homonymous hemianopia Loss of the visual fields serving the visual space on one side, in both eyes; this involves the nasal half of the visual field in one eye and the temporal half of the visual field in the other eye.

Horner's syndrome Miosis (constriction of the pupil), anhidrosis (dry skin with no sweat) and ptosis (drooping of the upper eyelid) due to a lesion of the sympathetic pathway to the head. Lesions can occur anywhere along the course of the sympathetic pathway such as in the descending portion in the brainstem, in the paraspinous sympathetic ganglia in the chest and along the course of the carotid artery to Mueller's muscle in the upper eyelid.

Hydrocephalus Enlargement of the ventricles due to excessive accumulation of CSF. If the CSF pathway obstruction is within the ventricular system, the disorder is termed noncommunicating (obstructive) hydrocephalus; communicating hydrocephalus means that the obstruction is in the extracerebral subarachnoid space.

Hypo/hyperreflexia Decrease (hypo) or increase (hyper) of the stretch (deep tendon) reflex.

Hypo/hypertonia Decrease (hypo) or increase (hyper) of the tone of muscles, manifested by decreased or increased resistance to passive movements.

Hypokinesia Markedly diminished movements (spontaneous).

Hypothalamus A region of the diencephalon that serves as the main controlling centre of the autonomic nervous system and is involved in several limbic circuits; also regulates the pituitary gland.

Infarction Local death of an area of tissue due to loss of its blood supply.

Infratentorial This refers to the space and structures contained in the area below the tentorium cerebelli or cerebellar tentorium such as the brainstem and cerebellum.

Infundibulum (funnel) Infundibular stem of the posterior pituitary (neurohypophysis).

Innervation Nerve supply, sensory or motor.

Insula (island) Cerebral cortical area not visible from outside inspection and situated at the bottom of the lateral fissure (also called the island of Reil).

Internal capsule White matter between the lentiform nucleus and the head of the caudate nucleus (anteriorly) and the thalamus (posteriorly); consists of anterior limb, genu and posterior limb; carries motor, sensory and integrative fibers connecting the thalamus and cortex.

Internuclear ophthalmoplegia (INO) Loss of adduction of one eye due to damage to the medial longitudinal fasciculus connecting CN VI to CN III.

Ion channel Pore in a neuronal cell membrane that selectively allows a specific ion to pass into or out of the cell; allows for the development of an electric charge within the neuron.

Ipsilateral On the same side of the body (e.g. ipsilateral to a lesion).

Ischemia A condition in which an area is not receiving an adequate blood supply.

Kinesthesia The conscious sense of position and movement.

Lacune The pathological small 'hole' remaining after an infarct in central cerebral hemispheric structures such as the internal capsule; also irregularly shaped venous 'lakes' or channels draining into the superior sagittal sinus.

Lateral ventricle CSF cavity in each cerebral hemisphere; consists of anterior horn, body, atrium (or trigone), posterior horn and inferior (temporal) horn.

Lemniscus A specific pathway in the CNS; medial lemniscus for discriminative touch, conscious proprioception and vibration; lateral lemniscus for audition.

Lenticulo-striate arteries Small vessels originating from the circle of Willis or the proximal middle cerebral arteries and supplying the basal ganglia; obstruction of or hemorrhage from one of these vessels is a common cause of stroke.

Lentiform Lens-shaped; lentiform nucleus, a part of the corpus striatum; also called the lenticular nucleus; composed of putamen (laterally) and globus pallidus.

Leptomeninges Arachnoid and pia mater, part of the meninges.

Lesion Any injury or damage to tissue, vascular, traumatic.

Leukodystrophy Diffuse degenerative disease of the central nervous system white matter, typically of hereditary origin; some types may be accompanied by peripheral nerve demyelination.

Limbic system Parts of the brain, cortical and subcortical, that are associated with emotional behaviour.

Locus ceruleus A small nucleus located in the uppermost pons on each side of the IVth ventricle; contains melanin-like pigment, visible as a dark bluish area in freshly sectioned brain; source of noradrenergic projections to the cerebral hemispheres, cerebellum and spinal cord.

Lower motor neuron Anterior horn cell of the spinal cord and its axon and also the cells in the motor CN nuclei of the brainstem; also called the alpha motor neuron. Its loss leads to atrophy of the muscle and weakness, with hypotonia and hyporeflexia; also, fasciculations are often noted.

Lumbar puncture (LP) This test is also called a spinal tap. The purpose is to obtain samples of CSF to determine infection, inflammation, malignancy and bleeding within the subarachnoid space. The test is performed by inserting a spinal needle into the subarachnoid space of the extension of the dural sac, the lumbar cistern. Usually performed between the spinal (vertebral) interspaces at the L3–L4 level. The commonest complication is headache, or LP headache. This is caused by the reduction of the hydrostatic pressure in the subarachnoid space caused by the lumbar puncture. This headache is characteristic in that it is worse when the patient is sitting or standing and immediately better when recumbent.

Mammillary Mammillary bodies; nuclei of the hypothalamus, which are seen as small swellings on the ventral surface of the diencephalon (also spelled *mamillary*).

Mass Effect The process of displacement or shutting of normal brain structures due to swelling caused by a mass lesion pushing against adjacent brain tissue.

Medial lemniscus Brainstem portion (crossed) of sensory pathway for discriminative touch, conscious pro-prioception and vibration, formed after synapse (relay) in nucleus gracilis and nucleus cuneatus.

Medial longitudinal fasciculus (MLF) A tract throughout the brainstem and upper cervical spinal cord that interconnects visual and vestibular input with other nuclei controlling movements of the eyes and of the head and neck.

Medulla Caudal portion of the brainstem; may also refer to the spinal cord as in a lesion within (intramedullary) or outside (extramedullary) the cord.

Meninges Covering layers of the central nervous system (dura, arachnoid and pia).

Meningismus A condition in which there is irritation of the meninges causing extreme neck stiffness: Brudzinski's sign occurs with meningeal irritation; there is reflex flexion of the legs when the head is flexed on the neck by the examiner. Similarly, Kernig's sign is reflex flexion of the head when the lower leg is extended at the knee. Neither of these signs is specific but if present should guide the clinician to look for causes of meningeal irritation by performing a lumbar puncture (if there are no contraindications).

Meningitis Inflammation of the cerebral and spinal leptomeninges, usually of infectious origin.

Mesencephalon The midbrain (upper part of the brainstem).

Microglia The 'scavenger' cells of the CNS, that is macrophages; considered by some as one of the neuroglia.

Midbrain Part of the brainstem; also known as the mesencephalon.

MMSE Mini-Mental Status Examination or Folstein Test. This is a standardized test of mental status that includes evaluation of attention, immediate and delayed recall, verbal and written language. The test has a maximum score of 30. Scores of less than 27 without intercurrent medical causes are considered to be abnormal. (See Chapters 2 and 9.)

MoCA Montreal Cognitive Assessment. This is similar to the MMSE but is more extensive in terms of frontal lobe testing. It has a maximum score of 30. Scores of lower than 27 without intercurrent medical causes are considered to be abnormal. (See Chapters 2 and 9.)

Motor To do with movement or response.

Motor unit A lower motor neuron, its axon and the muscle fibers that it innervates; includes the neuromuscular junction (NMJ).

MRI Magnetic resonance imaging. A diagnostic imaging technique that uses an extremely strong magnet, not x-rays. In clinical applications, this technique uses the electromagnetic echo returned by excited hydrogen ions to give anatomical, functional and spectral information of the brain and spinal cord. This technique is extremely sensitive to changes in water density caused by tumors and inflammatory processes such as multiple sclerosis.

Muscle spindle Specialized receptor within voluntary muscles that detects muscle length; necessary for the stretch/myotatic reflex (DTR); contains muscle fibers within itself capable of adjusting the sensitivity of the receptor.

Muscular dystrophy Hereditary degenerative disease of muscle tissue.

Myelin Proteolipid layers surrounding nerve fibers, formed in segments; it is important for rapid (saltatory) nerve conduction.

Myelin sheath Covering of a nerve fiber, formed and maintained by oligodendrocyte in CNS and Schwann cell in PNS; interrupted by nodes of Ranvier.

Myelitis Inflammatory disease of the spinal cord, of infectious or autoimmune origin.

Myelography Imaging after performing a lumbar puncture; an iodinated dye is injected into the subarachnoid space. The spread of the dye in the subarachnoid space is followed up to the neck as the patient is tilted head down to make the dye flow with gravity.

Myelopathy Generic term for disease affecting the spinal cord.

Myopathy Generic term for muscle disease.

Myotatic reflex DTR elicited by stretching the muscle, causing a reflex contraction of the same muscle; monosynaptic, from muscle spindle afferents to anterior horn cell (also spelled *myotactic* reflex).

Myotome Muscle groups innervated by a single spinal cord segment; in fact, usually two adjacent segments are involved (e.g. biceps, C5 and C6).

Neocerebellum Phylogenetically newest part of the cerebellum, present in mammals and especially well developed in humans; involved in coordinating precise voluntary movements and also in motor planning.

Neocortex Phylogenetically newest part of the cerebral cortex, consisting of six layers (and sublayers) characteristic of mammals and constituting most of the cerebral cortex in humans.

Neostriatum The phylogenetically newer part of the basal ganglia consisting of the caudate nucleus and putamen; also called the striatum or corpus striatum.

Nerve conduction studies (NCS) A technique involving transcutaneous electrical nerve stimulation to determine the speed of conduction and amplitude of response from peripheral nerves. This aids in localization of lesions to specific peripheral nerves. Slowing of conduction suggests disruption of myelin; low amplitude suggests axonal dysfunction or destruction.

Nerve fiber Axonal cell process, plus myelin sheath if present.

Neuralgia Pain, severe, shooting, 'electrical' along the distribution of a peripheral nerve (spinal or cranial).

Neuraxis The straight longitudinal axis of the embryonic or primitive neural tube, bent in later evolution and development.

Neuroglia Accessory or interstitial cells of the central nervous system; includes astrocytes, oligodendrocytes, ependymal cells and microglial cells.

Neuron The basic structural unit of the nervous system, consisting of the nerve cell body and its processes – dendrites and axon.

Neuropathy Disorder of one or more peripheral nerves.

Neuropil An area between nerve cells consisting of a complex arrangement of nerve cell processes, including axon terminals, dendrites and synapses.

Neuropsychological testing Complex evaluations of mental status usually performed by certified neuropsychologists. These tests are usually ordered when the MMSE or MoCA does not provide enough sensitivity to determine if a given individual has experienced a decrement in cognitive functions. Examples of these tests include the Wechsler Adult Intelligence Scale (WAIS), Wisconsin card sorting task (WCST), Minnesota Multiphasic Personality Inventory (MMPI) and others. The neuropsychologist will often choose the battery of tests that best suits the type of difficulty that the patient is reporting.

Neurotransmitter Chemical compound released into a synaptic cleft by the terminal process of the presynaptic neuron; excites or inhibits the postsynaptic site, depending on the type of transmitter and the type of receptor.

Nociception Perception of a potentially injurious stimulus, typically via small myelinated and unmyelinated sensory nerve fibers; may or may not be associated with the sensation of pain.

Node of Ranvier Gap in myelin sheath between two successive internodes; necessary for saltatory (rapid) conduction.

Nucleus (plural = nuclei) An aggregation of neurons within the CNS; in histology, the nucleus of a cell.

Nystagmus An involuntary oscillation of the eye(s), typically slow in one direction and rapid in the other; named for the direction of the rapid component.

Oligodendrocyte A neuroglial cell; forms and maintains the myelin sheath in the CNS; each cell is responsible for several internodes on different axons.

Operculum (from the Latin 'to cover') Regions of cerebral cortex along the upper and lower edges of the lateral (Sylvian) fissure.

Optic chiasm(a) Partial crossing of optic nerves – the nasal half of the retina representing the temporal visual fields – after which the optic tracts are formed.

Optic disc Area of the retina where the optic nerve exits; also the site for the central retinal artery and vein; devoid of receptors, hence, the physiological blind spot.

Optic nerve Second CN (CN II); special sense of vision; actually a pathway of the CNS, from the ganglion cells of the retina until the optic chiasm as the optic nerve, and its continuation, the optic tract.

Palmomental reflex This is a reflex movement of the mouth and lips of the contralateral face that occurs when the palm is stroked briskly on one side. This indicates disinhibition of the frontal cortex due to injury from stroke, trauma or a degenerative process.

Papilledema Edema of the optic disc (also called a choked disc); visualized with an ophthalmoscope; usually a sign of increased intracranial pressure.

Paralysis Complete loss of muscular action.

Paraplegia Paralysis of both legs and lower part of trunk.

Paresis Muscle weakness or partial paralysis.

Paresthesia Spontaneous abnormal sensation (e.g. tingling; pins and needles).

Parkinsonism Impaired motor control syndrome characterized by resting tremor, muscle rigidity, akinesia and postural dyscontrol; classic example is Parkinson's disease.

Paroxysmal depolarization shift (PDS) An initial decline in resting membrane potential of a cortical neuron followed by a burst of action potentials, then a return to a normal resting membrane potential.

Pathway A chain of functionally related neurons (nuclei) and their axons, making a connection between one region of the CNS and another; a tract (e.g. visual pathway, posterior column – medial lemniscus sensory pathway).

Peduncle A thick stalk or stem; a bundle of nerve fibers (cerebral peduncle of the midbrain; also three cerebellar peduncles: superior, middle and inferior).

Penumbra The penumbra is defined as an area of ischemic brain that has partially lost its blood supply but not to the degree that cell death has occurred. This is usually a circumferential area surrounding an area of brain that has been irreversibly damaged (an infarct). The goal of stroke treatment is to identify patients with areas of penumbra and to target treatment to these areas before irreversible damage occurs.

Perikaryon The cytoplasm surrounding the nucleus of a cell; sometimes refers to the cell body of a neuron.

Peripheral nervous system (PNS) Nerve roots, peripheral nerves and ganglia (motor, sensory and autonomic) outside the CNS.

PET Positron emission tomography. A radionuclear technique using the properties of positron emission. This technique uses radionuclear analogs of biological substrates such as glucose, oxygen or neurotransmitters such as dopamine to visualize areas of the brain where these substances are metabolically active.

Pia (mater) The thin innermost layer of the meninges, attached to the surface of the brain and spinal cord; forms the inner boundary of the subarachnoid space.

Plexus An interweaving arrangement of vessels or nerves.

Pons (bridge) The middle section of the brainstem that lies between the medulla and the midbrain; appears to constitute a bridge between the two hemispheres of the cerebellum.

Posterior column Fasciculus gracilis and fasciculus cuneatus of the spinal cord, pathways (tracts) for discriminative touch, conscious proprioception and vibration; dorsal column.

Praxis Greek word meaning 'do' or 'act'; in neurological parlance refers to normal performance of complex procedural motor activities such as combing hair or brushing the teeth.

Projection fibers Bidirectional fibers connecting the cerebral cortex with structures below, including the basal ganglia, thalamus, brainstem and spinal cord.

Proprioception The sense of body position (conscious or unconscious).

Proprioceptor One of the specialized sensory endings in muscles, tendons and joints; provides information concerning movement and position of body parts (proprioception).

Prosody Vocal tone, inflection and melody accompanying speech.

Ptosis Drooping of the upper eyelid; can be unilateral or bilateral.

Pulvinar The posterior nucleus of the thalamus; functionally linked with visual association cortex.

Putamen The larger (lateral) part of the lentiform nucleus, with the globus pallidus; part of the neostriatum with the caudate nucleus.

Pyramidal system So-called because the corticospinal tracts occupy pyramid-shaped areas on the ventral aspect of the medulla; may include corticobulbar fibers. The term pyramidal tract refers specifically to the corticospinal tract.

Quadriplegia Paralysis affecting the four limbs (also called tetraplegia).

Radicular Refers to a nerve root (motor or sensory).

Ramus (plural = rami) The division of the mixed spinal nerve (containing sensory, motor and autonomic fibers) into anterior and posterior branches.

Rancho Los Amigos Levels of Cognitive Functioning (RLA) This scale was designed to measure and track an individual's progress early in the recovery period following brain injury. The RLA scale describes levels of functioning and is used to assess the efficacy of treatment programs.

Raphe An anatomical structure in the midline; in the brainstem, several nuclei of the reticular formation are in the midline of the medulla, pons and midbrain (these nuclei use serotonin as their neurotransmitter).

Red nucleus Nucleus in the midbrain (reddish colour in a fresh specimen).

Reflex Involuntary movement of a stereotyped nature in response to a stimulus.

Reflex arc Consisting of an afferent fiber, a central connection (synapse), a motor neuron and its efferent axon leading to a muscle movement.

REM behaviour disorder (RBD) A sleep disorder characterized by what appears to be the acting out of dreams. This phenomenon occurs during rapid eye movement (REM) sleep in individuals who have lost REM atonia; seen in various forms of synucleinopathies such as Parkinson's disease and multisystem atrophy.

Repetitive nerve stimulation A technique by which a repetitive electrical stimulus is applied to a peripheral nerve and the output is measured over a muscle innervated by that nerve; changes in compound motor action potential (CMAP) amplitude over time are assessed. A decremental response (decreasing CMAP amplitude with repetitive stimulation) suggests a fatiguable NMJ and, therefore, a postsynaptic problem. An incremental response (increasing CMAP amplitude with repetitive stimulation) suggests a presynaptic problem.

Reticular Pertaining to or resembling a net; reticular formation of the brainstem.

Reticular formation Diffuse nervous tissue nuclei and connections in brainstem, quite old phylogenetically.

Rhinencephalon Refers in humans to structures related to the olfactory system.

Rigidity Abnormal muscle stiffness (increased tone) with increased resistance to passive movement of both agonists and antagonists (e.g. flexors and extensors), usually seen in Parkinson's disease; velocity independent.

Root The peripheral nerves – sensory (dorsal) and motor (ventral) – as they emerge from the spinal cord and are found in the subarachnoid space.

Rostral Toward the snout, or the most anterior end of the neuraxis.

Rubro (red) Pertaining to the red nucleus, as in rubrospinal tract and corticorubral fibers.

Saccadic (to jerk) Extremely quick movements, normally of both eyes together (conjugate movement), while changing the direction of gaze.

Schwann cell Neuroglial cell of the PNS responsible for formation and maintenance of myelin; there is one Schwann cell for each internode segment of myelin.

Secretomotor Parasympathetic motor nerve supply to a gland.

Sensory (afferent) To do with receiving information, from the skin, the muscles, the external environment or internal organs.

Septal region An area below the anterior end of the corpus callosum on the medial aspect of the frontal lobe that includes the cortex and the (subcortical) septal nuclei.

Septum pellucidum A double membrane of connective tissue separating the frontal horns of the lateral ventricles; situated in the median plane.

Single fiber EMG A test of the integrity of the neuromuscular junction. The ability of NMJs to transmit impulses within a given motor unit is tested and expressed as a parameter called 'jitter'. The test is based on measuring the variance of jitter between two NMJs within the same motor unit.

Somatic Used in neurology to denote the body, exclusive of the viscera (as in somatic afferent neurons from the skin and body wall); the word *soma* is also used to refer to the cell body of a neuron.

Somatic senses Touch (discriminative and crude), pain, temperature, proprioception and the sense of 'vibration'.

Somatosensory evoked potentials (SSEP) A neurophysiological test that measures the response to an electrical stimulus applied to either upper or lower limbs in order to determine the speed of conduction from peripheral nerves through the spinal cord to the contralateral somatosensory cortex. An electrical stimulus is usually applied to the median nerves at the wrist or to the posterior tibial nerves at the ankles; each side and limb is done separately. This test assists in localization by assessing, where there is slowing of transmission of the evoked potential in the various structures through which the signal passes.

Somatotopic The orderly representation of the body parts in CNS pathways, nuclei, thalamus and cortex; topographical representation.

Somesthetic Consciousness of having a body; somesthetic (somatic) senses are the general senses of touch, pain, temperature, position, movement and 'vibration'.

Spasticity Velocity-dependent increased tone and increased resistance to passive stretch of the limb muscles; in humans, most often the elbow, wrist, knee and foot flexors as well as the hip adductors; usually accompanied by hyperreflexia.

Special senses Sight (vision), hearing (audition), balance (vestibular), taste (gustatory) and smell (olfactory).

Sphingolipid Complex glycolipid component of neuronal and glial cell membranes; may accumulate intracellularly in hereditary disorders of sphingolipid and recycling.

Spinal shock Complete 'shut down' of all spinal cord activity (in humans) following an acute complete lesion of the cord (e.g. severed cord after a diving or motor vehicle accident), below the level of the lesion; usually up to two to three weeks in duration. This can also occur after indirect trauma such as a blast injury.

Spinocerebellar tracts Ascending tracts of the spinal cord, anterior and posterior, for transmission of 'unconscious' proprioceptive information to the cerebellum.

Spinothalamic tracts Ascending tracts of the spinal cord for pain and temperature (lateral) and nondiscriminative or light touch and pressure (anterior).

Split brain A brain in which the corpus callosum has been severed in the midline, usually as a therapeutic measure for intractable epilepsy of frontal lobe origin.

Stereognosis The recognition of an object using the tactile senses and also central processing, involving association areas especially in the parietal lobe.

Strabismus (a squint) Lack of conjugate fixation of the eyes; may be constant or variable.

Striatum The phylogenetically more recent part of the basal ganglia (neostriatum) consisting of the caudate nucleus and the putamen (lateral portion of the lentiform nucleus).

Stroke A sudden severe attack of the CNS; usually refers to a sudden focal loss of neurologic function due to death of neural tissue; most often due to a vascular lesion, either infarct (embolus, occlusion) or hemorrhage.

Subarachnoid space Space between the arachnoid and pia mater, containing CSF.

Subcortical Not in the cerebral cortex, that is at a functionally or evolutionary 'lower' level in the CNS; usually refers to the white matter of the cerebral hemispheres and also may include the basal ganglia or other nuclei.

Substantia gelatinosa A nucleus of the gray matter of the dorsal (sensory) horn of the spinal cord composed of small neurons; receives pain and temperature afferents.

Substantia nigra A flattened nucleus with melanin pigment in the neurons located in the midbrain and having motor functions – consisting of two parts: the pars compacta, with dopaminergic neurons that project to the striatum and which degenerate in Parkinson's disease, and the pars reticulata, which projects more diffusely to forebrain structures.

Subthalamus Region of the diencephalon beneath the thalamus, containing fiber tracts and the subthalamic nucleus; part of the functional basal ganglia.

Sulcus (plural = sulci) Groove between adjacent gyri of the cerebral cortex; a deep sulcus may be called a fissure.

Supratentorial The space and structures contained in the area above the tentorium cerebelli or cerebellar tentorium, such as the cerebral hemispheres and ventricular system.

Synapse Area of structural and functional specialization between neurons where transmission occurs (excitatory, inhibitory or modulatory), using neurotramsmitter substances (e.g. glutamate, GABA); similarly at the NMJ (using acetylcholine).

Syringomyelia A pathological condition characterized by expansion of the central canal of the spinal cord with destruction of nervous tissue around the cavity.

Tectum The 'roof' of the midbrain (behind the aqueduct) consisting of the paired superior and inferior colliculi; also called the quadrigeminal plate.

Tegmentum The 'core area' of the brainstem, between the ventricle (or aqueduct) and the corticospinal tracts; contains the reticular formation, CN and other nuclei, and various tracts.

Telencephalon Rostral part of embryonic forebrain; primarily cerebral hemispheres of the adult brain.

Tensilon test A procedure that temporarily increases the amount of acetylcholine present in the synaptic cleft to overcome a block at the neuromuscular junction. This test uses a medication called Tensilon or edrophonium, a mild reversible acetylcholinesterase inhibitor. The medication (usually a 9 mg bolus) is given intravenously (IV) after an initial dose of 1 mg is given to test for bradycardia. The test is positive if the Tensilon reverses the neurological deficit for 5 to 10 minutes. This test should always be performed with cardiac monitoring and with IV atropine drawn and available should bradycardia or hypotension develop.

Tentorium The tentorium cerebelli is a sheet of dura between the occipital lobes of the cerebral hemispheres and the cerebellum; its hiatus or notch is the opening for the brainstem – at the level of the midbrain.

Thalamus A major portion of the diencephalon with sensory, motor and integrative functions; consists of several nuclei with reciprocal connections to areas of the cerebral cortex.

Third (IIIrd) ventricle Midline ventricle at the level of the diencephalon (between the thalamus of each side), containing CSF.

Tic Brief, repeated, stereotyped, semipurposeful muscle contraction; not under voluntary control although may be suppressed for a limited time.

Tinnitus Persistent intermittent or continuous ringing or buzzing sound in one or both ears.

Tomography Radiological images, CT or MRI, done sectionally.

Tone Referring to muscle, its firmness and elasticity – normal, hyper or hypo – elicited by passive movement and also assessed by palpation.

tPA Tissue plasminogen activator. Catalyzes the conversion of plasminogen to plasmin resulting in the lysis of newly formed clots. If given within three to four hours of the onset of symptoms, this treatment can lyse acute clots and reverse deficits due to acute cerebral ischemia. It is now considered the standard of care for acute ischemic stroke.

Tract A bundle of nerve fibers within the CNS, with a common origin and termination (e.g. optic tract, corticospinal tract).

Transient ischemic attack (TIA) A nonpermanent focal deficit, caused by a vascular event; by definition, usually reversible within a few hours, with a maximum of 24 hours.

Tremor Oscillating, rhythmic movements of the hands, limbs, head or voice; intention (kinetic) tremor of the limb commonly seen with cerebellar lesions; tremor at rest commonly associated with Parkinson's disease.

Two-point discrimination Recognition of the simultaneous application of two points close together on the skin; varies with the area of the body (e.g. compare finger tip to back).

Uncus An area of cortex – the medial protrusion of the rostral (anterior) part of the parahippocampal gyrus of the temporal lobe; the amygdala is situated deep to this area; important clinically as in uncal herniation.

Upper motor neuron Neuron located in the motor cortex or other motor areas of the cerebral cortex, giving rise to a descending tract to lower motor neurons in the brainstem (corticobulbar, for CNs) or spinal cord (corticospinal, for body and limbs); may also refer to brainstem neurons projecting to the spinal cord (e.g. reticular formation giving rise to the reticulospinal tract).

Upper motor neuron lesion A lesion of the brain (cortex, white matter of hemisphere), brainstem or spinal cord interrupting descending (corticospinal and reticulospinal) motor influences to the lower motor neurons of the spinal cord, characterized by weakness, spasticity and hyperreflexia and, often, clonus; accompanied by a Babinski response.

Ventral root Efferent motor component of a spinal nerve; situated between the spinal cord and the mixed spinal nerve, including its portion travelling in the subarachnoid space within the CSF.

Ventricles CSF-filled cavities inside the brain.

Vermis Unpaired midline portion of the cerebellum, between the hemispheres.

Vertigo Abnormal sense of spinning, whirling or motion, either of the self or of one's environment.

Visual evoked potential (VEP) A neurophysiological technique using EEG electrodes applied over the occipital cortex to determine the speed of conduction of impulses from the retina through the optic nerve, optic tracts, optic radiations to the occipital cortex; the stimulus may be an alternating checkerboard pattern on a TV monitor or, for young children and comatose patients, a light flash from light-emitting diode goggles. This test assists in localization by assessing whether there is slowing of transmission of the evoked potential in the various structures through which the signal passes.

White matter Nervous tissue of the CNS made up of nerve fibers (axons), some of which are myelinated; appears 'whitish' after fixation in formalin.

ANNOTATED BIBLIOGRAPHY

This is a select list of references to 'neuro' books with some commentary to help the learner choose additional learning resources about the structure, function and diseases of the human brain.

The perspective is for medical students and residents, and also for non-neurological practitioners, as well as those in related fields in the allied health professions.

The listing includes texts, atlases as well as neurology reference books (some with CDs or publisher Web sites), as well as Web sites and videotapes.

NEUROANATOMICAL TEXTS AND ATLASES

The listing includes both neuroanatomical textbooks and atlases, as well as a few neuroanatomical texts with a significant clinical emphasis; recent publications (since the year 2010) have usually been preferentially selected.

TEXTS

Augustine, J.R. *Human Neuroanatomy: An Introduction*. New York: Academic Press, 2008.

The human nervous system is explained in full detail in this book, including mention of injuries to the various systems and other clinical correlations. The illustrations are almost entirely in black and white. There is an extensive reference section at the end.

Conn, P.M., Ed. *Neuroscience in Medicine*, 3rd ed. Totowa, NJ: Humana Press, 2008.

The nervous system is well described from an anatomical and functional perspective in this multi-authored book, with mention of clinical disease states appropriate for each section. Some of the chapters are accompanied by a short section describing clinical disorders (e.g. dementia following the chapter on the cerebral cortex). The illustrations are almost entirely in black and white. The last chapter discusses degeneration, regeneration and plasticity in the nervous system.

Crossman, A.R., Neary, D. *Neuroanatomy: An Illustrated Colour Text*, 5th ed. Edinburgh: Elsevier Churchill Livingstone, 2014.

This book format is part of a series of 'Illustrated Colour Texts' (the companion *Neurology* text is by Fuller, G., Manford, M., 2nd ed., 2006). The text is accompanied by many illustrations, including diagrams and photographs of the brain. This book is a useful addendum for students in the health sciences learning about the nervous system for the first time.

Gertz, S.D. *Liebman's Neuroanatomy Made Easy and Understandable*, 7th ed. Austin, TX: Pro-ed Publishers, 2007.

The title promises a book that makes neuroanatomy understandable, which it sometimes succeeds in doing by simplifying the subject matter and by bringing in some clinical correlations. Whether or not the subject has been made 'easy' is debatable and has not been helped by the rather low quality of the illustrations. The book may be usable by students in a non-biological field of study but is not nearly sufficient for medical students.

Gilman, S., Newman, S.W. *Manter and Gatz's Essentials of Clinical Neuroanatomy and Neurophysiology*, 10th ed. Philadelphia, PA: F.A. Davis, 2003.

This slender book attempts to condense a lot of neuroanatomical information with clinical correlations throughout. The illustrations are sketch-like and full of connections. This is possibly a handy review book for those who have taken a full course on the nervous system and would not be adequate as a learner's textbook.

Haines, D.E. *Fundamental Neuroscience for Basic and Clinical Applications*, 4th ed. Philadelphia, PA: Churchill Livingstone, 2012.

This edited, multiauthored large text, with many colour illustrations, is an excellent reference book, mainly for neuroanatomical detail.

Kandel, E.R., Schwartz, J.H., Jessell, T.M. *Principles of Neural Science*, 5th ed. New York: McGraw-Hill, 2012.

This thorough textbook presents a physiological depiction of the nervous system, often with experimental details and information from animal studies. It is suitable as a reference book and for graduate students.

Kiernan, J.A. *Barr's the Human Nervous System: An Anatomical Viewpoint*, 10th ed. Baltimore, MD: Lippincott Williams and Wilkins, 2013.

This new edition of Barr's neuroanatomical textbook has additional colour, as well as clinical notes (in boxes) and magnetic resonance images (MRIs). It is clearly written and clearly presented, with a glossary. There is no longer an accompanying CD but a publisher's Web site (accessed with the purchase of the book) with some of the illustrations, sample exam questions and some clinical cases.

Kolb, B., Whishaw, I.Q. *Fundamentals of Human Neuropsychology*, **6th ed. New York: Worth Publishers, 2008.**

A classic in the field and highly recommended for a good understanding of the human brain in action. Topics discussed include memory, attention, language and the limbic system.

Martin, J.H. *Neuroanatomy: Text and Atlas*, **4th ed. New York: McGraw-Hill, 2015.**

A very complete text, with a neuroanatomical perspective and accompanied by some fine (two-colour) explanatory illustrations, written as a companion to Kandel et al. The material is clearly presented, with explanations of how systems function. A detailed Atlas section is included at the end, as well as a glossary of terms.

Nieuwenhuys, R., Voogd, J., van Huijzen, C. *The Human Central Nervous System*, **4th ed. Berlin: Springer, 2008.**

The authors have transformed a book with unique three-dimensional drawings of the central nervous system (CNS) and its pathways, in half-tones (see Atlas section), into a complete textbook of neuroanatomy. Over 200 new diagrams have been added, some with colour. The terminology has likewise been altered, from the previous Latin to more contemporary terminology. The text is clearly written and heavily referenced, with some mention of clinical conditions. This book should now be considered as another reference source for information on functional neuroanatomy.

Nolte, J. *Elsevier's Integrated Neuroscience*, **Philadelphia, PA: Mosby Elsevier, 2007.**

This neuroscience (neuroanatomy) primer is part of a series from Elsevier Press integrating the basic sciences – anatomy, histology, biochemistry, physiology, genetics, immunology, pathology and pharmacology. These books are also linked via the Elsevier 'student consult' Web site (available only to purchasers of these books). The illustrations are well designed to elucidate the structure and function of the nervous system. The only clinical link is a number of exemplary case studies (and their answers) at the end of the book.

Nolte, J. *The Human Brain: An Introduction to Its Functional Anatomy*, **7th ed. St. Louis, MO: Mosby, 2015.**

This is a new edition of an excellent neuroscience text, with anatomical and functional (physiological) information on the nervous system, complemented with clinically relevant material. The textbook includes scores of illustrations in full colour, stained brainstem and spinal cord cross-sections, along with three-dimensional brain reconstructions by John Sundsten. There is also added material via a publisher's Web site (accessed with the purchase of the book). This edition is probably too detailed to be considered a textbook for the abbreviated courses of a medical curriculum and may now be considered in the category of a reference book.

Paxinos, G., Mai, J.K., Eds. *The Human Nervous System*, **3rd ed. Oxford: Elsevier Academic Press, 2011.**

Amongst the books reviewed, this is the most extensive in its discussion of the anatomical aspects of the human nervous system – including development, the peripheral and autonomic nervous systems and the structures of the CNS. This would be a reference book to consult for anatomical details.

Rubin, M., Safdieh, J.E. *Netter's Concise Neuroanatomy.* **Philadelphia, PA: Saunders Elsevier, 2007.**

This small-sized book with Netter's familiar illustrations has little in the way of text, but contains many tables with a lot of both anatomical and functional information. The illustrations are less useful in their smaller versions.

Williams, P., Warwick, R. *Functional Neuroanatomy of Man.* **Philadelphia, PA: W.B. Saunders, 1975.**

This is the 'neuro' section from *Gray's Anatomy*. Although somewhat dated, there is excellent reference material on the CNS, as well as the nerves and autonomic parts of the peripheral nervous system. The limbic system and its development are also well described. (See also Mancall under Neuroanatomical/Clinical Texts.)

Wilson-Pauwels, L., Akesson, E.J., Stewart, P.A. *Cranial Nerves: Anatomy and Clinical Comments.* **Toronto, BC: Decker, 1988.**

A handy resource on the cranial nerves, with some very nice illustrations. Relatively complete and easy to follow.

ATLASES

DeArmond, S.J., Fusco, M.M., Dewey, M.M. *Structure of the Human Brain: A Photographic Atlas*, **3rd ed. New York: Oxford University Press, 1989.**

An excellent and classic reference to the neuroanatomy of the human CNS. No explanatory text and no colour.

England, M.A., Wakely, J. *Color Atlas of the Brain and Spinal Cord: An Introduction to Normal Neuroanatomy.* **Philadelphia, PA: Mosby Elsevier, 2006.**

A very well illustrated atlas, with most of the photographs and sections in colour. Little in the way of explanatory text.

Felten, D.L., Jozefowicz, R.F. *Netter's Atlas of Human Neuroscience*, 3rd ed. Teterboro, NJ: Icon Learning Systems, 2015.

The familiar illustrations of Netter on the nervous system have been collected into a single atlas, each with limited commentary. Both peripheral and autonomic nervous systems are included. The diagrams are extensively labelled.

Haines, D. *Neuroanatomy: An Atlas of Structures, Sections and Systems*, 9th ed. Baltimore, MD: Lippincott Williams and Wilkins, 2014.

A popular atlas that has some excellent photographs of the brain, some colour illustrations of the vascular supply, with additional radiologic material, all without explanatory text. The histological section of the brainstem is very detailed. There is a limited presentation of the pathways and functional systems, with text. This edition comes with a CD containing all the illustrations, with some accompanying text.

Hendelman, W.J. *Atlas of Functional Neuroanatomy*, 3rd ed. Boca Raton, FL: CRC Press, Taylor & Francis, 2016.

This is a learner-oriented atlas, where each illustration has accompanying explanatory text on the opposite page. Each of the sections has a brief introduction (Central Nervous System Organization; Functional Systems; Meninges, Cerebrospinal Fluid and Vascular Systems; Limbic System). The accompanying Web site has all the illustrations, with the added features of 'roll-over' (mouse-over) labelling and animation added to the various pathways.

Mai, J.K., Paxinos, G., Voss, T. *Atlas of the Human Brain*, 4th ed. Heidelberg: Elsevier Academic Press, 2015.

The human brain is mapped in stereotaxic detail, using a very large-sized format. In part I, there are photographs accompanied by labelled illustrations, sometimes accompanied by MRIs. The second part consists of high resolution myelin-stained sections on one side with labelled colour diagrams on the opposite page. There are also detailed reconstructions of brain structures. The Atlas includes a DVD of the images plus three-dimensional visualization software.

Netter, F.H. *The CIBA Collection of Medical Illustrations, Vol. 1, Nervous System, Part 1: Anatomy and Physiology.* Summit, NJ: CIBA, 1983.

A classic! Excellent illustrations of the nervous system, as well as the skull, the autonomic and peripheral nervous systems, and embryology. The text is interesting but may be dated. (See also Royden-Jones under Clinical Texts.)

Nieuwenhuys, R., Voogd, J., van Huijzen, C. *The Human Central Nervous System.* Berlin: Springer, 1981.

Unique three-dimensional drawings of the CNS and its pathways are presented, in tones of gray. These diagrams are extensively labelled, with no explanatory text.

Nolte, J., Angevine, J.B. *The Human Brain in Photographs and Diagrams*, 4th ed. Philadelphia, PA: Mosby Elsevier, 2015.

A well-illustrated (colour) atlas, with text and illustrations and neuroradiology. Functional systems are drawn onto the brain sections with the emphasis on the neuroanatomy; the accompanying text is quite detailed. Excellent three-dimensional brain reconstructions by J.W. Sundsten. There is a chapter with clinical imaging. The glossary includes (coloured) images. The accompanying CD has all the images (in various formats), with some explanatory text for each.

Woolsey, T.A., Hanaway, J., Gado, M.H. *The Brain Atlas: A Visual Guide to the Human Central Nervous System*, 3rd ed. Hoboken, NJ: Wiley, 2008.

A complete pictorial atlas of the human brain, with some colour illustrations and radiographic material.

NEUROANATOMICAL/CLINICAL TEXTS

Afifi, A.K., Bergman, R.A. *Functional Neuroanatomy: Text and Atlas*, 2nd ed. New York: McGraw-Hill, 2005.

This is a neuroanatomical text with functional information on clinical syndromes. A chapter on the normal is followed by a chapter on clinical syndromes (e.g. spinal cord). The book is richly illustrated (in two colours) using semi-anatomic diagrams and MRIs; there is an atlas of the CNS at the end, but it is not in colour. It is a pleasant book visually and quite readable.

Benarroch, E.E., Daube, J.R., Flemming, K.D., Westmoreland, B.F. *Mayo Clinic Medical Neurosciences: Organized by Neurologic Systems and Levels*, 5th ed. Florence, KY: Informa Healthcare, 2008.

The first part of this book is devoted to the basic sciences underlying the clinical diagnosis of neurologic disorders. The second part consists of a description of the various systems – sensory, motor, consciousness, cerebrospinal fluid and vascular – presented with a clinical perspective. In the third part, the nervous system is viewed at various horizontal (axial) levels, starting in the periphery and progressing upwards to the cortical region.

The book is very nicely illustrated in four-colour, with functional diagrams and other illustrations, including imaging. The illustrations underscore the functional aspects of the nervous system, which are being explained, from synapses to pathways. Each chapter includes clinical correlations and clinical problems (in boxes). This full-sized book is too extensive for a 'required' book but is perhaps most useful for medical students doing a clinical elective in neurology and also for neurology residents at the beginning of their training.

Blumenfeld, H. *Neuroanatomy through Clinical Cases,* **2nd ed. Sunderland, MA: Sinauer Associates, 2011.**

This book attempts to bridge the gap between basic neuroanatomy and clinical neurology. Each chapter (e.g. motor pathways, visual system) presents a thorough explanation of the relevant neuroanatomy and is well illustrated, in full colour. This is followed by a section presenting key clinical concepts, with examples of signs and symptoms of various clinical disorders relevant to the aspect of the nervous system under review. The margin includes some illustrations of clinical examinations, some review exercises, plus some mnemonics. Then come the clinical cases, usually several, accompanied by a discussion of the case and its clinical course, and the relevant imaging (unfortunately, it is sometimes difficult to keep track of which imaging belongs to which case).

This is a valuable book for the advanced learner, likely a neurology or neurosurgery resident. The book accomplishes what it sets out to do by its title, but perhaps too much so by trying to be too comprehensive in each domain. There is no CD-ROM or Web site with this edition of the book.

FitzGerald, M.J.T., Gruener, G., Mtui, E. *Clinical Neuroanatomy and Neuroscience,* **7th ed. Edinburgh: Elsevier Saunders, 2015.**

The authors have attempted to create an integrated text for medical and allied health professionals, combining the basic neuroscience with clinical material. The neuroanatomical presentation is quite detailed and is richly illustrated, in full colour, with large appealing explanatory diagrams, and some MRIs, but there are few actual photographs. The clinical syndromes are in coloured panels sometimes accompanied by illustrations. Each chapter ends with a coloured panel with core information. A glossary of terms is included at the end of the book.

Fuller, G., Manford, M. *Neurology: An Illustrated Colour Text,* **3rd ed. New York: Elsevier Churchill Livingstone, 2010.**

The format of this book is part of a series of Illustrated Colour Texts (the companion *Neuroanatomy* text is by Crossman, A.R., Neary, D., 5th ed, 2014). This book presents select clinical entities with concise explanations, accompanied by many illustrations (in full colour). It is not intended to be a comprehensive textbook. The large format and presentation make this an appealing but limited book.

Mancall, E.L. *Gray's Clinical Neuroanatomy: The Anatomic Basis for Clinical Neuroscience.* **Philadelphia, PA: Elsevier/Saunders, 2011.**

The nervous system section of *Gray's Anatomy, 39th edition* (2010) has been reorganized and made into a separate text. Added to the neuroanatomical explanations and classic illustrations are a host of clinical vignettes. These are contained in 'boxes' usually accompanied by photographs of the clinical condition. The clinical cases are exemplary and relevant to the underlying anatomical principles, emphasizing that knowledge of neuroanatomy is essential!

Note: A previous version of a similar edition by Williams and Warwick was published in 1975 (listed previously); clinical material was not part of that edition.

Marcus, E.M., Jacobson, S.J. *Integrated Neuroscience: A Clinical Problem Solving Approach.* **Boston, MA: Kluwer Academic Publishers, 2003.**

This book attempts to bridge the gap between basic neuroanatomy and clinical neurology. The neuroanatomy is presented quite succinctly, accompanied by simplified illustrations not in colour. Diseases relevant to the system being discussed are included within each chapter. Case histories are presented and there are problem-solving chapters for different levels of the nervous system, with associated questions. The CD-ROM accompanying the book has a variety of anatomical and MRI images, and some of the problems have additional commentary. Overall, the book does not integrate the anatomy and physiology necessary to resolve the clinical problem and hence does not live up to its title, although it may be useful for senior neurology residents preparing for examinations.

Tate, M., Cooper-Knock, J., Hunter, Z., Wood, E. *Neurology and Clinical Neuroanatomy on the Move.* **Boca Raton, FL: CRC Press, Taylor & Francis, 2015.**

This book is part of a series that provides quick flexible access, both in text format and electronically (smartphone, computer, tablet). The emphasis is clearly on common clinical neurological conditions, with few neuroanatomical illustrations. The information is highly condensed ('micro-facts' and key points). MICRO-cases (in coloured boxes) are also included. This presentation is probably handy for medical students and those in training who are only passing through a brief neurology exposure.

NEUROLOGIC EXAMINATION

Bickley, L.S., Ed. *Bates' Guide to Physical Examination and History Taking,* **11th ed. Philadelphia, PA: Lippincott Williams & Wilkins, 2012.**

This is the gold standard for medical students on how to conduct a physical examination, as well as for history-taking. An audio-visual (videotape or CD-ROM) has been prepared for some of the systems.

Campbell, W.W. *DeJong's The Neurologic Examination*, **7th ed. Philadelphia, PA: Lippincott Williams & Wilkins, 2012.**

A more advanced and detailed description of the neurologic examination is found in this book, more intended for residents in neurology and neurosurgery and perhaps also for rehabilitation physicians (physiatrists). It is appropriately illustrated with some neuroanatomy and with many examples of physical signs associated with diseases of the nervous system.

CLINICAL TEXTS

Aminoff, M.J., Greenberg, D.A., Simon, R.P. *Clinical Neurology*, **9th ed. New York: Lange Medical Books/McGraw-Hill, 2015.**

If a student wishes to consult a clinical book for a quick look at a disease or syndrome, then this is a suitable book of the survey type. Clinical findings are given, investigative studies are included, as well as treatment. The illustrations are adequate (in two colours), and there are many tables with classifications and causes.

Asbury, A.K., McKhann, G.M., McDonald, W.I., Goodsby, P.J., McArthur, J.C. *Diseases of the Nervous System: Clinical Neuroscience and Therapeutic Principles*, **3rd ed. New York: Cambridge University Press, 2002.**

A complete neurology text, in two volumes, on all aspects of basic and clinical neurology and the therapeutic approach to diseases of the nervous system.

Donaghy, M. *Brain's Diseases of the Nervous System*, **12th ed. New York: Oxford University Press, 2008.**

A very trusted source of information about clinical diseases and their treatment.

Fauci, A.S., Hauser, S.L., Jameson, J.L., Loscalzo, J., Braunwald, E., Kasper, D.L., Longo, D.L., Eds. *Harrison's Principles of Internal Medicine*, **19th ed. New York: McGraw-Hill, 2015.**

Harrison's is a trusted, authoritative source of information, with few illustrations. Part 2 in Section 3 (Volume I) has chapters on the presentation of disease; Part 15 (Volume II) is on all neurologic disorders of the CNS, nerve and muscle disease, as well as mental disorders. The online version of Harrison's has updates, search capability, practice guidelines and online lectures and reviews, as well as illustrations.

Gorelick, P.B., Testai, F.D., Hankey, G.J., Wardlaw, J.M., Eds. *Hankey's Clinical Neurology*, **2nd ed. Boca Raton, FL: CRC Press, Taylor & Francis, 2014.**

This text includes a compendium of diseases of the nervous system, both CNS and peripheral nervous system, including muscle. Each disease entity is discussed in detail from definition to prognosis and prevention. Included are many images, particularly neuroradiologic. The text is aimed at the practicing clinical neurologist.

Patten, J. *Neurological Differential Diagnosis*, **2nd ed. New York: Springer, 2004.**

This comprehensive text systematically reviews clinical disorders affecting each major echelon of the nervous system. Individual chapters are devoted to the clinical assessment of the pupils, visual fields, retina, cranial nerves controlling eye movement, cerebral hemispheres, brainstem, nerve roots (and so on) and the diseases seen in each region. Neuroanatomical information is introduced in each chapter, as required. An important element is the liberal use of brief clinical vignettes to highlight many of the disorders being considered. The anatomical and clinical illustrations are largely the work of the author himself and are in gray-scale. The text is too detailed for medical students but is an excellent resource for trainees in neurology.

Ropper, A.H., Samuels, M.A., Klein, J.P. *Adams and Victor's Principles of Neurology*, **10th ed. New York: McGraw-Hill, 2014.**

A comprehensive neurology text, part devoted to cardinal manifestations of neurologic diseases, and part to major categories of diseases.

Rowland, L.P. *Merritt's Neurology*, **13th ed. Baltimore, MD: Lippincott Williams and Wilkins, 2015.**

A well-known, complete and trustworthy neurology textbook, now edited by L.P. Rowland.

Royden-Jones, H. *Netter's Neurology*, **2nd ed. Teterboro, NJ: Icon Learning Systems, 2011.**

Netter's neurological illustrations (see Netter under Atlases previously) have been collected in one textbook, with the addition of Netter-style clinical pictures; these add an interesting dimension to the descriptive text. There is a broad coverage of many disease states, but not in depth, with clinical scenarios in each chapter. The CD accompanying the book contains the complete textbook and its illustrations.

This is definitely a book that both medical students and others in the allied health sciences might consult in the course of their studies and clinical duties.

Scadding, J.W., Losseff, N. *Clinical Neurology*, **4th ed. Boca Raton, FL: CRC Press, Taylor & Francis, 2011.**

This is quite a good clinical text for non-neurologists interested in neurology. Of necessity, it is superficial in its approach to most neurological disorders, and despite claiming to be a general text for neurologists as well as non-neurologists, it would be of little use to any neurologist other than trainees. The illustrations are not very 'artistic' and are often too small, but they certainly are comprehensive and, for the most part, effective. The fourth edition has been fully revised and updated.

Schapira, A.H.V., Ed. *Neurology and Clinical Neuroscience.* **Philadelphia, PA: Mosby Elsevier, 2007.**

This multi-authored book is a compendium of neurological disorders. The subject matter covers the full range, from apraxia to schizophrenia. The chapters include tables and charts, photographs, neuropathology and MRIs, with the judicious use of colour. Clearly a reference book.

Weiner, W.J., Goetz, C.G., Shin, R.K., Lewis, S.L., Eds. *Neurology for the Non-Neurologist,* **6th ed. Philadelphia, PA: Lippincott Williams & Wilkins, 2010.**

This text presents a thoughtful approach to common neurological symptoms (e.g. vertigo) and diseases (e.g. movement disorders). There are chapters on diagnostic tests and neuroradiology. There are also chapters on principles of neurorehabilitation and medical-legal issues in the care of neurologic patients. The book is sparsely illustrated and not in colour.

PEDIATRIC NEUROLOGY

Fenichel, G.M. *Clinical Pediatric Neurology: A Signs and Symptoms Approach,* **7th ed. Philadelphia, PA: Elsevier Saunders, 2013.**

This book is recommended to medical students and other novices by a highly experienced pediatric neurologist as a basic text with a clinical approach, its structure based on presenting signs and symptoms.

NEUROPATHOLOGY

Kumar, V., Abbas, A.K., Fausto, N., Robbins, S.L., Cotran, R.S., Eds. *Robbins and Cotran Pathologic Basis of Disease,* **9th ed. Philadelphia, PA: Elsevier Saunders, 2014.**

A complete source of information for all aspects of pathology for students, including neuropathology. Purchase of the book includes a CD with interactive clinical cases, and access to the Web site.

Kumar, V., Cotran, R.S., Robbins, S.L., Eds. *Robbins Basic Pathology,* **9th ed. Philadelphia, PA: W.B. Saunders, 2012.**

Not as complete as the 2014 text by Kumar et al. just cited.

WEB SITES

Web sites should only be recommended *after* they have been critically evaluated by the teaching or clinical faculty. If keeping up with various texts is not difficult enough, a critical evaluation of the various Web resources is an impossible task for any single person. This is indeed a task to be shared with colleagues and perhaps undertaken by a consortium of teachers and students.

Additional sources of reliable information on diseases are usually available on the disease-specific Web site maintained by an organization, usually with clear explanatory text on the disease often accompanied by excellent illustrations.

The following sites have been visited by the author (WH), and several of them are gateways to other sites – clearly not every one of the links has been viewed. Although some are intended for the lay public, they may contain good illustrations and/or other links.

The usual WWW precaution prevails – look carefully at who created the Web site and when. A high-speed connection is a must for this exploration!

SOCIETY FOR NEUROSCIENCE

http://web.sfn.org/

This is the official Web site for the Society for Neuroscience, a very large and vibrant organization with an annual meeting attended by more than 30,000 neuroscientists from all over the world. The Society maintains an active educational branch, which is responsible for sponsoring a Brain Awareness Week aimed at the public at large and, particularly, at students in elementary and high schools. The following are examples of their publications.

BRAIN FACTS

http://www.brainfacts.org/

Brain Facts is an online primer on the brain and nervous system that is published by the Society for Neuroscience with partners. It is a starting point for a general audience interested in neuroscience. Educational resources are now available online at www.BrainFacts.org. This site includes information about neuroscience, sections on brain basics; sensing, thinking and behaving; and diseases and disorders, as well as a section called 'across the lifespan', such as parenting and bullying. The new edition updates all sections and includes new information on addiction, neurological disorders (A–Z) and psychiatric illnesses and potential therapies. In addition, there are entries called 'discoveries' and links to 'recent neuroscience in the news', as well as blogs for opinion and conversation about what is new, notable, or inspiring in neuroscience.

This DVD series explores the human face of degenerative brain diseases, including amyotrophic lateral sclerosis (also called Lou Gehrig disease), Alzheimer's and Parkinson's. Patients and families describe the powerful physical, emotional and financial impact of these devastating disorders Researchers tell how they are working to find treatments and cures for these and other diseases, such as Huntington's disease. There is also a video on healthy brain aging.

DIGITAL ANATOMIST PROJECT

http://www9.biostr.washington.edu/da.html

BRAIN ATLAS

The material includes two-dimensional and three-dimensional views of the brain from cadaver sections, magnetic resonance imaging scans, and computer reconstructions.

Authored by John W. Sundsten.

NEUROANATOMY INTERACTIVE SYLLABUS

This syllabus uses the images in the Brain Atlas (the previous entry) and many others. It is organized into functional chapters suitable as a laboratory guide, with an instructive caption accompanying each image. It contains the following: three-dimensional computer graphic reconstructions of brain material; MRI scans; tissue sections, some enhanced with pathways; gross brain specimens and dissections; and summary drawings. Chapter titles include Topography and Development, Vessels and Ventricles, Spinal Cord, Brainstem and Cranial Nerves, Sensory and Motor Systems, Cerebellum and Basal Ganglia, Eye Movements, Hypothalamus and Limbic System, Cortical Connections, and Forebrain and MRI Scan Serial Sections.

Authored by John W. Sundsten and Kathleen A. Mulligan, Digital Anatomist Project, Department of Biological Structure, University of Washington, Seattle, WA.

BRAINSOURCE

http://www.brainsource.com/

BrainSource is an informational Web site aimed at enriching professional, practical and responsible applications of neuropsychological and neuroscientific knowledge. The Web site is presented by neuropsychologist Dennis P. Swiercinsky. The site includes a broad and growing collection of information and resources about the following:

normal and injured brains; clinical and forensic neuropsychology; brain injury rehabilitation; creativity, memory and other brain processes; education; brain–body health; and other topics in brain science. BrainSource is also a guide to products, books, continuing education, and Internet resources in neuroscience. Some information may be dated.

This Web site originated in 1998 for promotion of clinical services and as a portal for dissemination of certain documents useful for attorneys, insurance professionals, students, families and persons with brain injury, rehabilitation specialists and others working in the field of brain injury. The Web site is growing to expand content to broader areas of neuropsychological application.

HARVARD: THE WHOLE BRAIN IMAGING ATLAS

http://www.med.harvard.edu/AANLIB/home.html

This is a neuroradiological resource using a variety of neuroimaging modalities. Images include the normal brain, cerebrovascular disease (stroke), as well as neoplastic, degenerative and inflammatory diseases.

THE BRAIN FROM TOP TO BOTTOM

http://thebrain.mcgill.ca/

This interesting Web site is designed to let users choose the content that matches their level of knowledge. For every topic and subtopic covered on this site, you can choose from three different levels of explanation – beginner, intermediate or advanced. The major topics include brain basics (anatomy, evolution and the brain, development and ethics, pleasure and pain, the senses and movement), a section called brain and mind (memory, emotions, language, sleep and consciousness) and a section on some brain disorders; other subject areas are under development. This site focuses on five major levels of organization – social, psychological, neurological, cellular and molecular. On each page of this site, you can click to move among these five levels and learn what role each plays in the subject under discussion.

Note to the learner: This site can be viewed in both English and French.

THE DANA FOUNDATION

http://www.dana.org/

The Dana Foundation is a private philanthropic organization with a special interest in brain science, immunology and arts education. It was founded in 1950. The

Dana Alliance is a nonprofit organization of more than 200 pre-eminent scientists dedicated to advancing education about the progress and promise of brain research. The Brain Center of this site is a gateway to the latest research on the human brain.

NEUROSCIENCE FOR KIDS

http://faculty.washington.edu/chudler/neurok.html

Neuroscience for Kids was created for all students and teachers who would like to learn about the nervous system. The site contains a wide variety of resources, including images – not only for kids. Sections include exploring the brain, Internet neuroscience resources, neuroscience in the news and reference to books, magazines articles and newspaper articles about the brain.

Neuroscience for Kids is maintained by Eric H. Chudler and is supported by a Science Education Partnership Award (R25 RR12312) from the National Center for Research Resources. To receive this interesting monthly newsletter, contact Eric H. Chudler, PhD (e-mail: chudler@u.washington.edu).

TELEVISION SERIES

http://www.pbs.org/wnet/brain/index.html

The Secret Life of the Brain reveals the fascinating processes involved in brain development across a lifetime. This five-part series, which was shown nationally by the Public Broadcasting Service in the winter of 2002, informs viewers of exciting new information in the brain sciences, introduces the foremost researchers in the field and uses dynamic visual imagery and compelling human stories to help a general audience understand otherwise difficult scientific concepts.

The material includes a history of the brain, three-dimensional brain anatomy, mind illusions and scanning the brain. Episodes include The Baby's Brain, The Child's Brain, The Teenage Brain, The Adult Brain and The Aging Brain. Some of this material may now be available online.

The *Secret Life of the Brain* is a co-production of Thirteen/WNET New York and David Grubin Productions,

© 2001 Educational Broadcasting Corporation and David Grubin Productions, Inc.

VIDEOTAPES (BY THE AUTHOR)

These edited presentations are on the skull and the brain as the material would be shown to students in the gross anatomy laboratory. They have been prepared with the same teaching orientation as the *Atlas of Functional Neuroanatomy, 3rd edition* (see Atlases section) and are particularly useful for self-study or small groups. Each video lesson is fully narrated and lasts for about 20 to 25 minutes. These videotapes of actual specimens are particularly useful for students who have limited or no access to brain specimens.

INTERIOR OF THE SKULL

This program includes a detailed look at the bones of the skull, the cranial fossa and the various foramina for the cranial nerves and other structures. Included are views of the meninges and venous sinuses.

THE GROSS ANATOMY OF THE HUMAN BRAIN SERIES

PART I: THE HEMISPHERES

This is a presentation on the hemispheres and the functional areas of the cerebral cortex, including the basal ganglia.

PART II: DIENCEPHALON, BRAINSTEM AND CEREBELLUM

This is a detailed look at the brainstem, with a focus on the cranial nerves and a functional presentation of the cerebellum.

PART III: CEREBROVASCULAR SYSTEM AND CEREBROSPINAL FLUID

This is a presentation on cerebrovascular system and the cerebrospinal fluid.

Note to the learner: These video lessons are now available on the atlas Web site: www.atlasbrain.com.

ANSWERS TO CHAPTER QUESTIONS

CHAPTER 4

1. The right parasagittal region, precentral gyrus.

2. (a) Weakness in both legs (especially of hip abduction, foot dorsiflexion and foot eversion). (b) Increased muscle tone in response to rapid stretch in the plantar and knee flexors. (c) Hyperactive tendon reflexes in the legs; normal reflexes in the arms. (d) Bilateral extensor plantar responses.

3. (a) Proximal and distal muscle weakness in both legs. (b) Reduced muscle bulk in all muscle groups of the lower limbs. (c) Reduced muscle tone in response to rapid stretch in the plantar and knee flexors. (d) Reduced or absent tendon reflexes in the legs (normal reflexes in the arms). (e) If sufficient residual muscle power in the intrinsic foot muscles, flexor plantar responses bilaterally. (f) Impaired position sense, touch, vibration, temperature and pinprick sensations in both legs below the inguinal regions.

4. Left-sided convexity motor cortex → central hemispheric white matter → posterior limb of left internal capsule (bypassing basal ganglia and thalamus) → left cerebral peduncle (midbrain, bypassing substantia nigra and cranial nerve nuclei III, IV) → left basis pontis (bypassing cranial nerve nuclei V–VIII and pontocerebellar nuclei) → left medullary pyramid → across the midline to the right lateral corticospinal tract → enter right anterior horn of cervical spinal cord at C8–T1. Quite an astonishing accomplishment!

5. (a) Generalized disease of peripheral nerves, affecting longer nerve fibers more than shorter. AHC disease is unlikely because of the sensory findings. Neuromuscular junction disorders are unlikely for the same reason. Muscle disease is unlikely because proximal muscle bulk and power are normal.
(b) A degenerative disease process, given the prolonged time course, possibly of genetic origin. A chronic toxic nerve injury cannot be excluded on clinical grounds.

6. (a) Degenerative disease. The patient's deficits are too diffuse to consider a malignant disease; a chronic toxic neuropathy would be extremely unlikely in a small infant.

(b) AHC or peripheral nerve (with motor axonal degeneration). Muscle disease would not typically be associated with distal muscle weakness and wasting, or with complete areflexia. Neuromuscular junction disease tends to affect cranial musculature more than the limbs, would not produce muscle atrophy and would not be associated with abnormal tendon reflexes. Of the two possibilities, AHC disease is the most probable cause for the patient's presentation. Her clinical deficits appear to involve exclusively the peripheral motor system (proximal and distal weakness with muscle wasting; fasciculations; complete areflexia). A chronic progressive generalized polyneuropathy would also affect sensory nerve fibers while AHC disease would not. The practical problem, of course, is that assessment of sensory function beyond a gross response to pinprick is not feasible in a six-month-old infant. Nevertheless, a patient with a polyneuropathy would typically have muscle weakness and wasting largely confined to the distal musculature below the elbows and knees. In this patient, proximal muscles (including the tongue) are just as severely affected as the distal muscles.
(c) Nerve conduction studies would determine whether there is any sensory nerve degeneration. An electromyography assessment would confirm the presence of chronic denervation changes. In the case of AHC disease at this age, one would see the presence of so-called 'giant' motor units.

This clinical picture is very typical for type 1 infantile muscular atrophy (SMA; Werdnig–Hoffmann disease), an autosomal recessive genetic disorder affecting AHCs.

7. (a) According to the clinical description, the patient has at least grade 2/5 weakness of the right tibialis anterior (foot dorsiflexion), extensor digitorum communis (toe extension), extensor hallucis longus (great toe extension) and peroneus longus/brevis (foot eversion) muscles. All of these muscles are supplied by the common peroneal nerve. The tibialis anterior, extensor digitorum communis and extensor hallucis longus muscles are supplied by the deep branch of the nerve, while the peroneus muscles and the vast majority of the skin on the

dorsum of the foot are supplied by the superficial branch. This being the case, the most likely explanation of the important but restricted deficits in this boy is an acute lesion of the right common peroneal nerve. This nerve branches off the sciatic nerve in the popliteal fossa and travels laterally around the neck of the fibula, where it is subcutaneous and much exposed to mechanical injury. The most probable explanation for the clinical picture is an extrinsic compression of the nerve in the fibular neck region by one of the crib bars during the night.

(b) As was the case for the patient in question 6, the simplest test would be a motor nerve conduction study of the right common peroneal nerve. This would show slowing in conduction velocity in the proximal portion of the nerve, above the point of division into the deep and superficial branches. Compound muscle action potentials recorded from the extensor digitorum brevis muscle in the foot would be normal from stimulation points *below* the fibular neck, but much reduced from points above, e.g. in the lateral popliteal fossa.

(c) In a young child with still-maturing nerves and what was probably largely an ischaemic process within the nerve, the prognosis for complete recovery should be excellent. The nerve dysfunction will be, in all probability, largely a temporary loss of nerve conduction capacity due to energy failure, without any substantial permanent injury to either axons or Schwann cells.

CHAPTER 5

1. The spinothalamic tract starts in the spinal cord having synapsed with the dorsal nerve root, which provides the fibers for pain and temperature from the periphery. It then crosses in the anterior commissure in front of the spinal canal and ascends in the spinal cord in the anterolateral funiculus. It ascends through the lateral portion of the brainstem to terminate on the ventroposteriolateral (VPL) nucleus of the thalamus. The third segment of the tract then ascends from the thalamus to the parietal cortex.

   The posterior columns originate in the posterior funiculus of the spinal cord as a continuation of the dorsal nerve root supplying sensation of vibration, proprioception and fine touch. The tract ascends ipsilateral to the nerve root entry to terminate in the medulla on the nucleus gracilis for the leg fibers and the nucleus cuneatus for the arm fibers. The second relay consists of the medial lemniscus, which crosses to the opposite side and ascends to the contralateral VPL of the thalamus. The third segment originates in the thalamus and terminates in the parietal cortex.

2. Sacral sparing refers to the preservation of sensation to the saddle area with a central spinal cord lesion expanding outwards. Due to the homuncular distribution of the fibers of the spinothalamic tract with the innermost fibers being those from either the thoracic or arm inputs with the leg and saddle fibers being most lateral. Therefore, a destruction central cord lesion will affect the innermost fibers first.

3. Spinal cord lesions can affect segments of spinal cord central gray matter as well as the ascending and descending white matter tracts. The clinical findings for each white matter tract and gray matter segments are as follows:

   Corticospinal tract: Upper motor neuron weakness, hyperreflexia, spasticity

   Spinothalamic tract: Loss of pain and temperature from the opposite side of the body

   Posterior columns: Loss of vibration and proprioception on the same side

   Gray matter: Weakness, sensory loss for pain, temperature, vibration, proprioception, hyporeflexia only at the level of the gray matter damage

4. There are three major arterial supplies to the spinal cord.

   Anterior spinal artery is formed from the two vertebral arteries in the neck before they enter the skull. The artery supplies the cervical and most of the thoracic spine. Occlusion of this vessel can lead to cervical cord infarction.

   Artery of Adamkiewicz supplies the anterior spinal artery at a level between cord level T10 and L1. It supplies the lower thoracic and lumbar cord. It can provide collateral flow to the cervical cord from below. Occlusion of the artery of Adamkiewicz can cause infarction of the thoracic cord involving both gray and white matter structures in the anterior two-thirds of the cord.

   Posterior spinal arteries arise from the aorta at each radicular level and supply the posterior third of the spinal cord. Infarction due to occlusion of these vessels is rare. Occasionally, large dural arteriovenous fistulas can form in the spinal cord and can result in a steal phenomenon from the posterior spinal arteries at one or more levels.

5. The complete transection will result in loss of all the major tracts and gray matter at the level of the transection as well as damage to the blood supply to the spinal cord. She will have all of the clinical deficits listed in answer 3 below the T10 level.

6. A cauda equina syndrome occurs when a mass lesion of some sort forms over the nerves of the cauda equina causing compression and dysfunction of these nerve with neurological symptoms.

The symptoms usually include sudden onset of severe back and leg pain, bilateral leg weakness and sensory loss for all modalities as well as loss of bowel and bladder control. The major risk factors include recent lumbar surgery or injury, degenerative disk disease or trauma. Other possibilities include metastatic disease or medulloblastoma. The cauda equina syndrome is a neurological emergency that requires prompt imaging and surgical decompression.

CHAPTER 6

1. The correct answer is c – the proximal right facial nerve. At this point, shortly after leaving the lower lateral pons, the facial nerve includes the so-called nervus intermedius that contains, along with parasympathetic autonomic fibers for tear production and salivary production, special sensory fibers for taste. The taste fibers leave the main trunk of the nerve in the petrous temporal bone as the chorda tympani, well before the nerve exits the stylomastoid foramen and pass forward into the tongue.

2. No discernible hearing deficit would be identified with standard clinical testing. The reason is that the ascending auditory pathway from a given ear ascends in the brainstem bilaterally, synapsing in *both* medial geniculate nuclei. Thus, the primary auditory cortex (Heschl's gyrus) in the cerebral hemisphere with the intact medial geniculate nucleus would process auditory data from both ears.

3. (a) Either Heschl's gyrus (primary auditory cortex) in both superior temporal lobes or in the white matter tracts that connect the medial geniculate nuclei to Heschl's gyri (for a typical example of the latter, see case 12-5).
(b) Left posterior-superior temporal lobe cortex (Wernicke's area; dominant temporal lobe).

4. The involved eye is deviated laterally and downwards: laterally because of the unopposed action of the lateral rectus muscle (supplied by the abducens nerve, CN VI); downwards because of the unimpeded action of the superior oblique muscle (supplied by the trochlear nerve, CN IV). The pupil in the involved eye will be markedly dilated and unreactive to light stimulation due to the loss of the pupillary constrictor (parasympathetic) fibers in CN III and the unopposed action of sympathetic autonomic fibers from the superior cervical ganglion in the sympathetic chain.

5. Your eyes will be deviated predominantly to the left as you attempt to follow objects in your environment that are all appearing to move rapidly to the left as the chair spins to the right. At regular intervals, your eyes will quickly move to the right in order to pick up whatever prominent object is next in line in the visual environment. Thus, your eyes will rapidly track leftwards as you pursue each new item of interest, followed by an even more rapid eye-flick to the right. This process produces a rhythmic nystagmus with the 'slow' phase to the left and the faster phase to the right. The phenomenon just described is known as *optokinetic nystagmus*.

6. (a) Diminished position sense, vibration and discriminative touch in the right half of the body; pinprick, temperature and crude touch will be intact on both sides. This clinical picture would be the result of loss of function in the left medial lemniscus, the relay bundle in the medial medulla that received second-order sensory axons from the posterior column nuclei in the lower medulla.
(b) Such a lesion would affect, among other structures, the left spinothalamic tract and the descending nucleus and tract of the left trigeminal nerve. The former lesion would result in loss of pain and temperature sensation in the right side of the body below the upper neck. The latter lesion would result in loss of pain and temperature sensation in the left side of the face. For a patient with a similar lesion, see also case 8e-1.
(c) Diminished sensation for *all* modalities on the right side of the body. At the level of the upper pons, the medial lemnisci have moved laterally to join the ascending spinothalamic tracts that, in the medulla, were also located laterally – well away from the medial lemniscus.

7. (a) Weakness of the lower facial muscles on the left side; weakness, spasticity and increased tendon reflexes in the left arm and leg.
(b) Left spastic hemiparesis as in a but *sparing* the left side of the face. Instead, there would be a lower motor neuron pattern of weakness in the *right* side of the face.

CHAPTER 7

1. (a) Indirect pathway
   (b) Both the direct and indirect pathways
   (c) Direct pathway – fine facial movements are analogous to fine finger movements and are mediated by single axons connecting the peri-sylvian portion of the precentral gyri and the appropriate parts of the facial nerve nuclei
   (d) Indirect pathway would predominate as you would not be employing the distal musculature (forefoot, toes) – unless you were barefoot!
   (e) Direct pathway

2. Inability to move the fingers of the contralateral hand in a coordinated or independent fashion; impaired or absent pincer grasp; inability to wriggle the toes of the contralateral foot and impaired ability to tap the foot on the floor. Muscle tone in the involved arm and leg would be normal or low rather than increased. Proximal arm and leg movements would be largely intact. There would nevertheless be an extensor plantar response.

3. (a) Left spastic hemiparesis involving the lower facial muscles and the arm and leg. There would be weakness of wrist extension, hand grip, hip abduction and foot dorsiflexion. Fine coordinated movements of the hand, fingers and toes would be severely compromised or completely absent. The right hand might be fisted (i.e. inability to extend the fingers) and the individual would tend to walk on the toes of the right foot.
   (b) Right spastic hemiparesis sparing the face

4. (a) A hyperkinetic movement disorder of the right side of the body resulting from disruption of the indirect pathway and removal of the 'foot from the brake'. The movement disorder resulting from a lesion in this structure typically consists of wild flinging movements of the opposite side of the body, a phenomenon known as *hemiballismus*.
   (b) Unilateral bradykinesia and rigidity of the left side of the body.

5. (a) An unsteady, staggering gait; coordination of upper limb movements is typically preserved.
   (b) Incoordination of the right arm and leg (i.e. the *ipsilateral* limbs)

6. Sleeping on an arm for a sustained period of time sometimes leads to transient ischemia and loss of function in motor and sensory axons within the proximal portions of the limb's various mixed nerves. The largest diameter axons – muscle spindle and joint capsule afferents – are particularly vulnerable as they have the highest metabolic requirements. Thus, while loss of function in large motor axons may cause temporary muscle weakness, the marked incoordination of movement in these circumstances is believed to result from lack of sensory input from the limb (mediated by both the posterior columns and the spinocerebellar tracts).

7. The 'older' antipsychotic drugs have potent antidopaminergic effects, i.e. they are dopamine antagonists. This results in impairment of activation of the 'accelerator' pathway – and impairment of suppression of the 'brake' pathway – leading to deficits that resemble those seen in Parkinson's disease. Alternatively, chronic use of these antidopaminergic drugs may lead to up-regulation of dopamine receptors in the striatum and the appearance of involuntary movements, particularly in the buccolingual region, a phenomenon known as a *tardive dyskinesia*.

CHAPTER 8

1. Refer to Section 8.4 and Figures 6.3 to 8.6
2. Refer to Sections 8.4.6 and 8.4.7 and Table 8.2
3. The principle measures of prevention for ischaemic stroke include
   a. Control of hypertension
   b. Smoking cessation
   c. Control of hypercholesterolemia
   d. Control of diabetes
4. The major risk factors for cortical hemorrhages include
   a. Amyloid deposition
   b. Trauma
   c. Aneurysms
   d. Cavernomas
   e. Tumors
   f. Systemic anticoagulation
5. The common genetic risk factors include family history of
   a. FMD
   b. CADASIL
   c. Marfan's syndrome
   d. Homocystinuria
   e. MELAS

CHAPTER 9

1. Skin and subcutaneous tissue → external periosteum → bone tissue → internal periosteum → epidural space → dural membrane → subdural

space → arachnoid membrane → subarachnoid space → pial membrane → cerebral cortex

2. Right lateral ventricle → right interventricular foramen (also known as the foramen of Monro) → third ventricle → aqueduct → fourth ventricle → 2 lateral foramina (of Luschka) and 1 median foramen (of Magendie) → cerebellopontine cistern (cisterna magna) → extra-cerebral subarachnoid space → midline subarachnoid granulations → superior sagittal sinus

3. (a) Nasal retina → optic disc of the right eye → optic nerve → across midline through the optic chiasm → left optic tract → synapse in left lateral geniculate body (nucleus) → optic radiation → primary visual cortex of left occipital lobe
   (b) Same route as far as the optic tract → bypass the left lateral geniculate nucleus of the thalamus → brachium of the superior colliculus (connects lateral geniculate nucleus region with the left superior colliculus in the superior midbrain tectum) → superior colliculus. At this point, the axon synapses with tectal neurons that project to both of the (parasympathetic) Edinger–Westphal nuclei (see pupillary light reflex pathway – Figure 2.2)

4. The midbrain aqueduct

5. With a history of progressively more severe headache and vomiting accompanied by papilledema, this boy has clear evidence of raised ICP. Considering the normal cranial CT study and the absence of any localizing signs on neurological examination, the origin of his symptoms is not immediately obvious. It is possible, of course, that – like Cheryl – he also has IIH. The clue to the answer lies in the engorged superficial face and neck veins. Their presence implies that there must be a *bilateral* obstruction in the venous drainage system close to its exit from the cranial cavity, i.e. both transverse and sigmoid sinuses as well as, possibly, the internal jugular veins in the neck. In such circumstances – typically the result of an acquired hypercoagulable state – central venous blood gains access to superficial facial veins via collateral vessels connecting them through the orbits with the cavernous sinuses. This pattern of venous blood divergence would not occur with either a complete superior sagittal sinus or unilateral transverse sinus thrombosis because venous blood could still escape the head by another route: respectively the cerebral deep venous drainage system and the opposite transverse sinus.

6. (a) Macular area pathology in the retina of the left eye
   (b) Left optic radiation or occipital lobe visual cortex
   (c) Optic chiasm
   (d) Left optic nerve
   (e) Right temporal lobe affecting the inferior portion of the optic radiation that briefly diverts into the temporal lobe as Meyer's loop

CHAPTER 10

1. (a) Yes. (b) No. (c) Yes. (d) No. (e) No. (f) No. (g) Yes.

2. The basilar artery at its terminal bifurcation divides into the posterior cerebral arteries. These, in turn, supply both thalami as well as the occipital and inferior temporal lobes. As noted in the answer to 1a, ischemic injury to both thalami will disrupt the non-specific thalamo-cortical projection systems and result in a disturbance of consciousness.

3. The most likely location of the lesion is the upper midbrain. A bilateral lesion in this location would result in coma due to compromise of the midbrain tegmentum. Bilateral cranial nerve III lesions as well as interruption of connections between the frontal eye fields and the horizontal gaze regions of the inferior pons would explain the dilated pupils and absent doll's eye movements (see case 6e-1). Damage to the cerebral peduncles (corticospinal, corticoreticular and corticobulbar tracts) and to the red nuclei (rubrospinal tracts) explains the decerebrate posturing (see Section 10.9).

   The upper midbrain is very vulnerable to selective damage in high-velocity head trauma. The reason for this vulnerability is that the upper midbrain effectively represents the upper end of a narrow stalk (the brainstem) at its attachment to a very heavy superstructure (the cerebral hemispheres). When the head strikes an object at high speed (for example, when hitting an automobile windscreen), the cerebral hemispheres suddenly veer anteriorly in reference to the midbrain, which is held in place by the remainder of the brainstem, the cerebellum and the cranial nerves III–XII. The resulting mechanical sheering forces in the rostral midbrain cause widespread axonal disruption and focal hemorrhages, leading to immediate, persistent loss of consciousness.

4. This unfortunate woman sustained infarction of the entire territory of the left middle cerebral artery, amounting to a large volume of brain tissue (see Figures 8.3 and 8.4). In principle, such a large dominant hemisphere could result in a semicomatose state from the onset. In her particular case, however, post-infarction edema would gradually lead to massive swelling of the left cerebral hemisphere followed by trans-tentorial herniation of the mesial left temporal lobe and compression of the adjacent midbrain (leading to coma) and of the left oculomotor nerve (leading to left pupillary dilation – see also case 10e-1).

5. The most probable location of the lesion in this case is the pontine tegmentum bilaterally. The slow respiratory rate with inspiratory pauses is suggestive of an apneustic breathing pattern, commonly encountered with pathology in the pons (see Section 10.9). The small, poorly responsive pupils are a result of disruption of the descending sympathetic pathways in the lateral pontine tegmentum; these connect the hypothalamus to the sympathetic spinal cord gray matter components of the thoracolumbar cord (see Figure 2.3). The preganglionic pupillary dilation fibers of the sympathetic system originate at the T1 cord level and ascend via the sympathetic chain along the common and internal carotid arteries to eventually reach the eyes. Disruption of this first link in the three-neuron sympathetic pathway within the pons results in constriction of the pupils due to the unopposed action of the pupillary parasympathetic autonomic neurons in the Edinger–Westphal nucleus of the upper midbrain tegementum (part of the oculomotor nucleus of cranial nerve III). Absence of the oculocephalic (doll's eye) reflexes is explained by injury to the pontine conjugate gaze centres bilaterally (see case 6e-1). Finally, the stiff, extended limbs equate to decerebrate posturing, resulting from disruption of the rubrospinal, corticospinal and corticoreticulospinal pathways in the base of the pons.

The most likely mechanism is a basilar artery thrombotic ischemic stroke at the pontine level, obstructing the midline and lateral pontine perforating branches of the parent vessel and leading to infarction of both sides of the pons. An elderly person with uncontrolled hypertension and type 2 diabetes mellitus is unfortunately a prime candidate for such an event.

6. (a) His Glasgow Coma Scale score is 8/15:
   Best ocular response: opens eyes to pain = 2
   Best verbal response: incomprehensible sounds = 2
   Best motor response: withdraws to pain = 4
   (see Table 2.4)
   (b) This previously well boy was found in a comatose state at the end of a presumably one- to two-hour afternoon nap. He is afebrile and has no abnormalities on neurological examination other than his poorly responsive state. If we consider the Etiology Matrix (Figure 3.4), we should consider the following disease categories:
   Paroxysmal: Coma following a first, unprovoked, prolonged generalized convulsion (positive family history of epilepsy in his brother – although there is a good reason for the latter's seizures).
   Trauma: Unlikely, but – in a two-year-old – one must consider the possibility of a non-accidental head injury. Under such circumstances, one would often see bruises or retinal hemorrhages, as well as radiological evidence of new and old skeletal fractures.
   Vascular: Ischaemic stroke or cerebral haemorrhage (e.g. from a vascular malformation) – but would be unlikely in the absence of abnormal motor system findings.
   Toxic: i.e. an accidental antiepileptic drug overdose. Two-year-olds are good climbers and are not always under continuous supervision by care-givers; he could have gotten at his brother's medication prior to his nap. The five-year-old brother would probably be taking carbamazepine chew-tablets, which taste like candy.
   Infection: Meningitis or encephalitis – The time course is very short for such diseases and the patient is afebrile.
   Metabolic: Hypoglycaemia is always a possibility but, in a two-year-old, more likely to be present in association with an intercurrent illness. The same would be true for previously asymptomatic hereditary metabolic disorders leading to hypoglycaemic or hyperammonaemic coma.
   Inflammatory/autoimmune: e.g. ADEM (see Chapter 9 and case 6e-5) – but the time course is too short.
   Given the abrupt onset of symptoms in a previously normal child, a reasonable first step would be to immediately draw blood for glucose and carbamazepine levels (as well as a more general toxic screen). If a measurable (presumably very high) carbamazepine level

were to be present, no further investigations would be necessary. If the toxic screen were negative and the blood sugar normal, an urgent magnetic resonance imaging study of the head would be required. If the patient had head bruises or retinal haemorrhages, an immediate cranial imaging study would be obligatory. If he had been found to be febrile on arrival, in addition to the previous investigations, a lumbar puncture would be necessary, preferably following the cranial imaging study.

CHAPTER 11

1. (a) Left parasagittal motor cortex
(b) Right precentral gyrus (motor cortex) as it dives into the sylvian (lateral) fissure
(c) Right postcentral gyrus (somatosensory cortex) about halfway between the interhemispheric fissure and the sylvian (lateral) fissure
(d) Auditory association cortex in the superior lateral temporal lobe, either hemisphere
(e) Right parieto-occipital region in the visual association cortex
(f) The amygdala in either mesial temporal lobe region
(g) Insular cortex, either cerebral hemisphere
(h) The primary olfactory (entorhinal) cortex in the anteromesial portion of either temporal lobe – the point at which the olfactory nerves enter the brain (see Chapter 2, Section 2.1.1.1)
(i) Either the right parietal lobe (the cortical region involved in automatic visual pursuit) or the right frontal eye field (see Figure 1.10c)
(j) Either frontal lobe
(k) Either hippocampal formation in the mesial temporal lobe

2. (a) *Migrainous* – This is a typical story of a visual migraine equivalent phenomenon: the result of spreading depression (of Leão) across the right occipital lobe cortex. In most instances, this symptom sequence is followed by the development of a pounding headache, in this case localized to the right side of the head. On occasion, however (as for this patient), the visual symptoms develop without any subsequent head pain. The shimmering zigzag line is often termed a *fortification spectrum*, as its appearance resembles the walls of an eighteenth or nineteenth century fort as seen from the air. The duration of symptoms in this patient is far too prolonged to suggest a partial simple seizure originating in the occipital cortex.

(b) *Syncopal* – The initial, premonitory symptoms of dizziness and darkened vision are reminiscent of those experienced by Mrs. V, the protagonist in case 11e-3. In the present case, the syncopal episode occurred in the context of a febrile illness and a period of sustained bed rest, rather than having been triggered by the pain of a venipuncture. Neurovascular syncope triggered by standing up occurs quite often in otherwise normal teenagers (especially females). Perhaps because of rapid growth and the relatively new phenomenon of cyclical hormonal fluctuations, the autonomic nervous system at this age may take longer to adapt to a sudden change in body position, leading to a transient period of hypotension and benign cerebral ischemia.

(c) *Syncopal* – In this instance, however, the cause of the sudden loss of consciousness may not be benign. The story is suspicious of so-called cardiac syncope, a phenomenon often triggered by vigorous exercise and due to a transient, abrupt tachyarrhythmia in which the heart is beating so rapidly that it is temporarily unable to generate sufficient blood pressure to perfuse the brain. The phenomenon occurs most often in individuals with a congenital anomaly (often hereditary) in the cardiac conduction system and, if the tachyarrhythmia does not quickly resolve spontaneously, may result in sudden death. While an episode of (benign) neurovascular syncope cannot be excluded on the basis of the history, the story is sufficiently disturbing that this patient will require an electrocardiogram and, if necessary, a period of Holter monitoring.

(d) *Epileptic* – As is typical for epileptic seizures, this patient's episodes are brief, recurrent and stereotypic. The seizures presumably originate in the left perisylvian sensory cortex, thus not far from the facial motor cortex and Broca's area – perhaps the reason why her speech may be transiently slurred while the symptoms are present. At age 24, there is a high probability that the seizures originate in an area of structural disturbance, e.g. focal cortical dysplasia or a neoplastic process. The headaches could be suggestive of a space-occupying lesion but, in the absence of any information concerning findings on a neurological examination, could also be migrainous in nature and therefore incidental to the problem at hand.

(e) *Migrainous* – This is a typical story for prodromal migraine (or migraine with aura), i.e. relatively rare episodes in which there is an initial localized sensory symptom – in this instance bilateral and accompanied by a motor deficit – resulting from spreading depression affecting primary somatosensory cortex, followed by a typical pounding headache accompanied by nausea. The head pain is the result of vasodilation and 'sterile inflammation' involving the meningeal and superficial scalp arteries. With an occipital region headache and a prodrome of dizziness as well as sensory disturbances, this woman may be suffering from a form of so-called 'basilar migraine', a bilateral form of prodromal migraine that stems from brain regions supplied by the basilar and posterior cerebral arteries.

(f) *Epileptic* – This case is more difficult than the other five in that, superficially, it resembles the scenario given in C. The obvious difference is that this young man has had two clear-cut generalized convulsions in the context of vigorous exercise – rather than episodes of syncope. While, as we have seen in case 11e-3, a brief 'convulsive' phenomenon may accompany neurovascular (or, for scenario C, possible cardiac) syncope, the present patient's convulsions have been much more prolonged, and very suggestive of primary generalized epilepsy. A history of learning difficulties is also more likely to be encountered in a patient with epilepsy than would be the case for someone with neurovascular syncope. This patient will nevertheless need the same cardiac work-up as the one suggested for question C – along with appropriate investigations to address the possibility of epilepsy.

CHAPTER 12

1. The following are the common forms of dementia:
   a. Alzheimer's disease: accumulation of amyloid and Tau protein
   b. FTD: accumulation Tau protein
   c. DLB: accumulation of synuclein protein
   d. Vascular dementia: small and large vessel disease
   e. Alcoholic dementia: mammillothalamic, cortical degeneration

2. An organic dementia depending on the type is usually a progressive disorder of cognition in multiple domains, with one domain usually memory, which predominates. Pseudodementias often have significant psychomotor slowing, but the cognitive domains if tested with patience are often intact.

   There is often a significant mood disorder that comes on concurrently with the perceived cognitive disorder. Treatment of a mood disorder can help with both, but the pseudodementia can show significant improvement of cognitive function.

3. When faced for the first time with a patient with moderately severe dementia, the number one concern has to be for the safety of the individual and close family members. There is often a lot of denial both by the patient and family members, but the treating physician must be firm about issues such as driving, cooking alone and personal affairs such as banking. Depending on the jurisdiction, mandatory reporting to the appropriate driving authority must be done according to the local laws. In extreme cases, family members must be told to remove the vehicle or car keys. Cooking alone on the stove is a significant hazard. Measures can be taken for the individual not cook alone or to disable the stove so that the individual can use a microwave and kettle with automatic cut-off. The family member or members appointed as the Power of Attorney should make arrangements how to proceed legally to assist the patient with legal and financial matters.

4. The commonest familial dementias are as follows:
   a. Cerebral amyloid angiopathy, Itm2b related
   b. DLB, Loci 1q22, 4q22.1, 5q35.2
   c. Familial Alzheimer disease, 1,2,3,4,5, multiple loci, Consult Online Mendelian Inheritance in Man Web site
   d. FTD, protein tau, Locus 17q21
   e. Huntington's disease, gene encoding Huntingtin, Locus 4p16
   f. Presenilin 1 Locus 14q24.2
   g. Spinocerebellar ataxia type 2, 4, 17

5. The following is a list of dementias caused by prions:
   a. Creutzfeldt–Jakob disease sporadic
   b. FFI related to asp178-to-asn mutation of the PRNP gene
   c. Gerstmann–Straussler disease, heterozygous mutation in the prion protein gene, Locus 20p13
   d. Kuru caused by funeral rite of cannibalism of dead relative

e. Variant CJD caused by eating infected cows, which were infected by eating sheep offal in Britain in 1980s

CHAPTER 13

1. Both prefrontal lobe areas

2. (a) This unfortunate man has sustained severe traumatic injury to both medial temporal lobe regions, including the hippocampal formations. Without at least one functioning hippocampus, permanent registration of new memories is impossible, hence his restricted but profound disability. With a high-velocity impact head injury, the most vulnerable regions of the cerebral hemispheres are the frontal and occipital poles and the anterior temporal lobes. The reason for this predilection is that, during a high-velocity impact of the head against the ground, the inertial forces in the heavy cerebral hemispheres – floating as they are in subarachnoid space CSF – cause the brain to shift violently against adjacent bony structures: the inner table of the frontal, sphenoid and occipital bones.

(b) The patient's pineal tumor was located just posterior to the fornices of the hippocampi, where they ascend dorsally and medially from the posterior regions of the hippocampi in the temporal lobes before almost meeting in the midline and proceeding anteriorly beneath the corpus callosum

and then inferiorly to the septal area before terminating in the hypothalamus (see Figure 13.5). In the early part of this trajectory, the patient's fornices were vulnerable to injury while the adjacent pineal tumor was being removed piecemeal from the bottom of a long surgical approach pathway. Such injuries are well-described complications of pineal region surgery and, fortunately, usually have a good prognosis.

(c) This woman's tendency to fabricate memories of events that never took place is a phenomenon referred to as *confabulation* and is one manifestation of a complex neurological syndrome occasionally encountered in chronic alcoholics known as *Wernicke–Korsakoff syndrome*. The disorder primarily results from severe thiamine (vitamin B1) deficiency with secondary neuronal loss in the medial thalamic region, the mammillary bodies of the *hypothalamus* (the termination of the hippocampal – forniceal pathway noted in the answer to the previous question – see Figure 13.5) and the periaqueductal gray matter in the midbrain. It is the mammillary body pathology that is believed to be the lesion responsible for the amnestic aspects of the syndrome. In addition to the severe memory disturbance noted in our patient, individuals with Wernicke–Korsakoff syndrome commonly also present with acute delirium, ataxia and generalized polyneuropathy.

INDEX

Page numbers followed by f and t indicate figures and tables, respectively.